ENGINEERING PSYCHOLOGY AND HUMAN PERFORMANCE

Christopher D. Wickens
University of Illinois at Champaign-Urbana

CHARLES E. MERRILL PUBLISHING COMPANY
A Bell & Howell Company
Columbus • Toronto • London • Sydney

Published by
Charles E. Merrill Publishing Company
A Bell & Howell Company
Columbus, Ohio 43216

This book was set in Melior.
Production Coordination: Constantina Geldis
Cover Design: Tony Faiola

Library of Congress Catalog Card Number: 83-62522
International Standard Book Number: 0-675-20156-X
Printed in the United States of America

1 2 3 4 5 6 7 8 9 10—88 87 86 85 84

To my wife, Linda, and my children, Nina, John, Allison, and Laura

Preface

I wrote this book because I saw a need to bridge the gap between the problems of system design and much of the excellent theoretical research in cognitive experimental psychology and human performance. Many human-machine systems do not work as well as they could because they impose requirements on the human user that are incompatible with the way a person attends, perceives, thinks, remembers, decides, and responds; i.e., the way in which a person processes information. Over the past four decades, tremendous gains have been made in understanding and modeling human information processing. My goal is to show how these theoretical advances have been, or might be, applied to improving human-machine interactions.

While engineers encountering system design problems may find some answers or guidelines either implicitly or explicitly stated in the book, this is not a handbook of human factors or engineering psychology. References in the text provide a more encyclopedic tabulation of such guidelines. I have organized the book directly from the perspective of human information processing. The chapters correspond to the flow of information as it is processed by a human being and are not generally organized from the perspective of different system components or engineering concerns, such as displays, illumination, controls, computers, and keyboards. Furthermore, although the following pages contain recommendations for certain system design principles, many of these are based only upon laboratory research and theory; they have not been tested in real world systems.

It is my firm belief that a solid grasp of theory will provide a strong framework from which the specific principles of good human factors can be more readily derived. My intended audience, therefore, is threefold: (a) the student in psychology, who will begin to recognize the relevance to many areas of the real world and applications of the theoretical principles of psychology which he or she may have encountered in other courses; (b) the engineering student, who, while learning to design and build systems with which humans interact, will come to appreciate not only the nature of human limitations—the essence of human factors—but also the theoretical principles of human performance and information processing that underlie those limitations; and (c) the actual practitioner in engineering psychology, human performance, and human factors engineering, who should understand the close cooperation that should exist between principles and theories of experimental psychology and issues in system design.

The twelve chapters of the book span a wide range of human performance components. Chapters 2 through 7 deal with the areas of perception, decision making, memory, and attention, with an emphasis on their potential applications rather than cognitive psychology. Chapters 8 through 11 cover topics of time-sharing, reaction time, action, and manual control, thereby addressing areas that are more traditionally associated with the engineering field. Finally, Chapter 12 is systems-oriented, discussing process control and automation. This chapter shows how many of the principles explained in earlier chapters are pertinent to one specific application area of rapidly growing importance.

While the twelve chapters are interrelated (just as the components of human information processing are), I have constructed them in such a way that any chapters may be deleted from a course syllabus and still leave a coherent body. Thus, a course on applied cognitive psychology might include Chapters 1 through 7; one emphasizing human performance theory might include Chapters 1, 2, 5, 7, 8, 9 and 10; and another emphasizing strictly engineering applications might include Chapters 1, 2, 8, 9, 10, 11, and 12. Some chapters contain supplements. This material falls into two classes: (a) relatively technical or quantitative in nature; and (b) the psychological theory behind some aspect of performance, expanding upon treatments presented in the main body of the chapter.

In any project of this kind, one is indebted to numerous people for assistance. In my case, the list includes my family, as well as several colleagues who have read and commented on various chapters or have stimulated my thinking. Among the colleagues are Bill Derrick, Dan Fisk, Danny Gopher, Art Kramer, Amir Mané, Diane Sandry, Pamela Tsang, Mike Vidulich, and Yei-Yu-Yeh. My thanks also to the following reviewers for their valuable comments on selected chapters of the book: Lyle Bourne, University of Colorado; Norman Gordon, State University of New York at Oswego; William C. Howell, Rice University; David Martin, New Mexico State University; Michael Masson, University of Victoria; and Philip J. Smith, The Ohio State University. I am especially grateful to Neville Moray, University of Toronto, for his thorough and integrative remarks about the entire text. In addition, I appreciate the efforts of Margaret Plantz, who provided beneficial suggestions on manuscript style, and Jane Sudbrink, Connie Geldis, and Phyllis Crandall at Charles E. Merrill Publishing, who guided the manuscript through the production and publication processes. I also want to acknowledge support from the engineering psychology program of the

U.S. Office of Naval Research. Much of the research carried out under their sponsorship contributed to the formulation of many ideas expressed in this book. Lorna Hunt was especially helpful in organizing the references.

Four specific individuals have contributed to the development of my interest in engineering psychology: my father, Delos Wickens, stimulated my early interests in experimental psychology; Dick Pew introduced me to academic research in engineering psychology and human performance; Stan Roscoe pointed out the importance of good research applications to system design; and Emanuel Donchin continues to maintain my awareness of the importance of solid theoretical and empirical research. Finally, it is impossible to credit Mary Welborn's contributions to this book. Without her hours of dedication at the word processor of a sometimes hostile computer, the project never would have succeeded.

Christopher D. Wickens

Contents

*The material within each supplement is of two kinds. Fairly technical mathematical or quantitative treatments are indicated by M. Fairly theoretical analyses of certain topics with less direct potential applications are indicated by T.

ENGINEERING
PSYCHOLOGY
AND
HUMAN
PERFORMANCE

Introduction to Engineering Psychology and Human Performance

At 4:00 A.M. on March 28, 1979, near Harrisburg, Pa., a temporary clog in the feedwater lines of Turbine No. 1 at the Three Mile Island nuclear power plant caused a rapid automatic shutdown of the feedwater pump and the turbine. A fraction of a second later, the redundant safeguards that are built into such systems functioned normally to supply an alternate source of feedwater. Immediately, four critical errors converged, demonstrating as never before how vulnerable is the human link in the performance of complex systems.

The first error had been committed before the clog developed. The pipe for the alternate feedwater supply had been blocked off during maintenance by personnel who had since gone off duty. As a result, the radioactive core, no longer receiving a continuous supply of cold water to remove its heat, began to increase in temperature, turning the surrounding coolant to vapor, and the pressure increased rapidly.

Automatic safeguards, however, continued to function properly. Cobalt rods descended into the core to slow down the process, while a pressure relief valve opened to "bleed off" some of the high-pressure overheated vapor from the primary cooling loop. Once pressure was reduced below the critical level, the automatic relief valve received a signal to close in the same manner that a furnace receives a signal from the thermostat to shut down when it has heated the room sufficiently. At this point the second error occurred. Because of a malfunction, the valve did not close.

Within a minute after the original shutdown, the supervisory crew at Three Mile Island was attempting to understand what was going on from a myriad of alarms, lights, and signals on massive display panels. Though their training allowed them to capture a fairly accurate picture, they were led astray by one signal. The display for the pressure relief valve was designed to indicate what the valve was commanded to do rather than what it actually did, and the display indicated that the pressure relief valve had closed. This was the third critical error in the sequence.

Meanwhile, the pattern of redundant automated safeguards that are a mainstay of such complex systems continued to operate, and an emergency pump switched on to supply the system with a now badly needed source of coolant. At this point, the supervisors made a decision that turned what might have been a relatively minor incident into a major catastrophe. With instruments showing that pressure was already high and that the relief valve had closed, the operators decided to override the controls manually and shut off the emergency pump. This was the fourth error in the sequence: a decision based on their inference that coolant level was excessive rather than too low. The core was thereby deprived of the vital cooling it needed, and the incident soon accelerated to the point of no return.

This description of the Three Mile Island incident highlights only certain of the primary contributing factors. Rubinstein and Mason (1979) provide a fascinating, detailed, event-by-event account of the disaster. In the exhausting inquiries, investigations, and hearings that followed the crisis, three things became clear. First, no single fault, mistake, event, or malfunction *caused* Three Mile Island. Rather, responsibility was distributed across a number of sources. Second, human error was involved at several levels, from the incorrect decision to shut down the emergency coolant to the human decisions in the design of a relief valve system that tells operators what the valve was commanded to do, not what it did. Third, and most important, the overwhelming complexity of information presented to the human operators and the confusing format in which it was displayed was probably sufficient to guarantee that, somewhere along the line, the intrinsic limitations of human abilities to attend, perceive, remember, decide, and act—that is, to process information—would be overloaded (Electrical Power Research Institute [EPRI], 1977; Rasmussen, 1981; Sheridan, 1981). Thus, in Three Mile Island as in so many other accidents involving complex systems (Hurst, 1976), while human error may be attributed as a cause, the human operators themselves are not at fault.

A major thesis of this book is that blame must be placed instead on the design of systems that overload human information-processing capabilities. Our own experiences offer many other less dramatic examples of "system-induced human error." One common irritation is mistakenly turning on a car's headlights instead of its windshield wipers because the controls are identically shaped and side-by-side. Making a mistake in operating a household appliance because of poorly worded instructions or looking up a 10-digit area code and phone number and then not remembering it long enough to finish dialing also are examples of problems arising from system designs that exceed human capacities.

One major purpose of this book is to examine human capabilities and limitations in the specific area of information processing. The second purpose is to demonstrate how knowledge of these limitations can be applied in the design of complex systems with which humans interact.

ENGINEERING PSYCHOLOGY
AND HUMAN FACTORS

Designing machines that accommodate the limits of the human user is the concern of a field referred to as *human factors*. The field is a very broad one—broader than this book, which is focused specifically on designing systems that accommodate the information-processing capabilities of the brain. Many of the principles that are important for designing machines that humans use, and that therefore belong in the province of human factors, are not related to this particular facet of human capacity. For example, designing an automobile's dashboard in such a way that all controls can be reached easily and all displays are visible without straining the neck is a human factors concern. This design problem, however, must respond to the physical properties and constraints of the driver's body, not the brain's information-processing capabilities. Another concern of human factors is designing controls to minimize muscle fatigue. Both these problems are real and legitimate concerns, but they are not within the purview of this text.

A focus on the information-processing capacities of the human brain, then, is a key characteristic delimiting the scope and contents of this book. Study of the processes of the brain falls within the realm of psychology. It is the discipline of engineering psychology, specifically, that applies a psychological perspective to the problems of system design.

Among the notable attributes of engineering psychology as it has emerged as a discipline in the last three decades are its solid theoretical basis (Howell & Goldstein, 1971) and its close affinity to the discipline of experimental psychology. While engineering psychology is akin to both human factors and experimental psychology, its goals are nevertheless unique. The goal of experimental psychology is to uncover the laws of behavior through experiments. However, the design of these experiments is unconstrained by a requirement to apply the laws. That is, it is not required that experiments generate immediately useful information. The goal of human factors, on the other hand, is to apply knowledge in designing systems that work, accommodating the limits of human performance *and* exploiting the advantages of the human operator in the process. Engineering psychology arises from the convergence of these two domains. "The aim of engineering psychology is not simply to compare two possible designs for a piece of equipment [which is the role of human factors], but to specify the capacities and limitations of the human [generate an experimental data base] from which the choice of a better design should be directly deducible"[1] (Poulton, 1966, p. 178). That is, while research topics in engineering

---

[1]From "Engineering Psychology" by E. C. Poulton, 1966, *Annual Review of Psychology, 17*, p. 178. Copyright 1966 by Annual Reviews, Inc. Reproduced by permission.

psychology are selected because of applied needs, the research transcends specific one-time applications and is conducted with the broader objective of providing a usable theory of human performance. In keeping with this attribute of the field, some topics treated in the present book have not been directly translated to applications in system design but are included because of their importance in the theory of human information-processing limitations and their consequent potential for applications.

One further limitation on the scope of this book should be noted. A major component of research in experimental psychology has been the subject of human learning. In applications to system design, this topic has translated into the area of training (Anderson, 1981; Stammers & Patrick, 1975). Here the question is what procedures and techniques should be used to teach the human operator to interact fluently with the machine. In engineering psychology, these issues for the most part are avoided. It is assumed that an operator already is well trained. The focus is on the limits of his or her *performance* of the tasks at hand. Hence the second part of the title, *human performance*.

It should be reemphasized that the decision to exclude certain areas of human factors does not reflect a view that these are areas of lesser importance. In fact, much of the research in these areas is potentially more informative about actual design modifications and decisions than is research in human performance. The reader is referred to textbooks by Bailey (1982), McCormick and Sanders (1982), Kantowitz and Sorkin (1983), and VanCott and Kinkade (1972) for more detailed treatments of these areas. The present book is intended to complement, rather than supplement or replace, these treatments.

The rest of this introductory chapter will provide additional background information to prepare the reader for the discussions to come. We shall first present a brief historical overview of the more general domain of human factors, including the forces that motivated the development of the field and some of the basic considerations that are taken into account in designing systems to accommodate the human users. This will be followed by a description of the different kinds of human factors research. The chapter will conclude with a general model of the human as an information processor who attends to, perceives, translates, stores, and responds to events in the environment. This model provides a schematic framework for the chapters that follow, most of which examine component processes within the model in detail and consider their implications for system design.

The Human Factor in System Design

Impetus for the development of human factors and engineering psychology as disciplines has arisen from three general sources: practical needs, technological advancements, and linguistic developments. Prior to the birth of human factors, or *ergonomics*, in World War II, emphasis had been placed on "designing the human to fit the machine." That is, the emphasis was on training. Experience in World War II, however, revealed a number of instances in which systems, even with well-trained operators, simply weren't working. Airplanes were flying into the ground with no apparent mechanical failures; enemy contacts were missed on radar by highly motivated monitors. As a consequence, experimental psychologists from both sides of the Atlantic were brought in to analyze the operator-machine interface, to diagnose what was wrong, and to recommend the solution (Fitts & Jones, 1947). This represented the practical need underlying the origin of human factors engineering.

A second motivation has come from evolutionary trends in technology. With increased technological development in this century, systems have become increasingly complex, with more and more interrelated elements, forcing the designer to consider the distribution of tasks between human and machine. This problem has led system designers to consider what functions to allocate to humans and what to machines. This in turn necessitates a close analysis of human performance in different kinds of tasks. At the same time, with increased technology, the physical parameters of all systems have grown geometrically. As examples, consider the increases in maximum velocity of vehicles, progressing from the oxcart to the spacecraft; in the temperature range of energy systems, from fires to nuclear reactors; and in the physical size of vehicles, from wagons to supertankers and wide-bodied aircraft such as the Boeing 747. Particularly with regard to speed, this increase forces psychologists to analyze quite closely the operator's temporal limits of processing information. To the oxcart driver, a few milliseconds of delay in responding to an environmental event will be of little consequence. But to the pilot of a supersonic aircraft, a delay of the same absolute magnitude may be critical in avoiding a collision.

Finally, a linguistic influence for the growth of human factors has come from the new language of information theory and cybernetics that after World War II began to replace the stimulus-response language of behavioral psychology as a description of human activities. Terms such as *feedback, channel capacity,* and *bandwidth* began to enter the descriptive language of human behavior. This new language enabled some aspects of human performance to be described in the same terms as the mechanical or electronic systems with which the operator was interacting. This shared terminology in turn facilitated the integration of humans and machines in system design and analysis.

As indicated above, the initial applications of engineering psychology during and after World War II were focused primarily on correcting systems that were obviously faulty. How should an altimeter be redesigned, for example, to prevent a pilot from being confused about his altitude and flying into the ground (Fitts & Jones, 1947)? Gradually engineers realized that many such faults could be avoided if the designer considered the inherent limitations of the human operator before, not after, the system was built and allocated functions between operator and machine in such a way that those limitations were avoided. Many lists have been generated that contrast strengths and limitations of humans and machines, and the following one presented by McCormick (1976, pp. 461–462) is typical:

> *Humans* are generally *better* in their abilities to:
>
> Sense very low levels of certain kinds of stimuli: visual, auditory, tactile, olfactory, and taste.
>
> Detect stimuli against high-noise-level background, such as blips on a cathode-ray-tube radar display with poor reception.
>
> Recognize patterns of complex stimuli that may vary from situation to situation, such as objects in aerial photographs and speech sounds.
>
> Sense unusual and unexpected events in the environment.
>
> Store (remember) large amounts of information over long periods of time (better for remembering principles and strategies than masses of detailed information).
>
> Retrieve pertinent information from storage (recall), frequently retrieving many related items of information; but reliability of recall is low.
>
> Draw upon varied experience in making decision; adapt decision to situational requirements; act in emergencies. (Does not require previous "programming" for all situations.)
>
> Select alternative modes of operation, if certain modes fail.
>
> Reason inductively, generalizing from observations.
>
> Apply principles to solutions of varied problems.
>
> Make subjective estimates and evaluations.
>
> Develop entirely new solutions.
>
> Concentrate on most important activities, when overload conditions require.
>
> Adapt physical response (within reason) to variations in operational requirements.
>
> *Machines* are generally *better* in their abilities to:
>
> Sense stimuli that are outside the human's normal range of sensitivity, such as x-rays, radar wavelengths, and ultrasonic vibrations.
>
> Apply deductive reasoning, such as recognizing stimuli as belonging to a general class (but the characteristics of the class need to be specified).
>
> Monitor for prespecified events, especially when infrequent (but machines cannot improvise in case of unanticipated types of events).
>
> Store coded information quickly and in substantial quantity (for example, large sets of numerical values can be stored very quickly).
>
> Retrieve coded information quickly and accurately when specifically requested (although specific instructions need to be provided on the type of information that is to be recalled).
>
> Process quantitative information following specified programs.
>
> Make rapid and consistent responses to input signals.
>
> Perform repetitive activities reliably.
>
> Exert considerable physical force in a highly controlled manner.
>
> Maintain performance over extended periods of time (Machines typically do not "fatigue" as rapidly as humans).
>
> Count or measure physical quantities.

Perform several programmed activities simultaneously.

Maintain efficient operations under conditions of heavy load (humans have a relatively limited channel capacity).

Maintain efficient operations under distractions.[2]

Stated in more general terms, this presentation of strengths and limitations suggests that humans can respond perceptually to a changing environment and to relations in the environment. They can go beyond the information immediately given, respond to low-probability occurrences, and adopt alternative strategies and alternate modes of performance when necessary. In addition, they are inherently interested in their own survival. In short, humans are *flexible*. Unfortunately, this flexibility is purchased at a cost. Humans also are *variable* (they produce errors), and they may become "creative" in changing their responses when it is not optimal to do so.

Of course, the dichotomy presented by McCormick is not a rigid one. In particular, the tremendous advances in computer technology that have taken place over the last decade have led machines to encroach more and more into the "human" domain. Automation and its implications for human performance, a critical issue in engineering psychology, will be dealt within some detail in Chapter 12.

Research Procedures in Engineering Psychology

The field of engineering psychology employs a variety of research procedures, which differ in the extent to which they employ real-world situations in their experimentation and testing. There are five general categories of investigation that can be placed on a continuum from high to low realism: observation of the system in action in the "real world," field studies of the fully developed system, simulation, laboratory experiments, and mathematical models.

As an example of movement along this continuum in the other direction—that is, from low to high realism—consider the design of an aircraft. Initially, mathematical models of the pilot's "tracking" capabilities may be used. These may be followed by laboratory investigations of a human's ability to perceive and respond to events in the airborne environment. A simulator of the aircraft will be developed in the laboratory and used to test the placement of controls and displays. Then the aircraft will be given a number of field studies by test pilots. Finally, even after it is bought by the contractor, its performance in the real world is evaluated continually.

Three important factors co-vary with this continuum of research techniques. First is the degree of fidelity or validity of conclusions drawn from the research: Will the system work? This fidelity decreases as one moves away from the high realism end of the continuum.

---

[2]From *Human Factors in Engineering and Design* (pp. 461–462) by E. J. McCormick, 1976, New York: McGraw-Hill. Reprinted by permission.

Second is the flexibility of modification if something is diagnosed as not working. This flexibility, of course, increases as realism is lost. It is quite trivial to erase and change a term in an equation or a mathematical model of human-machine system performance. It is a good bit more difficult and drastically more expensive to redesign an existing flying aircraft when performance is shown, through flight testing, to be less than satisfactory. Third is the generalizability to other systems of conclusions drawn from research on a specific system. This generality decreases as realism is increased. A mathematical model of performance or a laboratory experiment may be generalizable to a wide variety of systems that make use of the kind of processing being investigated. At the other end of the continuum, conclusions from observing a particular system in action may not generalize to other systems because too many variables differ between the systems.

HUMAN PERFORMANCE

A primary goal of engineering psychology is to predict system performance. For this, performance of the human component must be predicted as well, and this is the objective of human performance models.

Models of Performance

In general, models are theoretical representations of systems that specify the major components involved and the relationships among them. If the model is a good representation of reality, then by altering any component or its theorized relationship to other components, a researcher can predict how such a change would alter the system's overall functioning or output. In engineering psychology, human performance models generally are formulated to serve one of two functions, either predicting operator performance or examining the distinction between expert and novice.

Predictive modeling.   The role of models for performance prediction is critically important because of the tremendous savings in cost that can be realized if a system designer can predict how well a system will function before it is actually built. Will the airplane with its pilot fly adequately given the limits in the speed and precision with which the pilot can control? Will the power-plant operator perform his tasks without making incorrect decisions? A great deal of money is lost when predictions cannot be made in advance. Once into production, systems are tremendously expensive to modify because the change of any component almost invariably forces reconsideration and modification of related components. The more complex the system, the more intricate is this network of interrelated components.

A typical example of this concern relates to the size of the crew on certain aircraft. Can a given crew adequately accomplish all functions that the aircraft must perform in its various missions? Or is an extra person required? If a model of pilot performance can be used to determine early in the design process that the extra person is needed, then full engineering production can be devoted solely to the larger design. If no model to predict needed crew size exists and a smaller design is built and then proven impossible, tremendous costs may be imposed. In a high-speed military aircraft, the entire aircraft, including its handling and engine characteristics, would have to be redesigned to accommodate the extra size, weight, and environmental support for the additional person. In order to make predictions of system performance, then, the designer must be equipped with accurate mathematical models of the human performer, expressed in a common language with the system. From such models, hypothetical system performance can be evaluated with varying values to describe the capabilities of crew members.

The domain of predictive performance modeling varies dramatically from fairly accurate, detailed models of specific component processes such as signal detection (Green & Swets, 1966), decision making (Edwards, Lindman, & Phillips, 1965), fault diagnosis (Rouse, 1978, 1981), tracking (McRuer, 1980), and workload (Wickens, 1983) to very general models of the entire ensemble of components of human performance in combination (Card, Moran, & Newel, 1983; Siegel & Wolf, 1969; Wherry, 1976). The more detailed component models, because of their specificity, are expected to be quite precise and accurate in predicting variance in performance of the particular component task. In contrast, for models of a more general character, detailed accuracy of the component processes is sometimes sacrificed for the sake of avoiding unwieldy complexity. This sacrifice is compensated by the model's applicability to a much broader range of activities. For example, the human operator simulator (HOS) model of Wherry (1976) is intended to predict human performance in a wide variety of tasks that confront the aircraft pilot. Card et al.'s (1983) "model human processor" predicts performance across numerous tasks involving human interaction with video keyboards by taking into account such factors as the time to locate an item on the display, remember its meaning, and to depress an appropriate key.

Expert modeling.   The second use of performance modeling, somewhat less well developed in terms of human factors application, concerns the modeling of expert performance (Anderson, 1981). The goal is to describe and simulate on a computer the manner in which the expert performs a given task. This expert performance then may be compared to that of the novice performer. In instances where it is possible to identify specific differences between expert and novice performance, training may focus explicitly on reducing the discrepancies. Models of this sort have been implemented to describe expert behavior in such fields as circuit-wiring troubleshooting (Egan & Schwartz, 1979), physics problem solving (Larkin, McDermott, Simon, & Simon, 1980), computer programming (Jeffries, Turner, Polson, &

Atwood, 1981; Schneiderman, 1976), and skilled aspects of memory (Chase & Ericsson, 1981). A variant of this procedure occurs when a model of *idealized optimal* performance is devised with which performance of either the novice or the expert can be compared. Where there are discrepancies, attention may be focused on the dimensions of difference. (It is important to realize that expert performance may indeed be far from perfect. This is a theme that will be addressed later in some detail, particularly in the domain of decision making, discussed in Chapter 3.)

Performance models, whether of the novice, the expert, or the optimal ideal, are useful and important because they portray more than the final outcome of performance. They portray the mechanism and means whereby that performance is achieved. It is this information that is of critical importance. It is one thing to know that A is better than B. It is quite a bit more informative to know *why* A exceeds B.

Dimensions of Human Performance: Efficiency, Bias, and Optimality

There are clearly a large number of dimensions along which performance may vary, depending on the task performed by the operator and the options available. These dimensions may be divided into two very broad categories: those that pertain to the overall efficiency, or "goodness," of performance and those that pertain to a bias for one style of performance over another.

Human performance often is measured by its effectiveness, or *efficiency*. In a variety of performance paradigms, it is possible to identify a specific dimension on which one can assess "good" and "bad" performance. For example, in most tasks that require speeded actions, such as typing, good performance is fast and accurate; bad performance is slow and error-laden. Thus in this case the dimension of efficiency is a combination of speed and accuracy. In the realm of detection, a good nuclear power plant supervisor always detects failures when they occur but never "cries wolf" to cause an unnecessary shutdown. A poor supervisor misses failures and also has a high rate of unnecessary shutdowns, or "false alarms." Thus the efficiency dimension in detection is defined by a combination of miss rate and false-alarm rate.

We assume that each operator is capable of functioning at some maximum level of efficiency. This level can be achieved with sufficient practice only if it does not exceed "hardware" limits of the human processing system. For the nuclear power plant monitor, for example, less than maximum efficiency in failure detection may be a practice problem: the operator has not yet learned the various combinations of instrument readings that discriminate a malfunctioning reactor from a normal one. The existence of human hardware limitations is suggested if, for example, an operator finds it impossible to concentrate on essential instrument readings at two or more separate locations simultaneously no matter what level of training and practice he has achieved.

In addition to efficiency, there is a second general characteristic, or dimension, of human performance that is useful to consider. This dimension, which we label *bias*, normally encompasses some tradeoff between the quantities contributing to efficiency. In the typing task, for example, when changes in efficiency occur, speed and accuracy are positively correlated. A performance bias, however, may lead the typist to be fast but sloppy, or slow but accurate. In this case of bias change, the correlation between speed and accuracy is negative. In the realm of detection, returning to the example of the nuclear power monitor, efficiency is reflected by more true detections and fewer false alarms. On the bias dimension, detected failures and false positives are positively correlated. A "risky" bias will produce many detections but also many "false alarms." A conservative bias will produce few of each.

Many of the following chapters will present models of specific aspects of human performance (e.g., signal detection in Chapter 2, decision making in Chapter 3, time sharing in Chapter 8, reaction time in Chapter 9, tracking in Chapter 11). These models all have in common the partitioning of performance into dimensions of *efficiency* and *bias*, usually measured as correlations of two basic dimensions such as speed and accuracy. Some of these models also contain some explicit prescription of the *optimal* setting of the bias parameter when certain external conditions (costs and payoffs) are specified. This prescription allows the investigator both to determine how close the operator is to optimal in the selection of strategies of performance and to consider the implications for system design of important departures from optimality. In this regard, an important point should be emphasized: Optimal performance does not mean perfect performance. Rather, it is performance that maximizes some complex function of expected benefits (or minimizes a function of costs) given certain external conditions and the basic training or hardware limits in processing efficiency. In the case of weather forecasting, for example, the optimal forecaster is not 100 percent correct. However, given the probabilistic characteristics of meteorological data and her own cognitive limitations, the optimal forecaster does the most accurate job possible with the data at hand.

A MODEL OF HUMAN INFORMATION PROCESSING

The previous section has described the way performance outcomes—observable measurements—are represented and modeled. These outcomes are the consequence of complex mental operations that the brain performs on information perceived from the environment.

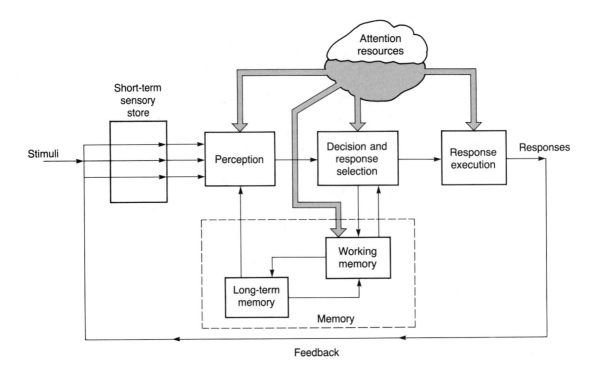

Figure 1.1 A model of human information processing.

It has been useful both to the experimental psychologist and to the engineering psychologist interested in applications, to conceptualize these operations as a sequence of information-processing stages. The model of information processing presented in Figure 1.1 is a composite of those presented by a number of previous investigators (e.g., Broadbent, 1958; Smith, 1968; Sternberg, 1969; Welford, 1976). It assumes that each stage of processing performs some transformation on the data and demands some time for its operation.

While this conceptualization is not to be taken literally, it nevertheless facilitates investigation of the components of human performance and so provides a useful technique for examining potential limitations in performance. These processing stages will be considered extensively in subsequent chapters. In the following pages they are described briefly in the "left to right" order shown in Figure 1.1.

Sensory processing.   Of primary importance in engineering psychology are the visual and auditory senses of the eyes, ears, and the proprioceptive or kinesthetic senses of body and limb position. Unique limitations on each sensory system influence the quality and quantity of information that may be initially registered and so, potentially, all processes that follow. The characteristics of rod and cone receptors in the retina of the eye, for example, influence the conditions under which color can be employed to display information. Characteristics of receptors and neural connections in the ear affect the ability of one sound to "mask" another in a noisy environment (Deatherage, 1972; Gerald, 1965). Thus the basic properties of the senses are of critical importance in the investigation of information processing in human factors design. These properties are not considered extensively in the present text, however. The reader is referred to VanCott and Kinkade (1972), Baty (1982), Kantowitz and Sorkin (1983), Kaufman and Boff (forthcoming), or McCormick and Sanders (1982) for further work in this area.

Short-term sensory store.   Each sensory system, or *modality*, appears to be equipped with a central mechanism that prolongs a representation of the physical stimulus for a short period of time after the stimulus has physically terminated. When attention is diverted elsewhere, the short-term sensory store (STSS) permits environmental information to be preserved temporarily and dealt with later. Three general properties are characteristic of STSS: (1) It is *preattentive*; that is, no conscious attention is required to prolong the image during the natural "time constant" of the store. (2) It is relatively *veridical*, preserving most of the physical details of the stimulus. (3) It is *rapidly decaying*. The particular time constant of the decay varies somewhat, normally being less than a second for short-term visual store (STVS or iconic memory; Neisser, 1967; Sperling, 1960) and a bit longer for both short-term auditory store (STAS or echoic memory; Neisser, 1967), and short-term kinesthetic store (Posner & Konick, 1966). Estimates for the last two range between 2 and 8 seconds, depending upon the techniques employed in measurement (Lachman, Lachman, and Butterfield, 1979). The experience of iconic memory may be likened to the rapidly decaying image on a photographic print when a light is suddenly turned on in a darkroom or to a Polaroid picture in reverse time. Echoic memory possesses the characteristic of the decaying "internal echo" of a voice after it has ceased. Iconic and echoic memory will be considered in more detail in Chapter 6.

Perceptual encoding.   STSS preserves the details of the stimulus image only briefly and without attention. The information is then processed by progressively higher centers of the nervous system. This information is assumed to make contact with a unique neural code that was previously *learned* and *stored* in the brain. At this point the stimulus is said to be perceived or recognized. The result is a *perceptual decision* in which the physical stimulus is assigned to a single perceptual category. The effect on perception of the previously learned neural code is represented in Figure 1.1 by the association between perception and *long-term memory*, the repository of relatively permanent information.

The perceptual process is a many-to-one mapping; that is, a large number of different physical stimuli may be assigned to a single perceptual category. For example, in perception of the letter a, different type styles (A, **a**, or a) all generate the same categorical perception, as does the sound "a" spoken by any voice. At the same time that there is a common response to all these forms of the letter a, their *differences* also are preserved at higher levels of processing. Mechanisms of focal attention allow us, if we choose, to attend to other dimensions of the stimulus as well, such as whether it is spoken by a male or female voice, or written in upper or lower case. Even with this flexibility, however, the common response to the various physical forms of a will remain.

In our analysis of perception, it will be important to distinguish between a number of levels of complexity in the categorization task imposed by both the nature of the stimulus and the task confronting the operator. At a basic level, the task may call for the simple judgment of whether a stimulus is present. Only two perceptual categories are used, yes and no, and the operation is one of *detection* (Chapter 2). Thus the monitor of a radar scope is called upon initially merely to detect the existence of a "blip" on her scope. At levels of greater complexity, the task may call for not just detection but also *recognition, identification,* or *categorization* of the stimulus into one of several possible groups.

The complexity of these tasks is affected by whether one physical dimension alone is to be used in perceptual categorization of the stimulus, or whether instead several dimensions are to be considered in concert. As one example of the former case—the *absolute judgment task*—an operator may be asked to judge the loudness of a tone, the size of a crowd, or the smoothness of a plate of steel and assign the stimulus to a particular categorical level (Chapter 2).

In more complex tasks, at least two dimensions must be considered to match a particular stimulus category. Such multidimensional considerations typify the task of *pattern recognition* (Chapters 4 and 5). Each pattern is uniquely specified by a combination of levels (called features) along the several physical dimensions. For instance, recognition of a particular malfunction of a complex system will occur when the operator can associate a unique combination (pattern) of dial readings (features) on several different instruments (dimensions) to the perceptual category associated with a particular malfunctioning state. Pattern recognition of a disease by a physician is elicited by certain unique combinations of features called symptoms. A particular collection of thermometer and barometer readings, satellite cloud-cover photos, and wind speeds may signal "storm coming" to a trained meteorologist.

Decision making. Once a stimulus has been perceptually categorized, the operator must decide what to do with it (Chapter 3). For example, recognizing that a traffic light has turned yellow (an absolute judgment task), the driver may decide to accelerate or to brake. In this case, the decision involves the selection of a response (Chapters 9 and 10). Alternatively, as indicated in Figure 1.1, the driver may decide to store the information in memory while searching the intersection ahead for the presence of a police car. If the decision is made to commit the information to memory, then the information flow in Figure 1.1 follows the path leading to the memory box (Chapter 6). At this point, a second decision may be made to retain the information for a short time by actively rehearsing it in "working memory" or attempt to store it permanently (learn it) and so enter the information in long-term memory. It is apparent that the point of decision making and response selection is a critical junction in the sequence of information processing. A large degree of choice is involved, and heavy potential costs and benefits depend upon the correctness of the decision (Chapter 3).

Response execution.   If a decision is made to generate a response, then an added series of steps is required to call up, and release with the appropriate timing and force, the necessary muscle commands to carry out the action (Chapters 10 and 11). The decision to initiate the response is logically separate from its execution. The baseball batter, for example, may decide to swing or not swing at a pitch (response selection) and, if he does swing, may execute one of a variety of successful or unsuccessful swings of the bat (response execution). One way to verify the complexity of the response execution process is to consider the extremely complex programming efforts required to enable robots to perform simple reaching, grabbing, and turning manipulations.

Feedback.   It is evident that we typically monitor the consequences of our actions, forming the closed-loop feedback structure depicted in Figure 1.1 (Chapter 11). While feedback most often is considered in terms of a visual feedback loop (i.e., we see the consequences of our responses), feedback through the auditory, proprioceptive, and tactile (skin senses) modalities may be at least as important under some circumstances. For example, when changing gears in an automobile, the proprioceptive feedback from the hand is of considerable importance in evaluating when the gears have been engaged successfully.

Attention.   Much of the processing that occurs following STSS appears to require attention to function efficiently. In this context, we may model attention both as a searchlight that chooses information sources to process and as a commodity or *resource* of limited availability. If some processes require more of this resource, then less is available for other processes, whose performance therefore will deteriorate. We need only note how one stops talking in a car (response selection and execution) if there is a sudden need to scan a crowded freeway for a critical road sign (perception). Yet learning and practice decrease the demand for the limited supply of resources. We can walk while talking because walking, a well-practiced skill, requires little attention. The selective aspects of attention are discussed in Chapter 7. The aspects of attention that will be addressed in Chapter 8 concern how attentional resources improve performance of tasks and how well humans can distribute attentional resources as they are required.

Qualifications of the Model

The model depicted in Figure 1.1 is a useful simplifying framework for interpreting human performance in complex tasks. However, it is important to realize that it *is* simplified, and that not all aspects of the representation should be taken literally. In particular, three important qualifications should be highlighted.

First, the "flow" of information has been portrayed in Figure 1.1 as moving from initial reception to response execution. There are many instances, however, in which the processes operate in reverse order, particularly in cases where our interpretation of sensory data is greatly influenced by our *expectancies*, generated from short- and long-term memory, as well as by the outcome of our previous decisions (Chapters 2, 4, and 5). Bobrow and Norman (1975) have referred to the distinction between "data-driven," or "bottom up," processes (those operating left to right in Figure 1.1) and "conceptually driven," or "top down," processes. Decisions that have already been made or information already stored in memory may influence the perceptual categorizations made. The effect of expectancies on perception will be considered in some detail in Chapters 4 and 5.

Second, these operations, or stages, should not be thought of as literal boxes or even necessarily as physical locations within the brain. Rather, they represent functional transformations performed on the information as it is processed. It may be that some of the mental operations can, in fact, be structurally localized to particular regions or physical networks within the brain. Definable areas in the cortex, for example, seem to be responsible for early stages of sensory and perceptual analysis. Yet in most instances, correlations between information-processing paradigms and physiological manifestations of brain activity are still too sparse to make definitive localization possible.

Third, the distinction between processing stages may not be as clear as indicated, although there does appear to be both logical and experimental grounds for distinguishing between the three processes of perception, action selection, and response execution. More importantly, although the diagram suggests a discrete temporal sequencing (i.e., perception is terminated before response selection begins), this sequencing may not be absolute. In fact, recent research, described in Chapter 9, suggests that information is not handed along stage by stage like a car on an assembly line. Rather, it undergoes a continuous "flow," with considerable overlap in time between the operation of different processing stages (Eriksen & Schultz, 1978; McClelland, 1979). For example, the response selection process may begin for a particular stimulus even before that stimulus has been absolutely and categorically recognized.

A NOTE ON ORGANIZATION

A book on human performance can be organized in at least two ways: by human performance *components* such as detection, perception, spatial judgment, memory, and action selection; or by *tasks* such as monitoring, navigation, manual control, decision making, and process control. The two organizing frameworks are highly, but not perfectly, correlated. For example, the task of process control involves the performance component abilities of monitoring, decision making, and motor control. This book is organized by performance components. However, this strategy imposes certain constraints on the order in which abilities are discussed. Accordingly, when a task application is discussed in the context of one ability, it sometimes will be necessary to refer to other abilities that will be described in future chapters. With one particular topic—process control—an exception to the organizing principle is made, and this task will be treated in a chapter of its own (Chapter 12). The component abilities of process control span such a wide range that the importance of the paradigm would be diluted if it received only piecemeal treatment throughout the book. Its place as the final chapter also serves to integrate the topics covered before.

References

Anderson, J. R. (1981). *Cognitive skills and their acquisition.* Hillsdale, NJ: Erlbaum Associates.

Bailey, R. W. (1982). *Human performance engineering: A guide for system designers.* Englewood Cliffs, NJ: Prentice-Hall.

Bobrow, D. G., & Norman, D. A. (1975). Some principles of memory schemata. In D. G. Bobrow & A. M. Collins (Eds.), *Representation and understanding: Studies in cognitive science.* New York: Academic Press.

Broadbent, D. E. (1958). *Perception and communications.* New York: Pergamon Press.

Card, S., Moran, T. P., & Newel, A. (1983). *The psychology of human-computer interactions.* Hillsdale, NJ: Erlbaum Associates.

Chase, W. G., & Ericsson, A. (1981). Skilled memory. In J. Anderson (Ed.), *Cognitive skills and their acquisition.* Hillsdale, NJ: Erlbaum Associates.

Deatherage, B. H. (1972). Auditory and other sensory forms of information presentation. In H. P. Van Cott & R. G. Kinkade (Eds.), *Human engineering guide to system design.* Washington, DC: U.S. Government Printing Office.

Edwards, W., Lindman, H., & Phillips, L. (1965). Emerging technologies for making decisions. In T. M. Newcomb (Ed.), *New directions in psychology* (Vol. 2). New York: Holt, Rinehart, and Winston.

Egan, D. E., & Schwartz, B. S. (1979). Chunking in recall of symbolic drawings. *Memory & Cognition, 7,* 149–158.

Electrical Power Research Institute. (1977, March). *Human factors review of nuclear power plant design: Final report.* (Project 501, NP–309–SY). Sunnyvale, CA: Lockheed Missiles and Space Co.

Eriksen, C. W., & Schultz, D. (1978). Temporal factors in visual information processing. In J. Requin (Ed.), *Attention and performance VII.* Hillsdale, NJ: Erlbaum Associates.

Fitts, P. M., & Jones, R. E. (1947, July). *Analysis of factors contributing to 460 "pilot error" experiences in operating aircraft controls* (Memorandum Report TSEA4-694-12, Aero Medical Laboratory). Dayton, OH: Wright Patterson Air Force Base. Reprinted in W. Sinaiko (Ed.), *Selected papers on human factors in the design and use of control systems.* New York: Dover, 1981.

Gerald, F. (1965). *The human senses.* New York: Wiley.

Green, D. M., & Swets, J. A. (1966). *Signal detection theory and psychophysics.* New York: Wiley.

Howell, W. C., & Goldstein, I. L. (Eds.) (1971). *Engineering psychology: Current perspectives in research.* New York: Appleton-Century-Crofts.

Hurst, R. (1976). *Pilot error.* London: Granada.

Jeffries, R., Turner, A. A., Polson, P. G., & Atwood, M. E. (1981). The processes involved in software design. In J. R. Anderson (Ed.), *Cognitive skills and their acquisition.* Hillsdale, NJ: Erlbaum Associates.

Kantowitz, B. H., & Sorkin, R. D. (1983). *Human factors: Understanding people-system relationships,* New York: Wiley.

Kaufman, L., and Boff, K. (Eds.). (forthcoming). *Handbook of perception and performance.* New York: Wiley.

Lachman, R., Lachman, J. L., & Butterfield, E. C. (1979). *Cognitive psychology and information processing.* Hillsdale, NJ: Erlbaum Associates.

Larkin, J., McDermott, J., Simon, D., & Simon, H. (1980). Expert and novice performance in solving physics problems. *Science, 208,* 1335–1342.

McClelland, J. (1979). On the time relations of mental processes: An examination of processes in cascade. *Psychological Review, 86,* 287–330.

McCormick, E. J. (1976). *Human factors in engineering and design.* New York: McGraw-Hill.

McCormick, E. J., & Sanders, M. S. (1982). *Human factors in engineering and design.* New York: McGraw-Hill.

McRuer, D. T. (1980). Human dynamics in man-machine systems. *Automatica, 16,* 237–253.

Neisser, U. (1967). *Cognitive psychology.* New York: Appleton-Century-Crofts.

Posner, M. I., & Konick, A. F. (1966). Short-term retention of visual and kinesthetic information. *Organizational Behavior and Human Performance, 1,* 71–88.

Poulton, E. C. (1966). Engineering psychology. *Annual Review of Psychology, 17,* 177–200.

Rasmussen, J. (1981). Models of mental strategies in process plant diagnosis. In J. Rasmussen and W. B. Rouse (Eds.), *Human detection and diagnosis of system failures.* New York: Plenum Press.

Rouse, W. B. (1978). A model of human decision making in a fault diagnosis task. *IEEE Transactions on Systems, Man, and Cybernetics, SMC–8,* 357–361.

Rouse, W. B. (1981). Experimental studies and mathematical models of human problem-solving performance in fault diagnosis tasks. In J. Rasmussen and W. B. Rouse (Eds.), *Human detection and diagnosis of system failures,* New York: Plenum Press.

Rubinstein, T., & Mason, A. F. (1979, November). The accident that shouldn't have happened: An analysis of Three Mile Island. *IEEE Spectrum,* pp. 33–57.

Schneiderman, B. (1976). Exploratory experiments in programmer behavior. *International Journal of Computer and Information Sciences, 5,* 123–143.

Sheridan, T. (1981). Understanding human error and aiding human diagnostic behavior in nuclear power plants. In J. Rasmussen & W. B. Rouse (Eds.), *Human detection and diagnoses of system failures.* New York: Plenum Press.

Siegel, A. I., & Wolf, J. A. (1969). *Man-machine simulation models.* New York: Wiley.

Smith, E. (1968). Choice reaction time: An analysis of the major theoretical positions. *Psychological Bulletin, 69,* 77–110.

Sperling, G. (1960). The information available in brief visual presentations. *Psychological Monographs, 74* (498).

Stammers, R., & Patrick, J. (1975). *The psychology of training.* London: Methuen.

Sternberg, S. (1969). The discovery of processing stages: Extension of Donders' method. *Acta Psychologica, 30,* 276–315.

VanCott, H. P., and Kinkade, R. G. (Eds.). (1972). *Human engineering guide to system design.* Washington DC: U.S. Government Printing Office.

Welford, A. T. (1976). *Skilled performance.* Glenview, IL: Scott, Foresman.

Wherry, R. J. (1976). The human operator simulator (HOS). In T. Sheridan & G. Johanssen (Eds.), *Monitoring behavior and supervisory control.* New York: Plenum Press.

Wickens, C. D. (1983). Processing resources in attention. In R. Parasuraman & R. Davies (Eds.), *Varieties of attention.* New York: Academic Press.

Signal Detection and Absolute Judgment

OVERVIEW

Information processing in most systems begins with the detection of some environmental event. In some cases, such as the Three Mile Island incident, the event is so noticeable that immediate detection is assured. The information-processing problems in these circumstances are those of recognition and diagnosis. However, there are many other circumstances in which detection itself represents a source of uncertainty or a potential bottleneck in performance, because it is necessary to detect events that are near the threshold of perception. Will the security guard monitoring a bank of television pictures detect the abnormal movement on one of them? Will the radiologist detect the abnormal x-ray as it is scanned? Will the monitor of the process control plant detect a significant rise in temperature in one of the indicators?

This chapter will first deal with the critical question of detection in which perceptual categorization is made into one of two states: a signal is present or it is absent. The detection process will be modeled within the framework of signal-detection theory, and we shall show how the model can assist engineering psychologists in understanding the complexities of the detection process, in diagnosing what goes wrong when detection fails, and in recommending corrective solutions.

The process of detection may in fact involve more than two states of categorization. It may, for example, require the operator to choose between three or four levels of uncertainty about the presence of a signal or to detect more than one kind of signal. At this point our consideration moves into the realm of identification and recognition. The final pages of the chapter will deal with the simplest form of multilevel categorization, the absolute judgment task.

Together, the treatments of signal detection and absolute judgment presented in this chapter provide the foundation for considering more complex extensions of detection and recognition in the three following chapters: decision making in Chapter 3, in which the *uncertainty*, or probabilistic element, of the detection task is emphasized in a wide variety of situations, and pattern recognition in Chapters 4 and 5, in which the rules for perceptual categorization are far more complex than the simple 2–5-level categorization considered here. In the supplement of this chapter, we will consider some of the more quantitative modeling approaches to detection and absolute judgment: the ROC curve as a means of describing signal detection performance and information theory as a model for describing absolute judgment performance.

SIGNAL DETECTION THEORY

The Signal Detection Paradigm

Signal detection theory is applicable in any situation in which there are two discrete *states of the world* (call these signal and noise) that cannot easily be discriminated. Signals must be detected by the human operator, and in the process two response categories are produced: "yes" (I detect a signal) and "no" (I do not). This situation may describe activities such as the detection of a contact on a radar scope (N. H. Mackworth, 1948), a malignant tumor on an x-ray plate by a radiologist (Parasuraman, 1980; Swets & Pickett, 1982), a malfunction of an abnormal system by a nuclear plant supervisor (Lees & Sayers, 1976), a critical event in air traffic control (Bisseret, 1981), typesetting errors by a proofreader (Anderson & Revelle, 1982), an untruthful statement from a polygraphy (Szucko & Kleinmuntz, 1981), or a communications signal from intelligent life in the bombardment of electromagnetic radiation from outer space (Blake & Baird, 1980).

The combination of two states of the world and two response categories produces the 2 × 2 matrix shown in Figure 2.1, generating four classes of joint events which are labeled hits, misses, false alarms, and correct rejections. It is apparent that perfect performance is that in which no misses or false alarms occur. However, since the signals are not very salient in the typical signal detection paradigm, misses and false alarms do occur, and so there are normally data in all four cells. In signal detection theory (SDT) these values are typically expressed as probabilities by dividing the number of occurrences in a cell by the total number of occurrences in a column. Thus if there were 5 hits and 15 misses, we would write $P(\text{hit}) = 5/20 = 0.25$.

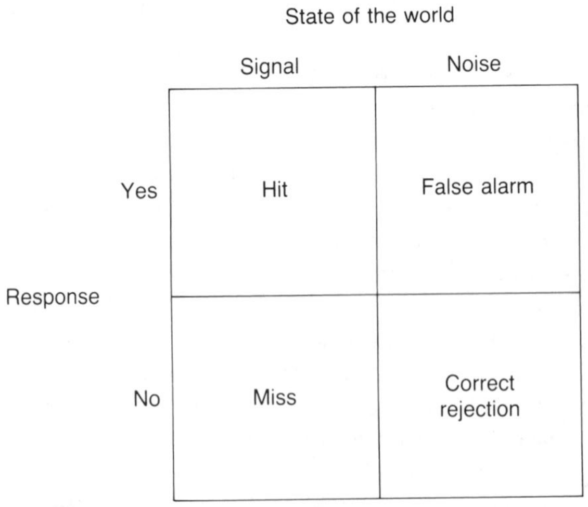

Figure 2.1 The four outcomes of signal detection theory.

The SDT model assumes that there are two stages of information processing in the task of detection: (1) sensory evidence is aggregated concerning the presence or absence of the signal and (2) a decision is made whether this evidence constitutes a signal or not. According to the theory, external stimuli generate neural activity in the brain. Therefore, there is assumed to be on the average more sensory or neural evidence in the brain when a signal is present than when it is absent. This neural evidence, X, may be conceived as the rate of firing of neurons at a hypothetical "detection center." The rate increases in magnitude with stimulus intensity. We refer to the quantity X as the *evidence variable*. Therefore, if there is enough neural activity, X exceeds a critical threshold X_C, and the operator decides "yes." If there is too little, the operator decides "no."

Because the amount of energy in the signal is typically low, the average amount of X generated by signals in the environment is not much greater than the average generated when no signals are present (noise). Furthermore, the quantity of X varies continuously even in the absence of a signal because of random variations in the environment and in the operator's own "baseline" level of neural firing (e.g., the neural "noise" in the sensory channels and the brain). Therefore, even when no signal is present, X will sometimes exceed the criterion X_C as a result of random variations alone, and the subject will say "yes" (generating a false alarm). Correspondingly, even with a signal present, the random level of activity may be low, causing X to be less than the criterion, and the subject will say "no" (generating a miss). The smaller the difference in intensity between signals and noise, the greater these error probabilities become, because the amount of variation in X resulting from randomness increases relative to the amount of energy in the signal. Figure 2.2 shows a hypothetical example of this random variation in X. The overall level of X is increased slightly in the presence of a weak signal and greatly when the strong signal is presented.

As an example, consider the monitor of a noisy radar screen. Somewhere in the midst of the random variations in stimulus intensity caused by reflections from clouds and rain, there is an extra increase in intensity that represents the presence of the stimulus—an aircraft. The amount of noise will not be constant over time but will fluctuate; sometimes it will be high, completely masking the stimulus, and sometimes low, allowing the plane to stand out. In this example, "noise" varies in the environment. Suppose instead you were standing watch on a ship searching the horizon on a dark night for a faint light. It becomes difficult to distinguish the flashes that might be real lights from those that are just "visual noise" in your own sensory system. In this case, the random noise is internal.

The relations between presence or absence of the signal, random variability of X, and X_C can be seen in Figure 2.3. The figure plots the probability of observing a specific value of X, given that a noise trial (left curve) or signal trial (right curve) in fact occurred. In Figure 2.2, these data might have been tabulated by counting the relative fre-

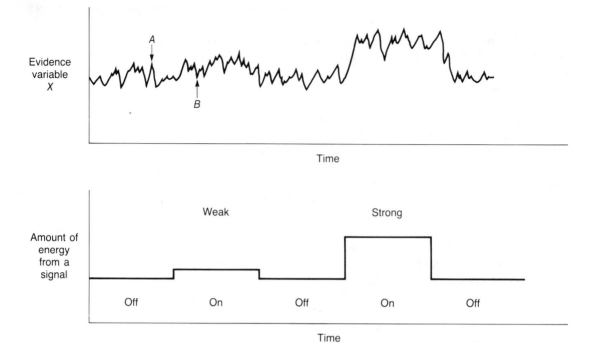

Figure 2.2 The change in the evidence variable X caused by a weak and a strong signal. Notice that with the weak signal there can sometimes be less evidence when the signal is present (point B) than when the signal is absent (point A).

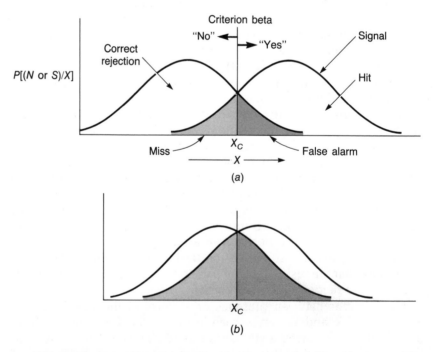

Figure 2.3 Hypothetical distributions underlying signal detection theory: (a) high sensitivity, (b) low sensitivity.

quency of different X values during the intervals when the signal was off, creating the probability curve on the left of Figure 2.3, and making a separate count of the probability of different X values while the weak signal was on, generating the curve on the right. As the value of X increases, it is relatively more likely to have been generated while a signal was present than while it was absent.

The point at which the absolute probability that X was produced by the signal equals the probability that it was produced by only noise is the value of X where the two curves intersect. The criterion value X_C chosen by the operator is shown by the vertical line. All X values to the right $(X > X_C)$ will cause the operator to respond "yes." All to the left generate "no" responses. The different shaded areas represent the occurrences of hits, misses, false alarms, and correct rejections. Since the total area within each curve is one, the two shaded regions within each must also add to one. That is, $P(H) + P(M) = 1$ and $P(FA) + P(CR) = 1$.

It is important to bear in mind that the distributions in Figure 2.3 are purely hypothetical. This representation is assumed to be present somewhere in the brain but cannot be observed directly. The assumption of the specific form of the two normal distributions will be critically important to the actual measurement of detection performance.

Setting the Response Criterion: Optimality in SDT

In any signal detection paradigm, whether an experimental one in the laboratory or a real-world situation, operators may be described in terms of their *response bias*. They may be prone to say "yes," thereby detecting most of the signals that occur but incurring many false alarms ("risky responders"), or they may be "conservative," saying "no" most of the time and making few false alarms but missing many of the signals.

Different circumstances may dictate whether a conservative or a risky strategy is best. For example, when the radiologist scans the x-ray of a patient who has been referred to her because of other symptoms of illness, it is probably appropriate to be more biased to say "yes" (detecting a tumor) than when examining the x-ray of a totally healthy patient, for whom there is no reason to suspect any malignancy (Swets & Pickett, 1982). Consider, on the other hand, the monitor of the power-generating station who has been cautioned repeatedly by the supervisor to guard against any unnecessary shutdowns of a turbine because of the resulting loss of revenue to the company. The operator will probably become conservative in monitoring the dials and meters for indication of malfunction and may be prone to miss (or delay responding to) a malfunction when it does occur.

According to Figure 2.3, an operator's conservative or risky behavior is determined by placing the decision criterion X. If X_C is placed to the right, then very much evidence is required for it to be exceeded, and most responses will be "no" (conservative responding). If it is placed to the left, then little evidence is required, most responses will be "yes," and the strategy is risky. An important variable closely related to the hypothetical X_C is the quantity *beta*, which is the ratio

$$\frac{P(X/S)}{P(X/N)}$$

This is the ratio of the height of the two curves of Figure 2.3, at a given level of X_C. Therefore, X_C and beta are correlated. Intuitively we see that higher values of beta (and X_C) will generate fewer "yes" responses and therefore fewer hits, but also fewer false alarms. Lower beta settings generate more "yes" responses, more hits, and more false alarms. Thus the beta parameter belongs to the category of *bias* measures described in Chapter 1, which have costs and benefits at both ends of the scale.

More formally, the actual probability values appearing in the matrix of Figure 2.1 that would be calculated from obtained data relate to the areas under the two probability distribution functions of unit area shown in Figure 2.3, to the left and right of the criterion. Thus, for example, the probability of a hit, with the criterion shown, is the relative area under the "signal" curve (a signal was presented) to the right of the criterion (the subject said "yes"). One can determine by inspection that if the two distributions are of equal size and shape, then the setting of beta = 1 occurs where the two curves intersect and will provide data in which $P(H) = P(CR)$ and $P(M) = P(FA)$, that is, a truly "neutral" criterion setting.

Signal detection theory, as a normative model, is able to prescribe exactly where the optimum beta should fall, given environmental conditions related to (1) the likelihood of observing a signal and (2) the costs and benefits of the four possible outcomes (Green & Swets, 1966; Swets & Pickett, 1982). We shall consider this optimal setting first as it is dictated by signal probability alone and then as the costs and benefits of different outcomes are considered.

1. *Equal probabilities.* Consider the situation in which signals occur just as often as they don't and, furthermore, there is neither differential cost to the two bad outcomes nor differential benefits to the two good outcomes of Figure 2.1. In this case *optimal* performance minimizes the number of errors (misses and false alarms). It can be shown that the particular symmetrical geometry of Figure 2.3 dictates that optimal performance will occur when X_C is placed at the intersection of the two curves, that is, when beta = 1. Any other placement, in the long run, would reduce the probability of being correct.

2. *Effect of probabilities.* It is intuitively reasonable that if a signal is more likely, the criterion should be lowered. For example, if the radiologist has available other symptomatic information to suggest that a patient is likely to have a malignant tumor or has received the patient on referral, she should be more likely to categorize a possible abnormality on the x-ray as a tumor than to ignore it as mere noise in the x-ray process. On the other hand, if signal probability is reduced, then beta

should be adjusted conservatively. For example, suppose a quality control inspector searching for defects in assembled circuit boards is told that the batch under inspection has a very low estimated fault frequency because the manufacturing equipment has just received maintenance. In this case, the inspector should be more conservative in searching for defects. Formally, this adjustment of the optimal beta to changes in signal and noise probability is represented by the prescription

$$\beta_{opt} = \frac{P(N)}{P(S)} \qquad (2.1)$$

This quantity will be reduced (made more risky) as $P(S)$ increases, thereby moving X_C to the left of Figure 2.3. If this setting is adhered to, performance will still maximize the number of correct responses (hits and correct rejections). It is important to emphasize that the setting of optimal beta will not produce perfect performance. There will still be false alarms and misses as long as the two curves overlap. However, optimal beta is the best that can be expected for a given signal strength and a given level of sensitivity.

3. *Effect of payoffs.* The optimal setting of beta is also influenced by *payoffs*. In this case, "optimal" is no longer defined in terms of minimizing errors but is now maximizing the total expected financial gains (or minimizing expected losses). Again, intuitively, if it were important that signals never be missed, the operator might be provided high reward for hits and high penalties for misses, leading to a low setting of beta. This payoff would be in effect for a power station monitor who, unlike his colleague described above, is admonished by his supervisor as to the severe costs in company profits (and his own paycheck) if an undetected malfunction of a turbine damages the turbine blades. Conversely, in different circumstances, if false alarms are to be avoided, they might be heavily penalized. These costs and benefits can be translated into a prescription for the optimum setting of beta by expanding Equation 2.1 to

$$\beta_{opt} = \frac{P(N)}{P(S)} \times \frac{V(CR) + C(FA)}{V(H) + C(M)} \qquad (2.2)$$

where V is the value of the specified desirable event (hit, H, or correct rejection, CR), and C is the cost of the specified undesirable event (miss, M, or false alarm, FA). In Equation 2.2 an increase in any quantity in the denominator will decrease beta and lead to risky responding. Conversely, an increase in numerator values will lead to conservative responding. Notice also that the value and probability portions of the function combine independently. Thus, although such an event as the malfunction of a turbine may occur only very infrequently, raising the optimum beta, the consequence of a miss in detecting it might be so severe (cost of miss) as to reduce beta to a relatively low value.

Human performance in setting beta.   Since the actual value of beta that an operator uses can be computed on the basis of the known probabilities of hits and false alarms obtained from a series of detection trials, we may ask how well humans actually perform in setting their criteria in response to changes in payoffs and probabilities. In fact, humans do adjust beta as dictated by changes in these quantities. However, laboratory studies suggest that beta is not adjusted as much as it should be. That is, subjects manifest a "sluggish beta," a relationship shown in Figure 2.4. They are less risky than they should be if the ideal beta is low and less conservative than they should be when the optimum beta is high. As shown in Figure 2.4, this sluggishness is found to be more pronounced when beta is manipulated by probabilities than by payoffs (Green & Swets, 1966).

A number of explanations have been proposed to account for why beta is sluggish in response to probability manipulations. It is possible that it is simply a reflection of the operator's need to respond "creatively," by introducing the rare response more often than is optimal, since extreme values of beta optimally dictate long strings of either yes (low beta) or no (high beta) responses. Another explanation may be that the operator misperceives probabilistic data. There is some convincing evidence that subjects tend to overestimate the probability of very rare events and underestimate that of very frequent events (Peterson & Beach, 1967; Sheridan & Ferrell, 1974). This behavior, which will be discussed in more detail in Chapter 3, would produce the observed shifting of beta toward unity.

The sluggish beta phenomena can be demonstrated most clearly in the laboratory, where precise values of probabilities and values can be

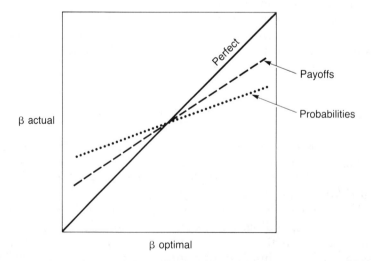

Figure 2.4   The "sluggish beta" phenomenon. From *Signal Detection Theory and Psychophysics* (p. 92) by D. M. Green and J. A. Swets, 1966, New York: Wiley. Copyright 1966 by John Wiley & Sons, Inc. Adapted by permission.

specified to the subjects. In real-world environments there is evidence that operators do indeed adjust beta with changes in probabilities although not necessarily as far as optimally prescribed. For example, quality control inspectors examining sheet metal for defects will adjust their response criterion according to the estimated defect rate of a batch (Drury & Addison, 1973). Bisseret (1981) has applied signal detection theory to the air traffic controller's task of judging whether an impending conflict situation between two aircraft will (signal) or will not (noise) require a course correction. He finds that controllers will adjust beta to lower values (be more willing to detect a conflict and therefore command a correction) as the difficulty of the problem and therefore the uncertainty of the future increases. Bisseret also compared performance of experts and trainees and found that the former group as a whole were lower in their setting of beta, being more willing to call for a correction. Bisseret attributes this finding to the fact that trainees are more uncertain as to how to implement the correction and therefore more reluctant to take the action that would require implementation. Bisseret argues that a portion of their training should be devoted to the issue of criterion placement.

In medicine, Lusted (1976) reports evidence that doctors do adjust the response criterion in diagnosis according to the disease prevalence rate—essentially an estimate of P(signal)—but adjust less than the optimal amount specified by the changing probabilities. However, the difficulty of specifying precisely the costs or benefits of all four of the joint events in medical diagnosis, as in air traffic control, makes it quite difficult to determine the exact level of optimal beta (as opposed to the direction of its change). How, for example, can the physician specify the precise cost of an undetected malignancy (Lusted, 1976), or the air traffic controller specify the costs of an undetected conflict that produced a midair collision? How can the power plant operator specify the cost of a turbine that will be off the line for an unpredictable amount of time, with ill-defined damages, before the shutdown has actually occurred? The problems associated with specifying costs define the limits of applying signal detection theory to determine optimal response criteria.

Sensitivity

One of the most important contributions of signal detection theory is that it has made a conceptual and analytical distinction between the response bias parameters described above and the measure of the operator's *sensitivity*, the keenness or resolution of the detection mechanisms. We have seen above that the operator may fail to detect a signal if his response bias is conservative. Correspondingly, the signal may be missed simply because the resolution of the detection process is low in discriminating signals from noise, even if the response bias is neutral or risky.

In terms of the theoretical representation of Figure 2.3, sensitivity refers to the separation of the internal noise and signal-plus-noise distributions along the X axis. If the separation is great, sensitivity is great and a given value of X is quite likely to be generated by either S or N but not by both. If it is low, sensitivity is low. Since the curves are assumed to be internal, their separation could be influenced either by physical properties of the signal (e.g., a change in its intensity or salience) or by properties of the subject (e.g., a loss of hearing for an auditory detection task or a lack of training of a medical student for the task of detecting complex tumor patterns on an x-ray). Figure 2.2 presented an example of two signals, one for which sensitivity is low, one for which it is high.

The sensitivity measure falls into the category of "efficiency" parameters described in Chapter 1: those in which changes in values clearly have a "good" and a "bad" interpretation. As may be noted by the change in separation of the curves in Figure 2.3b, decreases in sensitivity produce an increase in errors, both misses and false alarms. In the formal theory of signal detection, the sensitivity measure is called d' and corresponds to the separation of the means of two distributions in Figure 2.3 expressed in units of their standard deviations. This is a value that in most applications of signal detection theory ranges roughly between 0.5 and 2. Appendix A presents values of d' generated by different hit and false alarm rates.

Like bias, sensitivity also has an optimal value (which is not perfect). The computation of this optimal is more complex and is based upon an ability to characterize precisely the statistical properties of the physical energy in signal and no-signal trials. While this can be done in carefully controlled laboratory studies with acoustic signals and white-noise background, it is difficult to do in more complex environments. Nevertheless, data from auditory signal detection investigations showing the ways in which human sensitivity departs from optimal sensitivity (Green & Swets, 1966) have extremely important practical implications. These data suggest that the major cause for the departure results from the operator's lack of memory of the precise physical characteristics of the signal. When memory aides are provided that continuously remind the operator of what the signal sounds like or looks like, d' approaches the optimal levels. This point will be important when we consider the nature of vigilance tasks later in the chapter.

The ROC Curve

It should be apparent that all detection performance that has the same sensitivity is in some sense equivalent, no matter what its level of bias. A graphical method of representation known as the receiver operating characteristic (ROC) curve is used to portray this equivalence of sensitivity across changing levels of bias. The ROC curve plots the probability of a hit against the probability of a false alarm on a single graph, such as that shown in Figure 2.5. The ROC curves for two observers are shown. Each has tried to detect a series of signals at each of three levels of bias. Curve A is the performance of a very sensitive observer. There

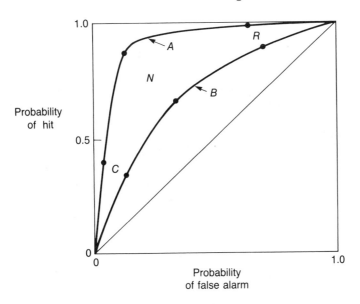

Figure 2.5 Example of two ROC curves. Curve *A*: high sensitivity; curve *B*: low sensitivity. Points *C*, *N*, and *R* represent conservative, neutral, and risky settings of beta, respectively.

are always many more hits than false alarms, no matter whether *A* is responding conservatively (point *C*), with a neutral criterion (point *N*), or a risky one (point *R*). These three points of course correspond to high, medium, and low settings of beta, respectively. Curve *B* represents the performance of a much less sensitive observer with a lower *d'*. For this observer, the probability of a false alarm is always fairly close to the probability of a hit, no matter what level of beta is used.

The ROC curve is a useful means of plotting different styles of performance and of comparing the performance of two observers (or one observer using two different systems). The chapter supplement discusses in greater detail the actual quantitative relation between the ROC curve and the theoretical curves of Figure 2.3. It also shows how the ROC curve is formally related to measures of sensitivity and response bias. Finally, it discusses the different methods of actually estimating sensitivity and bias from a set of detection data.

APPLICATIONS OF SIGNAL DETECTION THEORY

Signal detection theory has had a tremendous impact on the field of experimental psychology, and its concepts are highly applicable to many problems of human engineering as well. Its benefits can be divided into two general categories: (1) It provides the ability to compare sensitivity and therefore the quality of performance between conditions or between operators that may differ in response bias. (2) By partitioning performance and therefore performance change into bias and sensitivity components, it provides a diagnostic tool that implicitly recommends different corrective actions depending on whether a deterioration of performance results from a loss of sensitivity or a shift in response bias.

The implications with regard to the first category are clear. The performance of two operators or the hit rate obtained using two different pieces of inspection equipment are compared. If A has a higher hit rate but also a higher false alarm rate than B, which is superior? Unless the explicit mechanism for trading off hits for correct rejections and false alarms for misses is available, this comparison is impossible. Signal detection theory provides the mechanism.

The importance of the second point—the diagnostic value of signal detection theory—will be evident as we consider some actual examples of applications of signal detection theory to real-world tasks. In the myriad of possible environments in which the operator must detect an event and does so imperfectly, generating either misses or false alarms, the existence of these errors presents a challenge for the engineering psychologist: Why do they occur, and what corrective actions can prevent them? Two areas of application, medical diagnosis and eyewitness testimony, will be considered, leading to a more extensive discussion of vigilance.

Medical Diagnosis

The realm of medical diagnosis is a fruitful environment for the application of signal detection theory (Lusted, 1971, 1976). Abnormalities (diseases, tumors) are either present in the patient or they are not, and the physician's initial task is often to make a yes or no decision. The strength of the signal (and therefore, the sensitivity of the human operator) is related to factors such as the salience of the abnormality or the number of converging symptoms, as well as the training of the physician to focus upon relevant cues. Response bias, meanwhile, can be influenced by both signal probability and payoffs. In the former category, influences include the disease prevalence rate, and whether the patient is examined in initial screening (probability of disease low, beta high) or referral (probability higher, beta lower). Lusted (1976) has argued that physicians' detections generally tend to be less responsive to variation in disease prevalence rate than optimal. Factors that influence payoffs include the difficult-to-quantify consequences of hits (e.g., a detected malignancy that will lead to its surgical removal with associated hospital costs and possible consequences), false alarms (an unnecessary operation), and misses. Placing values upon these events based upon financial costs of surgery, malpractice suits, and intangible costs of human life and suffering is clearly difficult. Yet there is little doubt that they do have an important influence on a physician's detection rate. Lusted (1976) and Swets and Pickett (1982) have shown how diagnostic performance can be quantified using an ROC curve. In a very thorough treatment, Swets and Pickett go into considerable detail describing the appropriate methodology that should be employed when using signal detection theory to examine performance in medical diagnosis.

Several investigations have examined the more restricted domain of tumor diagnosis by radiologists, in which performance, at least theoretically, is far from optimal. Rhea, Potsdaid, and DeLuca (1979) have estimated the rate of omission in the detection of abnormalities to run between 20 and 40%. Swennsen, Hessel, and Herman (1977) have examined the effect on detection, of directing the radiologist's attention to a particular area of an x-ray plate in which an abnormality is likely to occur. They find that such focusing of attention will indeed increase the likelihood of the tumor's detection, but will do so by reducing beta rather than increasing sensitivity. Swennsen et al. (1979) have also compared search for x-ray abnormalities between conditions in which the plates contain only the abnormality (in some proportion) and conditions in which the critical target abnormality is mixed with plates containing other pathologies. While the former condition produced a higher hit rate, it was accomplished by a reduction in beta. This shift in beta was such that the sensitivity was actually reduced compared to the mixed plate conditions.

Parasuraman (1980), comparing detection performance of staff radiologists and residents, found major differences between the two populations in terms of sensitivity (favoring the staff radiologists) and bias (radiologists showing a more conservative criterion in general). A third dimension of contrast between these groups, with important implications for training, related to the adjustment of beta. The staff radiologists were much more responsive than residents in adjusting beta according to the disease prevalence rate.

Recognition Memory and Eyewitness Testimony

The domain of recognition memory represents a somewhat different application of signal detection theory. Here we are not assessing the correctness of a decision concerning whether or not a physical signal is present but rather a decision concerning whether or not a physical stimulus (the person or name to be recognized) "matches" a trace in memory. One important application of signal detection theory to memory is found in the study of eyewitness testimony (Buckhout, 1974; Ellison & Buckhout, 1981; Wells, Lindsay, & Ferguson, 1979). The witness to a crime may be called upon to recognize or identify a suspect as the perpetrator. The four kinds of joint events in Figure 2.1 can readily be specified. The suspect examined by the witness either is (signal) or is not (noise) the same individual actually perceived at the scene of the crime. The witness in turn can either say "that's the one" (Y), or "no it's not" (N).

In this case, the joint interests of criminal justice and protection of society are served by maintaining a high level of sensitivity while keeping beta neither too high (many misses, with criminals more likely to "go free") nor too low (a high rate of false alarms with an increased likelihood that innocent individuals will be prosecuted). Signal detection theory has been most directly applied to a witness's identification of suspects in police lineups (Buckhout, 1974; Ellison & Buckhout, 1981). In this case, the witness is presented a lineup of five or so individuals, one of whom is the suspect detained by the police, while the others are "foils." The decision is now a two-stage process: Is the suspect in the lineup, and if so, which one is it?

Ellison and Buckhout have expressed concern that witnesses generally have a low response criterion in lineup identifications and will therefore often say yes to the first question. This bias would present no difficulty if their recognition memory was also accurate, enabling them to identify the suspect accurately. However, considerable research on staged crimes shows that visual recognition of brief events is notoriously poor (Buckhout, 1974; Loftus, 1979; Wells, Lindsay, and Ferguson, 1979). Poor recognition memory coupled with the risky response bias in turn allows those conducting the lineup to use techniques that will capitalize on witness bias to ensure a positive identification. These techniques include such things as ensuring that the suspect is differently dressed, is seen in handcuffs by the witness prior to the lineup, or is quite different in appearance from the foils. In short, techniques are used that would lead even a person who had *not seen* the crime to select the suspect from the foils or would lead the witness to make a positive identification from a lineup that *did not contain the suspect* (Ellison & Buckhout, 1981). This is not testing the sensitivity of recognition memory. It is emphasizing response bias.

In applying signal detection theory to this procedure, investigators are interested in characteristics of the lineup process that might reduce the magnitude of response bias toward the optimum. Three procedures have been suggested: (1) Buckhout (1974) has found that those who express greater confidence in their positive identifications ("I'm sure that's the one") are actually less sensitive observers than those who are less confident. (2) Ellison and Buckhout suggest the simple procedure of informing the witness that the suspect may not be in the lineup. Like reducing the apparent probability of a signal, this procedure will drive beta upward toward a more optimal setting. (3) Ellison and Buckhout argue that individuals in the lineup should all be equally similar to each other (such that a nonwitness would have an equal chance of picking any of them). Although such greater similarity will reduce hit rate slightly, it will reduce the false alarm rate considerably more. The result will be a net increase in sensitivity.

VIGILANCE

The vigilance paradigm is one of the most common applications of signal detection theory, although occasionally a frustrating one. The general defining characteristics of this paradigm are that the operator is required to detect signals over a relatively long period of time (commonly referred to as the watch) and that the signals are intermittent, unpredictable, and infrequent. Examples include the radar monitor, who must observe infrequent contacts; the supervisory monitor of complex systems, who must detect the infrequent malfunctions of system components; and the quality control inspector who examines a stream of products (sheet metal, circuit boards, fruits) to detect and remove defective or flawed items.

Two very general conclusions emerge from the analysis of operators' performance in these paradigms: (1) Operators are far from optimal, in particular often showing a higher miss rate (or detection at longer latencies) than desirable. (2) In some situations, operators' detection performance as measured by P(hit) declines during the first half hour or so of the watch. This phenomenon was initially noted in radar monitors during World War II (N. H. Mackworth, 1948) and has subsequently been referred to as the *vigilance decrement* (loss in performance over time). It is distinguished from the *vigilance level*, which is the steady-state level of vigilance performance.

Vigilance Phenomena

It is important to distinguish two classes of vigilance situations or paradigms. The *free response* paradigm, such as that confronting the power plant monitor, is one in which a target event may occur at any time and nonevents are not defined. Event frequency in this case is defined by the number of targets per unit time. The *inspection paradigm*, the quality control inspector's task, is one in which *events* occur at fairly regular intervals. A few of these are *targets* (defects) and most are *nontargets* (normal items). "Event frequency" is therefore an ambiguous term in the inspection paradigm, because it may be defined either by the number of targets per unit time or by the ratio of targets to total events [$T/(T + NT)$]. The latter measure, of course, may be held constant as the number of targets per unit time is increased, simply by increasing the total event rate, by speeding up a conveyor belt, for example.

A large number of investigations of the vigilance decrement have been conducted over the last four decades with a myriad of experimental variables in various paradigms. Books by Broadbent (1971), Davies and Tune (1969), Davies and Parasuraman (1980), J. F. Mackworth (1970), and Mackie (1977) provide a comprehensive compilation and summary of these studies. Some variables influence vigilance level; others influence the decrement. The absolute vigilance *level* (the hit rate) is reduced when signals are short, of low energy, or complex and when the target to nontarget ratio is decreased. The level does not appear to be affected as much by *event rate* (events per unit time), as long as the target to event ratio remains constant across event rates. The magnitude of the vigilance decrement (over time) is increased by short signals and by low energy signals as above, as well as by decreasing the target/event ratio or by increasing the total event rate with the target/event ratio held constant. Discussions by Welford (1968) and Broadbent (1971) provide a fairly comprehensive summary of these effects.

Specifying vigilance performance in terms of signals detected and missed has important implications for the design of systems. However, it should be apparent that defining vigilance performance only by P(hit) is ambiguous, since in terms of signal detection theory P(H) may decline either through a loss of sensitivity or through an increase in beta with no change in sensitivity (i.e., the subject makes fewer false alarms but also makes fewer hits).

In inspection tasks, when nontarget events are clearly defined, the application of signal-detection-theory analysis is straightforward, since false alarms and the false-alarm rate may be easily computed. In the free response paradigm, however, when there is no "nontarget event," further assumptions must be made in computing a false-alarm rate so that the opportunity to make a false alarm can be specified numerically. In the laboratory this is typically accomplished by defining an appropriate response interval after each signal within which a subject's response will be designated a hit. The remaining time during a watch is partitioned into a number of "false-alarm intervals" equal in duration to the response intervals. P(FA) is simply the number of false alarms divided by the number of false-alarm intervals (Watson & Nichols, 1976; Wickens & Kessel, 1979). However, there are certain cautions that need to be considered when signal detection theory is applied to vigilance, particularly when the false alarm rates are quite low. Articles by Jerison, Pickett, and Stenson (1965), Craig (1977), and Long and Waag (1981) discuss some of these cautions.

Theories of Vigilance

An exhaustive listing of all of the experimental results of vigilance studies is far beyond the scope of this chapter. In any case, such a list would also be less than fully informative, because methodological and procedural differences between laboratories have often generated seemingly contradictory results. Instead, it is more instructive to present three major explanatory theories of vigilance (in particular, the vigilance decrement) that predict when a loss in vigilance performance is likely to occur. No single theory is likely to be entirely correct, to the exclusion of others. Rather, under some circumstances the influence of factors underlying one theory will be stronger than others. Also, it is valuable to realize that while the theories have been offered to account for a change in vigilance performance *over time* (i.e., the decrement), their underlying constructs may account for vigilance differences between conditions that differ as a result of any other factor. These might include, for example, differences between displays, between tasks, or between environmental stressors. The advantage of such theories is that they provide somewhat parsimonious ways of accounting for vigilance loss and thereby suggest techniques to improve vigilance performance. After outlining three theories of vigilance, we shall then show how these suggest corrective improvements.

The reduction in P(hit) over the watch has been observed in different laboratories to be a consequence of either a conservative shift in beta, a loss in sensitivity, or both. Of the three theories that are considered, the first will concern the cause of sensitivity loss, and the last two will account for criterion shifts.

Sensitivity loss: Fatigue and memory load. The earliest laboratory demonstrations of the vigilance decrement were reported by N. H. Mackworth (1948) in the classical "clock task." In this task, the subject monitored a clock hand which would tick at periodic intervals (events) and occasionally undergo a "double tick" (signal), moving twice the angle of the normal events. This task was chosen as the laboratory analog of a radar monitor's task, in which each sweep of the scope defines an event with an occasional contact on the scope defining the target. In another version, the clock hand would move continuously and occasionally pause. The former version is an inspection task, the latter a free response task. In this paradigm, as well as in subsequent ones that have primarily employed visual signals, a loss in sensitivity usually occurs over time. Broadbent (1971) has argued that the sustained attention necessary to fixate the clock hand or other visual signals continuously extracts a toll in *fatigue*. Because of this fatigue, the subject looks away more often as the watch progresses and therefore signals are missed.

While Broadbent and others have argued that fatigue theory (and the d' drop) in vigilance is primarily associated with visual signals, more recent evidence compiled by Parasuraman (1979) suggests that there is a different factor responsible for the sensitivity decrements. Parasuraman argues on the basis of his own data and results obtained by other investigators that there are certain conditions in *both* auditory and visual paradigms in which a sensitivity decrement is typically observed. These are conditions in which a high event rate is imposed while the detection of the signal event requires a comparison of the stimulus against a standard or template maintained in working memory. The sensitivity decrement will not occur when the signals can be compared against a standard that is physically present at the time of the signal.

An example of the memory condition in which a d' decrement would occur is provided by the detection of a target tone that varies in pitch, intensity, or duration from a standard nontarget. The sound of the nontarget would have to be continuously maintained as a comparative reference in memory. In contrast, detection of a pair of different tones mixed with identical pairs presents an example of the physical comparison condition, since no memory load is required to identify two tones of different pitch. It is only the former case, imposing the continuous load on the operator's working memory, in which fatigue will reliably reduce the efficiency of detection. Parasuraman argues, furthermore, that it is only an accident of choice that has caused most visual tasks to require comparisons with memory standards and most auditory tasks to allow physical comparisons. This has led to the consequent association of visual monitoring tasks with sensitivity decrements.

Criterion shifts: Arousal theory.   In most vigilance tasks in which hit rate declines, the decline is paralleled by a reduction in false alarms. On some occasions the shift in false-alarm rate is sufficient to suggest that there is no loss in sensitivity at all but only a conservative adjustment of the response criterion (Broadbent & Gregory, 1965). *Arousal theory,* articulated by Welford (1968), postulates that in a prolonged low event environment the "evidence variable" X (see Figure 2.3) shrinks while the criterion stays constant. This change is shown in Figure 2.6. This shrinking results from an overall loss in total

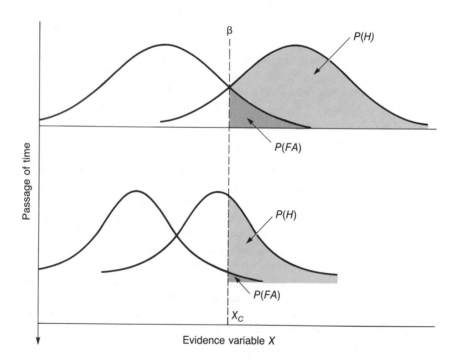

Figure 2.6   Welford's arousal theory of the vigilance decrement. From *Fundamentals of Skill* (p. 268) by A. T. Welford, 1968, London: Metheun. Copyright 1968 by Methuen & Co. Reprinted by permission.

activity (both signal and noise) in the nervous system with decreasing arousal. An examination of Figure 2.6 reveals that such an effect will reduce both hit rate and false-alarm rate (a change in beta) while keeping the separation of the two distributions, as expressed in standard scores, at a constant level (a constant d').

The arousal view has been supported by several findings: physiological manifestations of arousal correlate over the watch with the criterion shift (e.g., Dardano, 1962); drugs that increase arousal will reduce the vigilance decrement (N. H. Mackworth, 1950); and presenting background noise (which increases arousal) will decrease the decrement, as will any extraneous interruption like a phone call (e.g., McGrath, 1963; see Welford, 1968, for a summary of this evidence).

Despite the success of arousal theory in accounting for some aspects of vigilance performance, there is good evidence that this theory is insufficient to explain all occurrences of criterion shifts. This evidence is provided by instances in which a manipulated variable that would be expected to influence arousal level does not produce the expected effect on the criterion. For example, increasing total event rate, while keeping the intertarget interval constant (thereby decreasing the target/event ratio), will *increase* the decrement (Baddeley & Colquhoun, 1969; Jerison, et al., 1965). Yet arousal theory should predict the opposite effect, since the more frequent events should increase arousal, not decrease it. *Expectancy theory* therefore has been proposed to account for these and other nonarousal effects on the criterion shift.

Criterion shifts: Expectancy theory.   The expectancy theory proposed by Baker (1961) attributes the vigilance decrement to an upward adjustment of the response criterion in response to the perceived frequency (and therefore expectancy) of target events. Part of the cause of the decrement due to expectancy often reflects an artifact of the preexperimental training sessions. During these sessions subjects typically hear (or see) a series of targets at a fairly high rate in order to give them several examples of what to listen (look) for. Then the experiment begins and target rate is reduced to the lower frequency of the vigilance condition. In the early part of the experiment, beta then "drifts" upward to a new steady state value, actually approaching the optimum value provided by the Equation 2.1 (Craig, 1978). Williges (1976), therefore, has argued that the vigilance decrement may sometimes be considered a "*vigilance increment,*" in the sense that performance seeks an optimal setting of beta: high when signal frequency is low. Colquhoun and Baddeley (1967) have shown that if the pretraining is provided at a very low signal rate, then the decrement is reduced. Some investigators have even questioned whether the entire vigilance decrement phenomenon represents anything more than an artifact of laboratory pretraining procedures (Craig, 1978).

However, there are examples in which either performance shows a criterion shift over time that is not attributable to pretraining or steady-state performance is observed at a miss rate that is higher than optimal, given the costs and values of the situation. In these circumstances, expectancy theory explains a criterion shift in the following fashion. It is assumed that the subject sets beta on the basis of a subjective perception of signal frequency, $P_S(S)$. Then if a signal is missed for any reason, subjective probability $P_S(S)$ is reduced because the subject believes that one less signal occurs. This reduction in turn causes an upward adjustment of beta which further increases the likelihood of a miss. The consequent increase in miss rate decreases $P_S(S)$ further, and so on. Broadbent (1971) has labeled this phenomenon the "vicious circle" hypothesis, which would lead to an upward spiraling of beta and a downward spiraling of $P(H)$. While in theory this behavior could lead to an infinite beta and a negligible hit rate, in practice other factors will operate to level the criterion off at a stable but higher value. Since the vicious circle depends on signals being missed in the first place, it stands to reason that the kinds of variables that reduce sensitivity (short, low-intensity signals) should also increase the expectancy effect in vigilance. This, in fact, is the case (Broadbent, 1971).

Techniques to Combat the Loss of Vigilance

It is probably true that no single theory of the vigilance decrement is totally right and others are wrong. Instead, some aspects of all theories appear to be operating on performance at any given time, and total vigilance performance reflects some combination of the various effects described. Very often, for example, both d' and beta shifts are observed in a single vigil. It is also important to reemphasize that the theoretical mechanisms that have been proposed to account for the vigilance decrement are equally applicable to account for differences between conditions or tasks that are not related to time. Thus while Teichner (1974) has pointed out that the vigilance decrement is typically small (around 10%) in most laboratory studies, and Craig (1978) has suggested that it may be an artifact of pretraining levels, the point remains that corrective techniques that can reduce the decrement will also improve the absolute level of performance. Like the theories of vigilance, these corrective techniques may be categorized into those that affect the response criterion and those that enhance sensitivity.

Shift in response criterion.   Where possible, *knowledge of results* should be provided to allow an accurate estimation of the true $P(S)$ (N. H. Mackworth, 1950). Where this is not possible, Baker (1961) and Wilkinson (1964) have argued that introducing *false signals* will keep beta lower than it might be. This introduction will raise the subjective $P_S(S)$ and might raise arousal level as well. Furthermore, if the false signals are physically similar to the real signals, then by refreshing the standard of memory, this procedure should lower the burden on working memory. As discussed on page 35, this will reduce one source of departure of d' from its optimal. For example, as applied to the quality control inspector, a certain number of predefined defectives might be placed on the inspection line. These would be "tagged," so that if missed by the inspector they would still be removed. Their presence in the inspection stream should guarantee a higher $P_S(S)$ and therefore a lower beta than would be otherwise observed.

It seems essential, however, that operators be unaware of the presence of false signals. Otherwise, there is an inherent danger that if operators know certain signals are not real, or are not to be trusted, they may adopt a more stringent criterion to avoid detecting those "unreliable signals." This criterion shift would serve to reduce the possibility of detecting true signals as well and so negate the very intention of the false signal introduction. The problems of operator distrust of unreliable systems will be discussed again in Chapter 12, when we examine the effects of automation. There is also considerable danger in employing the technique if the actions which the operator would take after detection should have undesirable consequences for an otherwise stable system. An extreme example would occur if false warnings were introduced into a chemical process control plant and these led the operator to shut down the plant unnecessarily.

Other techniques to combat the decrement have focused more directly on arousal and fatigue. Welford (1968) has argued persuasively that any events (such as a phone call, drugs, or noise) that will sustain or increase arousal should reduce the decrement. Using biofeedback techniques, Beatty et al. (1974) have shown that operators trained to suppress theta waves (brain waves at 3–7 Hz, indicating low arousal) will also reduce the decrement.

Loss of sensitivity.   A logical outgrowth of Parasuraman's (1979) theory of vigilance loss is that any technique which will enhance the subject's memory of signal characteristics should reduce sensitivity decrements and preserve a higher overall level of sensitivity. Furthermore, as noted, any technique that combats the loss of sensitivity will also help reduce the decrements due to expectancy. Indeed, in the previous section it was noted that the introduction of false signals could improve sensitivity by refreshing memory. It would seem also that the availability of a "standard" representation of the target should assist. For example, Kelly (1955) reported a large increase in detection performance when quality control operators could look at television pictures of idealized target stimuli.

A study by Childs (1976) which found that subjects perform better when monitoring for only one target than when monitoring for one of several is also consistent with the importance of memory aids in vigilance. Childs also observed an improvement in performance when subjects were told specifically what the target stimuli were rather than what they were not. His recommendation is that inspectors should have access to visual representation of possible defectives and should not be asked simply to identify components that are not normal.

Various artificial techniques of *signal enhancement* are closely related to the reduction in memory load. A trivial example of such a technique is, of course, simply to amplify the energy characterizing the signal events. But this approach may magnify the noise as much as the signal and therefore will do nothing to change the overall signal-to-noise ratio. More ingenious solutions capitalize on procedures that will differentially influence signal and nonsignal stimuli. Luzzo and Drury (1980) developed a signal-enhancement technique know as "blinking" that also reduces memory load. When successive nontarget events are physically similar to each other (as in the inspection of wired circuit boards), Luzzo and Drury demonstrated how detection of miswired boards can be facilitated by a display technique that rapidly and alternately projects, at a single location, an image of two successive items. If both are "normal" the image will be identical, fused, and continuous. If one contains a malfunction (e.g., a gap in wiring), the gap location will appear as a stimulus that will turn on and off in a highly salient fashion as the displays are alternated.

A related signal enhancement technique in visual monitoring tasks is to induce coherent motion into targets but not into the noise, thereby taking advantage of the human's high sensitivity to target motion. For example, an operator scanning a radar display for a target blip among many similar noise blips encounters a very difficult detection problem. Scanlan (1975) has demonstrated how system design can capitalize on the fact that the radar target undergoes a coherent but slow motion while the noise properties are random. If the successive radar frames are stored, then a number of recent frames can be replayed in fast time, forwards and backwards. Under these conditions the target's coherent motion suddenly stands out as a highly salient and easily detectable feature, greatly improving detection performance.

Numerous other transformations or filters may be employed to influence the salience of signal versus noise events. These might include changing the raster scan on television displays, varying gray levels, jittering a display with rapid oscillating motion, or "cycling" the intensity of light points on a display in a technique known as pixel flicker (Genter & Weistein, 1980). In this technique the intensity of each point is increased by one log unit each display update until maximum brightness for a given point is achieved. Then the point recycles to zero and increases again. An alternative novel approach is to transcribe the events to the alternate sensory modality. This technique takes advantage of the *redundancy gain* that occurs when the signal is presented in two modalities at once. Employing this technique, Colquhoun (1975) found that sonar monitors detected targets more accurately when the target image was simultaneously displayed on a visual scope and as a transcribed auditory pattern than when either mode was employed by itself.

A technique closely related to the enhancement of signals through display manipulations is one that emphasizes operator training. Fisk and Schneider (1981) demonstrated that the magnitude of a vigilance decrement could be greatly reduced by training subjects to respond repeatedly to the target elements. This technique of developing *automatic processing* of the stimulus, which will be described in Chapter 4, tends to make the target stimulus "jump out" of the train of events just as one's own name is heard in a series of words. The conclusion of their study is quite consistent with Parasuraman's theory, since a characteristic of automatic processing is its limited dependence upon working memory (Schneider & Shiffrin, 1977).

Conclusions

Despite the plethora of vigilance experiments and the wealth of experimental data, system designers and human factors engineers have found little applicability of this basic research to problems of detection. This inadequacy may be due partly to the practitioner's inherent distrust of laboratory data, particularly in a field where much of the research *has* focused upon an artifactual source of decrement (i.e., the presession event rate). A second reason for distrust relates to the discrepancy between the fairly simple stimuli with known location and form employed in many laboratory tasks and the more complex stimuli existing in the real world. The monitor of the nuclear power plant, for example, does not know precisely what configuration of warning indicators will signal the onset of an abnormal condition, but it is unlikely that it will be the appearance of a single near-threshold light in foveal vision. Some laboratory investigators have examined the effects of signal complexity and uncertainty (e.g., Adams, Humes, & Stenson, 1962; Childs, 1976; Howell, Johnston, & Goldstein, 1966). These studies are consistent in concluding that increased complexity or signal uncertainty will lower the absolute vigilance level. However, their conclusions concerning its influence on other vigilance effects, and therefore the generalizability of these effects to complex signal environments, have not been consistent.

A final source of concern relates to the differences in motivation and signal frequency between laboratory data and real vigilance phenomena. In the laboratory, signal rates may range from one an hour to as high as three or four per minute—low enough to show decrements but far higher than rates observed in the performance of reliable aircraft, chemical plants, or automated systems, in which defects may occur at intervals of weeks or months. This difference in signal frequency may well interact with differences in motivational factors between the subject in the laboratory, performing a well-defined task and responsible only for its performance, and the real-time system operator confronted with a number of other competing activities and a level of motivation potentially influenced by large costs and benefits. This motivation level may be either lower or far higher than those of the laboratory subject, but it will not be likely to be the same.

A study by Robert Earing, for example, found that there was little effect of expectancy on operators' detection of system failures as long as the failures were relatively salient (Wickens & Kessel, 1981). But if they were subtle, then differences in signal expectancy, which might be anticipated between laboratory and real-world environments, exerted a strong effect on the response to those failures.

These differences do not mean that the laboratory data should be discounted. The basic variables causing vigilance performance to improve or deteriorate which have been uncovered in the laboratory will still probably affect detection performance in the real world, although the effect may be attenuated or enhanced. Rather, more data must be collected in real or highly simulated environments, such as those employed by Lees and Sayers (1976) in the process control task or Ruffle-Smith (1979) in aviation, to verify the generalizability of the conclusions.

While the argument for greater realism can be made with regard to other topics covered in this text, it seems to be more urgent in the field of vigilance for two reasons: (1) The paradigm requires operators to do very little most of the time and allows for much greater flexibility in operator behavior and motivation than is available in a highly stressed attention-demanding task. This flexibility produces an uncertain degree of experimenter control over subject behavior. Where subjects are allowed greater flexibility of behavior, the precise modeling efforts described in Chapter 1 become much more difficult. (2) Recreating the low-event characteristics of the real world renders the data that are collected far less reliable in a statistical sense. It is a paradox in vigilance research that this reliability can only be efficiently recovered by increasing the number of signals and so sacrificing fidelity to the very low frequency of events that is intrinsic to the paradigm.

ABSOLUTE JUDGMENT

The human senses, while not perfect, particularly with the underload condition of the vigilance paradigm, are still relatively keen when contrasted with the detection resolution of machines. In this light then it is somewhat surprising that the limits of absolute judgment—the task confronting the operator who must assign a stimulus into one of three or more levels along a sensory dimension—are relatively severe. This is the task, for example, that confronts an inspector of wool quality, who must categorize a given specimen into one of several quality levels, or the policeman who must rapidly estimate the size of a crowd. Our discussion of absolute judgment will first describe performance when stimuli vary on only a single dimension. The concept of information theory, used to quantify absolute judgment performance, will be briefly introduced and then described in much greater detail in the chapter supplement. We shall then consider absolute judgment along two or more physical dimensions that are perceived simultaneously and discuss the implications of these findings to principles of display coding.

Single Dimensions

In the typical laboratory paradigm used for investigating absolute judgment, a particular sensory stimulus continuum is selected for investigation (e.g., tone pitch, light intensity, texture roughness) and a number of discrete levels are selected along that continuum (e.g., four tones of different intensities). These stimuli are then presented randomly to the subject one at a time, and he is asked to associate a different discrete response to each one. From these data the experimenter can assess the extent to which each response matched the presented stimulus.

Typically, performance is described in terms of the amount of *information transmitted* by the subject from stimulus to response. A formal treatment of information theory is provided in the chapter supplement; here it is described only briefly. Information is potentially available in a stimulus any time there is some uncertainty as to what the stimulus will be. How much information is delivered by a stimulus depends in part on the number of possible stimuli that could occur in that context. If the same stimulus occurs on every trial, its occurrence conveys no information. In the absolute judgment task, the previous example described four equally likely alternatives. The amount of information conveyed by one of these stimuli when it occurs, expressed in *bits*, is simply equal to the base 2 logarithm of this number, for example, $\log_2 4 = 2$ bits. If there were but two alternatives, the information conveyed by the occurrence of one of them is $\log_2 2 = 1$ bit.

In human performance we are less interested in the amount of information in a stimulus than in the amount transmitted by the human operator to the response, a quantity designated H_T. While the formal technique for computing information transmission will be described in the supplement, an intuitive description will be given here. Obviously, if our hypothetical operator responds correctly to every one of our 2-bit stimuli, then 2 bits of information are transmitted. If the operator ignores the stimuli and responds randomly, zero bits are transmitted. If the operator makes some errors, performance ranges between these two limits. Thus the number of alternative stimuli, and therefore the amount of information in the input, places an upper bound on the maximum amount of information that an operator can transmit: $H_T \le H_S$.

In our discussion of a "typical" absolute judgment experiment, when four discriminable stimuli are presented, transmission is usually perfect—at 2 bits. Then the stimulus set is enlarged, and additional data are also collected with 5, 6, 7, and more discrete stimulus levels, while H_T is computed each time. Typically the results indicate that errors begin to be made when about 5 to 6 stimuli are used, and error rate increases as the number of stimuli increase further. These results indicate that stimulus sets of this magnitude have somehow saturated the subject's capacity to transmit information about the magnitude of the stimulus. We say the subject has a maximum *channel capacity*.

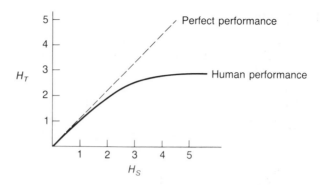

Figure 2.7   Human performance in absolute judgment tasks.

Graphically, these data can be represented in Figure 2.7, in which the actual information transmitted (H_T) is plotted as a function of the number of absolute judgment stimulus alternatives (expressed in informational terms as H_S). The 45° slope of the dashed line indicates perfect information transmission, and the "leveling" of the function takes place at the region in which errors began to occur (i.e., $H_T < H_S$). The level of the flat part of the function indicates the channel capacity of the operator: somewhere between 2 and 3 bits. George Miller (1956), in a classic paper entitled "The Magical Number Seven Plus or Minus Two," noted the similarity of this level of asymptote across a number of different absolute judgment functions with different sensory continua. Miller concluded that the limits of absolute judgment at 7 ± 2 stimulus categories (2–3 bits) is a fairly general one that transcends a number of different stimulus modalities.

This limit does however vary somewhat from one sensory continuum to another; it is less than 2 bits for saltiness of taste and about 3.4 bits for judgments of position on a line. The level of the asymptote does not however reflect a basic limit in sensory processing in the same manner that the signal detection parameter d' assesses a limit of sensory resolution. In the first place, to associate the crude limits of absolute judgment with shortcomings of sensory processing seemingly is at odds with the fact that the senses are extremely keen in their ability to make *discriminations* between two stimuli (are they alike or different?). In fact, the number of adjacent stimulus pairs on a sensory continuum, such as tone pitch, which the human can accurately discriminate is in the range of 1,800 (Mowbray and Gebhard, 1961). In the second place, Pollack (1952) has observed that the limits of absolute judgment are little affected by whether the stimuli are closely spaced on the physical continuum, producing fewer possible discriminations, or are widely dispersed. Pollack observed a limit of 1.8 bits when auditory stimuli in a pitch judgment task were restricted to a range of 400 Hz, but it only increased to 3 bits when 20 times the range was used. Apparently, then, the limit is not sensory but is in the resolution of the subject's *memory* for the representation of the 4–8 different standards (Siegel & Siegel, 1972).

If, in fact, absolute judgment limitations are related to memory, then there should be some association between this phenomenon and difference in learning or experience, since differences in memory are closely related to those of learning. Such evidence arises from two sources: differences in experience with different sensory continua and the phenomenon of *anchoring*.

Differences in experience. It is noteworthy that sensory continua for which we demonstrate good absolute judgment performance are those for which such judgments in real-world experience occur relatively often. For example, judgments of position along a line (3.4 bits) are made in measurements on rulers, and judgments of angle (4.3 bits) are made in clock reading. High performance in absolute judgment also seems to be correlated with professional experience with a particular sensory continuum in industrial tasks (Welford, 1968) and is demonstrated by the noteworthy association of absolute pitch with skilled musicians (Carpenter, 1951; Carroll, 1975; Siegel & Siegel, 1972).

The question of pitch judgment and the issue of "absolute pitch" (the ability to correctly identify any heard note on the musical scale) has been of particular interest to psychologists. Naturally, the correlation between musical ability and absolute pitch does not necessarily imply that performance was learned, since it is possible that superior abilities determined the musical proficiency rather than musical experience determining the superior absolute judgment of pitch. The critical data are those in which controlled differences in experience or training can be associated with differences in pitch judgment. However, these data are somewhat ambiguous. Hartman (1954) has tried to train naive subjects to a high level of absolute pitch. He succeeded in improving performance somewhat, but only with relatively extensive training. The level obtained rarely reached that of the skilled musician or conductor. In contrast, Carroll (1975), comparing the performance of individuals who claimed to possess absolute pitch innately with that of one subject who taught himself absolute pitch, found that the latter performed equivalently in all respects to the former group.

A different source of evidence concerning the role of learning in absolute pitch is provided by the cross-cultural comparisons between pitch judgments of Javanese musicians and American musicians. In Javanese music there is no "standard" scale; musicians tune instruments differently on each occasion. In the absence of such a standard, performance of the Javanese musicians on an absolute pitch task was found to be inferior to that of western musicians (Siegel & Siegel, 1972). In conclusion, the evidence clearly suggests that learning and experience play some role in the development of accurate absolute pitch judgment. While differences in basic ability undoubtedly contribute as well, the exact proportion of this contribution is still uncertain.

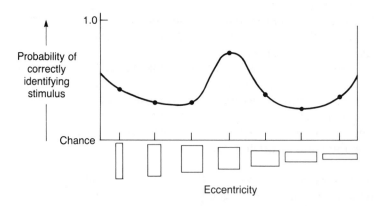

Figure 2.8   The phenomenon of anchoring. The two extreme stimuli and the familiar square in the middle show the highest probability of being recognized correctly.

Anchoring.   A second role of experience in absolute judgment is suggested by the improved accuracy of judgment that subjects make for particular stimuli along a continuum. These "anchors" include stimuli at the highest and lowest level and sometimes certain very familiar landmark stimuli in between. They serve as bases for improved judgments of adjacent stimulus levels. The role of learning in anchoring is suggested because the anchors in the middle of a continuum are usually stimuli that are particularly familiar. For example, squares along the continuum of rectangularity, circles along a continuum of ellipsoid eccentricity, or the vertical and horizontal on the continuum of orientation serve as anchors and are usually classified with exceedingly high accuracy. An example of the first of these continua is shown in Figure 2.8. It is presumably the greater degree of exposure in life's experience that provides a more well-defined category in memory for the anchor. This enhanced definition, in turn, helps a subject to judge these members of a given response category.

Multidimensional Judgment

If our limits of absolute judgment are so severe and seemingly can only be overcome by extensive training, how is it that we are able to recognize stimuli in the environment so readily? A major reason is that most of our recognition is based upon identification of some combination of two or more stimulus dimensions rather than levels along a single dimension. When a stimulus can vary on two (or more) dimensions at once, we make an important distinction between orthogonal and correlated dimensions. When dimensions of a stimulus are orthogonal, the level of the stimulus on one dimension can take on any value, independent of the other—for example, the weight and hair color of an individual. When dimensions are correlated, the level on one constrains the level on another—for example, height and weight, since tall people tend to weigh more than short ones.

Orthogonal dimensions.   The importance of multidimensional stimuli in improving absolute judgment has been repeatedly demonstrated (Garner, 1974). Egeth and Pachella (1969), for example, demonstrated that if 3.4 bits (10 levels) of information could be transmitted concerning position on a line, then when two lines were combined into a square, subjects could improve performance by transmitting 5.8 bits (57 levels) of information concerning the spatial position of a dot in the square. Note, however, that this improvement does not represent a perfect addition of channel capacity along the two dimensions. If processing along each dimension were independent and unaffected by the other, the predicted amount of information transmitted would be 3.4 + 3.4 = 6.8 bits, or around 100 positions in the square. Egeth and Pachella's results suggest that there is some loss of information along each dimension by virtue of the requirement to transmit information along each other.

Going beyond the two-dimensional case, Pollack and Ficks (1954), combined six dimensions of an auditory stimulus orthogonally. As each successive dimension was added, subjects showed a continuous gain in total information transmitted but a loss of information transmitted per dimension. These relations are shown in Figure 2.9, with seven bits the maximum total capacity transmitted. In a related observation,

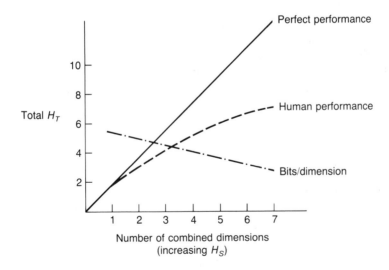

Figure 2.9   Human performance in absolute judgment of multidimensional auditory stimuli. As more dimensions are added, more total information is transmitted, but less information is transmitted per dimension. From "The Information of Elementary Multidimensional Auditory Displays" by I. Pollack and L. Ficks, 1954, *Journal of the Acoustical Society of America, 26,* 157. Reprinted by permission.

Carroll (1975) suggested that the reason why possessors of absolute pitch are superior to those without does not lie in greater discrimination along a single continuum. Rather, Carroll argued that those with absolute pitch make their judgments along two dimensions: the pitch of the octave and the value of a note within the octave. They have created a multidimensional stimulus from a stimulus that others treat as unidimensional.

Correlated dimensions.   In the examples cited above, information was conveyed by varying the levels along two dimensions *independently*. The data indicated that there is a gain in total information transmitted but a loss in transmitted information/dimension. When dimensions are combined *redundantly*, or in correlated fashion, different characteristics are observed. For example, if the stimuli employed are tones that vary in pitch and intensity, a redundant variation could make high-pitched tones always loud and low tones always soft. In this case, H_S, the information in the stimulus, is no longer the sum of H_S across dimensions, since this sum is reduced by the degree of correlation between levels on the two dimensions. If the correlation is unity, then total H_S is just the H_S on any single dimension (since all other dimensions are completely redundant). Clearly the maximum possible H_S for all dimensions in combinations is now less than its value could be in the orthogonal case. However, Eriksen and Hake (1955) found that by combining dimensions redundantly the information *loss* ($H_S - H_T$) was much less for a given value of H_S than it is when they are combined orthogonally, while information transmitted (H_T) is greater than it would be along any single dimension.

In their experiment Eriksen and Hake assessed information transmission performance when 20 color stimuli ($H_S = 4.32$ bits) were constructed in two different fashions.

1.   *Unidimensional.* In three different series of trials, one for each of three dimensions, 20 levels of a given stimulus dimension (hue, saturation, or brightness) were presented. The average value of H_T (across the three dimensions measured separately) was 2.75 bits. Therefore, $H_{loss} = 4.32 - 2.75 = 1.57$ bits.

2.   *Three correlated dimensions.* Twenty stimuli were constructed such that each level of brightness always had a corresponding level of hue and saturation. Again, 4.32 bits of stimulus information was presented. However, $H_T = 4.11$ bits, so H_{loss} was only 0.21 bits.

It is important to realize that a greater amount of information might have been transmitted (H_T) had the three dimensions been combined orthogonally ($H_S = 4.32 + 4.32 + 4.32 = 13$ bits), although Eriksen and Hake did not run this condition. However, based upon Pollack's results when dimensions are combined orthogonally, we would surely expect a much larger degree of information loss in this case. Thus we can see that orthogonal and correlated dimensions accomplish two different objectives in absolute judgment of multidimensional stimuli.

Orthogonal dimensions maximize H_T—the *efficiency* of the channel; correlated dimensions minimize H_{loss}—that is, maximize the security of the channel.

Implications of Absolute Judgment

The conclusions drawn from research in absolute judgment are highly pertinent to the performance of any task that requires operators to sort stimuli into levels along a physical continuum, particularly for industrial inspection tasks in which products must be sorted into various levels for pricing or marketing reasons (e.g., fruit quality) or into categories for different uses (e.g., steel or glass quality). The data from the absolute judgment paradigm provide some indication of the kind of performance limits that can be anticipated and suggest the potential role of training, of anchors, or of emphasizing multidimensional aspects of the stimuli to improve performance. It is also important to realize that a loss in resolution per dimension will occur if stimuli must be sorted on more than one dimension.

Display coding: One dimension. Absolute judgment data are equally relevant to the issue of *coding*. In many cases some physical or conceptual continuum of importance in the performance of a task will be coded for display by variation along a displayed sensory continuum. For example, the size of socket wrenches may be coded by color so that they can be easily differentiated even when the digital size indicator cannot be read (Pond, 1979). This is an example of one-dimensional absolute judgment. A more conceptual dimension, such as the hierarchical level of a personnel unit in an organization or the degree of danger of a particular environment, may be coded into a number of different levels. It is, of course, possible to use letters or digits to identify the various levels, but in conditions of low visibility, high visual clutter, or high stress these may not be read accurately. In this case basic data on the appropriate dimensions and the number of conceptual categories that can be employed without error are highly relevant to the development of such nonverbal display codes.

Moses, Maisano, and Bersh (1979) have cautioned that any conceptual continuum should not be arbitrarily assigned to a given physical dimension. They have argued that certain conceptual continua have a more "natural" association with some physical display dimensions than with others. The designers of codes should be wary of the potential deficiencies (decreased accuracy, increased latency) that are imposed by an arbitrary assignment. As an example, Moses et al. suggest that representation of danger and unit size should be coded by the color and size of a displayed object, respectively, and not the reverse.

Display coding: Several dimensions. When the principles of multidimensional absolute judgment are applied to issues of display coding, the designer must consider carefully the relation between the statistical properties of the information to be displayed and the physical characteristics of the display itself. With regard to the statistical properties of the information, we have described already how dimensions of information displayed to an operator may be independent or correlated. Consider, for example, the military commander who needs information about units under his command. These units are multidimensional stimuli. Some dimensions may be correlated. For example, unit type (infantry, artillery, etc.) may be correlated with unit size (e.g., infantry units may generally be larger than artillery). On the other hand, some dimensions may vary orthogonally. Unit size, for example, may be independent of unit experience or unit strength.

With regard to the physical properties of the display, Garner (1970, 1974) has drawn a distinction between what he refers to as *integral* and *separable* dimensional combinations. Integral and separable dimensional pairs differ in the degree to which levels on the two dimensions in a pair can be independently specified. Separable dimensions occur when the levels along each of the two dimensions can be specified without requiring the specification of the level along the other. For example, the length of the horizontal and vertical lines radiating from the stimulus in Figure 2.10*a* are two separable dimensions; one can be specified without specifying the other. For integral dimensions, this independence is impossible. The height and width of a single rectangle are integral, because to display the height of a rectangle its width must be given implicitly; otherwise, it would not be a rectangle (Figure 2.10*b*). Correspondingly, the color and brightness of an object are inte-

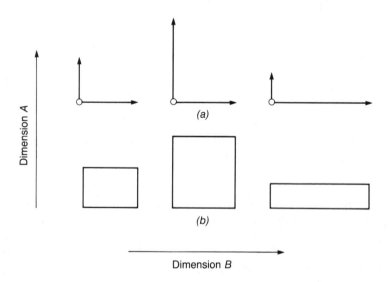

Figure 2.10   *(a)* Separable dimensions (height and width of a line segment); *(b)* integral dimensions (height and width of a rectangle). Dimension *A* is height; dimension *B* is width.

gral dimensions. Color cannot be physically represented without some brightness.

While numerous experimental paradigms and methodologies of measurement have been defined to contrast integrality and separability (Garner & Fefoldy, 1970; Treisman & Gelade, 1980), there are two important differences between integral and separable dimensional pairs that are of greatest importance for issues of multidimensional display coding. Both differences are defined in terms of the consequence of using integral rather than separable pairs. (1) When stimulus dimensions vary orthogonally, there is a greater information loss when integral dimensions are used. (2) When stimulus dimensions vary redundantly, there is a greater transmission of information when integral dimensions are used. The three implications of these principles to issues of display coding are as follows:

1. *Orthogonal variation.* When information dimensions vary independently, then a symbolic display should avoid using integral coding. The symbols in Figure 2.10a would seemingly be better than those in 2.10b. Since there is orthogonal variation, both formats will lead to some information loss/dimension, but this will be less with the separable display.

2. *Security and redundancy gain.* At the other extreme, sometimes the designer wishes to guarantee absolute security and minimize information loss. In this case the designer will choose redundant coding so that two physical dimensions represent only one dimension of information. This choice produces a *redundancy gain*. In this case Garner's work suggests that integral dimensions should be chosen. The color, hue, and brightness of stimuli used by Eriksen and Hake (1955) were examples of integral dimensions with correlated information. Color and shape are also integral (since any shape must have a color). Kopala (1979) found that Air Force pilots were better able to encode information concerning the level of threat of displayed targets when such information was presented redundantly by shape and color than when it was presented by either dimension alone.

3. *Partial correlation.* There is a range between complete independence and perfect redundancy. In our battlefield example the correlation between unit type and size was not 0, but neither was it perfect. Pressure and temperature are two variables in physical systems that often show some, but not perfect, correlation. At these intermediate levels of correlation it is impossible to say precisely whether integral or separable dimensions are superior. What Garner's research does suggest however is that as the level of correlation between variables increases, the relative merits of integral displays will increase, and those of separable displays will decline. The critical level of correlation at which the crossover occurs cannot be determined, however, and probably varies somewhat from one situation to the next.

The theme of dimensional combinations and the structure of correlated information is an important one that will be revisited in our discussions of decision making (Chapter 3), spatial perception (Chapter 5), and selective attention (Chapter 7). This chapter has introduced the critically important issue of display compatibility: the interface between the human operator, the physical display, and the information that is to be displayed.

Transition

The material in this chapter has set the stage for much of what is to follow in the next three chapters. The decisions that humans make are often based upon uncertain probabilistic stimuli like those discussed in our treatment of signal detection theory, but are also far more complex, with more possible states of the world and more complex cues than a simple continuum of energy. These decisions will be discussed in the next chapter. Correspondingly, the stimulus classifications that operators make with certainty when dealing with the more deterministic characteristics of the world are more complex and have more variations on more dimensions than the one-, two-, or three-dimensional stimuli discussed in the absolute judgment task. These more complex categorizations will be the topic of Chapters 4 and 5. Thus, while the world is more complex than we have described here, the principles underlying that description will continue to be important in the more complex processes of decision making and perception discussed in the next three chapters.

The ROC Curve
and Information Theory

THE ROC CURVE

Theoretical Representation

In the main chapter the receiver operating characteristic (ROC) curve was briefly discussed as a means of graphically representing the joint effects of sensitivity and response bias on performance. ROC curves were presented in Figure 2.5 that contrasted two observers with different sensitivities. Here we shall describe the curve in more detail by emphasizing its relation both to the typical signal detection data matrix in Figure 2.1 and to the theoretical signal and noise curves in Figure 2.3.

Figure 2.1 presented the raw data that might be obtained from a signal detection theory (SDT) experiment. Of the four values, only two are critical. These are normally $P(H)$ and $P(FA)$, since $P(M)$ and $P(CR)$ are then completely specified, as $1 - P(H)$ and $1 - P(FA)$, respectively. Figure 2.3 showed the theoretical representation of the neural mechanism within the brain that generated the matrix of Figure 2.1. As the criterion is set at different locations along the X axis of Figure 2.3, a different set of values will be generated in the matrix of Figure 2.1. Figure 2.11 presents a more detailed picture of the third way of representing signal detection data: the ROC curve. As we discussed in the main chapter, the ROC curve plots on a single graph the joint value of $P(hit)$ and $P(FA)$ as each is obtained at a number of different settings of the response criterion.

Each signal detection condition generates one point on the ROC. If the signal strength and the observer's sensitivity remain constant, then by decreasing beta (either through changing payoffs or increasing signal probability) from one block of signal detection trials to another, a series of points are produced that move from conservative responding in the lower left of Figure 2.11 to risky responding in the upper right. When connected, these points make the ROC curve. Figure 2.11 shows the relationship between the raw data, the ROC curve, and the assumed underlying hypothetical distributions, collected at three different beta sets. One can see that sweeping the criterion placement from left (low beta) to right (high beta) along the evidence variable produces progressively more "no" responses and moves points on the ROC curve from upper right to lower left.

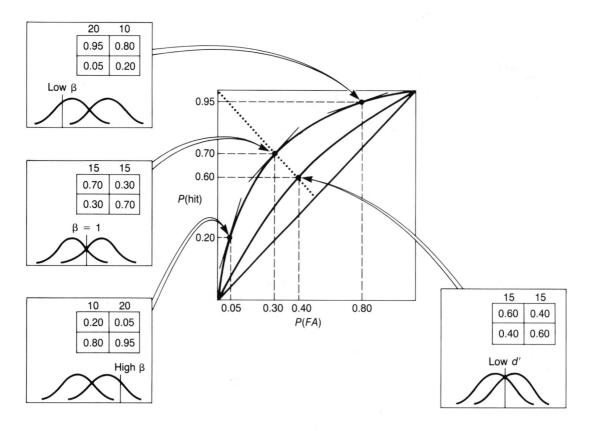

Figure 2.11 The ROC curve. The figure shows how three points on an ROC curve of high sensitivity relate to the stimulus-response matrix data and the underlying signal and noise curves. At the right, the figure also shows one point of lower sensitivity.

A more efficient means of collecting data from several criterion sets is to have the subject provide a rating of confidence that a signal was present (Green & Swets, 1966). If three confidence levels are employed (e.g., 1 = confident that a signal was present, 2 = uncertain, and 3 = confident that only noise was present), then the data may be analyzed twice in different ways. During the first analysis, categories 1 and 2 would be considered a "yes" response, and 3 a "no" response. This would produce data corresponding to a risky beta setting since roughly two-thirds of the responses would be called "yes." The second time, two responses (2 and 3) would be assigned instead to the "no" category, representing the data matrix of a conservative beta. Thus, two beta settings are available from only one set of detection trials. This economy of data collection is realized because the subject is asked to convey more information on each trial.

It is important at this point to scrutinize in more detail some of the general characteristics of the ROC curve and the way in which these characteristics relate to detection performance. Notice in Figure 2.11

that the ROC curve for a more sensitive observer is more bowed, being located closer to the upper left. Formally, the value of beta (the ratio of ordinate heights in Figure 2.3) at any given point along the ROC curve is equal to the slope of a *tangent* drawn to the curve at that point. This slope (and therefore beta) will theoretically always be equal to 1.0 at points that fall along the negative diagonal (shown by the dashed line). If the hit and false-alarm values of these points are determined, we will find that $P(H) = 1 - P(FA) = P(CR)$. Performance here is equivalent to performance at the point of intersection of the two distributions in Figure 2.3. Note also that points on the positive diagonal of Figure 2.11 represent chance performance: No matter how the criterion is set, $P(H)$ always equals $P(FA)$, and the signal cannot be discriminated at all from the noise. A representation of Figure 2.3 giving rise to chance perform-ance and analogous to the points on the positive diagonal would be one in which the signal and noise distributions were perfectly superim-posed.

While Figure 2.11, plotted with a linear probability scale, shows a typically bowed curve, an alternative way of plotting the curve is on probability paper (Figure 2.12). This representation has the advantage that the bowed lines of Figure 2.11 now become straight lines parallel to the chance diagonal. Constant units of distance along each axis represent constant numbers of standard scores of the normal distribu-tion. For a given point, d' is then equal to $Z(H) - Z(FA)$, reflecting the number of standardized scores that the point lies orthogonal to the chance diagonal. A measure of response bias that correlates very closely with beta, and is easy to derive from Figure 2.12, is simply the Z score of the false-alarm probability for a particular point (Swets & Pickett, 1982). It is also possible to look up values of d' and beta in published tables in which these parameters are listed as functions of $P(FA)$ and $P(H)$ (Swets, 1964); see Appendix A.

Empirical Data

It is important to realize the distinction between the theoretical, ide-alized curves presented in Figures 2.3, 2.11, and 2.12 and the actual

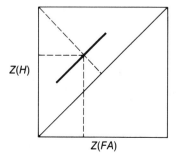

$Z(H)$

$Z(FA)$

Figure 2.12 The ROC curve on probability paper.

empirical data collected in a signal detection experiment or a field investigation of detection performance. The most obvious contrast is that the representations in Figures 2.11 and 2.12 are continuous smooth curves, while the data that would be collected would of course constitute a set of discrete points. More importantly, empirical results in which data are collected from a subject as the criterion is varied often provide points that do not fall precisely along a line of constant bowedness (Figure 2.11) or a 45-degree slope (Figure 2.12). More often the slope is slightly shallower. This situation arises because the distributions of noise and signal-plus-noise energy are not in fact precisely normal and of equal variance, as the idealized curves of Figure 2.3 portray, particularly if there is variability in the signal itself. This tilting of the ROC curve away from the ideal presents some difficulties for the use of d' as a measure of sensitivity. If d' is measured as the distance of the ROC curve of Figure 2.11 from the chance axis, and this distance varies as a function of the criterion setting, what is the appropriate setting for measuring d'? One approach is to measure the distance at unit beta arbitrarily (i.e., where the ROC curve intersects the negative diagonal). This measure is referred to as d_a and may be employed if data at two or more different beta settings are available, so that a straight-line ROC can be constructed on the probability plot of Figure 2.12 (Green & Swets, 1966).

While it is therefore desirable that two or more points on the ROC curve be generated, there are some circumstances in which it may be infeasible to do so, particularly when evaluating detection data in many real-world contexts. In such cases, the experimenter often has neither the luxury nor the feasibility of manipulating beta or using rating scales and so must use the data available from only a single stimulus-response matrix. In this case, an estimate of d' or d_a is exceedingly difficult unless beta is equal to one [i.e., $P(H) = 1 - P(FA)$], because the slope of the ROC curve cannot be assessed from a single point. Furthermore, when conditions are present in the data in which no errors are committed, the value of d' approaches infinity, making it quite difficult to average this value with others.

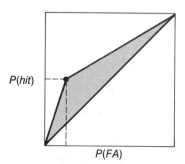

Figure 2.13 Example of the sensitivity measure $P(A)$, the area under the ROC curve, derived from one point.

Under these circumstances, the measure P(A) or the area under the ROC curve is an alternative measure of sensitivity that can be employed (Calderia, 1980; Craig, 1979; Green & Swets, 1966). Such a measure represents the area to the right and below the line segments connecting the lower left and upper right corners of the ROC space to the measured data point (Figure 2.13). Craig (1979) and Calderia (1980) have argued that the advantage of this measure is that it is "parameter free." That is, its value does not depend upon any assumptions concerning the shape or form of the underlying signal and noise distributions. For this reason, it is a measure that may be usefully employed even if two or more points in the ROC space are available but do not meet the equal variance assumptions (i.e., they do not fall along a 45-degree line). The measure P(A) may be calculated from the formula

$$P(A) = \frac{P\{H + [1 - P(FA)]\}}{2} \qquad (2.3)$$

Values of A' are represented in Appendix B. The reader is referred to Calderia (1980), Craig (1979), Swets and Pickett (1982), and Green and Swets (1966) for discussions of the relative merits of different sensitivity and bias measures.

INFORMATION THEORY

The Quantification of Information

A considerable portion of human performance theory revolves around the concept of transmitting information. In fact, in any situation in which the human operator either perceives changing environmental events or is responding specifically to events that have been perceived, the operator is encoding or transmitting information. The aircraft pilot, for example, must process a multitude of visual signals bearing on the status of the aircraft while listening to auditory messages from air traffic control concerning flight plans and the status of other aircraft. A fundamental issue in engineering psychology is quantifying this flow of information in terms that allow the diverse tasks confronting the human operator to be compared. Information theory accomplishes this goal. Given that we can associate processing efficiency with the amount of information that an operator can process per unit time and given that task difficulty can be associated with the rate at which information is presented, then information theory provides a metric with which these measures can be compared across a wide number of different tasks.

Formally, information is defined as the *reduction of uncertainty* (Shannon & Weaver, 1949). If, prior to the occurrence of an event, you are less sure of the state of the world (possess more uncertainty) than after, then that event conveyed information to you. The statement, "The sun rose this morning," conveys little information because, prior to the occurrence of the event, you could anticipate it. On the other hand, the statement, "Burma is at war with Liechtenstein," conveys quite a bit of information. Your knowledge and understanding of the world is probably quite different after hearing the statement than it was before. Information theory formally quantifies the amount of information conveyed by a statement. This quantification is influenced by three variables: the number of possible events that could occur N, the probabilities of those events P, and their sequential constraints or the context in which they occur. We shall now describe how each of these three variables quantitatively influences the amount of information conveyed by an event.

Number of events.   Prior to an event (which conveys information), the operator possesses a state of knowledge that is characterized by uncertainty about some aspect of the world. After the event, that uncertainty is normally less. The amount of uncertainty reduced by the event occurrence in *bits* (binomial digits) is formally defined to be the average *minimum* number of true-false questions that would have to be asked in order to reduce the uncertainty. For example, the information conveyed by the statement, "Reagan won," after the 1980 election is 1 bit, because the answer to one true-false question (e.g., "Did Reagan win?"—"True" or "Did Carter win?"—"False") is sufficient to reduce the prior uncertainty. If, on the other hand, there were four candidates, all running for office, two questions would have to be asked and answered to eliminate uncertainty. In this case one question, Q1, might be, "Was the winner from the liberal (or conservative) pair?" After this question was answered then Q2 would be, "Was the winner the more conservative (or liberal) member of the pair?" Thus, if you were simply told the winner, that statement would formally convey 2 bits of information. This question-asking algorithm assumes that all alternatives are equally likely to occur. Formally, then, when all alternatives are equally likely, the information conveyed by a stimulus H_S, in bits, can be expressed by the formula

$$H_S = \log_2 N \qquad (2.4)$$

where N is the number of equally likely alternatives.

Because information theory is based upon the minimum number of questions and therefore arrives at a solution in a minimum time, it has a quality of optimal performance. It is this optimal aspect that makes the theory attractive in its applications to human performance.

Probability.   In fact, events in the world do not always occur with equal frequency or likelihood. If you lived in the Arizona desert, much more information would be conveyed by the statement, "It is raining today," than the statement, "It is sunny." Your certainty of the state of the world is changed very little by knowing that it is sunny, but it is changed quite a bit (uncertainty is reduced) by hearing of the low-probability event of rain. In the example of the four election candidates, less information would be gained by learning that the favored candidate won than by learning that the candidate of the Socialist Workers or Libertarian party won. The probabilistic element of information is quantified by making rare events convey more bits. This in turn is accomplished by revising Equation 2.4 for the information conveyed by stimulus event i to be

$$H_S = \log_2 \left(\frac{1}{P_i} \right) \qquad (2.5)$$

where P_i is the probability of occurrence of event i. This formula will increase H for low-probability events. Note that if N events are equally likely, then each event will occur with probability 1/N. In this case, Equations 2.4 and 2.5 are equivalent.

As we noted above, information theory is based upon a prescription of optimal behavior. This optimum can be prescribed in terms of the order in which the true-false questions should be asked. If some events are more common or expected than others, we should ask the question about the common event first. In our four-candidate example we will do the best (ask the minimum number of questions in the long run) by first asking, "Is the winner Reagan?" or, "Is the winner Carter?" assuming that Reagan and Carter have the highest probability of winning. If instead the initial question was, "Is the winner the Libertarian candidate?" or, "Is the winner from one of the minor parties?" we have clearly "wasted" a question, since the answer is likely to be no, and our uncertainty will be reduced by only a small amount. Experimental evidence obtained by Sayeki (1969) suggests that subjects do tend to be relatively optimal in seeking information: the first questions they ask are the ones most likely to reduce uncertainty by the greatest amount.

The information conveyed by a single event of known probability is given by Equation 2.5. However, psychologists are often more interested in measuring the *average* information conveyed by a series of events with differing probabilities that occur over time—for example, a series of warning lights on a panel or a series of communication commands. In this case the average information conveyed is computed as

$$H_{ave} = \sum_{i=1}^{n} P_i \left[\log_2 \left(\frac{1}{P_i} \right) \right] \qquad (2.6)$$

In this formula, the quantity within the outer brackets is the information per event as given in Equation 2.5. This value is now weighted by the probability of that event, and these weighted information values are summed across all events. Accordingly, frequent low-information events will contribute heavily to this average, whereas rare high-information events will not. If a given number of events are equally likely, this formula will reduce to Equation 2.4.

An important characteristic of Equation 2.6 is that if a given number of events are not equally likely, H_{ave} will always be less than its value if the same number of events are equally probable. Consider four events, A, B, C, and D, with probabilities of 0.5, 0.25, 0.125, and 0.125. The computation of the average information conveyed by each event in a series of such events would proceed as follows:

Event:	A	B	C	D
P_i:	0.5	0.25	0.125	0.125
$\dfrac{1}{P_i}$:	2	4	8	8
$\log_2 \dfrac{1}{P_i}$:	1	2	3	3

$$\Sigma P_i \left(\log_2 \frac{1}{P_i} \right) = \quad 0.5 + 0.5 \quad + 0.375 + 0.375 = 1.75 \text{ bits}$$

This value is less than $\log_2 4 = 2$ bits, which is the value derived from Equation 2.4 when the four events are equally likely. In short, while low-probability events convey more information because they occur infrequently, the fact that they do occur infrequently causes their high information content to be down-weighted in their contribution to the average.

Sequential constraints.   In the preceding discussion, probability has been employed to reflect the long-term frequencies, or *steady-state* expectancies, that stimuli will occur. However, there is also a third contributor to information that reflects the short-term sequences of stimuli, or their *transient* expectancies. A particular stimulus may occur fairly rarely in terms of its absolute frequency. However, given a particular *context* it may be highly expected, and therefore its occurrence conveys very little information. In the example of rainfall in Arizona, we saw that the absolute probability of rain is low. But if we heard that there was a large front moving eastward from California, our expectancy of rain given this prior information would be much higher. That is, information can be reduced by the context in which it appears. As another example, the letter u in the alphabet is not terribly common and therefore normally conveys quite a bit of information when it occurs; however, in the context of a preceding q, it is almost totally predictable and therefore its information content, given that context, is nearly 0 bits.

Contextual information is frequently provided by *sequential constraints* on a series of stimuli. In the binominal series of stimuli ABABABABAB, for example, $P(A) = P(B) = 0.5$. Therefore, by Equation 2.6 each stimulus conveys 1 bit of information. But clearly the next letter in the above sequence is almost certainly an A. Therefore, the sequential constraints reduce the information content in the same manner as a change in stimulus probabilities reduces information from the equiprobable case. Formally, the information provided by an event given a context may be computed in the same manner as in Equation 2.5, except that the absolute probability of the event P_i is now replaced by a *contingent* probability P_i/X, which is read: "The probability of event i given that X has occurred." The variable X refers to the context.

Redundancy. To recapitulate, we note that three variables influence the amount of information that a series of stimuli can convey. The number of possible events N sets an upper bond on the maximum number of bits if all stimuli are equally likely, while changes in stimulus probability away from the equiprobable case and increases in sequential constraints both serve to reduce information from this maximum. The term *redundancy* formally defines this potential loss in information. Thus, for example, the English language is highly redundant, because of two factors: all letters are not equiprobable (e's vs. x's) and sequential constraints such as are found in common digraphs like *qu*, *th*, or *nt* reduce uncertainty.

Formally, the *percent redundancy* of a stimulus set is quantified by the formula

$$\% \text{ Redundancy} = \left(1 - \frac{H_{ave}}{H_{max}}\right) \times 100 \tag{2.7}$$

where H_{ave} is the actual average information conveyed taking into account all three variables (approximately 1.5 bits per letters for the alphabet) and H_{max} is the maximum possible information that would be conveyed by the N alternatives if they were equally likely ($\log_2 26 = 4.7$ bits for the alphabet). This means that the redundancy of the English language is $100(1 - 1.5/4.7) = 68\%$. Wh-t th-s sug-est- is t-at ma-y of t-e le-ter- ar- not ne-ess-ry fo- com-reh-nsi-n. However, to stress a point that will be emphasized in Chapter 4, this fact does not negate the value of redundancy in many circumstances. We have seen already in our discussion of both vigilance and absolute judgment that redundancy gain can improve performance when perceptual judgments are difficult.

Information Transmission of Discrete Signals

In much of human performance theory, investigators are concerned not only with how much information is *presented* to an operator but also with how much is *transmitted* from stimulus to response, the *channel capacity*, and how rapidly it is transmitted, the *bandwidth*. Using these concepts, the human processor is sometimes represented as an information channel; an example of which is shown in Figure 2.14. Consider the typist. Information is first of all present in the stimulus, the printed page from which he is typing. This value of stimulus information H_S can be computed by the procedures described above, taking into account probabilities of different letters and their sequential constraints. Second, each response on the keyboard is an event, and so we can also compute H_R in the same manner. Finally, we ask if each letter on the page was appropriately typed on the keyboard. That is, was the

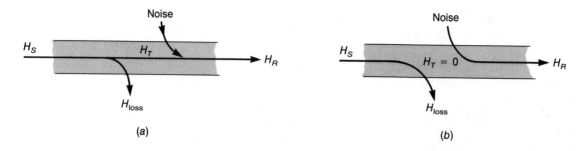

Figure 2.14 Information transmission and the channel concept: (a) information transmitted through the system; (b) no information transmitted.

information faithfully transmitted, H_T. If it was not, there are two types of mistakes: First, information in the stimulus could be lost, H_L, known as *equivocation*. This would be the case if a certain letter was not typed. Second, letters may be typed that were not in the original text. This is referred to as noise. Figure 2.14a indicates the relationship between these five information measures. Notice that it is theoretically possible to have a high value of both H_S and H_R, but to have H_T equal to zero. This would occur if the typist were totally ignoring the printed text (equivocation) while creating his own message as he went along. (Formally this would be considered "noise"; more charitably it might be called creativity.) A schematic example of this case is shown in Figure 2.14b.

We shall now go through an example of computation of H_T in the context of a four-alternative absolute judgment task rather than the more complex typing tasks. In the judgment task the subject is confronted by four possible stimuli, any of which may appear with equal probability, and must make a corresponding response for each.

When deriving a quantitative measure of H_T, it is important to realize that for an ideal information transmitter, $H_S = H_T = H_R$. In optimal performance of the absolute judgment task, for example, each stimulus (conveying 2 bits of information if equiprobable) should be processed ($H_T = 2$ bits) and trigger the appropriate response ($H_R = 2$ bits). As we saw, in information-transmitting systems, this ideal state is rarely obtained because of the occurrence of equivocation and noise.

The computation of H_T is performed by setting up a stimulus-response matrix, such as that shown in Figure 2.15a, and converting the various numerical entries into three sets of probabilities: the probabilities of stimuli shown along the bottom row, the probabilities of responses shown along the right column, and the probabilities of a given stimulus-response pairing. These latter values are the probability that an entry will fall in each cell, where a cell is defined jointly by a particular stimulus and a particular response. In Figure 2.15a there are four cells with $P = 0.25$ for each entry. Each of these sets of probabilities can be independently converted into the information measures by Equation 2.6.

Once the quantities H_S, H_R, and H_{SR} are calculated, the formula

$$H_T = H_S + H_R - H_{SR} \qquad (2.8)$$

allows one to compute the information transmitted. The rationale for using this formula is as follows: The variable H_S establishes the maximum possible transmission and so contributes positively to the formula. Likewise, H_R contributes positively. However, to guard against situations such as that depicted in Figure 2.14b in which stimuli are not coherently paired with responses, H_{SR}, a measure of the dispersion or lack of organization within the matrix, is subtracted. If each stimulus generates consistently only one response (Figure 2.15a), the entries in the matrix should equal the entries in the rows and columns. In this case, $H_S = H_R = H_{SR}$, which means that substituting the values in Equation 2.8, $H_S = H_T$. However, if there is greater dispersion within the matrix, then there are more bits within H_{SR}. In Figure 2.15b this is shown by eight equally probable stimulus-response pairs, or 3 bits of information in H_{SR}. Therefore, $H_{SR} > H_S$ and $H_T < H_S$. In terms of Venn diagrams the relation between these quantities is shown in Figure 2.16. In this representation it is possible to actually add and subtract areas to obtain results similar to Equation 2.8.

Often the investigator may be interested in the information transmission *rate* expressed in bits/second rather than the quantity H_T expressed in bits. To find this rate, H_T is computed over a series of stimulus events, along with the average latency for each transmission (i.e., the mean reaction time RT). Then the ratio H_T/RT is taken to derive a measure of the *bandwidth* of the communication system in bits/second. This becomes a critical measure for much of human factors research because it expresses a measure of processing efficiency that accounts for both speed and accuracy and transcends a number of

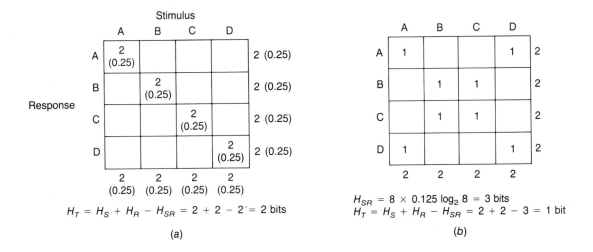

Figure 2.15 Two examples of the calculation of information transmission.

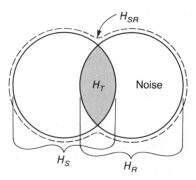

Figure 2.16   Information transmission represented in terms of Venn diagrams.

different tasks and measures. For example, a measure of typing or reaction time performance can be expressed in terms of bandwidth, while the same measure can also be used to express tracking performance. The following section will show how performance on continuous tasks such as tracking can also be described by information theory.

Information of Continuous Signals

Much of the information that is processed in complex systems (and in our daily lives) is continuous. Consider, for example, the curves in a road as we drive or the continuous deflections of a needle monitored by a pilot. While these are not discrete events, they clearly convey information to the operator in the same fashion as a warning light or a printed letter. When quantifying the bandwidth or stimulus information in continuous signals, it is first necessary to define the molecular event. This event is a reversal in the time series (a change in needle movement direction or the curve in the road). When expressed as a position function of time, the information conveyed by each reversal (bits) is derived from the number of possible positions at which a reversal can take place. In Figure 2.17a each downward reversal conveys 1 bit, since there are two locations at which the reversal could occur. The upward reversal is completely predictable in its location and so conveys no information. Thus there is, on the average, 0.5 bit/reversal across the continuous signal of Figure 2.17a.

When a signal is random, such as that shown in Figure 2.17b, it is necessary to make certain assumptions before the information content of each reversal can be assessed. If a reversal can occur at any point, the information, formally, is infinite. Therefore, it is necessary to assume the precision with which position must be localized by the operator. If a meter is deviating between scale markers of 0 through 16 and the operator must read only the nearest scale marker, then each deflection conveys $\log_2 16 = 4$ bits of information. From here it is a simple step to compute the information content of the continuous signal. This simply becomes the information/reversal multiplied by the number of reversals/second, sometimes referred to as the *upper cutoff frequency* of the continuous signal. The product will be the *bandwidth* of the continuous signal expressed in bits per second. Continuous information transmission may be calculated by the same S–R matrix technique shown in Figure 2.15, once the continuous signal and response are divided into their discrete levels. There are, however, also more elegant continuous techniques that apply *Fourier analysis* (see Chapters 4 and 11) to compute an equivalent measure of *transinformation* (Baty, 1971). These techniques are beyond the scope of this treatment, however, and the reader is referred to Sheridan and Ferrell (1974) for a description of their use. In this text, the utility of continuous information theory will be emphasized in Chapter 4 (supplement) on speech perception, Chapter 7 on attention, and Chapter 11 on continuous manual control.

Criticism of Information Theory

Information theory has been a valuable tool for investigating human performance because it can provide a relatively dimensionless unit of performance that is applicable across a wide variety of different depen-

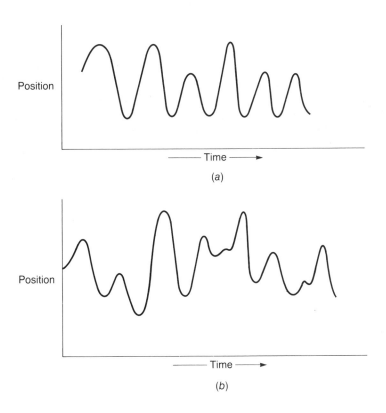

Figure 2.17   Continuous information: *(a)* one bit; *(b)* random.

dent variables. Furthermore, there appear to be certain limits of human information processing that remain relatively invariant when expressed in informational terms (Fitts & Posner, 1967). Despite its considerable success, however, the application of information theory to human performance has received some criticism, which may be divided into two general categories: limitations in the sensitivity of the information measure and limitations in its application to human performance. We will discuss the first category here, but the second is dealt with primarily in Chapter 9.

Consistency versus correctness.   Some critics charge that information theory measures response consistency, not correctness. This is a valid but not devastating criticism. For example, in the matrix representation of Figure 2.15, it is entirely possible to generate data in which only four S–R cells contain data but the responses do not correspond to the appropriate stimuli. While information theory would still calculate $H_S = H_T$ (i.e., perfect performance), it is clear that the subject is not performing appropriately. Although this criticism is a valid one, it is not terribly damaging. Such behavior would seemingly indicate a misunderstanding of instructions (or response assignment) on the part of the subject and could be easily discovered by the investigator.

Insensitivity to the magnitude of errors.   In the matrix of Figure 2.15 the formal measure of H_T will be indifferent to whether a response error is in an adjacent category or in a more distant category from the correct one. This insensitivity ("a miss is as good as a mile") means that the information transmission measure may not convey all of the information (used in the informal sense) about the quality of the subject's performance when the stimulus set represents a physically or conceptually ordered continuum. In this case, one who responds close to the appropriate category, as shown in Figure 2.18a, demonstrates greater resolution than one who does not (Figure 2.18b), but the H_T measure is not sensitive to this distinction. A measure such as a product-moment

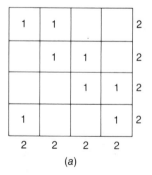
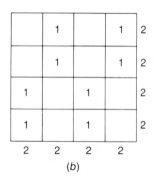

(a)   (b)

Figure 2.18   Two examples of information transmission. Both examples have $H_T = 1$ bit. Yet (a) is clearly better performance than (b).

correlation between the two ordered continua could be employed in this case if the investigator is greatly concerned about the size of an error.

Assumptions about prior uncertainty. In any quantification of H_S the investigator must make some assumptions concerning the subject's prior uncertainty of potential stimulus events. These assumptions may either underestimate or overestimate the actual subjective belief. For example, in a four-alternative reaction time task, suppose that the subject is informed that one of four lights on the right half of an eight-light linear array may appear ($H_S = 2$ bits). Even though the experimenter might caution the subject that the left four lights will never appear, a cautious subject might still believe that information could be delivered on the other set of four. Therefore, the subject's subjective prior uncertainty is in fact one of eight alternatives and each stimulus will convey 3 bits of information to the subject, even as the experimenter believes it to be 2 bits. The subject will take longer to process the information. The danger in this case is that a particular experimental manipulation may inadvertently affect the subject's expectancy of the potential stimulus set. The experimenter would attribute the resulting change in processing time directly to the experimental variable, when in fact it was due to the change in the actual information conveyed.

Relations of Information to Other Measures

It should be evident that the information transmission metric is not the only means of conveying accuracy information. As might be expected, it bears a close relation to three other accuracy measures: the product moment correlation, the percent of correct responses, and the d' measure of sensitivity from signal detection theory. To the extent that the stimulus and response dimensions of an information transmission task lie along an ordered continuum, then there is a monotonic relation between H_T and the product moment correlation for a given stimulus set size. Correspondingly, for a given set size, H_T will be monotonic with a simple measure of percent correct.

Finally, for the simple case of two stimulus alternatives and two responses, the measure of H_T is closely related to the d' measure of sensitivity. In their discussion of different sensitivity measures, Moray and Fitter (1973) have noted that the two measures are only conceptually equivalent when the response criterion in signal detection equals one. At nonunity levels, they depart because the d' efficiency measure is unaffected by changes in bias, while the H_T measure is affected. The variable H_T falls off from the maximum as beta becomes either more radical or more conservative. This departure results from the specific assumptions of the form of the signal and noise distributions employed by signal detection theory. These assumptions are not made in information theory.

References

Adams, J. A., Humes, J. M., & Stenson, H. H. (1962). Monitoring of complex visual displays III: Effects of repeated sessions on human vigilance. *Human Factors, 4,* 149–158.

Anderson, K. J., & Revelle, W. (1982). Impulsivity, caffeine, and proofreading: A test of the Easterbrook hypothesis. *Journal of Experimental Psychology: Human Perception & Performance, 8,* 614–624.

Baddeley, A. D., & Colquhoun, W. P. (1969). Signal probability and vigilance: A reappraisal of the "signal rate" effect. *British Journal of Psychology, 60,* 169–178.

Baker, C. H. (1961). Maintaining the level of vigilance by means of knowledge of results about a secondary vigilance task. *Ergonomics, 4,* 311–316.

Baty, D. L. (1971). Human transinformation rates during one-to-four axis tracking. In *Proceedings of the Seventh Annual Conference on Manual Control* (NASA SP–281). Los Angeles: University of Southern California.

Beatty, J., Greenberg, A., Deibler, W. P., & O'Hanlon, J. P. (1974). Operator control of occipital theta rhythm affects performance in a radar monitoring task. *Science, 183,* 871–873.

Bisseret, A. (1981). Application of signal detection theory to decision making in supervisory control. *Ergonomics, 24,* 81–94.

Blake, T., & Baird, J. C. (1980). Finding a needle in a haystack when you've never seen a needle: A human factors analysis of SET I. In G. Corrick, E. Hazeltine, & R. Durst (Eds.), *Proceedings of the Human Factors Society—24th Annual Meeting.* Santa Monica, CA: Human Factors.

Broadbent, D. E. (1971). *Decision and stress.* New York: Academic Press.

Broadbent, D. E., & Gregory, M. (1965). Effects of noise and of signal rate upon vigilance as analyzed by means of decision theory. *Human Factors, 7,* 155–162.

Buckhout, R. (1974). Eyewitness testimony. *Scientific American, 231* (6), 23–31.

Calderia, J. D. (1980). Parametric assumptions of some "nonparametric" measures of sensory efficiency. *Human Factors, 22,* 119–130.

Carpenter, A. (1951). A case of absolute pitch. *Quarterly Journal of Experimental Psychology, 3,* 92–93.

Carroll, J. B. (1975). Speed and accuracy of absolute pitch judgments: Some latter-day results. *The L. L. Thurston Psychometric Laboratory Research Bulletin.* University of North Carolina at Chapel Hill.

Childs, J. M. (1976). Signal complexity, response complexity, and signal specification in vigilance. *Human Factors, 18,* 149–160.

Colquhoun, W. P. (1975). Evaluation of auditory, visual, and dual-mode displays for prolonged sonar monitoring in repeated sessions. *Human Factors, 17,* 425–437.

Colquhoun, W. P., & Baddeley, A. D. (1967). Influence of signal probability during pretraining on vigilance decrement. *Journal of Experimental Psychology, 73,* 153–155.

Craig, A. (1977). Broadbent and Gregory revisited: vigilance and statistical decision. *Human Factors, 19,* 25–36.

Craig, A. (1978). Is the vigilance decrement simply a response adjustment towards probability matching? *Human Factors, 20,* 447–451.

Craig, A. (1979). Nonparametric measures of sensory efficiency for sustained monitoring tasks. *Human Factors, 21,* 69–78.

Dardano, J. F. (1962). Relationships of intermittent noise, intersignal interval, and skin conductance to vigilance behavior. *Journal of Applied Psychology, 46,* 106–114.

Davies, D. R., & Tune, G. S. (1969). *Human vigilance performance.* New York: American Elsevier.

Davies, D. R., & Parasuraman, R. (1980). *The psychology of vigilance.* London: Academic Press.

Drury, C. G., & Addison, S. L. (1973). An industrial study of the effects of feedback and fault density on inspection performance. *Ergonomics, 16,* 159–169.

Egeth, H., & Pachella, R. (1969). Multidimensional stimulus identification. *Perception & Psychophysics, 5,* 341–346.

Ellison, K. W., & Buckhout, R. (1981). *Psychology & Criminal Justice,* New York: Harper and Row.

Eriksen, C. W., & Hake, H. N. (1955). Absolute judgments as a function of stimulus range and number of stimulus and response categories. *Journal of Experimental Psychology, 49*, 323–332.

Fisk, A. D., & Schneider, W. (1981). Controlled and automatic processing during tasks requiring sustained attention. *Human Factors, 23*, 737–750.

Fitts, P. M., & Posner, M. I. (1967). *Human Performance:* Belmont, CA: Brooks/Cole.

Garner, W. R. (1970). The stimulus in information processing. *American Psychologist, 25*, 350–358.

Garner, W. R. (1974). *The processing of information and structure.* Hillsdale, NJ: Erlbaum Associates.

Garner, W. R., & Fefoldy, G. L. (1970). Integrality of stimulus dimensions in various types of information processing. *Cognitive Psychology, 1*, 225–241.

Genter, C. R., & Weistein, N. (1980, May). *Pixel flicker: A new form of object superiority.* Paper presented to the Optical Society of America. Atlanta, GA.

Green, D. M., & Swets, J. A. (1966). *Signal detection theory and psychophysics.* New York: Wiley.

Hartman, E. G. (1954). The influence of practice and pitch distance between tones on the absolute identification of pitch. *American Journal of Psychology, 67*, 1–14.

Howell, W. C., Johnston, W. A., & Goldstein, I. L. (1966). Complex monitoring and its relation to the classical problem of vigilance. *Organizational Behavior & Human Performance, 1*, 129–150.

Jerison, M. L., Pickett, R. M., & Stenson, H. H. (1965). The elicited observing rate and decision process in vigilance. *Human Factors, 7*, 107–128.

Kelly, M. L. (1955). A study of industrial inspection by the method of paired comparisons. *Psychological Monographs 69* (No. 394), 1–16.

Kopala, C. (1979). The use of color-coded symbols in a highly dense situation display. In C. Bensel (Ed.), *Proceedings of the 23rd annual meeting of the Human Factors Society.* Santa Monica, CA: Human Factors.

Lees, F. P., & Sayers, B. (1976). The behavior of process monitors under emergency conditions. In T. Sheridan & G. Johannsen (Eds.), *Monitoring behavior and supervisory control.* New York: Plenum Press.

Loftus, E. F. (1979). *Eyewitness testimony,* Cambridge, MA: Harvard University Press.

Long, C. M., & Waag, W. L. (1981). Limitations and practical applicability of d' and β as measures. *Human Factors, 23*, 283–290.

Lusted, L. B. (1971). Signal detectability and medical decision making. *Science, 171*, 1217–1219.

Lusted, L. B. (1976). Clinical decision making. In D. Dombal & J. Grevy (Eds.), *Decision making and medical care.* Amsterdam: North Holland.

Luzzo, J., & Drury, C. G. (1980). An evaluation of blink inspection. *Human Factors, 22*, 201–210.

Mackie, R. R. (1977). *Vigilance: Relationships among theories, physiological correlates, and operational performance.* New York: Plenum Press.

Mackworth, J. F. (1970). *Vigilance and attention.* Baltimore: Penguin.

Mackworth, N. H. (1948). The breakdown of vigilance during prolonged visual search. *Quarterly Journal of Experimental Psychology, 1*, 5–61.

Mackworth, N. H. (1950). *Researches in the measurement of human performance* (MRC Special Report Series No. 268). London: H. M. Stationery Office. (Reprinted in W. Sinaiko (Ed.), *Selected papers on human factors in the design and use of control systems.* New York: Dover, 1961.)

McGrath, J. J. (1963). Irrelevant stimulation and vigilance performance. In D. M. Buckner & J. J. McGrath (Eds.), *Vigilance: A symposium.* New York: McGraw-Hill.

Miller, G. A. (1956). The magical number seven plus or minus two: Some limits on our capacity for processing information. *Psychological Review, 63*, 81–97.

Moray, N., & Fitter, M. (1973). A theory and measurement of attention. In S. Kornblum (Ed.), *Attention and performance IV.* New York: Academic Press.

Moses, F. L., Maisano, R. E., & Bersh, P. (1979). Natural associations between symbols and military information. In C. Bensel (Ed.), *Proceedings of the Human Factors Society, 23rd annual meeting.* Santa Monica, CA: Human Factors.

Mowbray, G. H., & Gebhard, J. W. (1961). Man's senses vs. informational channels. In W. Sinaiko (Ed.), *Selected papers on human factors in the design and use of control systems.* New York: Dover.

Parasuraman, R. (1979). Memory load and event rate control sensitivity decrements in sustained attention. *Science, 205,* 924–927.

Parasuraman, R. (1980). Applications of signal detection theory in monitoring performance and medical diagnosis. In G. Corrick, E. Hazeltine, & R. Durst (Eds.), *Proceedings of the Human Factors Society, 24th annual meeting.* Santa Monica, CA: Human Factors.

Peterson, C. R., & Beach, L. R. (1967). Man as an intuitive statistician. *Psychological Bulletin, 68,* 29–46.

Pollack, I. (1952). The information of elementary auditory displays. *Journal of the Acoustical Society of America, 24,* 745–749.

Pollack, I., & Ficks, L. (1954). The information of elementary multidimensional auditory displays. *Journal of the Acoustical Society of America, 26,* 155–158.

Pond, D. J. (1979). Colors for sizes: An applied approach. In C. Bensel (Ed.), *Proceedings of the Human Factors Society, 23rd annual meeting.* Santa Monica, CA: Human Factors.

Rhea, J. T., Potsdaid, M. S., & DeLuca, S. A. (1979). Errors of interpretation as elicited by a quality audit of an emergency facility. *Radiology, 132,* 277–280.

Ruffle-Smith, H. P. (1979). *A simulator study of the interaction of pilot workload with errors, vigilance, and decision* (NASA Technical Memorandum 78482). Washington, DC: NASA Technical Information Office.

Sayeki, Y. (1969). Information seeking for object identification. *Organization Behavior & Human Performance, 4,* 267–283.

Scanlan, L. A. (1975). Visual time compression: Spatial and temporal cues. *Human Factors, 17,* 337–345.

Schneider, W., & Shiffrin, R. M. (1977). Controlled and automatic human information processing II: Perceptual learning, automatic attending, and a general theory. *Psychological Review, 84,* 127–190.

Shannon, C. E., & Weaver, W. (1949). *The mathematical theory of communications.* Urbana: University of Illinois Press.

Sheridan, T. B., & Ferrell, W. A. (1974). *Man-machine systems: Information, control, and decision models of human performance.* Cambridge, MA: MIT Press.

Siegel, J. A. (1972). The nature of absolute pitch. In E. Gordon (Ed.), *Experimental research in the psychology of music VIII.* Iowa City: Iowa University Press.

Siegel, J. A., & Siegel, W. (1972). Absolute judgment and paired associate learning: Kissing cousins or identical twins? *Psychological Review, 79,* 300–316.

Swennsen, R. G., Hessel, S. J., & Herman, P. G. (1977). Omissions in radiology: Faulty search or stringent reporting criteria? *Radiology, 123,* 563–567.

Swennsen, R. G., Hessel, S. J., & Herman, P. G. (1979). Radiographic interpretation with and without search: Visual search aids the recognition of chest pathology. *Radiology, 127,* 438–443.

Swets, J. A.(Ed.). (1964). *Signal detection and recognition by human observers: Contemporary readings.* New York: Wiley.

Swets, J.A., & Pickett, R. M. (1982). *The evaluation of diagnostic systems,* New York: Academic Press.

Szucko, J. J., & Kleinmuntz, B. (1981). Statistical vs. clinical lie detection. *American Psychologist, 36,* 488–496.

Teichner, W. (1974). The detection of a simple visual signal as a function of time of watch. *Human Factors, 16,* 339–353.

Treisman, A., & Gelade, G. (1980). A feature integration theory of attention. *Cognitive Psychology, 12,* 97–136.

Watson, D. S., & Nichols, T. L. (1976). Detectability of auditory signals presented without defined observation intervals. *Journal of the Acoustical Society of America, 59,* 655–668.

Welford, A. T. (1968). *Fundamentals of skill.* London: Methuen.

Wells, G. L., Lindsay, R. C., & Ferguson, T. I. (1979). Accuracy, confidence, and juror perceptions in eyewitness testimony. *Journal of Applied Psychology, 64,* 440–448.

Wickens, C. D., & Kessel, C. (1979). The effect of participatory mode and task workload on the detection of dynamic system failures. *IEEE Transactions on Systems, Man, and Cybernetics, SMC-13,* 24–34.

Wickens, C. D., & Kessel, C. (1981). The detection of dynamic system failures. In J. Rasmussen & W. Rouse (Eds.), *Human detection and diagnosis of system failures*. New York: Plenum Press.

Wilkinson, R. T. (1964). Artificial "signals" as an aid to an inspection task. *Ergonomics, 7*, 63–72.

Williges, R. C. (1976). The vigilance increment: An ideal observer hypothesis. In T. B. Sheridan & G. Johannsen (Eds.), *Monitoring and supervisory control* (pp. 181–191). New York and London: Plenum Press.

Decision Making

OVERVIEW

Midway in the sequence of processing, between the perception of environmental conditions and the execution of an action, arises the requirement to decide on the nature of the action. The general area of decision making that we consider in this chapter has the following characteristics: (a) the operator must select one of a number of possible choices or courses of action when presented with stimulus information bearing upon the choice. (b) The time frame for the choice is relatively long, greater than a second (distinguishing the present treatment from that of reaction time in Chapter 8). (c) The probability that the choice of action will be the correct or "best" one, over the long run, is considerably less than 1.0, either because of the probabilistic nature of the stimulus information or because of the operator's own cognitive limitations.

The attributes of this general paradigm describe a large number of real-world human operator environments. Examples include *medical diagnosis*: the physician considers a number of symptoms (stimuli) to diagnose a disease; *fault diagnosis*: the supervisor of a complex dynamic system (e.g., a nuclear plant) processes a number of warning indications to diagnose the nature of the underlying malfunction; *treatment*: the physician (or nuclear operator) weighs the costs and benefits of a number of possible treatments to remedy the diagnosed disease (or malfunction) before deciding upon a particular course of action; *weather forecasting*: the forecaster integrates a number of sources of information (time of year, pressure, temperature) to make a forecast for tomorrow; *pilot decision*: the aircraft pilot considers a number of factors (visibility, height, fuel consumption, schedule) to decide whether or not to continue an aircraft landing approach in bad weather; *factory production control*: the industrial manager decides which products should be emphasized given the availability of raw materials and personnel and the demands of the market; *selection*: the admissions committee weighs a number of attributes of an applicant before deciding to admit or reject; *judicial procedures*: the jury (or judge) weighs a number of sources of evidence about the suspect before deciding on a verdict of guilt or innocence or on the length of sentence; and *consumer behavior*: the buyer considers a number of products and compares several attributes of these products before deciding to purchase one. These manifestations of decision making are diverse, yet they all possess certain common elements. More important, all encounter limitations of human information-processing capabilities that prevent decisions from being correct as often as is optimally possible.

Initially in this chapter, a framework is provided for analyzing three different classes of decision-making problems related to prediction, diagnosis, and choice. The prominent dimensions of the decision-making task are highlighted in the process. Two major classes of limitations in human decision making will then be considered: those relating to the estimation of statistics from data and those related to actual diagnosis decisions and the inference process. The latter section involves a lengthy discussion of the manner in which human limitations in memory, attention, and logic affect behavior in a variety of decision-making and troubleshooting situations. A third section considers the role of costs and values in decision making, and a fourth addresses the role of learning and experience. The final section on decision-making aids identifies areas in which efforts have been made to improve decision-making capabilities.

Features of the Decision-Making Task

Decision making is complex, as the wide variety of examples cited above suggests. To provide some structure to the discussion in this chapter, therefore, it is important to identify certain dimensions that underlie all decision problems. Some of these are relevant to a problem's difficulty, some to its structure, and some to both.

Number of stimuli and outcomes.   In Chapter 2 we considered decision making in its most uncomplicated form, signal detection, in which the decision is simply a choice between two alternatives and only a single piece of evidence (the X value of a single stimulus) is considered. The complexity of decision theory grows with increases in number of stimuli and number of possible outcomes. For example, while the decision to admit or reject an applicant has only two possible outcomes, a larger number of stimuli (qualifications of the applicant) are involved, making the decision more complex. The diagnosis of a disease generally involves many possible stimuli (symptoms) and many possible outcomes (diseases that may be likely, given the set of observed symptoms). To cite an extreme example, the operator of a malfunctioning nuclear power plant may be confronted with more than 200 stimuli (warning indicators that show a change of state) that must be used to diagnose one of a very large number of potential malfunctions (Sheridan, 1981).

Choice, diagnosis, or prediction.   These terms refer to the specific type of decision task being confronted. In a *choice* task the decision maker is presented with a number of elements (objects, or courses of action) that are physically present or physically represented and that differ in certain ways. One of these elements must be selected. Examples include the consumer who is faced with a choice between purchases or the physician entertaining choices of treatments for a patient. In *diagnosis*, on the other hand, the operator is physically presented with a set of stimulus dimensions or cues (e.g., the symptoms of the patient) and must link these to one of two or more hypotheses held in memory in order to identify the present, existing state. *Prediction* is similar to diagnosis. The operator is again presented with a single set of observed variables. The task in prediction, however, is to map these onto a single quantitative or predicted value, as when an applicant's aptitude score is used to estimate how well she will perform in a position.

The similarities and differences between these three kinds of decision-making tasks are depicted in Figure 3.1. In each panel the human decision maker looks down upon the stimulus configuration that is physically presented. These physical stimuli (objects, symptoms, or measures) are labeled numerically. Above the decision maker are the hypotheses or mental constructs that the decision maker must maintain in some form of working memory.

In Figure 3.1a (choice) the consumer or operator observes two products or treatments (e.g., two automobiles) differing on two known attributes (e.g., gas mileage and cost). The value X of each object on each attribute is known and visible. The operator must then "multiply" the known value of these object attributes by a mental representation of how important the two attributes are to him—their utilities U_A and U_B—before deciding how the two products rate comparatively and making a decision. Assume in this example that cost is twice as important to the consumer as gas mileage. Car 1 has one-third the gas mileage but sells for half the price of car 2. Therefore car 1 would be chosen because of the greater weight placed upon the attribute (cost) that favored it. We describe this decision-making strategy as *compensatory* because an object that is slightly deficient in one attribute of great utility may compensate by having higher values on one or more attributes of lesser utility. Because the compensatory strategy takes into account all information on all attributes, it is therefore considered "optimal."

In diagnosis (Figure 3.1b) there are two observed symptoms. These are logically equivalent to the two attribute values of the choice task and might represent the blood pressure and temperature of a patient. If the physician is considering two possible hypotheses (disease A and disease B), then these correspond logically to the two objects in the choice example. To diagnose the most likely disease, the physician must mentally multiply the value X of each symptom by a *diagnostic weight W* that represents for a given symptom value the likelihood that disease A (or B) is present. A high temperature, for example, is more indicative of appendicitis than of colitis. Naturally, if the weights for a particular symptom are identical between the diseases, then whatever the symptom value may be, its information does not help the physician in diagnosis. If the weights differ substantially, then they are highly diagnostic. After performing this "multiplication" of symptom values

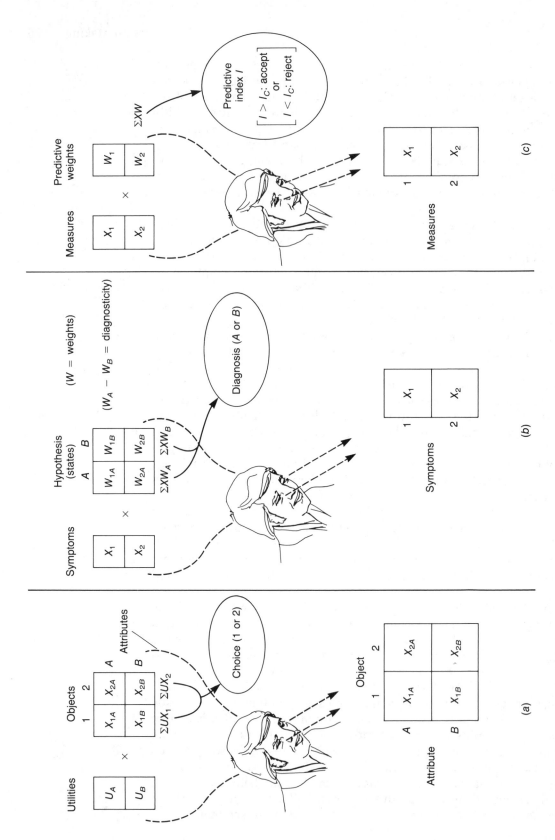

Figure 3.1 (a) Choice; (b) diagnosis; (c) prediction.

across diagnostic weights and aggregating the "products" across each of the diseases, the optimal physician diagnoses the disease as the one with the highest "score."

Figure 3.1b can also be used to describe the example of *forecasting*. If the two hypotheses are two alternative states to be predicted (e.g., rain or no rain) and the two symptoms are two pieces of meteorological data, then a similar aggregation process can be utilized to assess the relative probability of rain. The output of diagnosis may be either a statement of the most likely hypothesis in view of the data or a statement of the relative probability of one hypothesis over the other (i.e., the difference between the two aggregated products).

Finally, in Figure 3.1c, a single set of stimuli or cues is observed, such as the scores of an applicant on a number of tests. These are "multiplied" in turn by weighting functions that express how valid each is believed to be in predicting some criterion variable, such as the overall qualification of the candidate for a position. After all of these multiplications are completed, they are aggregated and the total is compared to a criterion. If it is exceeded, the candidate is accepted. If the aggregation process is done in the head, it is considered clinical prediction. If it is done by formal calculation on the basis of weights derived from actuarial data, it is a considered statistical prediction (Meehl, 1954).

The relative difficulty of the three decision tasks is determined by two factors: the ratio of number of values held in memory to number of values physically visible and the number of mental multiplications and summations to be performed. In this regard, prediction is probably simplest. Two weights are held in memory, two measures are observed, and a single sum of products is computed. Choice is more complex. In the example of Figure 3.1a, only the two utilities must be held in working memory, while the four attribute values (X's) are physically present. Additionally, two product sums must be computed and compared. Diagnosis is the most difficult decision task. In diagnosis (Figure 3.1b), as in choice, there are two product summations to be performed. However, of the six values to be manipulated, four (the diagnostic weights) must be held in working memory, while only two (the symptoms) are physically observable.

Diagnosis and choice often are closely interrelated. Consider the power plant operator who, confronting a failure, considers and must diagnose one of two possible malfunctions. Hypothesis 1 is a serious problem with a low probability and hypothesis 2 is relatively trivial. Upon diagnosis, a choice of actions must be made: shut the turbine down, which is the appropriate course of action if hypothesis 1 is true, or keep the turbine on the line, which is appropriate if hypothesis 2 is true. In this case, the choice of action is dictated as in Figure 3.1a. The two possible states of the system are equivalent to attributes A and B. The two actions or treatments are equivalent to objects 1 and 2 in the figure. The X values thus represent the costs or benefits of taking different actions under different circumstances (such as shutting a turbine off the line when nothing is wrong with it or keeping it running when it is malfunctioning). In performing the necessary arithmetic, the utilities of the choice task correspond formally to the probabilities associated with the two hypotheses.

Temporal sequencing.   Informational inputs to the decision task may become available to the operator nearly simultaneously, so that all may be processed at once (subject to the limits of human attention) or sequentially over a long course of time. Either format has major implications for the decision-making process.

Correlational structure.   Depending on the nature of the decision-making task, some cues may correlate in value with others. In prediction, for example, data generally indicate that applicants to a graduate program who have high grade-point averages also tend to score well on standardized tests (e.g., the Graduate Record Examinations). In other situations the correlation may be low. In choosing a product, we might find that its durability is unrelated to its appearance or attractiveness. The correlation may even be negative, so that the more attractive product is actually the less durable. Humans are found to use correlational structure to benefit decision-making performance and even will perceive correlations when they are absent.

Causal structure.   Causal structures are closely related to correlations. In diagnosis, the particular physical nature of the system under scrutiny imposes certain causal relations between symptoms. For example, in a hydraulic system, a sudden increase in pressure usually will soon cause an increase in temperature. Intercorrelations between cues are not necessarily attributable to a cause and effect relation, but if they are and if the operator understands this relation, the decision-making task appears to be less difficult (Einhorn & Hogarth, 1981). In fact, humans have a tendency to overestimate the strength of relationships for which they attribute a cause and effect relation (Einhorn & Hogarth, 1983; Tversky & Kahneman, 1982).

Optimality in Decision Making

As a general rule, across a wide range of experimental paradigms, the human is not optimal in decision making. The meaning of "optimal" here is similar to its meaning in Chapter 2; that is, it refers to the performance of the ideal "statistical person" or computer decision maker with full computational power at its disposal. While departures from optimal performance are not a startling discovery when extremely complex decisions are considered, in simpler paradigms these departures are somewhat disconcerting. What emerges from the research on decision making is that there are many fairly simple, logical reasons for the systematic departures that are observed. For example, humans employ "heuristics" (rules of thumb) that can simplify the cognitive complexity of a decision-making problem and unburden its demands

on attention and working memory. However, these heuristics sometimes induce systematic biases (Kahneman, Slovic, & Tversky, 1982;
Tversky & Kahneman, 1974). As one example to be described later,
people use the ease with which an example of a hypothesis can be
brought to mind as a means of estimating its probability—the heuristic
of *availability*. This heuristic may often approximate the probability of
the hypothesis, but sometimes it may not, and nonoptimal decisions
will be made. Certain of the features of the decision-making task can
enhance or diminish the influence of these heuristics and biases. Accordingly, solutions in terms of human-machine system aids often are
implicitly suggested.

In the following sections of this chapter, we shall consider the
elements of the decision-making task and the way in which these
elements interact with human limitations of attention, memory, and
computation to produce the systematic biases that are observed. The
research reported is aggregated both from fairiy abstract laboratory
tasks and from somewhat more valid real-world environments. It is
important to note that some of the conclusions drawn from laboratory
studies have yet to receive rigorous experimental tests in the non-
laboratory decision-making environment.

HUMAN LIMITS IN DECISION MAKING: STATISTICAL ESTIMATION

In discussing human limits in decision making, an analogy with the
classical procedure of statistical inference may be helpful. This procedure often is viewed as a two-stage process. First, the statistician or
psychologist computes some *descriptive statistics* of the data at hand
(e.g., mean, proportion, standard deviation). Then these estimated parameters are used to draw some *inference* about the sample under
consideration (e.g., whether it is from the same population as another
sample or a different population or whether two samples are correlated). When examining human limitations in decision making, two
issues that are analogous with these stages must be considered: the
human's ability to perceive and store probabilistic data accurately and
the ability to draw inferences (and thereby make decisions) on the basis
of those data. Peterson and Beach (1967) adopt this framework by
considering the human operator as an "intuitive statistician." In this
section, we shall examine data bearing on the human's ability to estimate four basic descriptive statistics: means, proportions, standard
deviations, and parameters of exponential growth. In the following
section we shall address human limits in statistical inference.

Perception of the Mean

When presented with a single number (e.g., 27), human perception
proceeds fairly automatically in accordance with principles to be outlined in Chapter 4. However, when the operator is presented with

several numbers or several marks on a measurement scale and asked to estimate the mean value of those numbers (without resorting to mental arithmetic), very different processes are engaged (Pitz, 1980). Experimental evidence suggests, however, that the estimation of the mean is done reasonably well (Peterson & Beach, 1967; Pitz, 1980; Sniezek, 1980). While the estimated mean does not necessarily correspond precisely to the true mean of a set of numbers, it does not appear to be systematically biased in one direction or another.

Perception of Proportions

When a quality control monitor is asked to make an intuitive estimate of a defect rate, he is estimating the value of a proportion from a sample of data. Estimates of proportions, unlike those of means, tend to show some small but systematic biases. Toward the midrange (e.g., from 0.10 to 0.90), estimates are fairly accurate. However, with more extreme values, subjects seem to "hedge their bets" conservatively away from extreme values (Sheridan & Ferrell, 1974). In a typical example (Erlick, 1961), subjects were presented a rapidly displayed series of two kinds of events (the letters a and c). Later, they were asked to report the proportion of letters of one kind. On a trial in which only 10 c flashes out of 100 occurred, $P(c) = 0.10$, subjects might report that $P(c) = 0.15$. The same sort of bias against perceiving (or reporting) extremely rare or frequent events was noted in Chapter 2 to be one potential cause of the "sluggish beta" in signal detection theory. Tversky and Kahneman (1981) note that rare events are overestimated in frequency not only when estimates are directly elicited from the subjects, but also when they are inferred on the basis of decision-making tasks such as purchasing insurance or choosing between pairs of gambles having different probabilities of winning. While it should be noted that some other investigations have reported the opposite effect—"risky" estimation of extreme probabilities (Pitz, 1965, 1966)—the conservative findings seem to be somewhat more frequent and are more in line with the data of signal detection theory.

Why is this conservatism observed? One explanation may be subjects' reluctance to report extreme values: "It is safer to err on the side of caution." A second may be that event probability is not just coded in terms of relative event frequency, but also is weighted by event *salience*. Psychologists investigating the "orienting response" long have recognized that novel stimuli attract attention and are more salient than those occurring with greater frequency (Sokolov, 1969). Frequent events, on the other hand, induce adaptation, or a failure to attend. The relatively greater salience of the rare event may cause the estimates of relative frequency to be biased upward. This finding illustrates an important general point that will recur later. A major source of bias in decision making and judgment results when operators direct attention to the most *perceptually* salient aspects of the environment rather than to those that are most relevant to the task or judgment at hand.

Estimating Variability

Human estimates of variability are used in different contexts. For instance, the task of estimating variability is important in situations where operators must assess the contribution of random noise to a process (e.g., to a meter reading) in order to determine how large a deflection of the process constitutes a noteworthy signal. In other contexts, the desirability of a product or outcome may be based upon its consistency (low variability) on some dimension. Finding that a batch of sheet metal consistently has a precise number of acceptable flaws, for example, may be more desirable than finding that the number of flaws in a batch is highly variable, even though the mean flaw rate in both batches is equivalent. In the former case, a constant corrective action can be taken, whereas in the latter it cannot.

When asked to estimate the variance or standard deviation of a set of numbers, humans do not perform as well as when they are estimating means. Given the more complex computation required to compute variance, this difference in accuracy is perhaps not surprising. In fact, two sorts of biases have been reported. Lathrop (1967) noted that the estimation of variability is inversely related to the mean value of the quantities. Of two sets of numbers or analog readings with equal variability, the set with the greater mean will be estimated to have the smaller variability. This finding seems to be a special case of Weber's law in psychophysics, which states that as stimulus magnitude increases, the amount of variability in stimulus magnitude that can just be perceived also increases. Pitz (1980) observed that estimates of variability are influenced greatly by the members of the data set that are most *salient* to the variability estimation task—namely, the two extreme values. The amount of dispersion of data points between these two is relatively discounted. In fact, for some subjects this dispersion is entirely ignored and the range of values is the only determination of the estimated variability.

Extrapolating Growth Functions

Humans often are required to predict or forecast future trends on the basis of a series of past and present data points. The chemical process control monitor must examine the past history of temperature recordings and decide if the process temperature is level or increasing. If it is increasing, is it doing so at a constant rate or is it accelerating? The economic forecaster or business investor must decide if an economic indicator is stable or is going out of control. A point that will recur elsewhere in the book is that humans do not generally perform well at this prediction task. Waganaar and Sagaria (1975) provide graphic evidence that a systematic *conservative* bias occurs when humans are asked to extrapolate the future course of an exponential or accelerating growth function (Figure 3.2). Their future predictions (dashed line) typically underestimate the growth that is predicted by mathematical

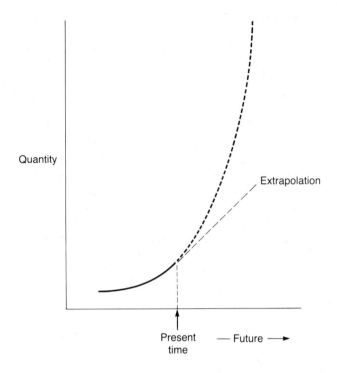

Figure 3.2 Conservatism in extrapolation.

extrapolation of the observed function into future (dotted line). Interestingly, Waganaar and Sagaria observed the same magnitude of conservative bias in a group of "experts," subjects from the Joint Conservation Committee of the Pennsylvania legislature who should be better trained in making such predictions than naive subjects. Gottsdanker (1955), Gottsdanker and Edwards (1957), and Runeson (1975) observed similar "conservative" trends in extrapolating the future motion of accelerating objects. Subjects tended to describe a constant velocity path that simply extended the velocity at the time of the most recent observation.

Three possible causes may be cited to account for this conservative bias in extrapolation: (1) Exponential growth functions are cognitively more complex, requiring more parameters to express verbally, or more analog manipulations to represent spatially, than linear ones. Therefore, the linear representation serves as a simplifying heuristic employed to reduce cognitive effort (Moray, 1980; Rasmussen, 1981; Tversky & Kahneman, 1974). (2) The underestimation may represent a "conservative" resistance to acknowledging the extreme values that an exponential extrapolation generates. This explanation is similar to one that was postulated to account for conservatism in extreme probability estimations. (3) There may, in fact, be a legitimate rational reason for conservatism in extrapolation. The subject simply goes beyond the data given and changes an estimation task into an inference task. An extrapolation task carries with it the assumption that the mathematical function generating the first portion of the curve will continue in effect for the latter portion. That is, an extrapolation task assumes a stable environment in which the growth function continues unaltered to infinity. In real-life experiences, however, most processes are not unchanging, but have built-in limits or self-correcting feedback loops that will arrest the process as it goes out of control. Countries where populations expand exponentially may adopt family-planning policies. Temperatures or pressures that increase exponentially will trigger fire extinguishers or safety valves. The expectation that such self-correcting procedures will "linearize" or at least reduce the speed of growth may be a sufficiently powerful and automatic bias (and also a legitimate and highly rational one) that it is imposed even in the pure laboratory environment when the subject is told to assume that an unchanging environment does exist (Einhorn & Hogarth, 1981).

This is an important general point that will be considered again. The human's assumptions in dealing with probabilistic data are formed on the basis of real-world experience. When decisions are required in a laboratory where the processes characteristic of the real world are often not in effect, the human may find it difficult to abandon these cognitive assumptions. Therefore, behavior will occur that is not rational from the perspective of the experimenter-task demands but is quite rational in the context of the real world (Ebbeson & Konecki, 1980; Hammond, 1980; Navon, 1979).

Whatever the underlying cause and degree, the essential finding remains that humans do not extrapolate growth functions according to the mathematical laws governing those functions. If, in an applied setting, it is necessary to extrapolate that function to determine what the future status will be *if self-correcting processes do not occur,* then some design innovations must be introduced to counter this human bias. One solution is for computer-generated "best fit" extrapolations of the function to be displayed explicitly. These "predictive" displays will be considered in more detail in Chapter 11. A second possible design change to attenuate the effect of conservatism is suggested by Waganaar and Sagaria (1975). They propose a graphical presentation in which the dependent variable in the extrapolated function is transformed so that the function is displayed linearly rather than as an accelerating curve. Projection of the linear function is done easily and without bias.

HUMAN LIMITS IN DECISION MAKING: STATISTICAL INFERENCE

We consider now the problem confronting the operator who must process a number of pieces of data and, rather than making a decision about the data themselves (as estimating the mean value), must make an *inference* about a hypothesis based upon these data. As Sniezek (1980) observed, optimality in estimation is no guarantee of optimality in inference. Figure 3.3 presents an information-processing conception of inferential decision making that provides a framework for our discussion. There is a state of the world which the operator must diagnose or predict. Yet this state is obscured by a "smoke screen" of unreliability and is reflected to the observer only by probabilistic cues. These cues are perceived, but if they are several in number, then perception is subject to limits of selective and divided attention (see Chapter 7). In the diagnostic process the information provided by the attended cues is evaluated against different competing hypotheses held in working memory. Limits on this capacity discussed in Chapter 6 hinder the decision-making process here. Eventually one hypothesis is selected as more likely than others, triggering a feedback process in which the hypothesis may be tested by sampling more cues. The diagnosis will trigger a choice of action. Here the costs and values of different outcomes are considered and balanced against the probabilities that the selected hypothesis or its alternatives may be true. As represented also in Figure 3.1, diagnosis and choice may follow each other in turn.

In the following discussion we will consider first the problems encountered when cues become available only sequentially over time. Then we will examine problems associated with diagnosis (and choice) when there are several different cues (or attributes) bearing on the state of the world. Finally, we will consider the iterative process of hypothesis selection and testing that produces the final candidate hypothesis (H_1 in Figure 3.3) that will guide a choice of action.

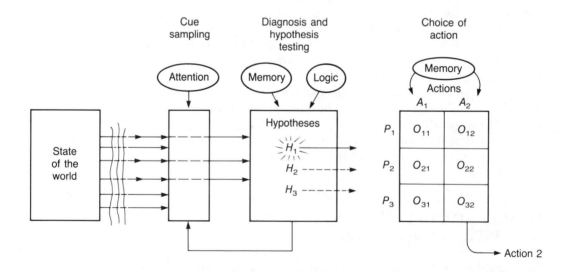

Figure 3.3 Information processing model of decision and choice. Three primary stages are shown across the top. Cognitive limitations are circled.

Integrating Information from a Single Source Over Time

A quality control inspector divides products delivered by an automatic manufacturing device into two groups, defective and normal. On the basis of the frequency of defectives (an estimated proportion), the inspector must decide if the manufacturing process is operating normally (Hypothesis H_1) or if it is malfunctioning in some respect (H_2). In a more abstract form, this scenario may be represented as one in which an operator samples pieces of information concerning the likelihood that one hypothesis or the other is true. The odds are revised after each sample. Formally, the odds may be represented as the ratio of the two probabilities. That is, the odds that the process is malfunctioning is the ratio of the probability of malfunction to the probability of normal operation, or $P(H_1$ is true$)/P(H_2$ is true$)$. (This assumes that the two hypotheses exhaust all possibilities.)

During the 1960s a large quantity of research was performed within this general experimental paradigm (e.g., Edwards, 1962, 1968; Edwards, Lindman, and Phillips, 1965; Edwards, Lindman, and Savage, 1963). A general observation that emerged from this research is that subjects give an undue amount of diagnostic weight to the early stimulus evidence in the sequence—a heuristic that Tversky and Kahneman (1974) labeled *anchoring*. The initial stimulus provides an anchor for the subjects' perception and understanding of the relevant hypotheses. Subsequent sources of evidence are not given the same amount of weight but are used only to shift the anchor slightly in one direction or the other (Einhorn & Hogarth, 1982; Wallsten, 1980). In short, the hypothesis suggested by the first stimulus has an intrinsic advantage because it was established first.

The research by Edwards and his colleagues explored this phenomenon within the framework of an optimal model of hypothesis revision known as Bayes's theorem. Briefly, this theorem states that the odds in favor of a given hypothesis after the acquisition of a piece of data should equal the prior odds (that is, the odds before the datum was collected) multiplied by the *likelihood ratio*. The likelihood ratio is the probability of observing that particular datum if the favored hypothesis is in effect, divided by the probability of observing the same datum if the other hypothesis is true. This relationship is described formally in Equation 3.1, which expresses the odds in favor of hypothesis 1, given the observation of a datum D:

$$\frac{P(H_1/D)}{P(H_2/D)} = \frac{P(H_1)}{P(H_2)} \times \frac{P(D/H_1)}{P(D/H_2)} \qquad (3.1)$$

Accordingly, a weather forecaster wishing to predict the odds for rain on a given day should multiply the prior odds for rain (the number of times it has rained on that day of the year divided by the number of times it has not) by the likelihood ratio of a particular piece of meteorological data (the probability of observing that datum if it is going to rain divided by the probability if it is not).

Edwards and his colleagues contrasted human performance when revising odds with the performance of an optimum statistical model following Bayes's theorem. The universal conclusion of their research is that the human operator is generally *conservative*. That is, in revising his hypothesis (or adjusting the odds), he does not extract as much information from each diagnostic observation of data as he optimally should do (Edwards, 1968; Edwards, Lindman, & Phillips, 1965). After a set of data have been viewed, the subjective odds in favor of one alternative or the other are not assessed to be as extreme nor given as much confidence as optimally they should be. This phenomenon demonstrates a conservatism not unlike that observed when extreme probabilities are to be estimated, or when the response criterion is adjusted in signal detection theory. The explanations for the phenomenon of conservative odds revision vary, but may be roughly categorized into three classes: aggregation deficiency, response bias, and nonindependence.

Aggregation deficiency. Edwards, Lindman, and Savage (1963) argue that humans are fairly good at estimating the independent diagnostic impact of each new piece of data (that is, the numerator and denominator of the likelihood ratio). However, they are deficient in the aggregation process of revising the odds, either because of cognitive limitations in the mental arithmetic required or because of a fading memory for the prior odds. (With a fading memory the odds shift back toward a neutral $0.5/0.5 = 1$ level.) To accommodate these human limitations, Edwards has proposed a cooperative form of human-computer integration in which humans estimate the likelihood ratio, while machines aggregate the evidence and keep track of the revised odds (Edwards, Phillips, Hays, and Goodman, 1968). Applying this technique to medical diagnosis, for example, the physician would estimate the probability of observing a certain set of symptoms if each of two diseases were in effect (H_1 and H_2), provide this information to the computer, and then allow the computer to determine the odds in favor of one disease over the other (Lusted, 1976). The prior odds (i.e., the odds prior to the first updating on the basis of symptomatic information) would simply be the ratio of the absolute frequency of occurrences of disease 1 and disease 2 in the population as a whole.

Response bias. Du Charme (1970) has argued that humans are simply reluctant to state (and be put on record as stating) the extreme odds that their mental calculations have derived with relative accuracy. According to this "response bias" explanation, they simply find it difficult to express extremely high (or low) odds.

Nonindependence. The nonindependence explanation, which is similar to the "rational" explanation for conservative extrapolation of exponential growth described on page 82, asserts that the laboratory assumptions for calculating odds according to Bayes's theorem are simply not valid in most real-world information acquisition tasks. Navon (1979) argues that the degree of *dependence of successive samples* is an important dimension along which the assumptions of Bayes's theorem depart from real-world situations. This difference actually makes conservatism an optimal form of behavior. In the real world, most successive samples of information about a hypothesis are not independent of each other. For example, two successive temperature readings of a process probably will be taken with the same thermometer, and therefore both are subject to any error and unreliability in the thermometer. To compensate for such interdependence, the operator optimally must reduce the diagnostic impact of each piece of data. Assuming interdependence in the real world, the skeptical subject automatically carries this rational and legitimate bias into the laboratory investigation and so does not extract as much information from each source as the optimal paradigm dictates.

Whatever the cause of "conservatism" in information aggregation and however much it represents rational or irrational behavior, its existence is generally indisputable. So also is the assertion that because of limits in memory, humans encounter a number of problems when aggregating evidence over time. These may be attributed to the tendency to give undue weight to early stimuli in a sequence (primacy) and the initially formulated hypothesis (anchoring), as well as to the tendency to overweight those stimuli that have occurred most recently and therefore are "fresh" in working memory (recency).

In arguing for such innovations as integrated graphics displays in process control (Moray, 1980, 1981; Rasmussen, 1981) or simultaneous display to consumers of unit/price information of a number of comparable products (Russo, 1977), researchers have made a convincing case that where possible evidence that is available simultaneously should be presented simultaneously and not sequentially (Einhorn & Hogarth, 1981). This format cannot guarantee that simultaneous processing will occur. That of course depends upon the limits of human attention and the operator's own processing strategies. At least, however, it provides the operator with the option of dealing with the information in parallel, if attentional limitations allow, or of alternating between different information sources if they do not. In this manner, one information source is not given extrinsic and automatic primacy over others.

Integrating Information from Several Sources

The preceding section indicated the difficulties of aggregating information sequentially. Research on decision making and choice suggests that added limitations are imposed when that information is derived from different sources, whether presented simultaneously or sequentially. (Examples include the physician who incorporates different symptoms in her diagnosis or the admissions committee who must attend to different test scores.) As the complexity of the diagnostic, choice, or prediction problem increases, furthermore, the departures from optimal performance become greater.

In this section we shall consider two kinds of problems encountered by the decision maker who must process several different cues: problems resulting from an increase in the number of cues and those resulting from a reduction in reliability.

Impact of number of sources.   The representations in Figures 3.1a and 3.1b depict only two stimulus sources. These may be two symptoms observed by the physician, two properties of a company presented to the stockbroker prior to an investment decision, or two potential consequences of a course of action to be taken following the diagnosis of a system failure. As the number of sources grows beyond two, humans generally do not use the greater information to make better, more accurate decisions (Dawes, 1979; Dawes & Corrigan, 1974; Hayes, 1964; Schroeder & Benbassat, 1975). Oskamp (1965), for example, observed that when more information was provided to psychiatrists, the confidence in their clinical judgment increased but the accuracy of those judgments did not. The limitations of human attention and working memory seem to be sufficiently imposing that an operator cannot easily integrate simultaneously the diagnostic impact of more than a few sources of information. In fact, Wright (1974) found that under time stress, decision-making performance deteriorated when more rather than less information was provided. Despite these limitations, humans have an unfortunate tendency to seek far more information than they can absorb adequately. The admiral or executive, for example, will demand "all the facts" (Samet, Weltman, & Davis, 1976).

To account for the finding that more information may not improve decision making, one must assume that the human operator employs a selective filtering strategy to process informational cues. When few cues are initially presented, this filtering is unnecessary. When several sources are present, however, the filtering process is required, and it competes for the time (or other resources) available for integration of information. Thus, more information leads to more time-consuming filtering at the expense of decision quality.

Any filter must have a tuning mechanism that determines which information is passed on to influence the decision and which is rejected. Payne (1980) argues that the filter is strongly tuned to the *salience* of the information or cue. Thus, Wallsten (1980) showed that subjects under time pressure in decision making selectively processed those cues that were presented at the top of an information display. Top locations presumably were more salient to Wallsten's subjects (as we read from top to bottom), despite the fact that the information presented there was of no greater diagnostic value than the information presented at lower locations.

These findings lead us to expect that, in any diagnostic situation, the brightest flashing light or the meter that is largest, is located most focally, or changes most rapidly will bias the operator toward process-ing its diagnostic information content over others. It is important for a system designer to realize, therefore, that the goals of altering (high salience) are not necessarily compatible with those of diagnosis, in which equal salience of a variety of information sources should be maintained.

Impact of reliability. The effect of salience in producing bias toward some information sources would be of no great harm if all sources were equally informative concerning the truth of one hypoth-esis or the other (diagnosis, Figure 3.1b) or the level of the predicted variable (prediction, Figure 3.1c). If this were the case, it would not matter which cues were selected and which ignored. However, this equality is not always present. Certain cues may be very diagnostic of a particular hypothesis (e.g., the air pressure 200 miles to the west pre-dicts tomorrow's weather with a high degree of accuracy), while other cues (e.g., the temperature 100 miles to the east) convey very little diagnostic information. Experimental results suggest that a salient but uninformative cue often will be weighted heavily relative to an un-salient but informative one, to the detriment of ultimate decision-making accuracy. In this light it is important to consider to what extent the operator attends to differences in the informativeness of cues.

Formally, a decision-making cue may be uninformative with regard to a hypothesis (or unpredictable with regard to a criterion) for one of two reasons: (1) The particular stimulus information may be equally likely under each of two hypotheses. In this case we say it is *undiag-nostic*. For example, low pressure on a gauge does not discriminate a leaking pipe from a failed pump. (2) The information itself may be *unreliable*. For example, an eyewitness testifying at a trial has a less than perfect probability of reporting a crime accurately as a result of memory failure (Ellison & Buckhout, 1981; Loftus, 1979). A remote sensor that provides information concerning weather patterns may occasionally be faulty and provide inaccurate readings to the weather forecaster. In prediction, test scores are somewhat unreliable in provid-ing the admissions committee with a true indication of an applicant's abilities.

The two causes of lack of informativeness—low diagnosticity and low reliability—are logically independent (Schum, 1975). However, both have the effect of reducing the information that *should* be gained about a hypothesis, given the observed stimulus data. Formal rules may be specified to indicate how much one's belief in a hypothesis should optimally be reduced, given the degree of unreliability (Johnson, Cava-nagh, Spooner, & Samet, 1973). Yet in both prediction and diagnostic decision making there is good evidence that people fail to make these optimal conservative adjustments in confidence. When processing sev-eral cues that may lack both perfect reliability and diagnosticity, they tend to apply an "as if" heuristic, treating all evidence "as if" it were equally reliable. This heuristic demonstrates a sort of *risky processing* bias in the sense that decision makers extract *more* information from the unreliable data than is warranted (Johnson et al., 1973; Schum, 1975; Snapper, O'Connor, & Einhorn, 1974).

Numerous examples may be cited in which the "as if" heuristic has been applied. In the task of prediction, Kahneman & Tversky (1973) have demonstrated that humans, even those well trained in statistical theory, do not down-weight unreliable predictions of a criterion variable when making "intuitive" predictions. In Figure 3.4, the optimal diagnostic weighting of a predictive variable is contrasted with the ascribed weights as indicated by subjects' predictive performance. Optimally, the information extracted, or diagnostic weighting, should vary as a linear function of the variable's correlation with the criterion. In fact, the weighting varies in more of an "all or none" fashion.

The insensitivity to differences in predictive validity or cue reliability should make humans ill-suited for performing tasks where prediction is involved. In fact, a large body of evidence (e.g., Dawes, 1979; Dawes & Corrigan, 1974; Kahneman & Tversky, 1973; Meehl, 1954) does indeed suggest that humans, compared to machines, make relatively poor intuitive or clinical predictors. In these studies, subjects are provided with information about a number of attributes of a particular case. The attributes vary in their weights and the subjects are asked to predict some criterion variable for the case at hand (i.e., the likelihood of success in a program). Compared with even a crude statistical system that knows only which way a given variable predicts a criterion (e.g.,

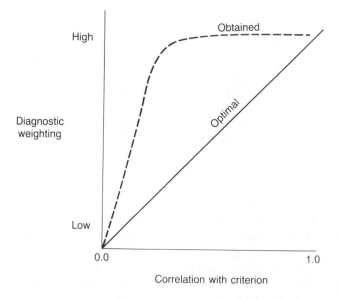

Figure 3.4 Demonstration of the "as if" heuristic. The function shows the relation of the validity of cues to the optimal and obtained weighting of cues in prediction.

higher test scores will predict higher criterion scores) and assumes equal weights for all variables, the human predicts relatively poorly. This observation has led Dawes (1979) to propose that the optimum role of the human in prediction should be to identify relevant predictor variables, determine how they should be measured and coded, and identify the direction of their relationship to the criterion. At this point a computer-based statistical analysis should take over and be provided the exclusive power to integrate information and derive the criterion value.

The "as if" heuristic has been amply demonstrated in decision-making and diagnosis tasks as well as in prediction. Kanarick, Huntington, and Peterson (1969) observed that subjects preferred to purchase cheap, unreliable information over more expensive but reliable information when performing a simulated military diagnostic task. This preference was exhibited despite the fact that a greater amount of total information per dollar spent could be attained by purchasing the reliable information. Rossi and Madden (1979) found that trained nurses were uninfluenced by the degree of diagnosticity of symptoms in their decision to call a physician. This decision was based only on the total number of symptoms observed. Schum (1975) discussed the problems in judicial proceedings arising when juries fail to weight different testimony adequately according to its reliability or to the witness's credibility.

Another potential cause of unreliable data results when the sample size of data used to draw an inference is limited. A political poll based upon 10 people is a far less reliable indicator of voter preferences than is one based upon 100. Yet these differences tend to be ignored by subjects when contrasting the evidence for a hypothesis provided by sources of both high and low sample size (Fischoff & Bar-Hillel, 1981; Tversky & Kahneman, 1971, 1974). In evaluating two polls, one favoring candidate A in 8 out of 10 voters sampled [$P(A) = 0.80$], and the other favoring candidate B in 30 out of 50 voters [$P(B) = 0.60$], the "intuitive statistician" might report the net evidence provided by the two polls as favoring A, despite the fact that the lower sample size (and therefore lower reliability) of the first poll should in fact lessen its impact. Obviously if the evidence of the two polls were lumped together and then tabulated, candidate B would be favored by 32 out of 60 voters. However, when the outcomes are computed separately and reliability must be inferred from sample size information, the impact of differential reliability is diminished. Tversky and Kahneman (1971) observe that even trained scientists who are familiar with principles of statistical inference are not immune to these biases.

Why do subjects demonstrate the "as if" heuristic in prediction and diagnosis? The heuristic seems to be another manifestation of cognitive simplification in which the operator reduces the load imposed on working memory by treating all data sources as if they are of equal reliability. When this is done, a person avoids the differential weighting or multiplication across cue values that would be necessary to

implement a compensatory strategy. When subjects are asked to estimate differences in reliability of a cue directly, they can clearly do so.
However, when this estimate must be used as part of a larger aggregation, then the values become distorted.

The same kind of cognitive simplification can also explain why
subjects do not make efficient use of information provided by nonlinear
cues (Brehmer, 1981; Slovic, Fischoff, & Lichtenstein, 1977). A nonlinear cue is one in which the probability of a hypothesis first increases
and then decreases as the value of the cue increases. For example, the
probability that a particular disease is present may increase with the
first several degrees rise in the patient's body temperature, but then
decrease after the temperature exceeds a certain level. Subjects have
difficulty accounting for nonlinear relations and tend to simplify
them—to treat them as if the relation between the cue value and the
probability of a particular hypothesis were monotonic and linear—in
making a diagnosis.

Overconfidence in judgment.   Overconfidence in judgment is one
manifestation of the risky tendency to overestimate diagnostic evidence. In one experiment, Wells, Lindsay, and Ferguson (1979) simulated a crime and called for eyewitness testimony concerning visual
details of the crime. They found that witnesses were far more confident
in the accuracy of their testimony (and therefore memory of the event)
than was warranted by the actual accuracy of their memory. Furthermore, estimates by mock jurors of the witnesses' confidence were
totally unrelated to the actual accuracy of the witnesses' memory.

Slovic et al. (1977) find that we have the same unwarranted overconfidence in the reliability of our own memory about facts of general
knowledge. Subjects reported themselves to be extremely confident in
answering general-knowledge questions such as the following: "Which
is the greater cause of deaths in the United States: abortion, pregnancy
and childbirth, or appendicitis?" (The answer is appendicitis.) The
disconcerting aspect of their findings is that subjects expressed such
overconfidence even on answers that were wrong far more often than
would be expected by chance. Closely related to these findings is an
observation by Mehle (1982) that subjects engaged in automotive troubleshooting are unjustly confident that they have entertained all possible diagnostic hypotheses. Here again, there is overconfidence in the
accuracy of memory.

Fischoff and MacGregor (1981) observed that overconfidence is also
demonstrated in forecasting. Subjects were asked to make predictions
about future local and national events and estimate their confidence in
those predictions. Later, the confidence of prediction was compared
with the frequency with which the predicted events actually did occur.
A typical set of results shown in Figure 3.5 indicates how consistently
the estimate of the confidence is greater than the actual chance of being
right. Fischoff and MacGregor cite an impressive body of investigations
that demonstrate how pervasive overconfidence is in a variety of pro-

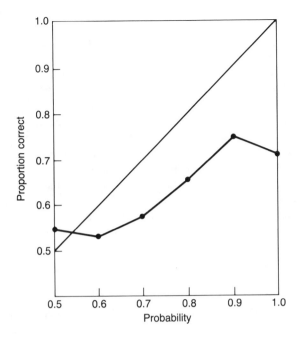

Figure 3.5 Overconfidence in forecasting. The straight line shows an optimal
forecast confidence. The actual function plots the forecast probability of
occurrence against the proportion of times the event actually did occur.
From *Subjective Confidence in Forecasts* (Perceptronics, Technical Report
PTR–1092–81–12) (p. 14) by B. Fischoff and D. MacGregor, 1981, Woodland Hills, CA: Perceptronics.

fessional forecasters. Their summary suggested that only weather forecasters (Murphy & Winkler, 1974) appear to be immune to such
overconfidence in prediction.

Entertaining Hypotheses

The process of entertaining hypotheses obviously is not independent of
the process by which an operator perceives multiple cues of information. In fact, as shown in Figure 3.3, the two operations are closely
related, and the limitations of perception affect hypotheses entertainment. This section will focus on that portion of the diagnosis problem
shown in Figure 3.1 that is "in the head" and consider how the hypotheses entertained influence the information-seeking behavior. An analogy can be drawn with perception, described in the next chapter, in
which expectations "in the head" guide our search for and interpretation of sensory data.

Filtering hypotheses.   Just as limitations of human memory and
attention restrict the perceptual cues that are processed, so they also
substantially curtail the hypotheses entertainment process. To begin

with, humans appear to have limited ability to entertain more than a few (three or four) hypotheses at once, whether these pertain to electronics troubleshooting (Rasmussen, 1981), automotive troubleshooting (Mehle, 1982), medical decision making (Lusted, 1976), or a variety of other decision-making tasks. The limitations of working memory and those of dividing attention between hypotheses and cue evaluation appear to be too restrictive to do otherwise. Because of these cognitive limitations, human decision makers seek to avoid, when possible, the optimal but cognitively demanding decision-making strategies. As described in the early pages of this chapter, these are the "compensatory" strategies in which the implications of all symptoms for all hypotheses are considered at once (in the diagnosis task) or the attribute levels of all objects or courses of action are considered at once (in the choice task). Operators thus avoid the optimal strategy in which favorable evidence for one hypothesis from one symptom may be canceled or "outvoted" by unfavorable evidence from another (Einhorn & Hogarth, 1981; Payne, 1982).

As a consequence of the limitation on the numbers of hypotheses that can be entertained simultaneously, when a large number of alternative hypotheses are initially available and plausible in a diagnosis task or when a large number of objects are presented initially in a choice paradigm, it is possible that relevant hypotheses will not get considered. Faced with an overabundance of alternatives, subjects often employ a simplifying "elimination by aspects" strategy which reduces the demand on cognitive space by cutting the hypotheses or options to the restricted number that can be considered more easily with one of the optimal multiplicative strategies portrayed in Figure 3.1 (Payne, 1982; Tversky, 1972).

In following this heuristic in choice, subjects only attend to the one or few critical attributes that are most heavily weighted in the choice task and eliminate from consideration all but the two to four objects with the highest values on the attributes. For example, in purchasing a car, if price is the most important attribute, the decision maker will eliminate from consideration all but the two to four cheapest automobiles despite the fact that those in the eliminated set may rate quite favorably on attributes that are initially neglected but still may be important. Only after the set has been paired down will other attributes be considered. In diagnosis, the operator might focus attention on the symptoms that most discriminate the hypotheses under consideration and eliminate from consideration any hypothesis that is not consistent with the observed value of that symptom. In diagnosing a process plant failure, for example, the operator might focus on the cue "low pressure at point X" as being most diagnostic and then eliminate from consideration all hypotheses that would not be likely to produce low pressure at X.

This simplifying heuristic may be distorting. On one hand, cues not processed in the initial reduction of the problem space may point very strongly to a hypothesis that was eliminated from consideration. The weight of evidence of these "secondary symptoms" may even be strong enough to outweigh the evidence of the attended attribute and make one of the eliminated hypotheses the most likely one. However, because the cue is not processed, the likely hypothesis is not entertained. On the other hand, with complex processes, some hypotheses eliminated from consideration may be associated with the value on the relevant attribute (low pressure at X), although with a lower likelihood than the hypotheses that are retained. Yet, as we shall see, once discarded from preliminary consideration, further biases in the decision-making process make it unlikely that a hypothesis will be reconsidered.

Choosing hypotheses: The heuristics of representativeness and availability.   Once the hypotheses (or objects) under consideration have been restricted to a manageable set, the human decision maker shows further biases in choosing a hypothesis in light of the symptoms. Following the logic of Bayes's theorem described earlier and presented formally in Equation 3.1, only two factors should enter into the choice of a hypothesis: its probability in light of the observed data (data information) and its prior probability, or the absolute frequency with which a given hypothesis appears to be true according to actuarial data. Examples of the latter include the frequency of a kind of system failure obtained from system reliability data or the prevalence rate of a disease obtained from medical annals. Furthermore, as reflected in Bayes's theorem, these two factors of data information and prior probability should be complementary so that if the prior probability of a particular hypothesis decreases, more data information should be required for the hypothesis to be chosen with an equal degree of confidence. This complementarity between probability and data evidence in affecting confidence is analogous to the complementarity described in Chapter 2 between signal probability and signal intensity in influencing the likelihood of responding "yes."

There are two important heuristics, however, that often dictate a decision in a way that contradicts the optimal balance between prior probability and data information. The two heuristics, *representativeness* (the degree to which the data "look like" those of a hypothesis) and *availability* (the ease of recalling the hypothesis), are simplifying techniques that the decision maker uses intuitively and automatically to approximate information concerning the data and prior probabilities, respectively (Kahneman, Slovic, & Tversky, 1982; Tversky and Kahneman, 1974). Often these heuristics are accurate, but as simplifying shortcuts they are occasionally inaccurate and thereby lead to some fairly systematic nonoptimal biases.

Representativeness refers to extent to which a set of observed symptoms is physically similar to or representative of the symptoms that would be generated if a particular, familiar hypothesis were true. If this similarity is present, then the hypothesis is selected. Thus a physician who observes a set of temperature and blood pressure readings, x-ray plates, and patient reports that are close to those typical of disease X is likely to diagnose the patient as having disease X. This is an intuitively appealing strategy and often there is nothing wrong with the representativeness approach. Carroll (1980) and Phelps and Shanteau (1978) have noted that the existence of correlated patterns of data, or "syndromes," in the real world that allow observed examples to be judged as representative of a category or hypothesis greatly increase the operator's capacity to process a large number of cues.

However, a major danger in the use of the representative heuristic is that it often is employed while information concerning the other critical element in hypothesis choice, prior probabilities, is ignored. The physician who diagnoses disease X on the basis of the similarity of the observed symptoms to the "template" of the disease syndrome may be totally ignoring the fact that disease X is an extremely rare one. Disease Y, which does not match the symptoms as closely (its probability *given the data* is less) but which occurs with 1,000 times greater frequency in the population, might be far more likely than X when both prior probabilities and symptomatic information are aggregated in an optimal compensatory fashion.

Representativeness reflects another example of the distorting effects of salience in decision making. As a real-world example of this phenomenon, Lusted (1976) has pointed out that physicians are insufficiently aware of disease prevalence rates in making diagnostic decisions. Balla (1980, 1982) confirmed the limited use of prior probability information by both medical students and senior physicians in a series of elicited diagnoses of hypothetical patients. Fischoff and Bar-Hillel (1981) report that base-rate information effectively enters into subjects' categorization judgments only when the pattern of symptoms presented by a particular case are not representative of any category. Finally, the "sluggish beta" adjustment in response to signal probability described in Chapter 2, in which decision-making criteria are not adjusted sufficiently on the basis of signal-frequency information, is another example of this failure to account for base-rate information.

Availability refers to "the ease with which instances or occurrences [of a hypothesis] can be brought to mind" (Tversky & Kahneman, 1974, p. 1127). This heuristic can be employed as a convenient means of approximating prior probability, in that more frequent events or conditions in the world generally are recalled more easily. Therefore, subjects typically entertain more available hypotheses. Unfortunately, other factors strongly influence the availability of a hypothesis that may be quite unrelated to their absolute frequency or prior probability. Recency is one such factor. An operator trying to diagnose a malfunction may have encountered a possible cause recently in a true situation, in training, or in a description just studied in an operating manual. This recency factor makes the particular hypothesis or cause more available, and thus it may be the initial one to be considered. Availability also may be influenced by hypothesis simplicity. For example, a hypothesis that is easy to represent in memory (e.g., a single failure rather than a compounded double failure) will be entertained more easily and naturally than one that places greater demands on working memory.

It should be reemphasized that the heuristics of *availability* and *representativeness* often work. That is, representative hypotheses often are the most likely in terms of the data, and more available hypotheses usually *do* have higher probabilities. In fact, if they did not work well, people probably would not employ them as heuristics. The attention of engineering psychologists, however, must focus on the situations in which they fail, because of the important consequences of choosing and acting on an incorrect hypothesis. For example, in the incident at the Three Mile Island nuclear plant described in Chapter 1, the incorrect initial formulation of a hypothesis was a major cause of the crisis (Rubenstein & Mason, 1979). The problems generated by the initial formulation of an incorrect hypothesis now will be addressed.

Testing Hypotheses: Problems in Logical Inference

Typically, the decision process is sequential (Payne, 1980). Once an initial hypothesis is formulated, further evidence is sought to confirm or refute it. Sequential processing is particularly typical of troubleshooting or fault diagnosis, when it is somewhat unlikely that a particular set of symptoms will instantly trigger a final diagnostic decision (Rasmussen, 1981). Unfortunately, the search process required to gain further information is itself characterized by three pronounced biases in human logic that tend to induce less than optimal decision making: the bias for confirmatory evidence, the failure to use negative evidence, and problems with biconditional association.

Confirmation bias.   The first of these biases, a manifestation of the *anchoring* heuristic discussed on page 85, produces an inertia which favors the hypothesis initially formulated. Operators tend to seek (and therefore find) information that confirms the chosen hypothesis and to avoid information or tests whose outcome which could disconfirm it (Einhorn & Hogarth, 1978; Mynatt, Doherty, & Tweney, 1977; Schustack & Sternberg, 1981; Wason, 1960; Wason & Johnson-Laird, 1972). This bias produces a sort of "cognitive tunnel vision" (Sheridan, 1981) in which operators fail to encode or process information that is contradictory to or inconsistent with the initially formulated hypothesis. Such tunneling seems to be enhanced particularly under conditions of high stress and workload (Sheridan, 1981; see also Chapter 7).

The confirmation bias can be described within the framework of Figure 3.6. In the figure a simplified set of two hypotheses are considered (although there may be others as indicated). In an industrial plant these may be a broken pump (H_1) and a clogged relief value (H_2). Evidence for a hypothesis is represented by a plus sign; evidence against by a minus sign; and irrelevant evidence by a zero. In this example, symptom 1 favors H_1 and provides evidence against H_2. Symptoms 2 and 4 provide the opposite information (this contradiction is possible since the symptoms are neither perfectly reliable nor perfectly diagnostic), while symptom 3 favors both hypotheses. Assume that the operator has originally chosen H_1 as the working hypothesis.

	H_1	H_2	H_3 $\cdots$
S_1	+	−	
S_2	−	+	
S_3	+	+	
S_4	0	+	
S_5	0	−	

Figure 3.6 The confirmation bias (symptoms 1–4) and the use of negative evidence (symptom 5). H_1 is the hypothesis chosen for consideration. S_1 and S_3 will be attended; S_2, S_4, and S_5 will not.

The confirmation bias will induce the operator to seek (and therefore be likely to find) evidence in favor of H_1, that is, S_1 and S_3. He would not attend to S_2 and S_4, since these support the alternative.

An example of this bias for the initial hypothesis is provided by Einhorn and Hogarth (1982), who reported that subjects' confidence in one of two hypotheses is influenced by the order in which favorable evidence for each is obtained. In Figure 3.6, if the order is S_1, S_2, subjects will choose H_1. If the order of cues were reversed, they would choose H_2. The hypothesis favored by the first evidence is "anchored" with greater strength. In point of fact, of course, assuming a stable world, the order should have no effect, since the aggregate evidence is the same no matter what order it is presented in.

The tendency to focus only on the initial hypothesis, akin to the "functional fixedness" demonstrated in many problem-solving environments (Adamson, 1952), magnifies the potential danger incurred when biases and heuristics generate an initial hypothesis that is incorrect. It is normally quite easy for the hypothesis tester to obtain information, such as symptom 3 in Figure 3.6, that is consistent with but does not prove a particular formulated hypothesis (i.e., the information also may be consistent with other hypotheses). In theory, it is just as easy to perform critical diagnostic tests that could refute the formulated hypothesis or to seek information contrary to the hypothesis (i.e., symptom 2). Yet this is not done.

Two possible reasons for this failure to seek disconfirmatory evidence may be proposed: (1) Humans encounter greater cognitive difficulty dealing with negative information than with positive information

(Clark & Chase, 1972), an issue that will be discussed further in Chapter 4. (2) To change hypotheses—abandon an old one and reformulate a new one—requires a higher degree of cognitive effort than does the repeated acquisition of information consistent with an old hypothesis (Einhorn & Hogarth, 1981). Given a certain "cost of thinking" (Shugan, 1980) and the tendency of operators, particularly when under stress, to avoid troubleshooting strategies that impose a heavy workload on limited cognitive resources (Rasmussen, 1981), operators adopt a natural bias to retain an old hypothesis rather than go to the trouble of formulating a new one. While maintaining a working hypothesis is valuable because it provides a guide to the search for new information that is more efficient than a random search, the issue of how to force a diagnostician simultaneously to entertain alternative hypotheses and to seek disconfirming evidence—in short, to break through the cognitive tunnel—represents a major challenge to the designer of systems in which troubleshooting will be required.

Investigations by Levine (1966), Bower and Trabasso (1963), and Arkes and Harkness (1980) have asked what information is being processed from cues that support an alternate hypothesis (i.e., S_2 and S_4 in support of H_2) while the decision maker is seeking to confirm the chosen hypothesis (H_1). This information is important if the operator ever does overcome the confirmation bias, abandons H_1, and must then choose among the remaining hypotheses. In laboratory studies of concept learning, Levine finds that subjects process far less of this information than is optimal. Bower and Trabasso reach the more pessimistic conclusion that subjects will sometimes fail to process *any* of the information concerning hypotheses that are not entertained. In a more realistic diagnostic setting, Arkes and Harkness also demonstrated the selective biasing of memory induced by the confirmation bias. They presented subjects with several symptoms related to a particular clinical abnormality (experiment 1) or to the state of a hydraulic system (experiment 2). Arkes and Harkness found that if the subject held a hypothesis or made a positive diagnosis, then symptoms they had observed that were consistent with that diagnosis were readily remembered, while inconsistent symptoms were more easily forgotten. Furthermore, subjects erroneously reported seeing symptoms that they actually had not seen but that were consistent with the diagnosis.

Negative evidence.   Symptom 5 in Figure 3.6 can be used to illustrate the second bias often encountered in troubleshooting and hypothesis testing: a bias against the use of negative information. Note that symptom 5 may or may not occur if H_1 is in effect but will be absent if H_2 is in effect. In this case, search for the *absence* of symptom 5 would provide indirect confirmatory evidence for H_1 because the fact of its absence provides evidence against H_2, thereby eliminating at least one competing hypothesis. For example, a process operator confronted by a malfunction in an energy conversion system who hypothesizes that the cooling-water level at some point is high could derive confirmatory evidence by observing that a meter is *not* indicating that temperature is excessive. Excessive temperature would be an expected symptom of low cooling-water level. Even though the meter might fail to differentiate normal from high water, it would eliminate one competing hypothesis and thereby reduce uncertainty in informational terms.

Wason and Johnson-Laird (1972) demonstrated this failure to use negative information in the laboratory. Rouse (1981) drew similar conclusions from a simulated troubleshooting task, finding that subjects derived very little diagnostic use from potentially valuable negative information that could be used to narrow the competing hypothesis set. Balla (1980) observed that neither medical students nor trained physicians were proficient in using the absence of symptoms to assist in diagnosis. Although Balla found that expertise does not affect the use of negative information, Hunt and Rouse (1981) found that people can be trained to use the absence of cues more efficiently in the simulated troubleshooting task employed by Rouse. Hunt and Rouse provided subjects with 10 sessions of practice and found a progressively greater reliance on negative information.

Causal inference.   A serious bias can arise in hypothesis testing when a causal relationship is inferred erroneously to exist between two cues or events. Such causal inference is most likely to be made when two events tend to occur together (that is, when their relationship can be described as "If A, then B"). Of course, the mere fact that they occur together does not necessarily mean that they are linked causally. "If A, then B," does not necessarily mean that A causes B. However, humans have a strong bias to attribute and emphasize causal factors relating two events. In their theory of causal inference, Einhorn and Hogarth (1983) propose that the simple ordering of two statements or facts on the page leads people to view a causal relation between them even when none may exist or when the actual causal relation may be reversed. Thus the statement, "The water pressure is high and the process is unstable," will be perceived as indicating that high water pressure caused the instability.

Tversky and Kahneman (1982) have demonstrated that different biases in causal inference are at work in diagnosis and prediction. "Prediction" refers to observing a present state and looking forward to predict a future set of cues. "Diagnosis" refers to observing a given set of cues and looking backward to infer the hypothesis or state that produced those cues. Thus, prediction goes from cause to effect, diagnosis from effect to cause. Kahneman and Tversky assert that humans are biased to perceive stronger relations from cause to effect than from effect to cause, even when the two may be equal. As an example, they considered the following pair of questions: What is the probability that a daughter will have blue eyes, given that the mother has blue eyes (prediction)? What is the probability that the mother has blue eyes given that the daughter has blue eyes (diagnosis)? Since the absolute probability of having blue eyes in both the parent and the daughter's generation is equal, it can be verified that there is no difference in the conditional probabilities of the two questions. Yet the subjects in Kahneman and Tversky's experiment consistently judge predictive statements like the first to be far more likely than diagnostic ones like the second, so strong is the tendency to conceive in a forward cause-effect direction.

Symmetrical strength of prediction and diagnosis can occur only when the base-rate probabilities of the predicted and diagnosed event are equal. When there is a marked difference in the absolute probability to two conditions A and B, then the statement, "A causes B" (prediction), will not have the same likelihood as the statement, "B was caused by A" (diagnosis). This situation of unequal probabilities would occur if, for example, B could be caused by several possible events and therefore had a higher absolute probability than A. Adopting a hypothetical process control example, the relation might be, "If the pump fails, then the water level will be low." While this causal relation is unidirectional, humans often err by assuming that it is bidirectional (Taplin, 1971; Taplin & Staudenmayer, 1973). That is, they also assume, "If B, then A" ("If the water level is low, then the pump must be broken"). This is not a valid inference, for simple logic informs us that there exist numerous other causes for low pressure (e.g., broken pipes, leaking valves). Yet subjects apparently have a difficult time generating a large set of plausible hypothesis that could have produced the observed symptoms. Mehle (1982), for example, observed this difficulty for subjects trying to troubleshoot problems in automobile engines.

Eddy (1982) has identified similar dangers in bidirectional inference in medical diagnosis when the absolute probabilities of cause and effect conditions differ. The probability that someone who has a particular disease will show a certain symptom is not the same as the probability that one who shows the symptom will have the disease. The former may be quite high even as the latter is low. Yet Eddy notes that physicians sometimes erroneously attribute the high probability of the symptom to the disease. As a consequence, physicians may be overconfident in diagnosing the disease in a patient who shows the symptom when such diagnosis is unwarranted, with the cost and potential danger of unnecessary surgery or treatments.

In summary, human limitations in hypothesis formulation and testing are well documented and pronounced. The reader is referred to excellent treatments by Kahneman et al. (1982) and Anderson (1979) for further discussion. We shall consider some of the implications of these limitations for system design at the end of the chapter.

VALUES AND COSTS

Systems and personnel both cost money. Because economic considerations often enter into decisions made by an operator, it is important to consider how perceptions of costs and benefits can be causes of potentially nonoptimal decision making. As described in Chapter 2 and shown in Figures 3.1 and 3.3, costs and benefits of various outcomes can and should influence the nature of decision-making biases. Yet subjects' behaviors in tasks such as diagnostic troubleshooting, in which actions should optimally be based both upon the information gained by actions and the costs or expected benefits of those actions in dollars, suggest that the choice of actions does not give adequate weight to financial considerations (Rouse, 1981; Towne, Fehling, & Bond, 1981). These observations are buttressed by a wide variety of laboratory findings on decision making and choice.

The Optimum Prescription

To provide a framework for analyzing these departures, we shall first consider the "optimum" manner by which costs and benefits should guide decision making and choice. In the optimal models of decision making, costs and values are assigned to different potential outcomes of a decision or choice, and the action chosen (based upon a diagnostic outcome) should be that which in the long run produces the highest expected gain or the minimum expected loss. Formally and optimally these relations may be expressed in the decision matrix shown in Figure 3.7, an expansion of the right side of Figure 3.3. Two courses of action (choices, treatments) are specified. Each can be taken in the face of the decision maker's belief that different states of the world exist. (This belief is the output of diagnosis.) Each state of the world (labeled

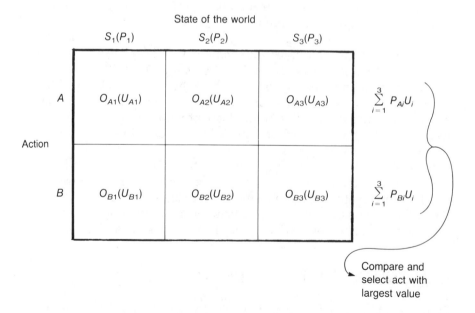

Figure 3.7 Optimal use of expected utility in the selection of action.

S) is associated with a subjective probability that it is true (labeled P). Each act may have a different outcome, given that a particular state of the world exists. These are shown within each cell. Finally, each outcome is associated with a value, cost, or utility U_{ij}. The action that is selected should optimally be that with the *highest expected utility*. This quantity is computed by multiplying the utilities and probabilities of each outcome and summing these products across actions. These expected utilities are shown in the right margin of the table. The optimal decision maker will select the action with the highest expected value.

There are, of course, some limitations to this optimal prescription, since all costs or utilities cannot be expressed in terms of dollar values. As we discussed in Chapter 2, the cost to the physician planning a certain diagnostic treatment that is associated with the possible loss of life obviously cannot be so expressed. This limitation, in fact, is one of the greatest difficulties in applying normative models of decision making to medical treatment (Lusted, 1976). Nevertheless, even within these constraints the typical decision maker still falls short of a optimal behavior. For example, according to the prescription, an action that reduces an expected loss by a constant amount should be preferred equally to one that enhances an expected gain by the same amount. Correspondingly, a given change in the expected utility should optimally have the same influence on decision behavior, no matter whether this change was produced by a change in probabilities or a change in utilities. We shall see that neither of these two "optimal equalities" are not in fact in force in human decision making.

Differences Between Perceived Gains and Losses

In contrast to the optimal prescription which equates loss reductions with gain enhancements, a number of investigators have demonstrated that people do not conceptualize losses as the mirror image of gains. Somewhat different principles seem to operate (Payne, Laughhunn, & Crum, 1982; Tversky & Kahneman, 1981). The major differences are threefold.

1. A potential loss of a given amount is viewed as having greater consequences and therefore exerts a greater influence over decision-making behavior than does a gain of the same amount. Thus, if given a choice between refusing or accepting a gamble that offers a 50% chance to win or lose $1.00, the subject would typically decline the offer. The potential $1.00 loss is viewed as more negative than the $1.00 gain is positive. As a result, the expected utility of the gamble (the sum of the probability of outcomes times their utilities) is a loss, and the subject will decline it. Furthermore, the negative utility of loss is found to be an accelerating function of the value of loss; so larger losses have disproportionally greater influences on decision making than smaller losses (Edwards, Lindman, & Phillips, 1965). As portrayed in Figure

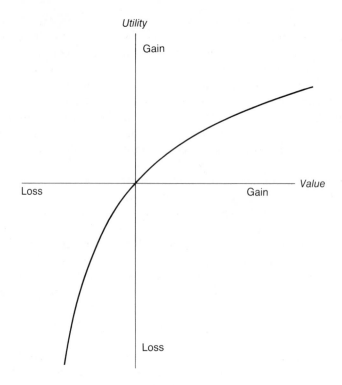

Figure 3.8   The hypothetical relation between value and utility.

3.8, this is the exact opposite of the effect on gains. As a consequence, the greater the difference between losses and gains in risky choices, the more risk-aversive subjects will be.

2.   Under time pressures, decision makers appear to give more weight to negative evidence concerning alternatives than to positive evidence (Wright, 1974).

3.   Individuals tend to be risk-aversive when choosing between potential gains of equal expected values ("Take the money and run") but risk-seeking when choosing between potential losses ("Throwing good money after bad") (Payne, 1980; Payne, Laughhunn, & Crum, 1982; Tversky & Kahneman, 1981). For example, if given the choice between two alternatives with the same expected values—winning $1.00 for sure (no risk) and taking a gamble with a 50/50 chance of winning $2.00 or nothing at all (risky)—subjects typically choose the former, certain option. However, suppose the word "winning" were replaced by "losing," so that the choice is between losses. This choice produces a so-called avoidance-avoidance conflict. In this case, subjects would now tend to choose the second, risky option. Payne et al. (1982) noted that this kind of bias described the behavior of 128 business executives given hypothetical investment decisions just as aptly as it described the behavior of typical laboratory subjects.

It should be noted that the third factor does not imply that subjects will always choose a risky loss over a sure loss if the amount of expected loss is not equal between the two options. A case in point is the purchase of insurance, in which the sure loss (the cost of the premium) is preferred over the risky loss. In this case the choice of the sure loss is dictated by the first factor, namely, the exponentially decreasing function of utility with negative value. Tversky and Kahneman (1981) have formalized a set of principles which they label "prospect theory" that describe the manner in which subject's perceptions of risks, gains, and losses collectively produce many of the systematic biases that are seen in decision-making and choice behavior.

The importance of these differences between perceived losses and gains is that a given change in value (or expected value) may be viewed either as a change in loss or a change in gain, depending upon what is considered as the reference or neutral point. For example, a tax cut may be perceived as a reduction in loss if the neutral point is "paying no taxes" or as a positive gain if the neutral point is "paying last year's taxes" (Tversky & Kahneman, 1981). As a consequence, different frames of reference in a decision problem may produce fairly pronounced changes in decision-making behavior (Carroll, 1980; Tversky & Kahneman, 1981).

For example, consider an operator choosing between two courses of action after diagnosing a potentially damaging failure in a large industrial process: continue to run while further diagnostic tests are performed or shut down the operation immediately. The first action he perceives will lead to a large financial cost (serious damage to the equipment) with some probability less than one (given that the problem is serious and immediate). The second action will produce a cost that is an almost certain cost but of lesser magnitude (lost production time and start-up costs). According to the argument stated above, when the choice is framed in this fashion as the choice between losses, the operator would tend to select the higher-risk first alternative over the low-risk second alternative so long as he perceived the expected values of the two actions to be similar. On the other hand, if the operator's perceptions were based upon a framework of profits to the company, the first alternative would be considered a probability mix of a full profit if nothing is wrong and a diminished (but still positive) profit if the disastrous event occurs. The second alternative will be considered a certain large but not maximum profit. In this case, the choice would more likely be the second, low-risk alternative.

It is apparent that perceived costs and benefits greatly influence most of the decisions that are made, and so differences in these utilities can influence decision outcomes. It is important therefore that the manager who supervises personnel who make decisions (e.g., process control monitors) should attempt to provide them with a relatively uniform set of costs and payoffs concerning the consequences of their various actions. The supervisor then must be aware of what kind of risk-seeking or risk-aversive actions those payoffs will induce.

LEARNING AND FEEDBACK: THE EXPERT DECISION MAKER

A number of investigators have questioned the generalizability of the conclusions reported above to "real-world" decision-making tasks. Some have merely questioned the validity of assertions that the human is not optimal when optimality is defined by normative (e.g., Bayesian) models operating on independent samples of data (Navon, 1979). As noted, there is some justification to the claim that the departures of human decision making from the model-based prescriptions of "optimal" are in fact closer to optimal than the models themselves given that the operator's valid assumptions about the way the world behaves (e.g., nonindependence of successive data samples) are brought into the laboratory. Furthermore, the argument can be made that the use of simplifying heuristics, such as representativeness and availability, is optimal when the decision to employ them, rather than more complex compensatory strategies, is based upon the utility of the cost of thinking (Shugan, 1980). Others, however (e.g., Ebbeson & Konecki, 1980), have gone farther and questioned whether any of the findings from laboratory experiments conducted under "idealized" conditions can generalize to the applied domain.

At the opposite extreme, some researchers (e.g., Balla, 1980; Brehmer, 1981; Eddy, 1982; Einhorn & Hogarth, 1978; Slovic et al., 1977) assert that even the true expert decision maker, whether medical doctor, investment broker, troubleshooter, or personnel selector, suffers from some fairly severe departures from optimality. This claim is initially surprising, since our general experience in most disciplines of human performance is that continued practice will cause performance to improve, approaching closer and closer to some "optimal level" (Fitts & Posner, 1967). The following discussion will consider first the data bearing on expert decision making and then the arguments concerning the generalizability of decision-making research.

The Myth of the Expert

Brehmer (1981) presents convincing evidence that decision makers in psychological and medical diagnosis, who are supposedly experts with their material, are not in fact terribly accurate and that their decision outcomes are highly variable. Two experts, provided the same data, are quite likely to arrive at very different outcomes. One cause of this variability is certainly that the cognitive limitations of working memory and attention impose the same constraints on the expert as they do on the amateur (Slovic et al., 1977). The diagnoses will be particularly inconsistent when a decision problem is not represented as a familiar "syndrome" of correlated attributes. However, three other causes of variability are directly attributable to the specific nature of the decision-making task, a task that is not generally structured in such a way as to encourage efficient learning.

Misleading feedback.   Because decison making is probabilistic, a correctly made decision (the right cues given appropriate weighting) will often yield an incorrect outcome because of chance factors. (We noted in Chapter 2 that optimal settings of d' and beta will still yield errors.) Correspondingly, one may obtain correct outcomes for the wrong reasons. Thus the "right-wrong" aspect of the feedback may often be misleading, thereby encouraging subjects to follow incorrect strategies that through chance yielded correct outcomes or to abandon appropriate strategies that yielded incorrect outcomes. Each correct feedback will serve to reinforce the rule that generated the outcome. If this rule is inappropriate but correct by chance, its strength becomes greater and greater, and it is harder to "unlearn" (Einhorn & Hogarth, 1978).

Limited attention to delayed feedback.   Often the feedback from a decision may be delayed—by a few days perhaps in the case of the physician or by several months or years in the case of an admissions committee (to assess whether an applicant succeeds in a program) or a parole officer (to assess whether his client has successfully adapted to the world outside of prison). This delay may cause many of the factors that went into the decision to be forgotten or distorted by the time the feedback (however limited) becomes available. Furthermore, the decision maker is often quite preoccupied with other matters at the time and thus gives little attention to the feedback. This tendency is exaggerated by a phenomenon that Fischoff (1977) labels *cognitive conceit*. Through a number of convincing experiments, he demonstrated how much we tend to underestimate the information gained from outcomes and therefore overestimate, retrospectively, the extent of our prior knowledge. If the apparent discrepancy between what we know now (after a decision outcome is observed) and what we thought we knew (retrospectively) before the decision is slight, then we see little wrong with our initial decision formulation and thus perceive little need to revise the decision-making process.

Fischoff and MacGregor (1981) identify the lack of attention to feedback as a major cause of overconfidence in forecasting, discussed on page 92. As they note, the exception appears to be in weather forecasting where forecasters are clearly judged on the long-run accuracy of their predictions (i.e., on 70% of the occasions when a "70 percent chance of showers" is predicted, it should actually shower). Here, when decision makers are forced to process feedback and thereby calibrate their forecasts, performance appears to be accurate (Murphy & Winkler, 1974).

Selective perception of feedback.   Einhorn and Hogarth (1978) have elegantly demonstrated how the manner of representing the prediction task fosters unwarranted confidence in one's decision-making abilities. This representation in turn leads experts to become progressively poorer, ironically even as their confidence in the correctness of their choice grows. Einhorn and Hogarth consider the situation in which a candidate is to be admitted to or rejected from a program. The basis of selection is the value of a composite predictive variable, aggregated from measures on a number of attributes as represented in Figure 3.1c. If the variable exceeds a criterion, the candidate is selected. Eventually, those candidates selected will be evaluated on the basis of their success in the program (or success of the treatment). Here again the authors assume a dichotomy of success and failure. These representations are shown in Figure 3.9a. The core of Einhorn and Hogarth's argument is that a host of factors combine to cause the decision maker to attend to (and therefore overestimate) the *positive hit rate* (applicants who exceeded the selection criterion and succeeded in the program).

In the first place, Einhorn and Hogarth argue that people tend to encode and thus remember the frequency of positive hits rather than the *probability* of a positive hit (i.e., positive hits divided by the total number of those exceeding the acceptance criterion). This distortion occurs in part because number is a more direct and salient quantity than is probability—an abstraction requiring an extra cognitive step (division). Also the bias against encoding by probability occurs in part because the denominator is dependent upon knowing the number of false positives (those who were accepted but failed). As indicated earlier in the chapter, these entries are data that disconfirm the decision-making strategy or selection rule and therefore, as disconfirming evidence, tend to be ignored.

Perceiving in this fashion the entries in the positive hit cell, the decision maker is spuriously reinforced by the knowledge that "n people exceeded the selection criterion based upon my decision rule and they succeeded; therefore, my rule must work." In addition to this bias of attention against the false positive numbers (disconfirmatory evidence), the decision maker will rarely possess any data concerning those who were rejected. Even if such data are available, they are unlikely to be in a usable probabilistic form. More likely they will be in the form of a single-case representation, for example, "the fish that got away."

Finally, the decision maker often has a vested interest in establishing the success of the rules that were employed. Therefore, treatment or "placebo effects" will increase the likelihood that those selected will be more likely to succeed than those who did not exceed the criterion but might have been admitted to the program for other reasons. This bias is shown by the upward shift in criterion scores to the right of the cutoff in Figure 3.9b. Through computer-simulation techniques, which assume that the rewarding effects of a positive hit reinforce the existing

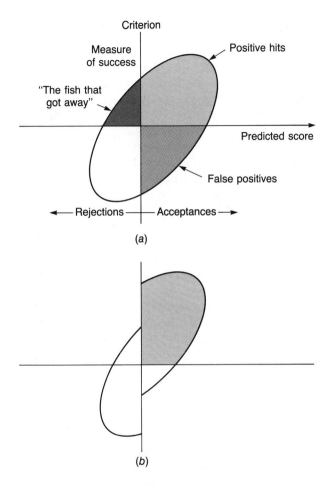

Figure 3.9 Source of unwarranted confidence in prediction. From "Confidence in Judgment: Persistence of the Illusion of Validity" by H. J. Einhorn and R. M. Hogarth, 1978, *Psychological Review, 85*, p. 397. Copyright 1978 by the American Psychological Association. Adapted by permission of the authors.

decision rules, Einhorn and Hogarth (1978) showed that both increasingly unwarranted confidence and less accurate prediction emerged with increased practice on the prediction task.

In summary, decisions of the expert in a given field may well be superior to those of the novice, as a result of the expert's greater familiarity with the material in the area, greater knowledge of correlations between cues, and better models of causal structure. Yet this knowledge alone is not sufficient to guarantee the accuracy of the decision process. The inherent biases built into the decision-making task will often be sufficient to guarantee that even the confident expert has an inflated estimate of his or her decision-making accuracy.

Validity of Decision-Making Research

A prominent argument against decision-making research and its generalizability is that the structures of problems are quite different in the natural world from those studied in the laboratory. A major source of this difference is the correlation between cues or symptoms. As pointed out on page 78, levels of attributes or values of symptoms do not vary independently, but certain values tend to co-occur. For example, the occurrence of high pressure in a chemical system will often precede a rise in temperature. In meteorology the presence of high winds will often precede the appearance of a high-pressure center.

By themselves, these correlations do little to assist the decision maker. However, as the patterns of correlation are learned by experience and training, an internal "model" of the environment is formed. Subjects can then use these correlated cues to advantage in decision making (Medin, Alton, Edelson, & Freko, 1982; Phelps and Shanteau, 1978). Furthermore, wherever possible subjects tend to perceive a causal relation between a pair of correlated cues (Einhorn & Hogarth, 1981, 1983). This internal model then provides a framework for easier integration of information that is consistent with it and removes much of the burden of working memory associated with processing large amounts of uncorrelated information (see Chapter 6). There is, of course, a cost associated with this model-based integration. Information that is inconsistent with the model is likely to be filtered or distorted to fit the expected pattern of correlations, a phenomenon noted in the experiment on diagnostic judgment conducted by Arkes and Harkness (1980). On the other hand, if there is enough information that is inconsistent (e.g., the symptoms don't co-occur in the manner that is expected), the decision maker will be forced to regress to the memory-loading strategy in which attributes are considered independently. Then the nonoptimal characteristics described in the preceding pages will again come into play.

There is probably some truth to claims that the human may be considerably more optimal than is suggested in some laboratory paradigms where uncorrelated cues are typically employed. However, it is important to reemphasize that the critical issue is not to establish how optimal the operator is in an absolute sense but to identify the variables, such as cue correlation, sequential information presentation, and negative information, that influence the departure from optimal performance.

Once these variables are identified, then the similarity of structure between the idealized laboratory paradigm and the different real-world environments can be carefully delineated. At this point the engineering psychologist can note that certain characteristics of the particular real-world environment under investigation lead the operator away from optimal performance, whatever the absolute level of that performance may be. The psychologist can then suggest corrective aids. In the final section of this chapter we will briefly discuss the nature of these aids, some implemented, some proposed, and others merely suggested by the data. Finally, in Chapter 12, the characteristics of process control that influence departures from optimality of decision making and diagnosis will be considered explicitly.

DECISION-MAKING AIDS

While several techniques have been proposed to assist the decision maker in real-world contexts, many of these have encountered difficulties on their way to implementation (Einhorn & Hogarth, 1981; Schrenk, 1969; Slovic et al., 1977). One problem is that the complexity of many of the aids renders them difficult to use. A second difficulty is that the aids are often based upon the principle of "divide and conquer" (Slovic et al., 1977). That is, their effectiveness often depends upon the ability of the user to decompose the overall decision problem into its constituent elements. Difficulties arise where this is not possible or where such decomposition substantially alters the nature of the original problem.

A third problem is that it is exceedingly difficult to evaluate the success of decision aids. If a decision is found to be correct with the use of an aid in a real-world context, would the outcome also have been correct if the decision were made without it? In the laboratory such data can easily be brought to bear as decision aids are demonstrated on simulated problems (Samet et al., 1976). However, because each real decision problem outside the laboratory is unique, the answer to the question is difficult to ascertain. The exact conditions under which the particular decision is made are impossible to re-create once, let alone often enough to obtain a statistically reliable estimate of the probability of a correct (or optimal) decision with and without the aid.

In spite of these difficulties in establishing the usefulness of decision aids, the numerous shortcomings of human decision making enumerated in the preceding pages implicitly suggest a number of techniques that could assist the decision maker in a wide variety of environments. The following section divides these aids into two categories: those based upon the limits of human memory and attention and those based upon training.

Memory and Attention Assists

It is evident that where possible decision-making tasks should be designed in such a manner as to present all diagnostic information concerning all alternatives so that it can be considered simultaneously. Integrated computer displays in complex process-monitoring stations will be of considerable use (Moray, 1980, 1981; Rasmussen, 1981) and these will be considered in more detail in Chapter 12. Naturally some selection may be required to reduce the displayed information to a manageable level. Samet et al. (1976) have suggested possible adaptive algorithms for this selection. They propose a system in which the computer monitors the operator's choice of information sources and infers the particular *attributes* of those sources that the operator finds most important. For example, does the monitor of a process-control plant typically refer to readings of pressure or temperature, and which ones weigh more heavily on her diagnostic judgment? After making this inference, the computer selectively filters information sources in order to present primarily those sources that contain the most information on the subjectively important attributes. In this manner the computer reduces some of the problems associated with information overload.

In a similar vein, computer-generated displays may be employed to assist the operator in a fault-diagnosis task by considering alternative hypotheses and keeping track of the outcome of sequential tests, so that recency and anchoring on the initial hypotheses do not dominate memory. Rouse (1981) reports that such memory aids substantially improve troubleshooting performance. A computer with sufficient intelligence could provide information relevant to the bearing of test outcomes on alternative hypotheses and could force the operator to entertain alternative hypotheses. This technique could provide an important safeguard against the bias for hypothesis confirmation (Einhorn & Hogarth, 1982; Moray, 1980; Rasmussen, 1981). With such an aid, however, there is a potential danger if the true state of the system is described by a hypothesis that neither the computer nor the operator considers. The computer after all is only as smart as its programmers. If they fail to foresee certain possibilities or implications, the operator who assumes a smart "fail-safe" computer system could encounter serious difficulties. These issues will be considered further in our discussion of automation in Chapter 12.

Other aids emphasize the computer's assistance in integrating diagnostic information. As described earlier, Dawes (1979) argues convincingly that computer assistance can be of great use in integrating predictive information to provide statistical prediction. Edwards (1962; Edwards et al., 1968) proposes a system of cooperation between humans and machines in sequential data-processing tasks. In order to overcome the inherent human conservatism relative to the "Bayesian optimal," he proposes that the human act as the likelihood-ratio estimator and the machine act as the Bayesian-odds aggregator. A more recent suggestion proposes that Bayesian techniques be used to assist the operator in his choice of modes of operations (automatic, semi-automatic, or manual) that are most effective in a gunnery target acquisition task (Zacharias & Levison, 1981).

Multiattribute utility theory has been proposed to assist people in adopting difficult compensatory strategies when choosing between several courses of action which vary along a number of attributes (Slovic et al., 1977). In fact, this technique, which formalizes and quantifies the representation of choice in Figure 3.1*a*, has been successfully employed in a number of diverse circumstances—for example, assisting land developers and environmentalists in reaching a compromise on coastal development policy in California (Gardiner & Edwards, 1975) and choosing the location of the Mexico City airport (Kenney, 1973).

Training

Training aids to the decision maker can be offered in three forms. One method is simply to make the decision maker aware of the nature of limitations and biases of which he or she may be totally unconscious. For example, it would seem that training operators to consider alternative hypotheses and to be aware of their resistance to doing it under stress might reduce the likelihood of cognitive tunnel vision. In an investigation relevant to this point, Koriat, Lichtenstein, and Fischoff (1980) found that forcing forecasters to entertain reasons why their forecasts might not be correct reduced their biases toward overconfidence in the accuracy of the forecast. As noted above, Hunt and Rouse (1981) succeeded in training operators to extract diagnostic information from the absence of cues. Lopes (1982) achieved some success at training subjects away from nonoptimal anchoring biases when processing multiple information sources. She called subjects' attention to their tendency to anchor on initial stimuli that may not be informative and had them anchor instead on the most informative sources. When this was done, the biases were reduced.

A second training aid is to provide more comprehensive and immediate feedback in predictive and diagnostic tasks, so that operators are forced to attend to the degree of success or failure of their rules. We noted above that the feedback provided to weather forecasters is successful in reducing the tendency for overconfidence in forecasting. Jenkins and Ward (1965) have demonstrated that providing decision makers simultaneously with data in all four cells of the contingency table in Figure 3.9, instead of simply the hit probability, improves their appreciation of predictive relations. Where selection tasks or diagnostic treatments are prescribed, box scores should be maintained to integrate data in as many cells of the matrix as possible (Einhorn & Hogarth, 1978; Goldberg, 1968). Tversky and Kahneman (1974) have suggested that decision makers should be taught to encode events in terms of probability rather than frequency, since probabilities intrinsically account for events that did not occur (negative evidence) as well as those that did.

A third way to improve training is to capitalize on the natural efforts of humans to seek causal relations between variables and their superiority in integrating cues when correlations between variables are known beforehand (Einhorn & Hogarth, 1981; Medin et al., 1982; Phelps & Shanteau, 1978). To capitalize in this manner, every effort should be made to identify "syndromes" and to emphasize the correlational structure existing in the cues that represent a certain hypothesis. Where possible it seems advantageous to explain these correlations in terms of a causal structure (if one exists) or a process model that can economically represent the nature of the intercorrelations (Moray, 1980, 1981). We shall cite some examples of the success of this instructional procedure in Chapter 12.

HOW OPTIMAL IS DECISION MAKING?

A review of this chapter suggests a fairly pessimistic view of human performance in decision making and diagnosis. The case studies of incorrectly made decisions, predictions, and diagnoses are numerous. They include the millions of dollars spent by consumers on unnecessary auto repairs and replaced parts, the proportion of the 4.5 million medical diagnostic tests made per year that are unnecessary (Lusted, 1976), the billion-dollar cost to the military of selecting pilots for flight school who fail to matriculate (North & Griffin, 1977), the tragic consequences of the errant predictions prior to the Japanese attack on Pearl Harbor, and the tremendous costs of incorrect decisions in nuclear incidents such as Three Mile Island. While many of these faulty decisions may indeed have simply been the result of the probabilistic nature of the environment (i.e., they were made correctly), many were undoubtably not. As long as this much room for improvement exists and human limitations can be easily demonstrated in the laboratory, improvements in the area of human decision making surely represent one of the most important potential contributions that engineering psychology can make.

In keeping with the spirit of this book, it is essential to emphasize that, although incorrect decisions are made by the individual human decision maker and may be attributable directly to human limitations, they are *rarely the operator's fault*. Instead, the faulty decisions are more likely to be the inevitable consequence of a decision problem, presented in such a manner as to accentuate rather than compensate for these limitations. The role of engineering psychology is to suggest the possible compensations; the role of human factors is to implement them.

Transition

It is appropriate to end this chapter on a note of optimism. While human decision making is often far from the optimal that could be achieved by the intelligent computer with all facts at hand, in most (but not all) situations humans far outperform decisions that would be made by the flip of a coin or a roll of the dice. Performance probably lies somewhere midway between chance and optimal, shifting further away from the optimum as complexity and stress increase. Yet there is a third anchor with which human decision making can be compared, and this is the computer, working on-line in real time with the same uncertainties and time pressure faced by the human. Under these circumstances one important characteristic enhances the human's performance relative to the computer. This is the ability to recognize familiar patterns of data and exploit correlated cues in such a way that the decisions appear automatic, without imposing upon the frailties of human attention and working memory. This kind of processing will be the topic of the next two chapters, the more automatic domain of pattern recognition.

References

Adamson, R. E. (1952). Functional fixedness as related to problem solving. *Journal of Experimental Psychology, 44,* 288–291.

Anderson, J. R. (1979). *Cognitive psychology.* New York: Academic Press.

Arkes, H., & Harkness, R. R. (1980). The effect of making a diagnosis on subsequent recognition of symptoms. *Journal of Experimental Psychology: Human Learning and Memory, 6,* 568–575.

Balla, J. (1980). Logical thinking and the diagnostic process. *Methodology and Information in Medicine, 19,* 88–92.

Balla, J. (1982). The use of critical cues and prior probability in concept identification. *Methodology and Information in Medicine, 21,* 9–14.

Bower, G., & Trabasso, T. (1963). Reversals prior to a solution in concept identification. *Journal of Experimental Psychology, 66,* 409–418.

Brehmer, B. (1981). Models of diagnostic judgment. In J. Rasmussen & W. Rouse (Eds.), *Human detection and diagnosis of system failures.* New York: Plenum Press.

Carroll, J. S. (1980). Analyzing behavior: The magician's audience. In T. S. Wallston (Ed.), *Cognitive processes in choice and decision making,* Hillsdale, NJ: Erlbaum Associates.

Clark, H. H., & Chase, W. G. (1972). On the process of comparing sentences against pictures. *Cognitive Psychology, 3,* 472–517.

Dawes, R. M. (1979). The robust beauty of improper linear models in decision making. *American Psychologist, 34,* 571–582.

Dawes, R. M., & Corrigan, B. (1974). Linear models in decision making. *Psychological Bulletin, 81,* 95–106.

Du Charme, W. (1970). Response bias explanation of conservative human inference. *Journal of Experimental Psychology, 85,* 66–74.

Ebbeson, E. D., & Konecki, V. (1980). On external validity in decision-making research. In T. Wallsten (Ed.), *Cognitive processes in choice and decision making.* Hillside, NJ: Erlbaum Associates.

Eddy, D. M. (1982). Probabilistic reasoning in clinical medicine: Problems and opportunities. In D. Kahneman, P. Slovic, & A. Tversky (Eds.), *Judgment under uncertainty: Heuristics and biases.* New York: Cambridge University Press.

Edwards, W. (1962). Dynamic decision theory and probabilistic information processing. *Human Factors, 4,* 59–73.

Edwards, W. (1968). Conservatism in human information processing. In B. Kleinmuntz (Ed.), *Formal representation of human judgment* (pp. 17–52). New York: Wiley.

Edwards, W., Lindman, H., & Phillips, L. D. (1965). Emerging technologies for making decisions. In T. M. Newcomb (Ed.), *New directions in psychology II.* New York: Holt, Rinehart, & Winston.

Edwards, W., Lindman, H., & Savage, L. J. (1963). Bayesian statistical inference for psychological research. *Psychological Review, 70,* 193–242.

Edwards, W., Phillips, L. D., Hays, W. L., & Goodman, B. C. (1968). Probabilistic information processing systems: Design and evaluation. *IEEE Transactions on Systems, Science, and Cybernetics, ssc-4,* 248–265.

Einhorn, H. J., & Hogarth, R. M. (1978). Confidence in judgment: Persistence of the illusion of validity. *Psychological Review, 85,* 395–416.

Einhorn, H. J., & Hogarth, R. M. (1981). Behavioral decision theory. *Annual Review of Psychology, 32,* 53–88.

Einhorn, H. J., & Hogarth, R. M. (1982). *Theory of diagnostic interference I: Imagination and the psychophysics of evidence.* (Technical Report No. 2). Chicago: University of Chicago, School of Business.

Einhorn, H. J., & Hogarth, R. M. (1983, January). *Diagnostic inference and causal judgment: A decision making framework* (research report). Chicago: University of Chicago, Center for Decision Research.

Ellison, K. W., & Buckhout, R. (1981). *Psychology & Criminal Justice*. New York: Harper & Row.

Erlick, D. E. (1961). Judgments of the relative frequency of a sequential series of two events. *Journal of Experimental Psychology, 62*, 105–112.

Fischoff, B. (1977). Perceived informativeness of facts. *Journal of Experimental Psychology: Human Perception and Performance, 3*, 349–358.

Fischoff, B., & Bar-Hillel, M. (1981). *Diagnosticity and the base-ratio effect* (Technical Report PTR–1092–81–11). Woodland Hills, CA: Perceptronics.

Fischoff, B., & MacGregor, D. (1981). *Subjective confidence in forecasts* (Technical Report PTR–1092–81–12). Woodland Hills, CA: Perceptronics.

Fitts, P. M., & Posner, M. I. (1967). *Human performance*. Pacific Palisades, CA: Brooks/Cole.

Gardiner, P. C., & Edwards, W. (1975). Public values: Multiattribute ability measurement for social decision making. In M. F. Kaplan & B. Schwarts (Eds.), *Human judgment and decision processes*. New York: Academic Press.

Goldberg, L. R. (1968). Simple models or simple processes? *American Psychologist, 23*, 483–496.

Gottsdanker, R. M. (1955). A further study of prediction-motion. *American Journal of Psychology, 68*, 432–437.

Gottsdanker, R. M., & Edwards, R. V. (1957). The prediction of collision. *American Journal of Psychology, 70*, 110–113.

Hammond, K. R. (1980, July). *The integration of research in judgment and decision theory* (Report No. 226). Boulder, CO: University of Colorado, Institute of Behavioral Science, Center for Research on Judgment and Policy.

Hayes, J. R. (1964). Human data processing limits in decision making. In E. Bennett (Ed.), *Information systems, science, and engineering: Proceedings of the First International Congress on the Information Systems Science*. New York: McGraw-Hill.

Hunt, R., & Rouse, W. (1981). Problem-solving skills of maintenance trainees in diagnosing faults in simulated power plants. *Human Factors, 23*, 317–328.

Jenkins, H. M., & Ward, W. C. (1965). Judgment of contingency between responses and outcomes. *Psychological Monographs: General and Applied, 79* (Whole No. 594).

Johnson, E. M., Cavanagh, R. C., Spooner, R. L., & Samet, M. G. (1973). Utilization of reliability measurements in Bayesian inference: Models and human performance. *IEEE Transactions on Reliability, 22*, 176–183.

Kahneman, D., Slovic, P., & Tversky, A. (Eds.). (1982). *Judgment under uncertainty: Heuristics and biases*. New York: Cambridge University Press.

Kahneman, D., & Tversky, A. (1973). On the psychology of prediction. *Psychological Review, 80*, 251–273.

Kanarick, A. F., Huntington, A., & Peterson, R. C. (1969). Multisource information acquisition with optimal stopping. *Human Factors, 11*, 379–386.

Kenney, R. L. (1973). A decision analysis with multiple objectives: The Mexico City airport. *Bell Telephone Economic Management Science, 4*, 101–117.

Koriat, A., Lichtenstein, S., & Fischoff, B. (1980). Reasons for Confidence. *Journal of Experimental Psychology: Human Learning and Memory, 6*, 107–118.

Lathrop, R. G. (1967). Perceived variability. *Journal of Experimental Psychology, 23*, 498–502.

Levine, M. (1966). Hypothesis behavior by humans during discrimination learning. *Journal of Experimental Psychology, 71*, 331–338.

Loftus, E. F. (1979). *Eyewitness testimony*. Cambridge, MA: Harvard University Press.

Lopes, L. L. (1982, October). *Procedural debiasing*. (Technical Report WHIPP 15). Madison, WI: Wisconsin Human Information Processing Program.

Lusted, L. B. (1976). Clinical decision making. In D. Dombal & J. Grevy (Eds.), *Decision making and medical care*. Amsterdam: North Holland.

Medin, D. L., Altom, M. W., Edelson, S. M., & Freko, D. (1982). Correlated symptoms and simulated medical classification. *Journal of Experimental Psychology: Learning, Memory, and Cognition, 8*, 37–50.

Meehl, P. C. (1954). *Clinical versus statistical prediction*. Minneapolis: University of Minnesota Press.

Mehle, T. (1982). Hypothesis generation in an automobile malfunction inference task. *Acta Psychologica, 52*, 87–116.

Moray, N. (1980). *Information processing and supervisory control* (MIT Man Machine Systems Laboratory Report). Cambridge, MA: Massachusetts Institute of Technology.

Moray, N. (1981). The role of attention in the detection of errors and the diagnosis of errors in man-machine systems. In J. Rasmussen & W. Rouse (Eds.), *Human detection and diagnosis of system failures*. New York: Plenum Press.

Murphy, A. H., & Winkler, R. L. (1974). Subjective probability forecasting experiments in meteorology: Some preliminary results. *Bulletin of the American Meteorological Society, 55*, 1206–1216.

Mynatt, C. R., Doherty, M. E., & Tweney, R. D. (1977). Confirmation bias in a simulated research environment: An experimental study of scientific inference. *Quarterly Journal of Experimental Psychology, 29*, 85–95.

Navon, D. (1979). The importance of being conservative. *British Journal of Mathematical and Statistical Psychology, 31*, 33–48.

North, R. A., & Griffin, G. R. (1977, October). *Aviator selection 1919–1977* (Technical Report L5-77-2). Pensacola, FL: Naval Aerospace Medical Research Laboratory.

Oskamp, S. (1965). Overconfidence in case-study judgments. *Journal of Consulting Psychology, 29*, 261–265.

Payne, J. W. (1980). Information processing theory: Some concepts and methods applied to decision research. In T. S. Wallsten (Ed.), *Cognitive processes in choice and decision behavior*. Hillsdale, NJ: Erlbaum Associates.

Payne, J. W. (1982). *Contingent decision behavior: A review and discussion of issues* (Technical Report 82-1). Durham, NC: Duke University School of Business Administration.

Payne, J. W., Laughhunn, D., & Crum, R. (1982). *Multiattribute risky choice behavior: The editing of complex prospects* (ONR Technical Report 82-2). Durham, NC: Duke University Graduate School of Business Administration.

Peterson, C. R., & Beach, L. R. (1967). Man as an intuitive statistician. *Psychological Bulletin, 68*, 29–46.

Phelps, R. H., & Shanteau, J. (1978). Livestock judges: How much information can an expert use? *Organizational Behavior in Human Performance, 21*, 209–219.

Pitz, G. F. (1965). Response variability in the estimation of relative frequency. *Perceptual & Motor Skills, 21*, 867–873.

Pitz, G. F. (1966). The sequential judgment of proportion. *Psychonomic Science, 4*, 397–398.

Pitz, G. F. (1980). The very guide of life: The use of probabilistic information for making decisions. In T. S. Wallsten (Ed.), *Cognitive processes in choice and decision behavior*. Hillsdale, NJ: Erlbaum Associates.

Rasmussen, J. (1981). Models of mental strategies in process control. In J. Rasmussen & W. Rouse (Eds.), *Human detection and diagnosis of system failures*. New York: Plenum Press.

Rossi, A. L., & Madden, J. M. (1979). Clinical judgment of nurses. *Bulletin of the Psychonomic Society, 14*, 281–284.

Rouse, W. B. (1981). Experimental studies and mathematical models of human problem solving performance in fault diagnosis tasks. In J. Rasmussen & W. Rouse (Eds.), *Human detection and diagnosis of system failures*. New York: Plenum Press.

Rubinstein, T., & Mason, A. F. (1979, November). The accident that shouldn't have happened: An analysis of Three Mile Island. *IEEE Spectrum*, pp. 33–57.

Runeson, S. (1975). Visual prediction of collisions with natural and non-natural motion functions. *Perception & Psychophysics, 18*, 261–266.

Russo, J. E. (1977). The value of unit price information. *Journal of Market Research, 14*, 193–201.

Samet, M. G., Weltman, G., & Davis, K. B. (1976, December). *Application of adaptive models to information selection in C3 systems* (Technical Report PTR-1033-76-12). Woodland Hills, CA: Perceptronics.

Schrenk, L. P. (1969). Aiding the decision maker—a decision process model. *IEEE Transactions on Man-Machine Systems, MMS-10*, 204–218.

Schroeder, R. G., & Benbassat, D. (1975). An experimental evaluation of the relationship of uncertainty to information used by decision makers. *Decision Sciences, 6*, 556–567.

Schum, D. (1975). The weighing of testimony of judicial proceedings from sources having reduced credibility. *Human Factors, 17,* 172–203.

Schustack, M. W., & Sternberg, R. J. (1981). Evaluation of evidence in causal inference. *Journal of Experimental Psychology: General, 110,* 101–120.

Sheridan, T. (1981). Understanding human error and aiding human diagnostic behavior in nuclear power plants. In J. Rasmussen & W. Rouse (Eds.), *Human detection and diagnosis of system failures.* New York: Plenum Press.

Sheridan, T. B., & Ferrell, L. (1974). *Man-machine systems.* Cambridge, MA: MIT Press.

Shugan, S. M. (1980). The cost of thinking. *Journal of Consumer Research, 7,* 99–111.

Slovic, P., Fischoff, B., & Lichtenstein, S. (1977). Behavioral decision theory. *Annual Review of Psychology, 28,* 1–39.

Snapper, K. J., O'Conner, M. F., & Einhorn, H. J. (1974). *Social indicators: A new method for indexing quality* (Soc. Res. Group Tech. Rep. 74-4). Washington, D.C.: George Washington University Press.

Sniezek, J. A. (1980). Judgments of probabilistic events: Remembering the past and predicting the future. *Journal of Experimental Psychology: Human Perception & Performance, 6,* 695–706.

Sokolov, E. N. (1969). The modeling properties of the nervous system. In I. Maltzman & K. Cole (Eds.), *Handbook of contemporary Soviet psychology.* New York: Basic Books.

Taplin, J. E. (1971). Reasoning with conditional sentences. *Journal of Verbal Learning and Verbal Behavior, 10,* 218–225.

Taplin, J. E., & Staudenmayer, H. (1973). Interpretation of abstract conditional sentences in deductive reasoning. *Journal of Verbal Learning and Verbal Behavior, 12,* 530–542.

Towne, P. M., Fehling, M. R., & Bond, N. A. (1981). *Design for the maintainer: Projecting maintenance performance from design characteristics* (Technical Report No. 95). Los Angeles: University of Southern California.

Tversky, A. (1972). Elimination by aspects: A theory of choice. *Psychological Review, 79,* 281–299.

Tversky, A., & Kahneman, D. (1971). The law of small numbers. *Psychological Bulletin, 76,* 105–110.

Tversky, A., & Kahneman, D. (1974). Judgment under uncertainty: Heuristics and biases. *Science, 185,* 1124–1131.

Tversky, A., & Kahneman, D. (1981). The framing of decisions and the psychology of choice. *Science, 211,* 453–458.

Tversky, A., & Kahneman, D. (1982). Causal schemas in judgments under uncertainty. In D. Kahneman, P. Slovic, & A. Tversky (Eds.), *Judgment under uncertainty: Heuristics and biases.* Cambridge: Cambridge University Press.

Waganaar, W. A., & Sagaria, S. D. (1975). Misperception of exponential growth. *Perception & Psychophysics, 18,* 416–422.

Wallsten, Thomas S. (1980). *Cognitive processes in choice and decision behavior.* Hillsdale, NJ: Erlbaum Associates.

Wason, P. C. (1960). On the failure to eliminate hypotheses in a conceptual task. *Quarterly Journal of Experimental Psychology, 12,* 129–140.

Wason, P. C., & Johnson-Laird, P. N. (1972). *Psychology of reasoning: Structure and content.* London: Batsford.

Wells, G. L., Lindsay, R. C., & Ferguson, T. I. (1979). Accuracy, confidence, and juror perceptions in eyewitness testimony. *Journal of Applied Psychology, 64,* 440–448.

Wright, P. (1974). The harassed decision maker: Time pressures, distractions, and the use of evidence. *Journal of Applied Psychology, 59,* 555–561.

Zacharias, G. L., & Levison, W. (1981). A supervisory control model of the AAA crew. *Proceedings, 17th Annual conference on Manual Control* (Technical Report JPL 81-95). Pasadena, CA: Jet Propulsion Laboratory.

Perception of Verbal Material

4

OVERVIEW

Chapters 4 and 5 focus upon the relationship between meaning and sensory stimuli. *Meaning* can be thought of as resulting from the activation of associations, or *nodes*, in long-term memory. A stimulus is likely to have meaning if it (a) has been experienced before, (b) is represented by a node in memory, and (c) has important potential consequences for immediate or future actions.

The stimuli that impinge upon our senses vary along a number of physical dimensions (for example, color, shape, size, and orientation). Each level of a stimulus dimension (for example, a specific line orientation or a given color) may be referred to as a *feature*. Certain *combinations* of features in the environment, because of their linkage in previous experience, are responded to in a consistent fashion. Thus, to the child, a certain combination of the shape, color, and size of an object is categorized or perceived consistently as a *fire truck*. For the pilot, a certain combination of the angles and length of four lines in a trapezoid is responded to as a *runway*, and more specific combinations as *the runway at the appropriate angle of approach*.

Previous chapters already have presented simple examples of this process. Chapter 2 dealt with the response to specific features on one physical dimension only (two levels in signal detection, several levels in absolute judgment). Particular features were assigned by the experimenter to generate invariant responses. In the discussion of decision making in Chapter 3, the categorization was more complex. We noted that the decision maker's task was simplified if correlations existed between the various attributes or dimensions of a decision-making problem. Thus swine judges (Phelps & Shanteau, 1978) and parole officers (Payne, 1980) were better able to classify correctly when repeatedly encountering correlated attributes. In these instances certain combinations of features triggered a discrete categorical response. We might also say that the particular combination of features defined by this correlation was represented by a perceptual response.

This superiority of categorization was obtained at a cost, however. Cues that did not quite fit into the expected correlated pattern were distorted perceptually to be forced into that category. Some information in the stimulus was lost. Thus Payne (1980) noted that the parolee who shares some but not all attributes of the "typical" drug offender will be readily categorized as one for parole purposes, probably to his detriment. Perception thus can be defined as the process of mapping a number of forms of a stimulus into a single perceptual category. This category corresponds to a unit or node in long-term memory that is activated when the stimulus is present. Repetition and experience with the correlated stimuli strengthen both these perceptual categories themselves and their linkage with stimuli.

Spatial and Verbal Perception

Within the last 15 years a large body of research has related the efficiency of processing different kinds of information to the two cerebral hemispheres (Moscovitch, 1979). The processing of verbal material has been more associated with the left cerebral hemisphere, and the processing of spatial material with the right. The primary methodology employed in this research has been to present the different kinds of information to different sides of the body (i.e., left ear, right ear, left visual field, right visual field) or to require responses with different hands. In the data from these experiments, interactions are noted between the type of material presented and the side of the body to which information is presented or from which a response is required. Two general principles have emerged from this research. The *direct access* principle states that material that is processed predominantly by one hemisphere or the other (i.e., verbal, left; spatial, right) will be perceived more efficiently when stimulus material is provided direct access to the predominant hemisphere—that is, when information is presented to the *contralateral* side of the body from the hemisphere which processes it (because the left hemisphere processes information from the right side of the body and vice versa). The *resource competition* principle states that two processes carried on within a hemisphere will interfere with each other to a greater extent than two processes carried on within separate hemispheres (Kinsbourne & Hicks, 1978).

Although there is by now convincing experimental evidence to support the direct access principle (see Moscovitch, 1979, for a comprehensive summary), the relevant effects as demonstrated in the laboratory are not large, accounting for only a small proportion of the experimental variance, primarily because the two hemispheres of the brain are somewhat flexible. Each hemisphere is able to compensate for the other's deficiencies, and each is able to assume some level of the other's primary processing mode. For example, the right hemisphere does have some linguistic ability, and the left hemisphere may be used in spatial problem solving. Because the laterality effect is not strong, these experimental results will probably not contribute a great deal to system-design principles. The most direct implication of the direct access principle is evident when operators must monitor complex multielement displays containing both verbal and spatial information. In this time-sharing environment, superior performance is obtained if the configurations place the verbal information to the right of the spatial (Wickens, Sandry, & Hightower, 1982). A discussion of the implications of the resource-competition principle to issues in system design will appear in Chapter 8.

In contrast to the "hardware" aspects of hemispheric laterality, considerably greater benefits of the research in this area derive from the characteristics of the two contrasting *codes* of processing that have been associated, through experimental manipulation, with one hemisphere or the other. The left hemisphere has been associated with processing of verbal, linguistic material, processing that is analytic (breaking elements into constituent features), and processing of high spatial frequencies (a fine "grain" of texture). The right hemisphere has been associated with processing of spatial relations, processing that is "holistic," and processing of low spatial frequency (Moscovitch, 1979; Sergent, 1982).

It is probably true that the difference between these two codes of processing is in fact more of a continuum than a dichotomy. At one extreme, recognition of the meaning of words and subsequent integration of words into meaningful sentences are clearly verbal processes, discussed in Chapter 4. At the other extreme, perceiving differences in texture and orientation of surfaces would readily be labeled *spatial*. We do not normally verbalize these properties as we perceive them, nor is verbalization necessary to respond to changes in orientation. Between these extremes, however, lie tasks like navigation which combine both verbal and spatial processing. As we shall discuss in Chapter 5, people encode directions both in terms of words (left, right, etc.) and in terms of spatial maps.

The spatial-verbal continuum is important because it correlates with other dimensions of human processing. These include implications that the continuum may define individual differences in spatial and verbal abilities (Gordon, Silverberg-Shalev, & Czernilas, 1982; Hunt, Frost, & Lunneborg, 1973; Lohman, 1979; Snow, 1980), different training methodologies (graphic-visual versus verbal; Snow, 1980; Yallow, 1980), different patterns of dual task interference (see Chapters 6 and 8), and different strategies of problem solving (logical versus intuitive) or of task performance (verbal versus spatial; Hammond, 1980; Landeweerd, 1979; see Chapter 12). These continua may well interact. For example, individuals with high verbal ability may benefit more from verbal than from graphic training, while spatial training may be more appropriate for others (Yallow, 1980).

It is certainly true that the practical implications of these continua do not depend upon knowing that verbal and spatial processing are lateralized in the brain. Nevertheless, the recent research on laterality and hemispheric differences, by providing a plausible anatomical basis for the dichotomies and continua described above, has brought their implications into a clearer focus. These implications will be discussed in greater detail in later chapters. The remainder of this chapter will address the dominant attribute of left hemispheric processing—namely, its verbal characteristics. Verbal processing in perception, in turn, will be treated by a detailed analysis of the perception of printed material, the implications for display design, and the integration of verbal information in terms of comprehension. The chapter supplement will present an analogous discussion of the perception of speech. Then in Chapter 5 we shall focus on perception of nonverbal material.

THE PERCEPTION OF PRINT

Stages in Word Perception

The perception of printed material is hierarchical in nature. When we read and understand the meaning of a sentence (a categorial response), we must in the process analyze its constituent words. Each word, in turn, depends upon the perception of constituent letters, while each letter is itself a collection of elementary features (lines, angles, curves). These hierarchical relations are shown in Figure 4.1, which is based on Neisser (1967). There is good evidence that each level of analysis of the hierarchy (feature, letter, word) is defined by a set of nodes. The brain manifests a unique, relatively automatic categorical response, characterized by the many-to-one mapping typical of perception. Thus in describing the perception of words we may refer to feature units, letter units, or word units. A given unit at any level will become active if the corresponding stimulus is physically presented and the perceiver has had repeated experience with the stimulus in question.

We shall consider, first, the evidence provided for the unit at each level of the hierarchy and the role of learning and experience in integrating higher-level units from experience with the repeated combina-

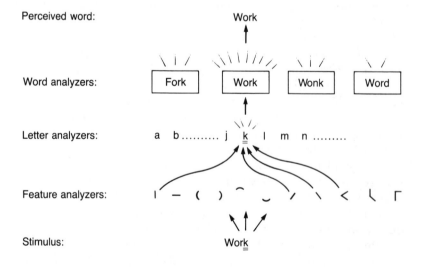

Figure 4.1 Hierarchical process of perception of the visual word "work."

tion of lower-level units. Then we shall consider the manner in which our expectancies guide perceptual processing from the "top down." After the theoretical principles of visual pattern recognition are described we shall then address their practical implications for system design.

The features as a unit: Visual search.   The constituent features that compose letters are represented as vertical or diagonal lines, angles, and curved segments of different orientations. Examples of these are shown at the bottom of Figure 4.1. Gibson (1969) has demonstrated that the 26 letters of the alphabet can be economically created by the presence or absence of a limited subset of these features. The importance of constituent features in letter recognition is most clearly demonstrated by the visual search task developed by Neisser (1967). In this paradigm, subjects scan a vertical column of random three- or five-letter sequences as rapidly as possible until they detect a particular target letter (or set of letters) that is specified prior to a trial. An example of two such search trials is shown at the top of Figure 4.2. Neisser observed a linear relation between the serial position of the letter in the list and the time to detect the target. This function is shown at the bottom of Figure 4.2. The slope of the function reflects the average time required to decide that each given letter is not the target (or not in the target set, if two or more targets are to be searched for).

The importance of features in letter identification is supported by the finding that the search rate is heavily dependent upon the feature similarity between targets and nontargets (Neisser et al., 1964). Thus, searching for the letter C in the background of S's, Q's, U's, and D's (all of which share the curved feature) is far more difficult than searching for C in K's, L's, Z's, and N's. Such data are quite nicely accounted for by a model that assumes that the subject develops a representation of the unique features of the target letter(s). Each letter in the search list is then processed only to a level at which a feature is found that is not a member of the target feature set. In this way, noise letters will only be fully processed if they share most features with the target item. Such a model explains why the target letters appear phenomenologically to "jump out" of the array when they are encountered. Only these are fully processed (Neisser, 1967).

A second conclusion reached by Neisser, Novick, and Lazar was that after practice subjects could search for two or more targets as rapidly as they could search for a single target. That is, as each letter in the list was encountered, it was mentally compared, in parallel, to all members of the target set. The investigators reasoned that the comparison must be parallel, for if it were serial, then the more targets to be searched for, the longer it would take to examine each character.

The letter as a unit: Automatic processing.   While features are a basic constituent unit of letter analysis, there is also strong evidence that a letter is more than simply a bundle of features. That is, the whole

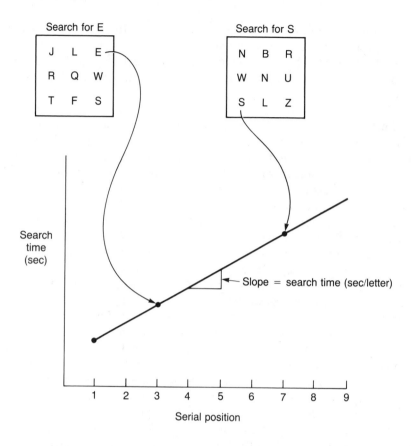

Figure 4.2 Neisser's letter-search paradigm. The top of the figure shows two lists with different targets. The bottom graph shows the search time for letters in each of the two lists as a function of their serial position on the list. Also presented are the data from other serial positions and the resulting linear slope. From "Decision Time without Reaction Time" by U. Neisser, 1963, *American Journal of Psychology, 76*, p. 377. Copyright 1963 by the University of Illinois Board of Trustees. Reprinted by permission.

letter is greater than the sum of its parts. An experiment of LaBerge (1973) provides an example of such evidence. The subjects in LaBerge's experiment were required to make rapid perceptual judgments about whether a pair of stimuli were identical or different. On some proportion of the trials subjects expected the particular stimulus pair, and therefore attention was focused on it. On the other trials they did not expect the stimuli to occur, and so the pair was not initially attended. Under both attended and nonattended conditions two different kinds of stimuli were presented: familiar letters (for example, *b, d*) and symbols that were as complex as letters but did not form letters (e.g., ∟ and ⌐). LaBerge found that when subjects expected the stimuli and so were attending to them when they arrived, there was no difference in the time to make identity decisions about letter or nonletter stimuli. However, when their attention was focused elsewhere, the nonletter stimuli took longer to process.

LaBerge interpreted these results to suggest that the letter stimuli, because their features have consistently occurred together, may be processed *automatically* or *preattentively*. In a sense their features are "glued together" to form a unit that automatically becomes active when the stimulus is presented, bypassing the feature analysis level. On the other hand, the nonletters are only processed after their constituent features are processed, and this processing cannot occur while attention is diverted elsewhere. Attention must "glue" the features together. The role of practice in developing automaticity was also demonstrated by LaBerge. After several days of experience, the nonletter stimuli were eventually processed preattentively as well.

Research conducted by Schneider and Shiffrin (1977; Shiffrin & Schneider, 1977) has explored in considerably greater detail the relation between response consistency, automatic processing, and the "unitization" of perception found in letter recognition. In their general paradigm, Schneider and Shiffrin required subjects to search for sets of target letters that appeared in a sequentially presented series of arrays of letters (see Figure 4.3). If, over a large number of repeated trials, subjects were consistently asked to detect one letter (or set of letters) as a target and these letters never appeared as nontarget stimuli,

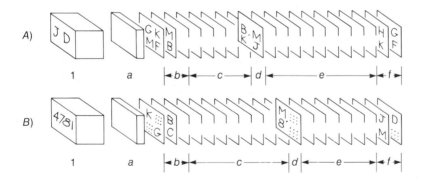

Figure 4.3 Two examples of a positive trial in the 20-frame search paradigm. (A) Varied mappings: memory-set size = 2 (J, D); frame size = 4. (B) Consistent mappings: memory-set size = 4 (4, 7, 8, 1); frame size = 2. (1) Presentation of memory set. (a) Fixation dot goes on for 0.5 sec when subject starts trial; (b) three dummy frames that never contain target; (c) distractor frames; (d) target frame; (e) distractor frames; (f) dummy frames that never contain target. Frame time is varied across conditions. Trial A is varied mapping because letters which are targets on one trial may be distractors on the next. Trial B is consistent mapping because digits are always targets and letters always distractors and the same digits are always targets. From "Controlled and Automatic Human Information Processing I: Detection, Search, and Attention" by W. Schneider and R. M. Shiffrin, 1977, *Psychological Review, 84*, p. 18. Copyright 1977 by the American Psychological Association. Reprinted by permission of the authors.

Schneider and Shiffrin's data suggested that the target letters were processed in a qualitatively different manner from the nontargets. On the other hand, when subjects searched for letters that sometimes appeared as targets and sometimes as nontargets, then processing of both targets and nontargets was qualitatively similar.

Schneider and Shiffrin labeled these two modes of processing as *automatic* and *controlled*. Automatic processing is characterized by a rapid parallel search through the stimulus array to locate the target; the target will "jump out" of the array even in the absence of focused attention. Controlled processing, on the other hand, is characterized by slower serial search that demands attention. The critical criterion for developing automatic processing appears to be the *consistent mapping* of a stimulus (or set of stimuli) into one response class, repeated over a number of exposures. In Schneider and Shiffrin's original research, some 20,000 trials were employed to demonstrate automatic processing. More recently a reduced amount of 200 trials has been employed.

The combined results of LaBerge's and Schneider and Shiffrin's results suggest that "automaticity" exists on more than one level. LaBerge's work indicates that all letters are automatically processed because of the consistency with which the co-occurring sets of features are mapped into the letter response. This is the result of a lifetime's experience. Schneider and Shiffrin find that arbitrary sets of letters may be processed even more automatically with just a few hours practice as a consequence of consistent mapping to a given response category.

The word as a unit: Word shape.   Under some circumstances words may be perceived through the analyses of their letters, just as letters were perceived through the analysis of their constituent features. Yet there is also evidence that familiar words can be perceived as units, just as LaBerge's experiment provided evidence that letters were perceived as units because of their familiarity. Thus the pattern of full-line ascending letters (d, b), descenders (p, g), and half-line letters (a, r) in a familiar word such as "the" forms a global shape that can be recognized and categorized as "the" even if the individual letters are obliterated to such an extent that each is illegible. Broadbent and Broadbent (1977, 1980) propose that a mechanism of spatial frequency analysis in the brain (see Chapter 6) is responsible for this crude analysis of word shape.

The analysis based upon word shape is more "holistic" in nature than the detailed feature analysis described above. The role of word shape, particularly with such frequent words as "and" and "the" for which unitization is likely to have occurred, seems to be revealed in the analysis of proofreading errors (Haber & Schindler, 1981; Healy, 1976). Haber and Schindler had subjects read passages for comprehension and proofing at the same time. They observed that misspellings of short, function words of higher frequency ("the" and "and") were difficult to detect. The role of word shape in these shortcomings was suggested because errors in these words were concealed most often if the letter change that created the error was one that substituted a letter of the same class (ascender, descender, or short) and thereby preserved the same word shape. An example would be "anb" instead of "and." If all words were only analyzed letter by letter, these confusions should be as hard to detect in long words as in short ones. As Haber observes, they are not. Corcoran and Weening (1967) noted that for words that were longer and less frequent (the two variables are, of course, highly correlated), acoustic or phonetic factors play a more prominent role than visual ones in proofreading errors. In this case, misspellings are concealed if the critical letters are not pronounced in the articulation of the word.

Top-down versus bottom-up processing: Context and redundancy. In the system shown in Figure 4.1, "lower-level" units (features and letters) feed into "higher-level" ones (letters and words). Sometimes lower-level units may be bypassed, if higher units are unitized. This process then is sometimes described as bottom-up or data-driven processing (Lindsay & Norman, 1972). There is also strong evidence that much of our perception proceeds in a "top-down," context-driven manner (Lindsay & Norman, 1972). More specifically, in the case of reading, hypotheses are formed as to what a particular word should be, given the context of what has appeared before, and this context enables our perceptual mechanism to guess the nature of a particular letter within that word, even before its bottom-up feature-to-letter analysis may have been completed. Thus the ambiguous word in the sentence, "I think it will rxxx today," can be easily and unambiguously perceived, not because of its shape or its features but because the surrounding context limits the alternatives to only a few (e.g., rain, snow, hail) and the apparent features of the first letter eliminate all but the first alternative.

In a corresponding fashion, top-down processing can work on letter recognition. Knowledge of surrounding letters may guide the interpretation of ambiguous features, as in the two words "THE CAT." The middle letters of the two words are physically identical. The features are ambiguously presented as parallel vertical lines or as converging lines. Yet the hypotheses generated by the context of the surrounding letters quite naturally force the two stimuli into two different perceptual categories. Top-down processing of this sort, normally of great assistance in reading, can prove to be a source of considerable frustration in proofreading, in which allowing context to "fill in the gaps" is exactly what is *not* required. All words must be analyzed to their full letter level in order to perform the task properly.

The foundations of top-down processing were established in the discussion of redundancy and information in Chapter 2. Top-down processing, in fact, is only possible (or effective) because of the contextual constraints in language that allow certain features, letters, or words to be predicted by surrounding features, letters, words, or sentences.

When the redundancy of a language or a code is reduced, the contribution to pattern recognition of top-down, relative to bottom-up processing, is reduced as well. This tradeoff of top-down, context-driven processing governed by redundancy against bottom-up, data-driven processing governed by sensory quality is nicely illustrated in an investigation by Tulving, Mandler, and Baumal (1964). They presented subjects with sentences of the form, "I'll complete my studies at the _____," and displayed the final word for very brief durations, producing a degraded sensory stimulus. The experimenters could adjust both the duration (and therefore the quality) of the stimulus and the amount of prior word context between 8, 4, and 0 letters. The results, shown in Figure 4.4, illustrate the almost perfect tradeoff in recognition accuracy

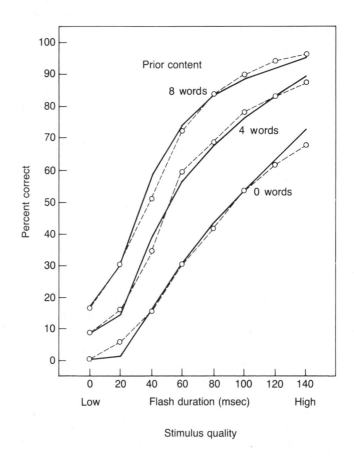

Figure 4.4   Tradeoff between bottom-up and top-down processing illustrated by experiment of Tulving, Mandler, and Baumal. From "Interaction of Two Sources of Information in Tachistoscopic Word Recognition" by E. Tulving, G. Mandler, and R. Baumal, 1964, *Canadian Journal of Psychology, 18,* p. 66. Copyright 1964 by the Canadian Psychological Association. Reproduced by permission.

between stimulus quality and redundancy. As one of these variables increases, the other may be degraded to maintain a constant level of recognition performance.

The word superiority effect.   As it has been presented so far, top-down processing has been seen as a kind of a guessing assistance that is used to compensate for poor data quality. The context reduces the number of possible alternatives for the degraded ambiguous stimulus, and so there is a greater likelihood that the degraded features will uniquely match one of the remaining alternatives. Current models of reading (e.g., McClelland & Rummelhart, 1981; Rummelhart, 1977) also propose that top-down processing operates to speed the comprehension of normal, undegraded text.

To some extent the top-down effects proposed in normal reading are those already described; context reduces the number of alternative words and therefore allows more abbreviated processing of the letters in each word. However, there is evidence for another form of top-down processing that works not by eliminating alternatives but rather by speeding up the analysis of letters that appear within words, because words are familiar units. An experiment by Reicher (1969) provided the initial demonstration of this phenomenon. Subjects presented stimuli for very brief durations were immediately offered two letters and asked to select which had been presented in the original stimulus. In the *single-letter* condition, only a single stimulus letter was presented (e.g., the letter N, followed by the test alternatives N and R). In the *whole-word* condition, an entire word containing the letter was presented. For example, the word "hand" was presented, followed by the test letters N or R. In this case again, the subject was to pick the letter N as the one seen in the word "hand." Reicher found that subjects performed more accurately in the whole-word than in the single-letter condition, despite the fact that in the whole-word condition four times as much sensory information needed to be processed in the same amount of time. The finding that letters are processed better within a word than by themselves is sometimes referred to as the word superiority effect.

It is important to realize that the word superiority effect found by Reicher cannot be attributed to top-down processing as we have so far described it. (That is, surrounding context reduces the alternatives that can be guessed.) Knowing the three letters H, A, and D cannot assist the subject in identifying the critical test letter, since both the alternatives N and R in the third position form a legitimate word ("hand" and "hard"). In fact, letters in words that have a large number of possible alternatives (i.e., -ink) are actually recognized slightly better than letters with few alternatives (McClelland & Rummelhart, 1981). Instead, the results argue that the "wordness" of the four-letter stimulus facilitates the bottom-up processing of the constituent letters so that this processing is performed at a faster rate than would be observed in the absence of the word context (e.g., in the single-letter condition).

Since Reicher's original demonstration, the word superiority effect has been replicated many times in various forms. It appears to operate consistently as long as subjects expect to see words (McClelland & Rummelhart, 1981). The existence of the effect suggests that the visual system processes the letters within a word in parallel more than sequentially, at least for words with up to six or seven letters (Wickelgren, 1979). The physical breaks between word boundaries serve as "parsing" mechanisms and the letters between those breaks are processed in a whole-word-like fashion. The importance of the physical breaks in defining the units was shown by Kolers and Lewis (1972). They observed that a single six-letter word could be perceived more accurately, under equivalent viewing conditions, than two three-letter words. The former contains a single unit, fostering parallel processing of all its members, while the latter contains two competing units.

The pattern of analysis of word perception described up to this point may be best summarized by observing that top-down and bottom-up processing are continuously ongoing at all levels in a highly interactive fashion (Navon, 1977; Neisser, 1967; Rummelhart, 1977). Sensory data suggest alternatives, which in turn provide a context that helps interpret more sensory data. This interaction is represented schematically in Figure 4.5. The conventional bottom-up processing sequence of features to letters to words is presented by the upward flowing arrows in the middle of hierarchy. The dashed lines to the left indicate that automatic unitization at the level of the letter and the common word

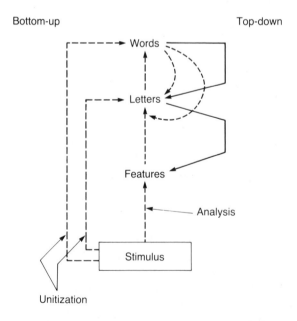

Figure 4.5 Bottom-up processing (analysis and unitization) versus top-down processing.

may occur as a consequence of the repeated processing of these units. Thus unitization may serve to identify a "blurred word" by word shape alone, even when neither feature nor context is available (Broadbent & Broadbent, 1980). Unitization does not necessarily replace or bypass the sequential bottom-up chain, but operates in parallel. Represented to the right of Figure 4.5 are the two forms of top-down processing: those that reduce alternatives through context and redundancy (solid lines) and those that actually facilitate the rate of lower-level analysis (dotted lines).

While all these factors may be operating simultaneously, two primary dimensions underlie the relative importance of one or the other. The first contrasts sensory quality against context and redundancy, as these trade off in bottom-up versus top-down processing. The second contrasts the relative contribution of higher-level unitization to hierarchial analysis in bottom-up processing. This contribution is determined by the familiarity and consistent mapping of the lower-level units. These two dimensions will be important as a framework for later discussion of the applications of pattern recognition.

From words to sentences.   Our previous analysis has focused upon the recognition of words. Yet more often than not word recognition is performed in the service of comprehending a string of words in a sentence. We have, of course, implicitly suggested that sentences must be processed in order to provide the higher-level context that in turn provides top-down processing for word recognition. In normal reading, sentences are processed by visually scanning across the printed page. Scanning occurs by a series of fixations, interspersed by discrete *saccadic* eye movements. The fixation has a minimum duration of around 200 msec. During each fixation, as we have seen, there appears to be some degree of parallel processing of the letters within the fixated word. While the meaning of an isolated word can normally be determined during fixations as short as the minimum fixation value of 200 msec, fixations made during continuous reading are sometimes considerably longer (Just & Carpenter, 1980; McConkie, 1983). The extra time is required to integrate the meaning of the word into the ongoing sentence context as well as to process more difficult words. Both the absolute duration of fixations and the frequency of fixations along a line of text vary greatly with the difficulty of the text (Anderson, 1979; McConkie, 1983).

While a given word is fixated, information understood from the preceding words provides context for top-down processing. A series of investigations conducted by McConkie and his colleagues (see McConkie, 1983) has asked how much information is actually processed from different locations along a line of print during fixations when a given letter is in the center of foveal vision. McConkie's research suggests that different kinds of information may be processed at different regions. As far out as 10 to 14 characters to the right of the fixated letter, very global details pertaining to word boundaries may be perceived for the purpose of directing the saccade to the next fixation. Some processing of word shape may occur somewhat closer to the fixated letter. Individual letters however are only processed within a fixation span of roughly 10 letters: four to the left and six to the right. This span itself is not fixed but varies in width according to the size of the word currently fixated and the position of the fixated letter within the word.

McConkie's research on the use of visual information in normal reading has employed an ingenious technique that we may describe as fixation-guided text display (McConkie & Raynor, 1974). In this technique, subjects read a computer-displayed text while the computer monitors the exact point of fixation of the eye on the printed line. The computer proceeds to "scramble" the text a certain number of characters, N, to the right of fixation. When the eye makes a saccade to the right, the scrambled text is rearranged to its coherent level, and the characters that are now N to the right of the new fixation point are scrambled. McConkie and Raynor found that if N was quite large, subjects were not even aware of the scrambling and they could read at a normal rate. However, when N reached a value of less than 10 (about 2.5 degrees of visual angle to the right, or roughly two words), their reading speed was disrupted. This distance indexes the span of processed peripheral information in normal reading.

The fact that subjects notice changes this far into the periphery does not of course guarantee that the information registered in the visual system at that degree of visual angle is actually "perceived" in the sense that it contributes to comprehending the material being read. In another experiment using the fixation-guided display technique, McConkie, Zola, Blanchard, and Wolverton (1982) changed a letter of a given word that fell within the fixation span on two successive and overlapping fixations. The letter change was fabricated such that the altered word gave the sentence an entirely different meaning. The investigators found that even though the meaning of the sentence was altered on the two successive fixations, subjects perceived no ambiguity. The sentence meaning that they perceived corresponded to the value of the word at the time the word was directly fixated and not when it was closer to the periphery.

In summarizing the results of his research, McConkie (1983) offers a conclusion that is quite consistent with the word superiority effect described previously. It is that the perception of letters within a word are heavily "word-driven." When a word lies at the center of fixation, its letters are processed cooperatively and guide the semantic interpretation of the sentence. Letters not in the fixated word may indeed be processed, but the results of this processing do not seem to play an active role in comprehension.

Applications

The research on recognition of print is, of course, applicable to system design in contexts in which warning signs are posted or maintenance and instruction manuals are read. These contexts will be discussed at the end of the chapter. Furthermore, the acquisition of verbal information from computer terminals and video displays is also an area of increasing importance. In designing such displays the goal is to present information in such a manner that it can be read rapidly and accurately. In addition, certain critical items of information (one's own identification code, for example, or critical diagnostic information) should be recognized automatically, with a minimal requirement to invest conscious processing. There are two broad classes of practical implications of the research that generally align themselves with the two dimensions of pattern recognition described above: applications that capitalize on unitization and applications that are related to the tradeoff between top-down and bottom-up processing.

Unitization.   Training and repetition lead to automatic processing. Some of this training is the consequence of a lifetime's experience (e.g., recognition of letters), but as LaBerge (1973) and Schneider and Shiffrin (1977) clearly demonstrated, the special status of automatic processing of critical key targets can also be developed within a relatively short period of practice. These findings suggest that when a task environment is analyzed, it is important to identify critical signals (and these need not necessarily be verbal) that should always receive immediate priority if they are present. Training regimes should then be considered that will develop automatic processing of these signals. In such training, operators should be presented with a mixture of the critical signals and others and should always make the same invariant responses to the critical signals.

In this regard, there would appear to be a distinct advantage in calling attention to critical information by developing automatic processing rather than by simply increasing the physical intensity of the stimulus. In the first place, as we will describe in our discussion of alarms in Chapter 12, loud or bright stimuli may be distracting and annoying and may not necessarily ensure a response. Second, physically intense stimuli are intense to all who encounter them. Stimuli that are "subjectively intense" by virtue of automatic processing may be "personalized" to be alerting only for those who require alerting.

At any level of perceptual processing it should be apparent that the accuracy and speed of recognition will be greatest if the displayed stimuli are presented in a physical format that is maximally compatible with the visual representation of the unit in memory. For example, the prototypal memory units of letters and digits preserve the angular and curved features as well as the horizontal and vertical ones. As a consequence, "natural" letters that are not distorted into an orthographic grid should be recognized with greater facility than dot matrix letters or letters formed with only horizontal and vertical strokes. These suggestions were confirmed in recognition studies comparing digits constructed in right-angle grids with digits containing angular and curved strokes (e.g., Ellis and Hill, 1978; Plath, 1970). Ellis and Hill, for example, found that 5-digit sequences were read more accurately when presented as conventional numerals than in 7-segment right-angled format. This advantage was enhanced at short exposure durations, as might be typical of time-critical environments.

A similar logic applies to the use of lowercase print in text. Since lowercase letters contain more variety in letter shape, there is more variety in word shape and so a greater opportunity to use this information as a cue for "holistic" word-shape analysis. Tinker (1955) found that subjects could read text in mixed case better than in all capitals. However, the superiority of lowercase over uppercase appears to hold only for printed sentences. For the recognition of isolated words, the stimuli appear to be better processed in capital than in lowercase (Vartabedian, 1972). These findings would seemingly dictate the use of capital letters in display labeling (Grether & Baker, 1972; McCormick, 1976) where only one or two words are required but lowercase in longer segments of verbal material.

As a result of unitization, words are both perceived faster and understood better than are abbreviations or acronyms. Therefore, it seems that words should always be used instead of abbreviations, except when space is at an absolute premium. Norman (1981) discusses the difficulties that naive users of computer text-editing systems encounter when confronted with abbreviations like "ln" for "link" or "cat" for "concatenate." He asks why the full word should not be employed instead. The "cost" of a few extra letters is surely compensated by the benefits of better understanding and fewer blunders. Where abbreviations are used, Norman suggests that, at a minimum, relatively uniform abbreviating principles should be employed (i.e., all abbreviations of common length) and that as much effort as possible be expended to make the abbreviated term as logical and meaningful to the user as possible. Moses & Ehrenreich (1981) have summarized an extended evaluation of abbreviation techniques and conclude that the most important principle is to employ consistent rules of abbreviation. In particular, they find that truncated abbreviations, in which the first letters of the word are presented, are processed better than contracted abbreviations, in which letters within the word are deleted. For example, "reinforcement" would be better abbreviated by "reinf" than by "rnfnt." This finding makes sense in terms of our discussion of reading, since truncation preserves at least part of any unitized letter sequence. Ehrenreich (1982) concludes that whatever rule is used to generate abbreviations, rule-generated abbreviations are, at least, always superior to subject-generated ones, in which the operator decides the best abbreviations for a given term.

A related recommendation derives from the unitizing influence of gaps between words (Wickelgren, 1979). It appears that this benefit may carry over to processing unrelated material such as alphanumeric strings by defining high-order visual "chunks" (see Chapter 6). Klemmer (1969) argues that there is an optimum size of such chunks for encoding unrelated material. In Klemmer's experiment strings of digits were presented to be entered as rapidly as possible into a keyboard. In this task the most rapid entry was achieved when the chunks between spaces were three or four digits long. Speed declined with either smaller or larger groupings. These findings have important implications for deciding on formats for various kinds of displayed material— license plates, identification codes, or data to be entered on a keyboard.

Of course the engineering psychologist must recognize that there are design tradeoffs which dictate that optimal configurations cannot always be made. "Natural" letters, for example, require considerably more points, or light sources, to project on a video display than do right-angle-grid letters. Limitations of space may force the use of abbreviations rather than whole words. These tradeoffs in design are considerably more pronounced when the implications of the bottom-up/top-down, context/data distinction are considered in the next section.

Context/data tradeoffs.   An example of the tradeoff of design considerations between bottom-up and top-down processing can be seen when a printed message is to be presented on an aircraft display in which space is clearly at a premium. Given certain conditions of viewing (high stress, vibration), the sensory qualities of the perceived stimulus may be far from optimal. A choice of designs is thereby offered as shown in Figure 4.6: (1) Present large print, thus taking advantage of

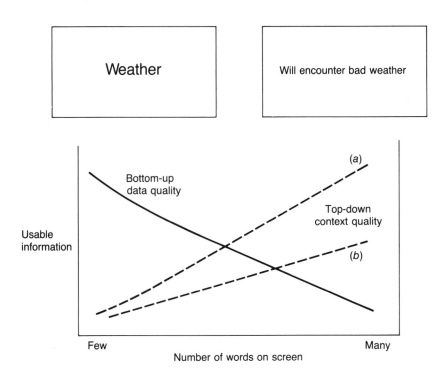

Figure 4.6   An illustration of the tradeoff between top-down and bottom-up processing in display of limited size. The two dashed lines represent different amounts of contextual redundancy: (*a*) high context of printed text; (*b*) low context of isolated word strings.

improving bottom-up "sensory" quality but restricting the number of words that can be viewed simultaneously on the screen (and thereby limiting top-down processing). (2) Present more words in smaller print and enhance top-down at the expense of bottom-up processing. Naturally the appropriate text size will be determined by an evaluation of the relative contribution of these two factors. For example, if there is more redundancy in the text matter, then smaller text size is indicated. If the display or viewing quality is extremely poor, then larger size is suggested. It is essential that the system designer be aware of the factors that influence the tradeoff between data-driven and context-driven processing, which determines the optimum point on the tradeoff to be selected.

Top-down processing may also be greatly facilitated through the simple technique of restricting a message vocabulary. With fewer possible alternatives to consider, top-down hypothesis forming becomes far more efficient. This technique is a major factor behind the strict adherence to standard terminology in many message-routing centers or communication systems (Kryter, 1972).

Economy versus security.   The tradeoff between top-down and bottom-up processing is demonstrated by the fact that messages of greater probability (and therefore less information content formally) may be transmitted with less sensory evidence. We have already encountered one example of this tradeoff in the compensatory relation between d' and beta in signal detection theory (Chapter 2). As a signal becomes more frequent, offering less information, and beta is lowered, it can be detected at lower sensitivity (i.e., with less evidence and lower d'). It is fortunate that the tradeoff in human performance corresponds quite nicely with a formal specification of the optimum design of codes, referred to as the *Shannon-Fano* principle (Sheridan & Ferrell, 1974). In designing any sort of code or message system in which short strings of alphanumeric or symbolic characters are intended to convey longer ideas, the Shannon-Fano principle dictates that the most efficient or *economic* code will be generated when the length of the physical message is proportional to the information content of the message. The principle is violated if all messages are of the same length. Thus high-probability, low-information messages should be short, and low-probability ones should be longer. For example, if the four events that make up a code and their associated probabilities are A (0.4), B (0.4), C (0.1), D (0.1), a code using binomial symbols that assigns A = 00, B = 01, C = 10, and D = 11 violates the Shannon-Fano principle. One that assigns A = 0, B = 1, C = 01, D = 10 does not.

It is interesting to observe that any natural language roughly follows the Shannon-Fano principle. Words that occur frequently ("a", "of," or "the") are short, and ones that occur rarely tend to be longer. This relation is known as Zipf's law. The relevant finding from the viewpoint of human performance is that adherence to such a code reinforces the natural tendencies to expect frequent signals and therefore require less sensory evidence for those signals to be recognized. For example, in an efficient code designed to represent engine status, the expected normal operation might be represented by N (one unit), while *HOT* (three units) should designate a less-expected, lower-probability overheated condition. Violations of the Shannon-Fano principle are observed in a coding system in which all events, independent of their probability, are specified by a message of the same length. Such a violation was evident in the Navy's system of computerized maintenance records. The operator servicing a malfunctioning component was required to enter a nine-digit malfunction code on a computerized form. This code was of uniform length whether the malfunction was highly probable (overheating engine) or extremely rare.

Numerous other properties of a useful code design have been nicely summarized by Bailey (1982). For example, code strings should be relatively short (fewer than six characters), and a code, like an abbreviation, should be meaningfully related to its referent. Alphabetic codes, because of the greater richness of the alphabet, generally meet this criterion better than alphanumeric ones.

In the context of information theory, there is, in addition to efficiency, a second critical factor that must be considered when a code or message system is designed. This factor illustrates again the tradeoffs frequently encountered in human engineering. The Shannon-Fano principle is one intended to produce maximum processing efficiency, and this is compatible with human processing biases. However, it may often be the case that relatively high-frequency (and therefore short) messages of a low information content are in fact very *important*. It is therefore essential that they be perceived with a high degree of *security*. In these instances the principle of economy should be sacrificed by including redundancy, as discussed in Chapter 2. Redundancy is accomplished by allowing a number of separate elements of the code to transmit the same information. The use of a communications-code alphabet in which "alpha," "bravo," and "charlie" are substituted for *a*, *b*, and *c* is a clear example of such redundancy for the sake of security. The second syllable in each utterance conveys information that is highly redundant with the first. Yet this redundancy is advantageous because of the need for absolute security (communication without information loss) in the contexts in which this alphabet is employed. It is possible to look on the tradeoff between efficiency and security in code design as an echo of the tradeoff discussed in Chapter 2 between maximizing information transmission and minimizing information loss. Certain conditions (orthogonal dimensions, adherence to the Shannon-Fano principle) will be more efficient, and other conditions emphasizing redundancy will be more secure.

Speed of information input. When the security of transmission is of less importance than its efficiency (amount of meaning per unit time), then concern must be directed to the speed of encoding printed information. For longer messages such information is perceived in a "self-paced" manner in which a line of text is scanned through repeated visual fixations. In normal text processing, such scanning is achieved at a rate that is highly variable but shows a typical value between three and six words per second. Allowing the rate of processing to remain under total control of the reader may present problems, however, if there are constraints on how slow a message can be processed. In this case, a slow reading rate will start to generate an increasingly long queue of unprocessed information. The technique of rapid serial visual presentation (RSVP) has been examined by Forster (1970) and by Potter, Kroll, and Harris (1980) as a means of increasing reading speed by forced pacing.

In RSVP the sequential words of a text are displayed one after the other at the same displayed location and a very rapid rate (up to 16 words per second). This procedure has the advantage of force-pacing the subject. Yet unlike other force-pacing schemes, such as the scrolling of texts (similar to the presentation of movie credits), RSVP also eliminates the need for visual scanning. Potter et al.'s results indicate that subjects can read single sentences for comprehension at rates up to 12 words per second with little or no loss in comprehension. However, when the material is in longer paragraph form, requiring that meaning be integrated over a longer time, comprehension is reduced. As we noted earlier, the latter results confirm that the 300–400 msec per fixation in normal reading, although longer than needed to extract the meaning of each isolated word, is not wasted time. Instead, it is time required to perform a more global integration across words and sentences that is associated with comprehension. When this time is eliminated in RSVP, comprehension suffers. These shortcomings notwithstanding, the technique of RSVP seems to be one that offers some promise of increasing perceptual efficiency of short messages in high-load conditions. This issue will be dealt with further as *pacing* is discussed in Chapter 10.

COMPREHENSION

Whether presented by voice or by print, words are normally combined into sentences whose primary function is to convey a message to the receiver. So far our discussion has considered how the meaning of the isolated words and word combinations is extracted. In this section we shall consider properties of the words themselves, and not just their physical representation, that influence the ease of comprehending a message (Broadbent, 1977). Instructions and procedures vary dramatically in the ease with which they may be understood. Chapanis (1965), in a delightful article entitled "Words Words Words," provides several classic examples of instructions that are poorly if not incomprehensibly phrased. Some examples are shown in Figure 4.7. A few very commonsensical principles seem to be involved in generating good

Elevator notices (top right):

PLEASE

WALK UP ONE FLOOR

WALK DOWN TWO FLOORS

FOR IMPROVED ELEVATOR SERVICE

What it says:
(13 words)

IF YOU ARE ONLY GOING

UP ONE FLOOR

OR

DOWN TWO FLOORS

PLEASE WALK

(IF YOU DO THAT WE'LL ALL HAVE
BETTER ELEVATOR SERVICE)

What it means:
(14 words)

TO GO UP ONE FLOOR

OR

DOWN TWO FLOORS

PLEASE WALK

Would this do?
(11 words)

Telephone instructions (center):

You can Dial LOCAL and TRUNK CALLS

LOCAL CALLS cost 3d. for 3 mins (cheap rate 6 mins)

for LONDON exchanges, dial the first three
letters of the exchange name followed by the number you want

LONDON exchanges are shown in the DIALLING CODE BOOKLET and Telephone
Directory with the first three letters in heavy capitals

For the following exchanges, dial the code shown followed by
the number you want

	Code		Code		Code
Byfleet	BY	Hoddesdon	HO3	St. Albans	LN
Crayford	CY	Hornchurch	HX	Slough	SL
Danford	DA	Ingrebourne	IL	Staines	SW
Erith	ET	Leatherhead	LE7	Uxbridge	UX
Farnborough(Kent)	FN	Northwood	NL	Waltham Cross	WS
Garston	GR7	Orpington	MM	Walton on Thames	WT
Gerrards Cross	GE4	Potters Bar	PR	Watford	WA
Hatfield	HL6	Romford	RO	Weybridge	WR

You can DIAL other LOCAL CALLS shown in the
DIALLING CODE BOOKLET (inside the A-D directory)

TRUNK CALLS

For BIRMINGHAM exchanges Dial 021
" EDINBURGH " " 031
" GLASGOW " " 041
" LIVERPOOL " " 051
" MANCHESTER " " 061

Then the first three letters of the
exchange name
then the number
e.g. for Birmingham Midland 7291
dial 021 MID 7291

You can DIAL other TRUNK CALLS shown in the
DIALLING CODE BOOKLET (inside the A-D directory)

To make a call first check the code (see above)

USE 3d bits, 6d or 1/- coins (Not Pennies)

HAVE MONEY READY, but do not try to put it in yet

LIFT RECEIVER, listen for dialling tone and

DIAL — see above — then wait for a tone

Ringing tone (burr-burr) changes, when the number answers, to

Pay tone (rapid pips) — Now PRESS in a coin and speak

(Coins cannot be inserted until first pay tone is heard)

Engaged tone (slow pips) — try again later

N.U. tone (steady note) — check number and redial

INSERT MORE MONEY to prolong the call
at any time during conversation
at once if pay tone returns

**Remember—Dial first and when you hear
pay tone (rapid pips) press in a coin**

For Directory Enquiries dial DIR   For other enquiries dial INF
For OTHER SERVICES and CALL CHARGES
see the DIALLING CODE BOOKLET (inside the A-D directory)

For the Operator – dial 100

Radio notices (left):

NOTICE

THIS RADIO USES A LONG LIFE PILOT
LAMP THAT MAY STAY ON FOR A SHORT
TIME IF RADIO IS TURNED OFF BEFORE
RADIO WARMS UP AND STARTS TO PLAY

What it said:
(29 words)

NOTICE

DON'T WORRY IF THE PILOT LAMP
SHOULD STAY ON FOR A LITTLE WHILE
AFTER YOU TURN THE RADIO OFF

What is meant:
(19 words)

NOTICE

IF YOU TURN THE RADIO ON, AND THEN
OFF RIGHT AWAY, THE PILOT LAMP MAY
STAY ON FOR A LITTLE WHILE

This is more
accurate:
(21 words)

NOTICE

THE PILOT LAMP SOMETIMES STAYS ON
FOR A LITTLE WHILE AFTER YOU TURN
THE RADIO OFF

But would this do?
(16 words)

Figure 4.7 Three examples of poorly worded instructions. From "Words Words Words" by A. Chapanis, 1965, *Human Factors, 7,* pp. 6, 7, 9. Copyright 1965 by The Human Factors Society, Inc.
Reproduced by permission.

instructions: State directly what is desired without adding excess words; use familiar words; and ensure that all information to be communicated is explicitly stated, leaving nothing to be inferred. Many of the techniques of good instruction derive from these intuitive principles, while many others are the less obvious consequence of findings in cognitive psychology. We shall now consider some findings of the latter category.

Provide Context

The essential role of context in comprehension is to bias the perceiver to encode the material in the manner which is intended. This is an effect that was considered in two different forms in Chapter 2: the influence of probability on response bias and the influence of context on information. Furthermore, context should provide a framework upon which details of the subsequent verbal information may be "hung." Bransford and Johnson (1972) have demonstrated the dramatic effect that the context of a descriptive picture or even a thematic title can exert on comprehension. In their experiment, subjects read a series of sentences that described a particular scene or activity (e.g., the procedures followed when washing clothes). The subjects were asked to rate the comprehensibility of the sentences and were later asked to recall them. Large improvements in both comprehensibility and recall were found for subjects who had been provided a context for understanding the sentences prior to hearing them. This context was in the form of either a picture describing the scene or a simple title of the activity. For those subjects who received no context, there was little means of organizing or storing the material, and performance was poor. In a more applied setting, Schneiderman (1980) argues that it is far easier for a computer programmer to learn a second programming language than a first, in part because the semantic aspects of the first language transfer and provide a *context* in which the syntactic details of the second language may be more easily encoded.

In order for context to aid in recall or comprehension, however, it should be made available prior to the presentation of the verbal material (Bower, Clark, Winzenz, & Lesgold, 1969). The important benefits of prior context would account for the results of an investigation by Norcio (1981) of computer program documentation (i.e., commentary on the meaning of the various logical statements). He noted that documentation only helped comprehension of the program if the documentation was provided at the beginning of the program and not when it was interspersed throughout. Like a good filing system, context can organize material for comprehension and retrieval if it is set up ahead of time. Even a highly organized filing scheme will be of little assistance if it is made available only after the papers are dumped loosely into a drawer.

Linguistic Factors

Logical reversals.   Whenever a reader or listener is required to logically reverse the meaning of a statement to translate from a physical sequence of words to an understanding of what is intended, comprehension seems to be hindered. One example is provided by the use of negatives. We comprehend more rapidly that a particular light should be "on" than that it should be "not off." A second example of logical reversals is falsification. It is faster to understand that a proposition should be true than that it should be untrue or false. Experiments by Clark and Chase (1972; Chase & Clark, 1972) and by Just and Carpenter (1971; Carpenter & Just, 1975) suggest that these differences are not simply the result of the greater number of words or letters that normally occur in reversed statements, but result from the cognitive difficulties in processing them as well.

The experimental paradigm employed by these investigators was the sentence-picture verification task. It is the experimental analog of an operator who reads a verbal instruction (i.e., "Check to see that valve X is closed") to verify it either against the physical state or against his own mental representation of the state of the system. In the actual experimental paradigm, as implemented by Clark and Chase, subjects are shown a verbal sentence describing the relation between two symbols (e.g., "The star is below the plus") along with a picture depicting the two symbols. The symbols are in an orientation that is either true or false relative to the verbal proposition. The subject is to indicate the truth of this statement as rapidly as possible. Sentences vary in their truth value relative to the picture (e.g., whether the correct answer is true or false) and in whether or not they contain a negative ("The star is above the circle" vs. "The circle is not above the star"). Four examples of the sentence-picture relation are shown in Figure 4.8. Beneath each sentence, describing the picture on the left, is shown the response time to verify sentences of that category obtained by Chase and Clark (1972).

The results of these experiments suggest three important findings:

1.   Statements that contain negatives invariably take longer to verify than those that do not, as shown by the greater response latency for the two sentences to the right in Figure 4.8. Therefore, where possible, instructions should contain only positive assertions (i.e., "Check to see that water level is normal") rather than negative ones ("Check to ensure that water level is not abnormally high").

2.   Whether a statement is verified as true or false influences verification time in a more complex way. If the statement contains no negatives (is positive), then true statements are verified more rapidly than false. This is shown to the left of Figure 4.8. However, if statements contain negatives, then false statements are verified more rapidly than true ones. This is shown to the right of Figure 4.8. The reason for this reversal relates to the principle of "congruence" described next.

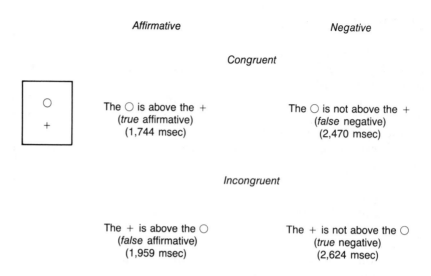

Figure 4.8 The picture-sentence verification task. The four sentences describe the picture on the left. The top two sentences are congruent with the picture relation "circle above plus." The two sentences on the right contain negatives. Response times are shown in parentheses. From "On the Process of Comparing Sentences Against Pictures" by H. H. Clark and W. G. Chase, 1972, *Cognitive Psychology, 3*, p. 482. Copyright 1972 by Academic Press. Adapted by permission of the authors.

3. Clark and Chase and Carpenter and Just find that there are very predictable differences in latency between the four kinds of picture-sentence relations. In response to this regularity they have modeled the processes involved as a series of very basic "constituent comparisons" (Carpenter & Just, 1975) of constant duration, each performed in sequence and each taking a constant time. These are performed in series until the final verification is obtained. Comparisons are made as to the equality, or "congruence," of the propositional form between picture and sentence, disregarding any negatives. For the two sentences at the top, this form is "circle above plus." For the two at the bottom it is "plus above circle." This may be roughly thought of as a comparison of the congruence of the word order on the page with the order of representation in memory of the picture. If there is disagreement in this congruence, then extra time is added. Since we normally read from top to bottom, the order of the picture in Figure 4.8, circle—plus, is incongruent for the bottom two sentences. A comparison is also made as to the existence of negatives (the two right sentences). Negatives also add time. As each of these constituent comparisons is made, units of time are added, and the "truth value" of the comparison is updated. After all comparisons are made the final response is given with a latency determined by the number of comparisons. The longest response to the "true negative" sentence occurs because it alone both is incongruent and contains negatives.

When this model is used to help the designer phrase proper instructions or to predict the time that will be required for operators to respond to instructions, the meaning of "true" and "false" must be reconsidered slightly. In Figure 4.8 the relations are always true or false because the picture never changes. However, in application, a "picture"—the actual state of a system—may take on different values with different probabilities. "True" must therefore be defined as the most likely state of a system. Therefore, if a switch is normally in an up position, the instruction should read, "Check to insure that the switch is up," or "Is the switch up?" Since this position has the greatest frequency, such a statement will normally be verified as a true positive. Furthermore, as long as negatives in wording are avoided, then the principle will always hold that affirmations will be processed faster than falsifications.

The results of the basic laboratory work on the superiority of positives over negatives have been confirmed in at least one applied environment, the formating of highway traffic-regulation signs. Experiments have suggested that prohibitive signs, whether verbal, "no left turn," or symbolic, ⊘ , are more difficult to comprehend than permissive signs such as "right turn only" (Dewar, 1976; Whitaker & Stacey, 1981).

Absence of cues.   The dangers inherent when operators must extract information from the absence of cues are somewhat related to the recommendation to avoid negatives in instructions. This problem, described in the discussion of troubleshooting in the previous chapter, is also quite relevant here. Fowler (1980) articulates this point in his analysis of an aircraft crash near the airport at Palm Springs, California. He notes that the *absence* of an R symbol on the pilot's airport chart in the cockpit was the only indication of the critical information that the airport did *not* have radar. Since terminal radar is something pilots come to depend upon and the lack of radar is highly significant, Fowler argues that it is far more logical to call attention to the absence of this information by the *presence* of a visible symbol than it is to indicate the presence of this information with a symbol. In general, the presence of a symbol should be associated with information that an operator *needs to know* rather than with certain expected environmental conditions.

Order reversals.   Many times instructions are intended to convey a sense of ordered events. This order is often in the time domain (procedure X is followed by procedure Y). When instructions are to convey a sense of order, it is important that the elements in those instructions are congruent with the order of events. For example, if subjects are to learn or to verify that the order of elements is $A > B > C$, it is better to say, "A is greater than B, and B is greater than C," rather than "B is greater than C, and A is greater than B," or "B is less than A, and C is less than B" (DeSoto, London, and Handel, 1965). In the first case, the physical ordering of information, *A B B C*, conforms with the intended "true" ordering (*A B C*). In the last two cases, it does not (*B C A B* or *B A C B*). Furthermore, in the third case the word "less" is used to verify an ordering that is specified in terms of "greater." This represents an additional form of cognitive reversal. This finding would seem to dictate that procedural instructions should read, "Do *A*, then do *B*," rather than "Prior to *B*, do *A*," since the former preserves the actual sequencing of events in the ordering of statements on the page.

The notion of congruence of ordering appears to be a specific case of the more general finding that people comprehend active sentences more easily than passive ones. Active sentences (e.g., "The malfunction caused the symptoms") preserve an ordering (first the malfunction, then the symptoms) that is more congruent with a mental causal model of the process than a passive sentence is (e.g., "The symptoms were caused by the malfunction"). Research has found that passive sentences both require longer to comprehend and require greater capacity to hold in working memory (Savin & Perchonock, 1965).

Embedded relations.   The problems of logical reversals and order reversals combine when embedded relations are considered. Research from psycholinguistics indicates that when we comprehend a proposition such as, "The light is not on," we do so by first comprehending the proposition, "The light is on," and then negating it. This negation requires extra time. The added difficulty with double negatives is not simply that two negatives are required: "Do not start component *X* if the light does not go on." If this were the only problem, we could simply cancel the two negatives mentally (Wickelgren, 1979). The greater source of difficulty occurs because we must first interpret the proposition, "light on," then go back in the flow of words to negate this proposition ("light not on"), and then move back again, to negate the previous operation ("start the component"). The sequence of logical operations is the opposite of the flow of words on the page. For example, the statement, "Some fish are not birds," is easier to comprehend than the logical equivalent, "Not all fish are birds" (Anderson, 1979). The former sentence negates birds after fish are presented. In the latter case, the entire sentence is first read, and then one jumps back to the beginning to negate it. This movement back and forth through embedded levels of logic imposes a serious cognitive load. Such a load may be reduced by a more linear flow that makes the order of words congruent with the order of operation (e.g., "If the light is on, start the component; if the light remains off, do not start the component").

Conclusion

The foregoing discussion indicates that our linguistic-logical system can process some relations more naturally than others. When these relations are violated in instructions or procedures, the likelihood of comprehension is reduced. While the effects reported in the laboratory studies are not generally large (with only a few hundred milliseconds differentiating the sentences in Figure 4.8), the differences are nonetheless significant in the applied environment. Greater latency of response under idealized conditions will generally translate into a greater potential for an error under less than optimal conditions. Further, Savin and Perchonock (1965) have shown how added sentence transformations, such as the inclusion of negatives and passives, impose upon the limited attentional resources used for performing concurrent tasks (see Chapter 8). Good engineering psychology should adhere to the principles of minimizing cognitive operations when instructions are provided.

SUPPLEMENT B

Speech Perception

In 1977 a tragic event occurred at the Tenerife airport in the Canary Islands when a KLM Royal Dutch Airlines 747 jumbo jet, accelerating for takeoff, crashed into a Pan American 747 taxiing on the same runway. Although poor visibility was partially responsible for the disaster, in which 538 lives were lost, the major responsibility lay with the confusion between the KLM pilot and air traffic control regarding whether clearance had been granted for takeoff. Air traffic control, knowing that the Pan Am plane was still on the runway, was explicit in denying clearance. The KLM pilot misunderstood and believed that clearance had been granted. In the terms described earlier, the failure of communications was attributed both to less-than-perfect audio transmission—poor data quality or bottom-up processing—and to less-than-adequate message redundancy, so that context and top-down processing could not compensate. The disaster, described in more detail in Bailey (1982) and fully documented by the Spanish Ministry of Transportation and Communications (1978), calls attention to the critical role of speech communications in engineering psychology.

In conventional systems, the human operator's auditory channel has been primarily used for transmitting verbal communication from other operators (e.g., messages to the pilot from air traffic control) and for presenting auditory warning signals (tones, horns, buzzers, etc.). Recently, however, rapid advances in microcomputer technology have produced highly efficient speech-synthesis units. These allow computer-driven displays of auditory verbal messages to be synthesized on line in a fashion quite analogous to visual information presentation on computer-driven video displays. This capability provides a considerably greater degree of flexibility than that inherent in tape-recorded verbal messages, such as those employed in the cockpit of some aircraft to warn of extreme emergencies. A large and flexible vocabulary of messages can be selected and played instantly without encountering the physical limitations of the tape recorder.

Human perception of speech shares some similarities but also manifests a number of pronounced contrasts with the perception of print. In one respect the perception of speech shares common attributes with Forster's (1970) work on rapid serial visual presentation, since words are uttered (and therefore must be perceived) in a serial force-paced fashion. In contrast, normal processing of print allows information to be extracted from surrounding words and also allows the reader to scan back and forth across a line.

In a second respect, the perception of speech is like the perception of cursive writing. The reason for this similarity is that the raw physical speech signal, like the cursive line but in contrast to print, is continuous, or analog, in format. The perceptual system must undertake some analog-to-digital conversion in order to translate the continuous speech wave form into the discrete units of speech perception. In order to understand the way in which these units are formed and their relation to the physical stimulus, it is necessary first to understand the representation of speech. We shall consider the difference between the time and frequency representations of continuous analog signals.

REPRESENTATION OF SPEECH

Physically, the stimulus of speech is a continuous variation or oscillation of the air pressure reaching the eardrum, represented schematically in Figure 4.9*a*. As with any time-varying signal, the speech stim-

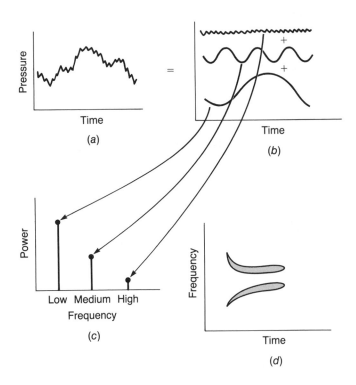

Figure 4.9 Different representations of speech signal: *(a)* time domain; *(b)* frequency components; *(c)* power spectrum; *(d)* speech spectrograph.

ulus can be analyzed using the principle of *Fourier analysis* into a series of separate sine wave components of different frequencies and amplitudes. Figure 4.9*b* is the Fourier-analyzed version of the signal in Figure 4.9*a*. We may conceptualize the three sinusoidal components in Figure 4.9*b* as three *features* of the initial stimulus. A more economical portrayal of the stimulus is in the *spectral representation* in Figure 4.9*c*. Here the frequency value (number of cycles per second, or Hertz) is shown on the abscissa, and the mean amplitude or power (square of amplitude) of oscillation at that particular frequency is on the ordinate. Thus the raw continuous wave form of Figure 4.9*a* is now represented quite economically by only three points in Figure 4.9*c*.

Unfortunately, the frequency content of articulated speech does not remain constant but changes very rapidly and systematically over time. Therefore, the representation of frequency and amplitude shown in Figure 4.9*c* must also include the third dimension of time. This is done in the *speech spectrograph*, an example of which is shown in Figure 4.9*d*. Here the added dimension of time is now on the abscissa. Frequency, which was originally on the abscissa of the power spectrum in 4.9*c*, is now on the ordinate, while the third dimension, amplitude, is represented by the width of the graph. Thus in the representation of Figure 4.9*d* one tone starts out at a high pitch and low intensity and briefly increases in amplitude while it decreases in pitch, reaching a steady-state level. At the same time a lower pitched tone increases in both pitch and amplitude to a higher and louder steady level. In fact, this particular stimulus represents the spectrograph that would be produced by the sound *da*. The two separate pitches are called formants.

UNITS OF SPEECH PERCEPTION

There are two main characteristics of speech perception that make the bottom-up, unit-level analysis of speech perception somewhat more complex than the bottom-up processing encountered in reading. These relate to the fact that the physical form of a given unit of speech might be different in different contexts. This is the *invariance problem*. The second problem relates to the fact that the units themselves are not physically separated or parsed like the words on a page. This is the *segmentation problem*. Each of these will be encountered as the discussion of the different units of speech is presented below.

Phoneme

The phoneme is analogous in many respects to the letter unit in reading. The phoneme represents the basic unit of speech because changing a phoneme in a word will change its meaning (or change it to a nonword). Thus the 38 English phonemes roughly correspond to the letters of the alphabet plus distinctions such as those between long and short vowels and between sounds such as *th* and *sh*. The letters *s* and soft *c* (as in "ceiling") are mapped into a single phoneme. Although the phoneme in the linguistic analysis of speech is quite analogous to the printed letter, there is a sense in which it is quite different from the letter in its actual perception. The problem is that the physical form of a phoneme is highly dependent upon the context in which it appears (the invariance problem). The speech spectrograph of the phoneme *k* as in "kid" is quite different from that of *k* as in "lick." As another example of this context dependency, Figure 4.10 shows the differences between the physical stimulus *d* as it appears in three different contexts defined by the following vowel. Note the contrast with the physical letter unit in print. The physical form of a letter may indeed vary, but this variation is not consistently determined by the value of the preceding and following letters (except to some degree with cursive writing, since a letter may be written slightly differently depending upon whether preceding letter ends, or a following letter begins, at a high or low position on a line).

Liberman, Cooper, Shankweiler, and Studdert-Kennedy (1967) have offered further evidence for the dependence of a consonant phoneme upon its neighboring vowel. Synthesizing consonant-vowel pairs such as the *da* depicted in Figure 4.10, they gradually shortened the steady-state vowel portion and asked subjects to report what they heard. When the steady-state portion was eliminated altogether, subjects no longer heard a pure *d* sound but rather a sort of a "chirp" sound. The phoneme lost its entire identity in the absence of its surrounding context.

Another example of the dissociation between the physical and subjective aspects of speech was provided by Eimas (1963). He continuously changed the physical formants of synthetic speech from one syllable to another, for example, *ba* to *da* (Figure 4.11a). The subject does not hear a gradual change of perception from a *ba*-like sound to a

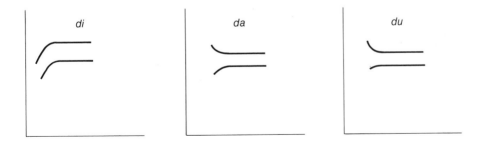

Figure 4.10 Speech spectrographs of the consonant *d* in different vowel contexts.

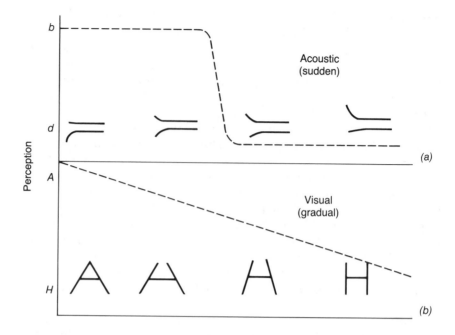

Figure 4.11 *(a)* Discontinuous change of acoustic perception from *b* to *d* as a function of continuous change in stimulus quality. *(b)* Both perception and stimulus are continuously changed from *A* to *H* in the visual stimulus.

da-like sound, but hears only a series of *ba*'s, all of which sound alike, followed abruptly by a series of *da*'s. This phenomenon of *categorical perception* is quite different from the perceptual experience when the analogous manipulation is performed on print. As shown in Figure 4.11*b*, the eye can easily discern the continuous gradation of physical change from *A* to *H*.

Articulatory Features

If the subjective discontinuity in phoneme perception does not correspond to the physical stimulus, to what does it correspond? Several investigators have suggested that it corresponds directly to the *articulatory features* of speech production (Liberman, 1957; Liberman et al., 1967). According to this reasoning, all stimuli to the left of the boundary in Figure 4.11*a* are heard as *ba* and not *da* because they would all be articulated by a voiced separation of the lips (*ba*) and not by a voiced withdrawal of the tongue from the roof of the mouth (*da*). The physical stimulus is not invariant across stimuli within a boundary but the production mechanism is. According to this viewpoint each phoneme is uniquely defined by a set of features. Dimensions defining these features are, for example, voiced (*m*, *n*) versus voiceless (*p*, *t*), lips (*p*, *m*) versus throat (*k*, *g*). The articulatory features defining a *d* sound (voiced, tongue, nonnasal) will always be the same, whether it is followed by an *i*, an *a*, or a *u*, even though the physical stimulus will vary.

Syllables

Articulatory features are defined in terms of production. An alternative approach is to define the *syllable* as the basic unit of speech perception (Massaro, 1975). This definition is in keeping with the notion that, while a following vowel (V) seems to define the physical form of the preceding consonant (C), the syllabic unit (CV) is itself relatively invariant in its physical form. The syllable in fact is the smallest unit with such invariance. One line of evidence in support of the syllable unit was provided by Huggins (1964). Huggins' subjects listened to continuous speech that was switched back and forth between the two ears at different rates. Comprehension was somewhat disrupted at all switching rates, but was most difficult at a rate of three per second. This is just the rate that would obliterate half of each syllable during the normal rate of speech production (Neisser, 1967). A faster rate obliterating half of each phoneme or a slower one obliterating half of each word was found to be far less disruptive. This finding suggests that the subject is particularly dependent upon the syllable unit in speech perception. Huggins also found that if the rate of speech was increased (by increasing tape speed), the frequency at which the critical, most disrupting interruption rate occurred increased proportionately.

Words

Although the word is the smallest cognitive or semantic unit of meaning,[1] like the phoneme it shows a notable lack of correspondence with the physical speech sound. This lack of correspondence defines the "segmentation" problem (Neisser, 1967). In a speech spectrograph of continuous speech (in contrast to the spectrograph of isolated consonants shown in Figure 4.10) there are identifiable breaks or gaps in the continuous record. However, these physical gaps show relatively little correspondence with the subjective pauses at word boundaries that we seem to hear. For example, the spectrograph of the four-word phrase, "She uses st*and*ard oil," would show the two physical pauses marked, neither one corresponding to the three word-boundary gaps that are heard subjectively. The segmentation issue then highlights another difficulty encountered by automatic speech-recognition systems that function with purely bottom-up processing. If speech is continuous, it is practically impossible for the recognition system to know the boundaries that separate the words in order to perform the semantic analysis without knowing what the words are already.

---

[1]Actually the morpheme is a slightly smaller cognitive unit than the word, consisting of word stems along with prefixes and suffixes such as "un-" or "-ing."

TOP-DOWN PROCESSING OF SPEECH

The description presented so far has emphasized the bottom-up analysis of speech. In fact, top-down processing in speech recognition is more essential than it is in reading. The two features that contrast speech perception with reading—the invariance problem and the segmentation problem—make it considerably more difficult to analyze the meaning of a physical unit of speech (bottom-up) without having some prior hypothesis concerning what that unit is likely to be. To make matters more difficult, the serial and transient nature of the auditory message precludes a more detailed and leisurely bottom-up processing of the physical stimulus. This restriction therefore forces a greater reliance upon top-down processing.

Demonstrations of top-down or context-dependent processing in speech perception are quite robust. In an experiment by Miller, Heise, and Lichten (1951) five-word strings were presented to subjects under varying levels of a masking noise, thereby varying the bottom-up stimulus quality. The strings were either of sentence form ("Who brought some wet socks?") or were the same words arranged in random order ("socks some brought wet who"). With equivalent levels of masking noise, the sentence-form stimuli were recognized far more accurately than the random. In fact, in order to achieve equivalent recognition accuracy between the two forms, the noise level in the random form had to be reduced to half the intensity of its value in the sentence form, even though the physical stimulus information in the two strings was equivalent. In the sentence case, the sentence structure provides context to assist in recognizing the constituent words. This added context compensates for a greatly degraded stimulus quality.

In a related experiment, Miller and Isard (1963) compared recognition of degraded word strings between random word lists ("loses poetry spots total wasted"), lists that provided context by virtue only of their syntactic (grammatical) structure but had no semantic content ("sloppy poetry leaves nuclear minutes"), and full semantic and syntactic context ("A witness signed the official document"). The three kinds of lists were presented under varying levels of masking noise. Miller and Isard's data suggested the same tradeoff between signal quality and top-down context that was observed by Tulving, Mandler, and Baumal (1964) with the recognition of print. Less context, resulting from the loss of either grammatical or semantic constraints, required greater signal strength to achieve equal performance.

It is apparent that the perception of speech proceeds in a manner similar to the perception of print, through a highly complex, iterative mixture of bottom-up and top-down processing. Neisser (1967) has described this process as one of analysis by synthesis. While lower-level analyzers at the acoustic-feature and syllable level progress in a bottom-up fashion, the context provided at the semantic and syntactic levels generates hypotheses and actually synthesizes plausible alternatives concerning what a particular speech sound should be. This synthesis guides and thereby facilitates the bottom-up analysis. The presence of subjective gaps that are heard between word boundaries of continuous speech also gives evidence for the dominant role of synthesis. Since such gaps are not present in the physical stimulus, they must result from the top-down processes that decide when each word ends and the next begins.

APPLICATIONS

Research and theory of speech perception has contributed to two major categories of applications. First, understanding of how humans perceive speech and employ context-driven top-down processing in recognition has aided efforts to design speech-recognition systems that perform the same task (Lea, 1978). Such systems are becoming increasingly desirable for conveying responses in complex environments, such as high-performance aircraft in which the hands and arms are continuously occupied.

The second major contribution of research on speech recognition has been to measure and predict the effects on speech comprehension of various kinds of distortion, which was a source of the Tenerife disaster. Such distortion may be extrinsic to the speech signal—for example, in a noisy environment like an industrial plant. Alternatively, the distortion may be intrinsic to the speech signal when the acoustic wave form is transformed in some fashion, either when synthesized speech is used in computer-generated auditory displays or when a communication channel for human speech is distorted. The following pages will describe how the disruptive effects of speech distortion are represented and will identify some possible corrective techniques.

Effects of Noise

As we discussed earlier, natural speech is conveyed by the differing amplitudes of the various phonemes distributed across a wide range of frequencies. Thus it is possible to construct a spectrum of the distribution of power at different frequencies generated by "typical" speech. Typical spectra generated by male and female speakers are shown in Figure 4.12. The effects of noise on speech comprehension will clearly depend upon the spectrum of the noise involved. A noise that has frequencies identical to the speech spectrum will disrupt understanding more than a noise that has considerably greater power but occupies a narrower frequency range than speech.

Engineers are often interested in predicting the effects of background noise on speech understanding. The *articulation index* (Kryter, 1972) accomplishes this purpose by dividing the speech frequency range into bands and computing the ratio of speech power to noise power within each band. These ratios are then weighted according to the relative contribution of a given frequency band to speech, and the weighted

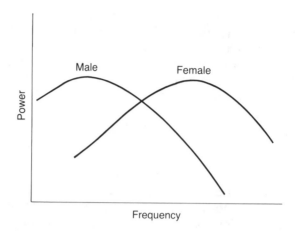

Figure 4.12   Typical power spectra of speech generated by male and female speakers.

ratios are then summed to provide the articulation index (AI). A simplified example of this calculation is shown in Figure 4.13. The spectrum of a relatively low-frequency noise is shown superimposed upon a typical speech spectrum. The speech spectrum has been divided into four bands. The ratio of speech to noise power and the logarithm of this ratio is shown below each band, and the contributions (weights) are shown below these. To the right is the sum of the weighted products, the AI, reflecting the extent to which the speech signal can be heard above the noise.

From our discussions of bottom-up and top-down processing it is apparent that the AI provides a measure only of bottom-up stimulus quality. A given AI may produce varying levels of comprehension depending upon the information content or redundancy available in the material and the degree of top-down processing that is employed by the listener. To accommodate these factors, measures of *speech intelligibility* are derived by delivering vocal material of a particular level of redundancy over the speech channel in question and computing the percentage of words understood correctly. Naturally, for a given signal-to-noise ratio (defining signal quality and therefore the articulation index) the intelligibility will vary as a function of the redundancy or information content of the stimulus material. A restricted vocabulary produces greater intelligibility than an unrestricted one; words produce greater intelligibility than nonsense syllables; high-frequency words produce greater intelligibility than low-frequency words; and sentence context provides greater intelligibility than no context.

The important implications of the above discussion are twofold: (1) Either the AI or the speech-intelligibility measures by themselves are inherently ambiguous, unless the redundancy of the transmitted material is carefully specified. (2) To reiterate a point made earlier in this chapter, data-driven bottom-up processing may trade off with context-driven top-down processing, and limitations of the former can be compensated by augmenting the latter. In noisy environments this may be accomplished by restricting the message set size (using standardized vocabulary) or by providing redundant "carrier" sentences to convey a particular message. The latter procedure is analogous to the use of the redundant carrier syllables of the communications-code alphabet (alpha, bravo, charlie) to convey information concerning a single alphabetic character. A high level of redundancy in the message from air traffic control to the KLM pilot would probably have stopped the premature takeoff and so averted the disaster.

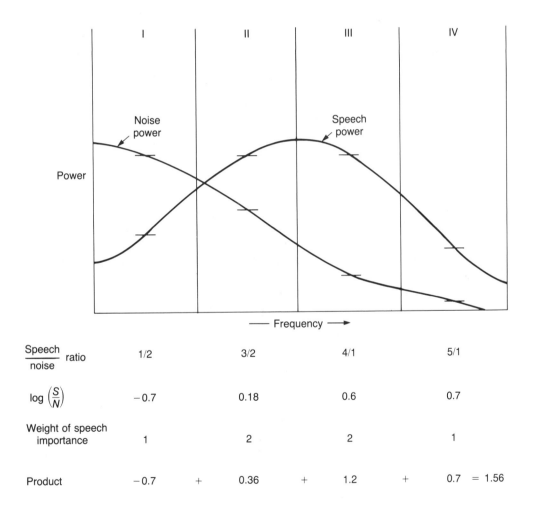

Figure 4.13 Schematic representation of the calculation of an articulation index. The speech spectrum has here been divided into four "bands" weighted in importance by the relative power that each contributes to the speech signal.

An experiment by Simpson (1976) demonstrated the effect of redundant carrier sentences on comprehension. Pilots listened to synthesized speech warnings presented in background noise. The warnings were either the critical words themselves ("fuel low") or the words embedded in a contextual carrier sentence ("Your fuel is low"). Recognition performance was markedly superior in the latter condition. Simpson also found that the beneficial effects of carrier sentences were greater for one-syllable words than for multi-syllabic ones. The greater level of redundancy in the multisyllabic words reduces the need for additional redundancy in the carrier sentence.

Effects of Distortion

The tradeoff between bottom-up and top-down processing is only one of a number of tradeoffs with which the designers of speech systems must deal. While any kind of speech distortion relayed by communications equipment is undesirable, there are times when it is inevitable. These are instances in which limitations on the bandwidth or information-processing rate of the communications channel impose a reduction in the information content of a message. A phone line, for example, has a limited bandwidth along which it may convey the acoustic signals of several conversations simultaneously. The time-varying acoustic wave carrying the speech information must therefore be filtered in some way in order to reduce the information content as described in Chapter 2 and still, ideally, preserve a maximum amount of voice quality. The tradeoff is defined in practice by the different techniques that may be employed to reduce the information content by equivalent amounts. With which technique can the greatest amount of voice quality and intelligibility be preserved relative to the amount of information reduction? Four techniques that can be alternatively employed are time compression, frequency compression, amplitude compression, and a combination of these accomplished by a device called the vocoder. While each of these will be described briefly in turn here, they are presented in greater detail in Kryter (1972). The engineering details of these techniques need not concern us here. Rather we are interested in their effects on performance.

At its most intuitive level, time compression may be achieved simply by talking less (reducing the redundancy of speech). However, time compression is mechanically achieved by deleting periodic intervals of the continuous speech signals. Deletion of more than 10% of the signal produces a measurable loss of intelligibility (Kryter, 1972). Frequency or bandwidth compression is achieved, conceptually, by translating the speech signal into the frequency domain (Figures 4.12 and 4.13) and filtering certain frequencies. Crudely, this may be accomplished by employing either a low- or high-pass filter and simply not transmitting the upper or lower range of frequencies, respectively. The problem with this procedure is that it systematically eliminates certain phonemes that are specific to either the high or low range. For example, s's and e's would be greatly disrupted by a low-pass filter (a filter that transmits only low frequencies), since these phonemes contain very-high-frequency oscillations. The o sound would be disrupted by a high-pass filter. A technique that allows considerably greater preservation of speech quality is one that deletes alternating bands of frequency across the spectral range.

A third technique that can be followed is to distort the raw speech signal in the amplitude dimension. This may be accomplished by one of two procedures. *Peak clipping* simply truncates the high and low excursions of the time signal shown in Figure 4.9a when these excursions pass above and below a given criterion. The closer these two criteria are brought together, the greater is the degree of compression. Peak clipping thereby reduces the information content of the signal by reducing the range of alternative levels. Peak clipping produces a harsh unpleasant speech quality when the signal is heard in quiet, but this loss of quality is less evident in noise (Kryter, 1972). Alternatively, the time-domain signal can be *digitized* to produce only a small number of discrete levels of amplitude, thereby reducing the *number* of levels within the original range. Digitization into at least 32 discrete levels (five bits) appears to be necessary in order to maintain a faithful resolution of voice quality (Kryter, 1972). Since the original speech signal contains theoretically an infinite amount of information because it is continuous, digitizing into even 32 levels still reduces information content significantly. However, even with considerably fewer levels, comprehension is still possible.

For a given amount of acoustic information transmitted, a comparison of the three techniques reveals that the greatest loss in intelligibility occurs with time compression or with high- or low-bandwidth compression (that is, frequency compression in which low or high frequencies are eliminated). This finding suggests that both the time and frequency dimensions are important to speech comprehension, and when either of these is distorted, comprehension is hindered. Somewhat greater intelligibility is preserved with amplitude compression or narrow-band frequency compression in which alternate frequency bands are eliminated. The first finding suggests the lesser degree of importance of amplitude variation in the raw speech signal: eliminating amplitude variation between the formants depicted in Figure 4.10 (the width of the strokes) would not increase the confusability of the different phonemes in their visual representation. Neither does it appear to do so in their auditory representation.

A fourth way of reducing speech bandwidth capitalizes directly upon the speech spectrograph representation. Since the two-dimensional shape of the formants in Figure 4.10 must be preserved in order to maintain comprehension, any technique that transmits this shape independent of other distortions will be adequate. The vocoder (Dudley, 1936; Kryter, 1972) effectively accomplishes this goal by "digitizing" the frequency axis of the speech spectrograph into a small number of levels (18). The vocoder then preserves discrete levels representing the amount of energy within each of these frequency bands at each sampled point in time. In this packaged, digitized form the information may be sent from transmitter to receiver. At the receiving end, the signal is resynthesized into speech by a series of 18 frequency generators that "play" at the intensities and times called for by the message. Of all the techniques described, the vocoder preserves the greatest intelligibility of speech per unit reduction in bandwidth.

When considering the various effects of speech distortion, it is important to realize that each alters the acoustic signal along certain dimensions. These dimensions are redundant to a considerable degree. (The formal information content of the speech signal is estimated at 59,000 bits per second. The information conveyed by language in normal speech is far less than that, with estimates in the tens of bits per second [Luce, 1960]). Therefore, distortion of a particular dimension will reduce the degree of redundancy and potentially may increase the vulnerability of the speech signal to other forms of disruption (e.g., noise). The particular form of the interaction between distortion and noise that may be observed will depend upon the degree of overlap between the dimensions affected by the distortion and those affected by the noise. Thus, for example, the disrupting effect of peak clipping is not enhanced, and in fact is attenuated, by noise (Kryter, 1972). On the other hand, combining a low-pass filter of a speech signal with low-frequency noise could be devastating, since the noise would obliterate selectively only that portion of the signal that remained.

References

Anderson, J. R. (1979). *Cognitive psychology.* New York: Academic Press.

Bailey, R. W. (1982). *Human performance engineering: A guide for system designers.* Englewood Cliffs, NJ: Prentice-Hall.

Bower, G., Clark, H., Wizenz, D., & Lesgold, A. (1969). Hierarchical retrieval schemes in the recall of categorical word lists. *Journal of Verbal Learning & Verbal Behavior, 8,* 323–343.

Bransford, J. D., & Johnson, M. K. (1972). Contextual prerequisites for understanding: Some investigations of comprehension and recall. *Journal of Verbal Learning and Verbal Behavior, 11,* 717–726.

Broadbent, D. E. (1977). Language and ergonomics. *Applied Ergonomics, 8,* 15–18.

Broadbent, D., & Broadbent, M. H. (1977). General shape and local detail in word perception. In S. Dornic (Ed.), *Attention & Performance VI.* Hillsdale, NJ: Erlbaum Associates.

Broadbent, D., & Broadbent, M. H. (1980). Priming and the passive/active model of word recognition. In R. Nickerson (Ed.), *Attention & Performance VIII.* New York: Academic Press.

Carpenter, P. A., & Just, M. A. (1975). Sentence comprehension: A psycholinguistic processing model of verification. *Psychological Review, 82* (1), 45–73.

Chapanis, A. (1965). Words words words, *Human Factors, 7,* 1–17.

Chase, W. G., & Clark, H. H. (1972). Mental operations in the comparison of sentences and pictures. In L. Gregg (Ed.), *Cognition in learning and memory.* New York: Wiley.

Clark, H. H., & Chase, W. G. (1972). On the process of comparing sentences against pictures. *Cognitive Psychology, 3,* 472–517.

Corcoran, D. W., & Weening, D. L. (1967). Acoustic factors in proofreading. *Nature, 214,* 851–852.

DeSoto, C. B., London, M., & Handel, S. (1965). Social reasoning and spatial paralogic. *Journal of Personal & Social Psychology, 2,* 513–521.

Dewar, R. E. (1976). The slash obscures the symbol on prohibitive traffic signs. *Human Factors, 18,* 253–258.

Dudley, H. W. (1936). The vocoder. *Bell Laboratories Records, 18,* 122–176.

Ehrenreich, S. (1982). The myth about abbreviations. *Proceedings, 1982 IEEE International Conference on Cybernetics and Society.* New York: Institute of Electrical and Electronic Engineers.

Eimas, P. D. (1963). The relation between identification and discrimination along speech and non-speech continua. *Language & Speech, 6,* 206–217.

Ellis, N. C., & Hill, S. E. (1978). A comparison of seven-segment numerics. *Human Factors, 20,* 655–660.

Forster, K. I. (1970). Visual perception of rapidly presented word sequence of varying complexity. *Perception & Psychophysics, 8,* 215–221.

Fowler, F. D. (1980). Air traffic control problems: A pilot's view. *Human Factors, 22,* 645–654.

Gibson, E. J. (1969). *Principles of perceptual learning and development.* Englewood Cliffs, NJ: Prentice-Hall.

Gordon, H. W., Silverberg-Shalev, R., & Czernilas, J. (1982). Hemispheric asymmetry in fighter and helicopter pilots. *Acta Psychologica, 52,* 33–40.

Grether, W., & Baker, C. A. (1972). Visual presentation of information. In H. P. Van Cott and R. G. Kinkade (Eds.), *Human engineering guide to equipment design.* Washington, DC: U.S. Government Printing Office.

Haber, R. N., & Schindler, R. M. (1981). Error in proofreading: Evidence of syntactic control of letter processing? *Journal of Experimental Psychology: Human Perception & Performance, 7,* 573–579.

Hammond, K. R. (1980, July). *The integration of research in judgment and decision theory* (Report No. 226). Boulder, CO: University of Colorado, Institute of Behavioral Science, Center for Research on Judgment and Policy.

Healy, A. F. (1976). Detection errors on the word "the." *Journal of Experimental Psychology: Human Perception & Performance, 2,* 235–242.

Huggins, A. (1964). Distortion of temporal patterns of speech: Interruptions and alterations. *Journal of the Acoustical Society of America, 36,* 1055–1065.

Hunt, E., Frost, N. & Lunneborg, C. (1973). Individual differences in cognition. In G. H. Bower (Ed.), *Psychology of Learning & Motivation* (vol. 7). New York: Academic Press.

Just, M. T., & Carpenter, P. A. (1971). Comprehension of negation with quantification. *Journal of Verbal Learning and Verbal Behavior, 10,* 244–253.

Just, M. T. & Carpenter, P. A. (1980). Cognitive processes in reading: Models based on reader's eye fixation. In C. A. Prefetti & A. M. Lesgold (Eds.), *Interactive processes and reading.* Hillsdale, NJ: Erlbaum Associates.

Kinsbourne, M., & Hicks, R. (1978). Functional cerebral space. In J. Requin (Ed.), *Attention and Performance VII.* Hillsdale, NJ: Erlbaum Associates.

Klemmer, E. T. (1969). Grouping of printed digits for manual entry. *Human Factors, 11,* 397–400.

Kolers, P. A., & Lewis, C. L. (1972). Bounding of letter sequences and the integration of visually presented words. *Acta Psychologica, 36,* 112–124.

Kryter, K. D. (1972). Speech communications. In H. P. VanCott & R. G. Kinkade (Eds.), *Human engineering guide to equipment design.* Washington, DC: U.S. Government Printing Office.

LaBerge, D. (1973). Attention and the measurement of perceptual learning. *Memory & Cognition, 1,* 268–276.

Landeweerd, J. A. (1979). Internal representation of a process: Fault diagnosis and fault correction. *Ergonomics, 12,* 1343–1362.

Lea, W. (Ed.). (1978). *Trends in speech recognition.* Englewood Cliffs, NJ: Prentice-Hall.

Liberman, A. M. (1957). Some results of research on speech perception. *Journal of the Acoustical Society of America, 29,* 117–123.

Liberman, A. M., Cooper, F. S., Shonkweiler, D., & Studdert-Kennedy, M. (1967). Perception of the speech code. *Psychological Review, 74,* 431–461.

Lindsay, P. H., & Norman, D. A. (1972). *Human information processing.* New York: Academic Press.

Lohman, D. F. (1979). *Spatial ability: A review and reanalysis of the correlational literature* (Aptitude Research Project T.R. No. 8). Palo Alto, CA: Stanford University, School of Education.

Luce, R. D. (1960). The theory of selective information and some of its behavioral applications. In R. D. Luce (Ed.), *Developments in mathematical psychology.* Glencoe, IL: Free Press.

Massaro, D. W. (1975). *Experimental psychology and information processing.* Chicago: Rand McNally College Publishing.

McClelland, J., & Johnston, J. (1977). The role of familiar units in perception of words and non-words. *Perception & Psychophysics, 22,* 249–261.

McClelland, J., & Rummelhart, D. C. (1981). An interactive model of the effect of context in perception, Part 1. *Psychological Review, 88,* 375–407.

McConkie, G. W. (1983). Eye movements and perception during reading. In K. Raynor (Ed.), *Eye movements in reading.* New York: Academic Press.

McConkie, G. W., & Raynor, K. (1974). *Identifying the span of the effective stimulus in reading* (Final Report OEG 2-71-0537). Washington, DC: U.S. Office of Education.

McConkie, G. W., Zola, D., Blanchard, H. E., & Wolverton, G. (1982). Perceiving words during reading: Lack of facilitation from prior peripheral exposure. *Perception & psychophysics, 32,* 271–281.

McCormick, E. J. (1976). *Human factors in engineering and design.* New York: McGraw-Hill.

Miller, G. A., Heise, G. A., & Lichten, W. (1951). The intelligibility of speech as a function of the text of the test materials. *Journal of Experimental Psychology, 41,* 329–335.

Miller, G., & Isard, S. (1963). Some perceptual consequences of linguistic rules. *Journal of Verbal Learning and Verbal Behavior, 2,* 217–228.

Moscovitch, M. (1979). Information processing and the cerebral hemispheres. In M. S. Gazzaniga (Ed.), *The handbook of behavioral biology: Volume on neuropsychology.* New York: Plenum Press.

Moses, F. L., & Ehrenreich, S. L. (1981). Abbreviations for automated systems. In R. Sugarman (Ed.), *Proceedings of the Human Factors Society 25th annual meeting.* Santa Monica, CA: Human Factors.

Navon, D. (1977). Forest before trees: The presence of global features in visual perception. *Cognitive Psychology, 9,* 353–383.

Neisser, U. (1963). Decision time without reaction time. *American Journal of Psychology, 76,* 376–385.

Neisser, U. (1967). *Cognitive psychology.* Englewood Cliffs, NJ: Prentice-Hall.

Neisser, U., Novick, R., & Lazer, R. (1964). Searching for novel targets. *Perceptual and Motor Skills, 19,* 427–432.

Norcio, A. F. (1981). *Human memory processes for comprehending computer programs* (Technical Report AS-2-81). Annapolis, MD: U.S. Naval Academy, Applied Sciences Department.

Norman, D. A. (1981). The trouble with unix. *Datamation, 27* (12), 139–150.

Payne, J. W. (1980). Information processing theory: Some concepts and methods applied to decision research. In T. S. Wallsten (Ed.), *Cognitive processes in choice and decision behavior.* Hillside, NJ: Erlbaum Associates.

Phelps, R. H., and Shanteau, J. (1978). Livestock judges: How much information can an expert use? *Organizational Behavior in Human Performance, 21,* 209–219.

Plath, D. W. (1970). The readability of segmented and conventional numerals, *Human Factors, 12,* 493–497.

Potter, M. C., Kroll, J. F., and Harris, C. (1980). Comprehension and memory in rapid sequential reading. In R. S. Nickerson (Ed.), *Attention and Performance VIII.* New York: Academic Press.

Reicher, G. M. (1969). Perceptual recognition as a function of meaningfulness of stimulus material. *Journal of Experimental Psychology, 81,* 275–280.

Rummelhart, D. (1977). *Human information processing.* New York: Wiley.

Savin, H. B., & Perchonock, E. (1965). Grammatical structure and the immediate recall of English sentences. *Journal of Verbal Learning and Verbal Behavior, 4,* 348–353.

Schneider, W., & Shiffrin, R. M. (1977). Controlled and automatic human information processing I: Detection, search, and attention. *Psychological Review, 84,* 1–66.

Schneiderman, B. (1980). *Software psychology.* Cambridge, MA: Winthrop.

Sergent, T. (1982). The cerebral balance of power: Confrontation or cooperation? *Journal of Experimental Psychology: Human Perception and Performance, 8,* 253–272.

Sheridan, T. E., and Ferrell, L. (1974). *Man-machine systems.* Cambridge, MA: MIT Press.

Shiffrin, R. M., & Schneider, W. (1977). Controlled and automatic human information processing II: Perceptual learning, automatic attending, and a general theory. *Psychological Review, 84,* 127–190.

Simpson, C. (1976, May). Effects of linguistic redundancy on pilot's comprehension of synthesized speeds. *Proceedings, 12th annual conference on manual control* (NASA TM-X-73, 170). Washington, DC: U.S. Government Printing Office.

Snow, R. E. (1980, September). *Aptitudes and instructional methods* (Final Report, Aptitudes Research Project). Palo Alto, CA: Stanford University, School of Education.

Spanish Ministry of Transportation & Communications. (1978). Report of Collision between PAA B-747 and KLM B-747 at Tenerife. *Aviation Week & Space Technology, 109,* November 20, pp. 113–121, and November 27, pp. 67–74.

Tinker, M. A. (1955). Prolonged reading tasks in visual research. *Journal of Applied Psychology, 39,* 444–446.

Tulving, E., Mandler, G., & Baumal, R. (1964). Interaction of two sources of information in tachistoscopic word recognition. *Canadian Journal of Psychology, 18,* 62–71.

Vartabedian, A. G. (1972). The effects of letter size, case, and generation method on CRT display search time. *Human Factors, 14,* 511–519.

Whitaker, L. A., & Stacey, S. (1981). Response times to left and right directional signals. *Human Factors, 23,* 447–452.

Wickelgren, W. A. (1979). *Cognitive psychology.* Englewood Cliffs, NJ: Prentice-Hall.

Wickens, C. D., Sandry, D. C., & Hightower, R. (1982, October). *Display location of verbal and spatial material: The joint effects of task-hemispheric integrity and processing strategy* (Technical Report EPL-82-2/ONR-82-2). Champaign, IL: University of Illinois, Engineering Psychology Research Laboratory.

Yallow, E. (1980). *Individual differences in learning from verbal and figural materials* (Aptitudes Research Project T.R. No. 13). Palo Alto, CA: Stanford University, School of Education.

Nonverbal Perception 5

OVERVIEW

As outlined in Chapter 4, three important properties characterize the perceptual processing of verbal material: (1) The units of perception are directly related to language and linguistic experience. (2) The units of analysis are symbolic, whether heard or seen, whether features, letters, phonemes, or words. That is, the final units activated in memory are uncorrelated with the physical representation of the stimulus. For example, neither the visual printed appearance nor the sound of the word "chair" bears any physical relation to a chair. (3) Perception of these units proceeds hierarchically and analytically, with the output of features feeding into letters (or phonemes) and these into words. Only with extensive practice did some sort of more holistic perception take place so that higher-order units were directly activated, a phenomenon called unitization. This process was typified by the advantage provided to whole words resulting from word-shape analysis.

This chapter on nonverbal perception will focus on perception of stimuli that do not generally share one or more of these three properties. One concern will be the perception of objects or texture. Here the physical stimulus does not bear an arbitrary symbolic relation to the subject's concept of that stimulus, but bears a direct *analog* relation to it. The term "analog" implies that some physical equivalence exists between the stimulus and its mental representation. The equivalence may be achieved by continuous transformations of rotation, magnification, distortion, or translation. Thus a pilot's mental representation of an aircraft in space should have a direct analog correspondence with the true orientation of the actual aircraft. As a second contrast, unlike verbal stimuli, the information to be dealt with in the present chapter is often perceived "holistically" rather than analytically. A box, for example, is not perceived first as a set of lines and angles joined together, but is perceived initially as a whole box. It is only after further analysis that our conscious perception can parse the box into its constituent elements.

We should reiterate that the distinctions drawn between verbal and nonverbal material, between analytic and holistic processing, and between symbolic and analog relations are not absolute. For example, as discussed in Chapter 4, the perception of familiar words is based partly upon a "holistic" processing of word shape, augmenting the hierarchical feature and letter analysis. Correspondingly, when one reads print but concentrates on the shape of the word or the size of the print, one is processing analog rather than symbolic properties of the printed stimulus. Conversely, perception of objects often engages heavily symbolic processes as well, as when one views a chair and attends not to its size or shape but to its *functional* properties (a place to sit, to work, to be comfortable) (Levy, 1974). Furthermore, it is also important to realize that objects are sometimes perceived analytically as well as holistically, as when a common object may be dissected by the visual system into its constituent attributes and components. As emphasized in Chapter 4, these two "modes of perceiving" define more of a continuum than a dichotomy.

The following sections of the chapter will deal first with two primary aspects of the "spatial" processing syndrome: its holistic properties, including individual differences in processing style and the perception of objects, and its analog properties, including the perception of texture and the issue of display compatibility. In the final section we discuss the practical application of spatial perception to the problems of maps and navigation.

HOLISTIC PROCESSING

The term *holistic* describes a mode of information processing in which the whole is perceived directly rather than as a consequence of the separate perceptual analysis of its constituent elements. This description does not mean that perceptual analysis of the whole precedes analysis of the elements. It only suggests that the conscious perceptual awareness is initially of the whole and that perception of the elements must follow from a more detailed analysis (Navon, 1977). To provide an intuitive example, when we perceive a face we are normally immediately aware of its overall configuration and not (at least initially) aware of the separate features (shape of the nose, separation of the eyes, etc.). This section provides an umbrella for discussing a number of perceptual phenomena that relate in one manner or another to this holistic form of perceptual analysis.

Individual Differences in Holistic Versus Analytic Processing

If holistic processing does not depend upon the analysis of a stimulus's constituent elements, then the complexity of those elements should not affect the time to process the stimulus holistically but should influence the time required for analytic processing. Research conducted by Cooper (1976) presents a prototypical example of this characteristic of holistic processing. In a series of investigations, she has demonstrated that certain individuals tend to process geometric stimulus information holistically, while others process the same kind of stimuli in a slower, analytic, feature-by-feature fashion. Cooper labeled the two groups Type I and Type II processors, respectively.

In Cooper's paradigm, subjects see a random-shaped geometric form. Then, after a short pause, a second form is presented and subjects decide as rapidly as possible whether the second form is the same as or different from the first. The critical independent variable is the degree of similarity of the two forms. Similarity is manipulated by varying the features (angles, sides) of the second form, as shown in Figure 5.1. When reaction time is examined as a function of form similarity, Cooper finds results like those shown in Figure 5.1. For Type I subjects, "same" responses are slightly faster than all "different" responses, and "different" responses, while slower, do not vary as a function of similarity. Overall responding is quite rapid compared with that of Type II subjects. For the latter subjects "same" responses are slower than for Type I, and "different" responses become progressively faster as the stimulus similarity decreases, so stimulus pairs that are most different are responded to more rapidly than "same" responses.

Type I subjects are thus presumed to process the stimuli holistically, like matching a template: If the second stimulus perfectly matches the memory-trace template of the first, they answer "yes." If it doesn't, the response is "no." Type II subjects, on the other hand, process analytically, feature-by-feature, until the perceptual comparison process encounters a feature that doesn't match. The more identical features the two stimuli share, the longer on the average this comparison will take, since the comparison is more likely to encounter identical features in

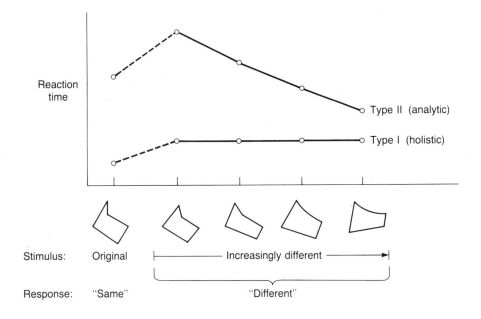

Figure 5.1 Stimuli and data of Cooper's figure-matching task. From "Individual Differences in Visual Comparison Processes," by L. Cooper, 1976, *Perception and Psychophysics, 19,* p. 436. Copyright 1976 Psychonomics Society, Inc. Adapted by permission of the author.

the search. In parallel with the search for different features, Type II subjects are assumed to have an "identity detector" that will trigger a "same" response if identity is detected (Bamber, 1969) before the feature analysis is complete. This assumption accounts for the fact that the same responses are faster than the different responses for the highest-similarity pairs. It is important to note that the two patterns of results (Type I and Type II) are *qualitatively* different from each other. This suggests that one group does not simply have more facile processing than the other but that its processing is quite different.

In the same-different task used by Cooper, holistic processing was clearly superior: faster and more accurate. In a subsequent experiment, Cooper (1980) changed the task to make the holistic processing of Type I subjects disadvantageous. Subjects were required to identify the location of discrepant features (not just to detect their existence). Despite the fact that the identification task now required an analytic approach, some of the Type I subjects could not easily abandon their (now harmful) holistic processing strategy. These results argue that the strategy differences represented something more than a simple preference on the part of subjects to process in one mode or the other. Rather it reflected a more fundamental difference in perceptual processing style.

Cooper argued on the basis of her data that there is a dichotomy in the population between Type I and Type II subjects. However, a subsequent test of 56 subjects on Cooper's task indicated that the distinction between analytic and holistic processes is defined more as a continuum between two end points than as a dichotomy. In this study, Dupree and Wickens (1982) observed a continuous range of slopes between the two extremes in Figure 5.1.

Cooper's work is important in part because it provides a clear example of holistic processing in the behavior of her Type I subjects. However, the work is also relevant because it contributes a potentially important dimension of information processing, or "cognitive style," to the growing catalog of such dimensions. These include such differences in information processing as those related to field dependence and field independence and to impulsivity and reflectivity (Goodenough, 1976). Dimensions such as these are of potential importance because they may assist in the selection of operators for jobs with different kinds of information-processing demands. The particular quality of holistic processing, for example, would be valuable in jobs such as photo interpretation in which the rapid, accurate perception of objects is required. Cooper's subjects of course perceived only arbitrary, unfamiliar geometric shapes. We shall now consider how people perceive more familiar objects.

Object Perception

The perception and representation of objects in the environment is another example of holistic processing. Although our perceptual system must analyze the separate dimensions that make up an object, we experience objects not as a disembodied "bundle of features" but as an integral whole whose attributes are processed in parallel, without mutual interference, and are somehow "glued together" by their very objectness (Kahneman & Treisman, in press; Lockhead, 1972, 1979; Wickelgren, 1979). Furthermore, this characteristic of holistic object perception appears to be a relatively primitive one, encouraged by innate perceptual mechanisms of size constancy and figure-ground segregation (Rock, 1975). There are three general characteristics that allow an object to be perceived holistically, although it is probable that none of the three by itself is a necessary ingredient for object perception. These characteristics are its surrounding contours, its correlated attributes, and its familiarity.

1. *Surrounding contours.* An object is normally defined by a shape, which in turn is represented by contours. These contours strongly trigger the perception of an object. This property is demonstrated in Figure 5.2. Even if the contours are not physically complete, our perceptual mechanism will complete the contours through top-down processing. As the examples shown in Figure 5.2 indicate, the contours thus define the *shape* of the object, one of its most important attributes.

2. *Correlated attributes.* The physical properties and constraints of objects dictate that their attributes are often correlated as viewed in normal perceptual experience. As an object is moved away from the observer, for example, its horizontal and vertical dimensions will covary. As it is tilted, its shape will covary with the pattern of light and shadow or with the sharpness of a textural gradient across one of its faces. Chapter 2 described the manner in which this correlation of dimensions leads to better performance in absolute judgment tasks (Eriksen & Hake, 1955), and Chapter 3 described the assistance that correlated attributes offered to decision-making performance. In this way, the unique perceptual responses to objects will benefit from the correlation. Indeed, it is quite possible that this causal relation is reversed, and the intrinsic advantage to processing integral dimensions when they are correlated (Garner, 1974) results from the lifetime of experience with such correlations when dealing with virtual objects: the dimensional combinations that tend to be more integral as defined in Chapter 2 are also those that tend to be attributes of a single object.

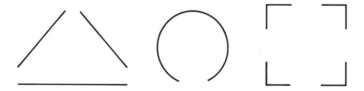

Figure 5.2   Role of top-down processing in object perception.

3. *Familiarity*. Familiarity is a common but not a necessary characteristic of an object. There is little doubt that familiarity facilitates object perception when contours are degraded or missing. Yet even totally unfamiliar contours can be recognized as holistic objects with little difficulty if they are clearly presented. This contrast illustrates the same interaction between bottom-up and top-down processing that was reviewed in the context of verbal perception in Chapter 4. A demonstration of the role of top-down processing in object perception was provided by Palmer (1975). In a procedure analogous to that described by Miller and Isard (1963) for speech perception and by Tulving, Mandler, and Baumal (1964) for printed sentences (both discussed in Chapter 4), Palmer presented subjects a context display of a visual scene (e.g., a kitchen). This was followed by tachistoscopic presentation of an object which could be either appropriate in the context of that scene (a loaf of bread), appropriate in physical form but out of context (a home mailbox, which looks like a loaf of bread but does not belong in the kitchen), or appropriate in neither form nor context (a drum). Palmer found that the visual recognition threshold was predicted directly by the amount of contextual appropriateness from the loaf of bread (highest appropriateness, lowest threshold) to the mailbox, to the drum (lowest appropriateness, highest threshold).

In a related study, Biederman et al. (1981) demonstrated top-down processing in rapid photo interpretation of objects. Their subjects had to detect objects in a complex visual scene from a rapid 200-msec exposure. The objects were in either appropriate or inappropriate contexts, where appropriateness was defined in terms of several expected properties of objects (e.g., the object must be supported, it should be of the expected size given the background). Biederman et al. found that if the object was appropriate, it was detected equally well at visual angles out to 3° of parafoveal vision. If it was not, performance declined rapidly with increased visual angle from fixation.

The role of familiarity in object perception shown by Palmer and by Biederman et al. suggests that familiar objects are coded and represented symbolically (in terms of abstract words and ideas) as well as in analog spatial form. We must assume that presentation of familiar objects rapidly activates both visual and symbolic codes in some cooperative manner (see Chapter 6). Evidence for the speed and efficiency with which the symbolic meaning of objects can be encoded and interpreted is provided by Potter and Faulconer (1975), who found that pictures could be understood at least as rapidly as words. This conclusion was further supported in an investigation by Potter, Kroll, and Harris (1980). Employing the technique of rapid serial visual presentation (RSVP) of text described in Chapter 4, they occasionally replaced words (concrete nouns) with pictures. Even at high presentation rates of 12 symbols per second, the comprehension accuracy did not appear to be hindered by the incorporation of the pictorial material.

The conclusion that pictures can often be processed at least as rapidly as words has particular relevance for the use of verbal versus pictorial/symbolic highways signs (Ells & Dewar, 1978; Green & Pew, 1978; Whitaker & Stacey, 1981). The results of such studies generally indicate that for familiar symbols, pictorial symbology is comprehended at least as rapidly as verbal messages. In particular, pictorial representations appear to be less disrupted by degraded viewing conditions (Ells & Dewar, 1978).

The object display.   The holistic property of object perception and the special unique perceptual response given to integral objects has a major straightforward implication for the design of displays. When the purpose of a display is to provide information to an operator concerning the status of an object (e.g., the position and attitude of an aircraft, vehicle, or remote manipulator robotic arm), there is an intrinsic advantage to displaying the information concerning the object in an object format rather than just in terms of its disembodied separate attributes—for example, as separate gauges or meters (Roscoe, 1968; Roscoe, Corl, & Jensen, 1981). While it may be important to display this attribute information separately for precise reading, the integral object display appears to represent the most compatible means of providing the supervisor with an overall representation or grasp of the status of the system under supervision.

In aviation, this holistic, analog depiction is partially provided by the *contact analog display* in which the two variables of roll and pitch are combined into a single, highly schematic representation of the aircraft itself, rather than presented on separate meters. In remote manipulator systems, stationary television cameras give users a direct perception of the grasping claw under their control. This kind of display is of considerable value in actual system operation. Alternatively, however, the object display could be useful as a means of training the learner to incorporate a high-fidelity "internal model" of the system by calling attention to the pattern of time correlation between its attributes (see Chapter 3). This pattern of correlations will be best perceived when the displayed elements are closely integrated in space.

The preceding discussion suggests that the supervision of variables that do not actually represent attributes of a single object but are nonetheless correlated in time (e.g., pressure and temperature in a chemical process) might also benefit if these variables are integrated in a connected contoured object format for the purposes of display.

An experiment performed by Jacob, Egeth, and Bevon (1976) provides a graphic demonstration of the utility of the object display in helping subjects make judgments on the basis of several correlated dimensions of a hypothetical system. The experiment also capitalizes on familiarity by using a particularly familiar object: the human face. In the experiment, subjects viewed a series of stimuli that varied along nine correlated dimensions. They were required to sort these rapidly into one of five categories each defined by correlated levels along the nine dimensions. This categorization might correspond to five different failure "syndromes" in a complex process or five different diseases. As noted in Chapters 2 and 3, such correlation between dimensions is expected in real-world systems (Moray, 1981) and assists subjects in their judgments.

The critical independent variable in Jacob et al.'s investigation was the manner in which the nine dimensions were displayed. These are shown across the columns in Figure 5.3 in an order that essentially defines their degree of "objectness." Number displays in the left column show each dimension separately in verbal numerical format. The "glyph" display, in the next column, portrays each dimension as an analog quantity, but the quantities are not integrated into any holistic configuration. The polygon display in the next column is created by drawing nine spokes from the center of an imaginary circle, proportional in length to the dimensional value, connecting the end points of the spokes and then deleting the spokes. This is the format of the "iconic" display proposed for safety-parameter information for nuclear power plant operators, which will be discussed further in Chapter 12.

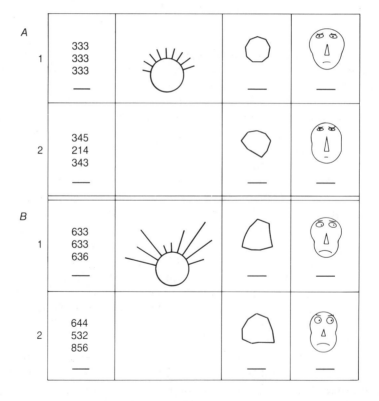

Figure 5.3 Four categories of stimuli. Within each column (except column 2) are two examples of each of two assigned categories (*A* and *B*). From "The Face as a Data Display" by R. J. K. Jacob, H. E. Egeth, and W. Bevon, 1976, *Human Factors, 18,* pp. 189–200, Figs. 1–5. Copyright 1976 by The Human Factors Society, Inc. Reproduced by permission.

The polygon representation capitalizes on both analog format and integrated contours. Finally, in the right column, the face display adds the final dimension of familiarity. Each dimension is created by a feature of the face (e.g., the mouth) and levels on that dimension are coded by the shape of the feature. A smile and a frown for example, would represent two levels of the "mouth feature." Within each column of Figure 5.3 are shown two representative categories A and B of the five to be sorted. Within each category (except the glyphs) are two examples of members 1 and 2. The similarity of values between the two members of the category reflects the correlation of the dimensions.

The results of Jacob et al.'s study clearly indicated that more objectlike representations to the right of Figure 5.3 assisted classification. This advantage was reflected in both speed and accuracy. The one exception to the trend was found with the glyph display, which generated poorer performance than even the digit display. This finding would suggest that contours are particularly valuable for the display of multivariate correlated information. Analog properties by themselves as shown in the glyph display are not valuable in this context.

The advantage of the face display over numerical values has been supported in more recent experiments by Hahn, Morgan, and Lorensen (1983) and Moriarity (1979). Moriarity presented subjects (both business students and professional accountants) with information about a series of corporations on 13 different dimensions of financial interest, the Dun and Bradstreet business ratios. As in the experiment by Jacob et al., these dimensions are typically correlated. Subjects were asked to process this information to decide if a company was in a position to file for bankruptcy. Moriarity found that performance of both groups was faster and more accurate when the 13 dimensions described variations of features in a face display than when they were presented as numerical values.

It is important to bear in mind that both of the previous experiments required that judgments be made on the basis of the configuration of all dimensions considered simultaneously, a reasonable requirement given the high correlation of dimensional values. However, the discussion of integral and separable dimensions in Chapter 2 suggests that the benefits of integral object representation of correlated variables might be partially neutralized by the cost resulting when a single variable must be *selectively* perceived and other dimensions must be selectively filtered out (Garner & Fefoldy, 1970). It seems that the integrated dimensions of an object "want" to be perceived together and "dislike" the partial perception of one to the exculsion of others (Kahneman & Triesman, in press).

The requirement for such selective filtering would occur in a situation in which the goal of perception is no longer a holistic one (to grasp or understand the overall state of the system as one might to diagnose the cause of a failure) but an analytic one (to read precisely the value of one variable as one might to identify the severity of a particular symptom). It is for this reason that system variables should always be made available in their separate representations, even when the integral representation is presented for configurational holistic perception. For example, the pilot of an aircraft wishing to perceive the overall state of the aircraft in order to predict its future behavior would benefit from a glance at an integrated object display (since most aerodynamic variables are highly intercorrelated). However, to check if one particular variable (e.g., airspeed) is within permissible bounds, the pilot should be able to view this information separately displayed.

Perceptual Schema

Normally objects must be perceived and assigned to perceptual categories even when the separate features that define all objects within a category may be quite variable. This was the case with the stimuli that Jacob et al. (1976) employed in their investigation of the face display. The two stimuli in the top of each column of Figure 5.3 were assigned to one category but differed a good bit in terms of their dimensional values. In fact, much of perceptual categorization of nonverbal stimuli requires that the information from several dimensions must be considered because, while dimensions may be correlated, no single defining feature is sufficient to assign a stimulus positively to a category.

To provide an extreme example, consider the perceptual category of "dog." It would be quite difficult to find a single stimulus feature possessed by all dogs and not possessed by any nondogs. Examples more relevant to engineering psychology of such ill-defined categories that do not possess simple distinctive features include the recognition of certain abnormalities from x-ray data or recognition of critical meteorological conditions or geological formations. The mental representation of psychiatric diseases is also argued to be more associated with correlated sets of features or symptoms than with the presence or absence of certain defining features (Cantor, French, Smith, & Mezzich, 1980).

The term *perceptual schema* has been employed to designate the form of knowledge or mental representation that people use to assign stimuli to ill-defined categories. The schema is a general body of knowledge about the characteristics of a perceptual category that does not contain a strict listing of its defining features (e.g., features which must all be present for a particular instance to be termed a category). Because of this somewhat fuzzy defining characteristic, the schema is normally acquired as a result of perceptual experience with examples rather than by learning a simple defining set of rules.

In a series of investigations, Posner and Keele (1968, 1970; Posner, 1973) developed a paradigm in which the different schema to be learned consisted of a set of four random 9-dot patterns. Four *prototypes* of typical members of each of the four schema were created by selecting four distinct dot patterns. Two examples of these are shown in Figure 5.4. From each of these prototypes, a number of *distortions* of each prototype were then created by randomly "jittering" the dots a

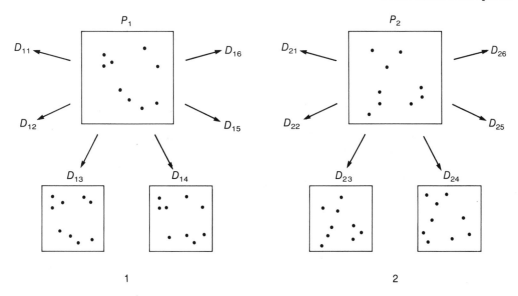

Figure 5.4   Typical stimuli employed by Posner and Keele (1968). Shown are
schemata 1 and 2. For each is shown the prototype P and examples of two
of the distortions D. From "Classification of distorted patterns" by M. I.
Posner, R. Goldsmith, and K. E. Welton, Jr., 1967, *Journal of Experimental
Psychology, 73,* pp. 28–38, Figs. 1 and 2. Copyright 1967 by the American
Psychological Association. Adapted by permission of the authors.

certain amount. Two typical distortions of each prototype are also
shown in Figure 5.4. (The distortions may also be called *exemplars* of
each schema.) After these distortions were synthesized, the subjects
were given the task of learning to associate the distortions with the
appropriate schema name. This training was accomplished by having
the subjects sort the distortions into the four piles, receiving corrective
feedback until they could do so accurately. It is essential to note that
the original prototype was *not* included in these training trials of
distortion sorting during which the schema was formed.

Once the subjects had learned to sort correctly, they were assumed to
possess some mental representation of the distinguishing properties of
the schema classes. Of what did this schema consist? Posner and Keele
argued that it consisted of two components: a general representation of
the mean, or "typical," value of the schema and some abstract representation of the variability. In a second phase of the experiment, they
presented subjects with a new set of stimuli from all four schema
classes to classify. In this test phase, the stimuli from each class consisted of three types: old distortions that the subjects had seen while
forming the schema, new distortions, and the prototype of each schema
from which all distortions (both new and old) had been generated. Like
the new distortions, the prototype was a stimulus never before encountered by the subjects.

Of greatest importance was the classification accuracy made in this new sort. The errors made here showed that subjects could sort the prototypes (novel) just as accurately as the old (familiar) distortions and with considerably greater accuracy than the equally novel new distortions. Moreover, after a week's time, when the classification test was repeated, memory (as measured by classification accuracy) for the old distortions declined, but accuracy for the new prototype remained unchanged. A final interesting observation made by Posner (1969) is that subjects not only classify the prototypes accurately, but also truly believe that they are familiar and have been experienced before (like the old distortions) even when they are in fact entirely novel.

The data that have been collected on schema formation suggest that the nature of the mental representation which people use to classify stimuli in ill-defined categories is not a strict list of characteristics of the prototype but that the mental representation also contains information concerning the variability around the template. The importance of this variability is suggested by studies of Dukes and Bevon (1967) and Posner and Keele (1968). These investigators found that exposure to a variety of instances of a schema induced better performance at recognizing new instances than did repeated exposure to a single instance. Posner and Keele investigated the specific amount of variability provided in training. They found that subjects given "tight" instance training (low variability from the prototype) performed relatively poorly at classifying highly distorted patterns. Those given "loose" training performed better in classifying high distortions. However, this improvement in categorization was accomplished with the cost of an increased false-alarm rate (categorizing a nonschema member into the schema class).

More recently, other theories of how people assign new stimulus instances to ill-defined categories have been put forth. For example, Medin and Schaffer (1978) propose that the assignment is made not by relating each new instance to a "central prototype" but rather by relating it to the exemplar to which it is most similar and then assigning the new instance to the residence category of that exemplar. While the theories make some different predictions in category learning (Medin & Schaffer, 1978), the more general importance of work in this area relates to how operators should be trained to recognize such ill-defined patterns in a variety of contexts: The medical technician scanning x-rays or electroencephalographic traces for certain patterns, the photo interpreter looking for evidence of certain phenomena, or the scientist examining naturalistic data for "prototypal conditions" are all attempting to classify stimuli with ill-defined, noisy features into discrete perceptual categories.

The practical implications of this research are straightforward but worthy of note. To the extent that the stimuli to be classified in a given task are variable, the training should also be of variable instances of the category rather than of the single example of the prototypal member or a strict memorization of the defining features. The degree of training variability should reflect the degree of actual instance variability. If either of these sources of variability is large however, the potential cost of false alarms should be taken into account. Finally, it seems evident that the best training for this sort of procedure can be achieved by requiring the learner to apply the classification rules repeatedly by actual physical sorting rather than by simply requiring the learner to study the examples (without a response) or to learn (verbally) the list of distinctive features that may characterize the prototype but not all of the exemplars. "Learning by doing," a principle emphasized in the preceding chapter in the discussion of the development of automatic processing, is reinforced here.

ANALOG PERCEPTION

As we discussed in Chapter 4, the perceptual categorization of verbal material is symbolic and discrete. The physical dimensions of the original stimulus are relatively unimportant in the internal code. However, much of perception requires that these physical parameters be retained in memory and internalized with as much precision as the stimulus has. To cite two contrasting examples, the shape of an *a* has little or no relevance to the meaning of the word in which the *a* appears. Therefore, in reading, its perception is not analog; the letter *a* is perceived categorically. However, the driver making steering corrections on the basis of perception of the road curvature cannot afford to perceive curvature categorically (e.g., curve left, curve right) but must respond to the precise degree of curve in either direction. As a second example, when adjusting the rate of chemical flow in a process industry on the basis of an analog indicator of process quality, the operator cannot afford to perceive this meter categorically but must be responsive to changes at all levels.

Thus the essence of analog perception is that the *ordering* and magnitude of physical changes in the stimulus are preserved in ordering and magnitude of the perceptual response and the mental representation. The focus of this section then will be on those aspects of perception in which there appears to be a continuous ordered mapping between variations in the physical stimulus and variations in the desired perceptual response or internal representation, that is, the mode of analog as opposed to categorical perception. This section will consider two aspects of analog perception: the perception of texture and the implications of analog perception and representation to the design of displays.

Perception of Texture

Most of the surfaces that occur within environments in which we must navigate and negotiate possess some texture. In tasks such as aviation in which the pilot's negotiation within airspace must be precise, the angle of orientation or aspect of the surface can provide very important cues. For example, the textural gradient is one cue available to the pilot concerning whether or not the plane is on the correct angle of approach to landing (Gibson, 1950). The term *textural gradient* refers to the change in texture grain per unit of visual angle that the pilot perceives in the ground surrounding the approach (Warren & Owen, 1981). This is a classic example of analog perception since changes in gradient of any magnitude ideally give rise to changes in the perceived state of the aircraft of corresponding magnitude. Physical stimulus changes become increasingly more relevant for action as their magnitude increases.

The perception of texture and textural gradients is a natural and automatic process that rarely represents a bottleneck of human performance. It is important to study texture processing in engineering psychology so that the system designer can build artificial displays of texture that preserve the necessary features and allow operators to perceive and respond to textural gradients in an unhindered "natural" fashion. If the mechanisms, or "features," of texture perception can be discovered, then they can be incorporated into the displays of aircraft simulators or remote-viewing visual terminals in an economical manner that provides sufficient information for the task at hand.

There is a good deal of evidence that texture is not perceived in the same way that letters are perceived—that is, by analyzing in detail each feature that makes up the textured surface in an analytic fashion employing curves, angles, and line detectors. Such a perceptual operation would be terribly time consuming and would impose an unnecessary burden on perceptual analysis (Welford, 1976). There is, however, emerging evidence that texture can be represented by a relatively small number of *spatial* features and may, in fact, be perceived analytically in terms of these spatial features, because the high degree of redundancy in texture allows much of its information content to be captured by just a few levels of *spatial frequency* (Harvey & Gervais, 1978; Sekular, 1974).

Spatial frequency analysis.   Spatial frequency is the analog of temporal frequency as discussed in the previous chapter supplement. However, spatial frequency is expressed in terms of cycles of brightness per unit of visual angle rather than cycles of pressure per unit time. Thus a row of pillars on a building viewed from close proximity provides a stimulus of low spatial frequency. The teeth of a comb provides one of high frequency.

Spatial frequency analyzers seem to operate in parallel with, and independent from, the analyzers of discrete features such as lines, curves, and angles. As noted in Chapter 4, Broadbent and Broadbent (1977, 1980) and Navon (1977) suggest that such analysis forms the basis of a very early "global" processing of word shape that is performed independently of the more detailed analysis at the feature level. Sergent (1982) suggests that the perceptual analysis of high and low spatial frequencies may to some extent be hemispherically lateralized.

High frequencies, those required for the analyses of fine details, are processed more efficiently by the left cerebral hemisphere, while low frequencies, those involved in the perception of global shapes, are processed more efficiently in the right hemisphere. Ginsburg (1971) also argues that spatial frequency analysis represents a major component of our ability to resolve individual targets and letters. He points out that tests of visual acuity assess only a small portion of an operator's true visual capabilities (and limitations), because these are sensitive to only the highest spatial frequency range (resolving gaps of very small visual angle). Yet degradation in operator performance may occur from a poor resolution of spatial frequency at the lower frequency ranges.

Spatial frequency and textural perception.   Richards (1978) has studied the contribution of spatial frequency analysis to texture perception. He asked subjects to compare synthetically generated textures consisting of a large number of spatial frequencies (and thereby reproducing "typical" textures viewed in the real world) with patterns of textures containing just four different spatial frequencies in each of four selected orientations. His objective was to determine if textures with the four-featured set would be produced that were perceptually equivalent to the more complex stimuli. If so, the former stimuli could lead to a great deal of economy in terms of the amount of information required to generate real-looking synthetic textures. By selecting the appropriate values of these four frequencies on each of the four appropriately chosen orientations, Richards found that textures could be synthesized that were perceptually indistinguishable from the real-world textures.

Richards' results with texture perception are analogous to results in color perception. Here the data suggest that appropriate levels of intensity of three primary colors can be combined to create a perceived hue that is indistinguishable from any wavelength (or mixture of wavelengths) of light stimuli occurring in natural visual experience. These color-perception data are important because of the savings that can be achieved when color displays (e.g., color televisions) are intended to convey high-fidelity color information. The data on color mixing in the visual system imply that this economy can be achieved by employing only three color generators in the screen. With regard to texture perception, an analogous situation is encountered. An issue in many information displays of real-world scenes (e.g., computer-generated landing displays of aircraft) is how to create realistic representations of texture at different orientations without consuming the entire bandwidth, or information-transmission capacity, of the display device. Richards' data suggest that this representation can indeed be achieved with considerable economy (approximately eight bits of information) by utilizing the "natural" spatial-frequency channels of the visual system.

Richards' data also illustrate a more general issue in the human factors of display technology that concerns the tradeoff between fidelity and economy. This tradeoff was discussed in Chapter 4 in the context of the speech distortions imposed by various techniques for reducing the amount of information in a channel. With regard to visual displays, fidelity (to the real scene that is to be displayed) obviously requires that as accurate a representation of the information as possible be displayed. This criterion usually dictates in turn a high information content, which often can be achieved only by adding cost or weight to the display generator, clearly undesirable options from the system designer's point of view. It is for this reason that any data bearing on reducing information content with minimum loss of display quality is of great importance. Such information may be obtained from the kind of theoretical investigation concerning the nature of human sensory channels that was undertaken by Richards.

An investigation by Sheridan and Ranadive (1981) on the performance of operators of remote undersea manipulators provides another example of the tradeoffs between information quantity and spatial-display quality. Because of the hazards of undersea work, it is preferable for operators to remain on the surface and control the remotely manipulated robots through feedback displayed on a television picture. The question at issue is the resolution (maximum spatial frequency) of the television display. Greater resolution yields a higher quality picture and generally more accurate control. Yet greater resolution also dictates an increase in the bandwidth of the display channel that tethers the undersea camera to the remote viewing screen. This, in turn, increases the cable diameter and the resulting drag on the remote system, thereby hampering its mobility.

Sheridan and Ranadive examined the loss of display quality induced by degrading three information parameters: spatial resolution, number of levels of gray scale, and display update rate. At issue was which of the three parameters, each influencing the bandwidth of the signal, could be degraded most and still preserve the greatest level of overall display quality. The ordering obtained suggested that lowering the update rate was least disruptive, decreasing the number of gray scales yielded an intermediate level of performance, and decreasing spatial frequency resolution was the most degrading.

In summary, to reiterate a point made in Chapter 1, one important goal of engineering psychology should be to obtain data banks representing these tradeoffs between display variables, so that system designers can select beforehand the display parameters that may be sacrificed or shortchanged in the interest of economy with minimum cost to system performance. The research of Richards on texture perception and of Sheridan and Ranadive on display resolution provide prototypical examples of these kinds of data.

Display Compatibility

The fact that humans perceive and represent the world in analog as well as symbolic fashion has important implications for display design. For example, there is good evidence that analog, continuously changing systems have continuous analog mental representations. These internal representations form the basis for understanding the system, predicting its future behavior, and controlling its actions. As a consequence, there exist three "levels of representation" which must be considered in designing display interfaces: (1) the physical system itself (which *is* analog), (2) the internal representation (which should be analog), and (3) the critical interface between these two, the display surface upon which changes in the system are presented to "update" the operator's internal model and to form the basis for control action and decision. It is important to maintain a high degree of congruence or *compatibility* between these three representations. Compatibility between the real system and the mental representation is clearly a matter of training (but may be influenced by ability differences as well). If an operator correctly understands the nature of the system dynamics, then the internal representations will be analog. If both the physical system and the mental representation are analog, then it is important that the third element—the display—be formatted in a manner that will be compatible with the other two. The following sections will discuss the issue of compatibility between display representation and mental representation. The related concept of compatibility between display and response, the issue of stimulus-response compatibility, will be discussed in Chapter 9.

It is important to distinguish two major components of display compatibility: static aspects of orientation and dynamic properties of motion. As we shall see, the two interact in terms of their implications for analog display design. However, in terms of processing mechanisms, there appears to be good evidence that separate perceptual channels exist to encode position and velocity information (Sekular, 1974; Sekular, Pantle & Levinson, 1978). Lappin, Bell, Harm, and Kottas (1975) obtained experimental evidence that subjects could extract information about the velocity of a moving stimulus that was independent of the two dimensions, distance and time, that combine to create that velocity. The static and dynamic components of compatibility will now be considered in detail.

The static component: Analog compatibility of orientation.   In the context of aviation displays, Roscoe (1968) has dealt with the issue of display compatibility at some length as an example of what he refers to as "the principle of pictorial realism." The representation of aircraft altitude is a typical instance. Altitude physically is an analog quantity. Conceptually, it is also represented by the pilot in analog form with large changes in altitude more "important" than small changes. Therefore, to achieve compatibility, a display of altitude (i.e., an altimeter) should also be of analog (moving needle) rather than digital format. The transformation of symbolic digital information to the analog conceptual representation imposes an extra processing step, which will lead to longer processing time or a greater probability of error (Grether, 1949).

There are of course other factors that influence the choice of analog or digital representations of altitude or of other continuously varying quantities. For example, a requirement to read the absolute value of the indicator with high precision would favor the digital format. On the other hand, the need to perceive rate-of-change information or to estimate the magnitude of the variable when it is rapidly changing favors the analog format.

While altitude represents an obvious example of an analog quantity with analog representation, numerous other variables may be identified whose internal representations are probably also analog (e.g., temperature, pressure, direction), as well as those conceptual dimensions for which the nature of internal representation is less well specified (degree of danger, readiness status). Basic research in engineering psychology is presented with the challenge to identify the nature of representation of these continua and so to indicate how important it is that their display representation should be analog. The issue of analog versus digital displays will be considered in greater detail in Chapter 12.

Displaying information in analog format that is mentally represented in analog fashion is a necessary but not sufficient condition to assure maximum compatibility. Since analog continua necessarily are associated with some physical ordering, it is essential also that the orientation and ordering of the display be compatible with the ordering of the mental representation. This compatibility defines another aspect of Roscoe's (1968) principle of pictorial realism. Display compatibility may be violated in shapes, for example, if a circular altimeter (pointer or dial) represents the linear conception of altitude (Grether, 1949) or if a linear display indicates compass heading, which is physically and conceptually circular.

Display compatibility may also be violated in direction. In the simple case of the altimeter, our "mental model" of altitude represents high altitude as "up" and low altitude as "down." Therefore, the altimeter also should present high altitudes at the top of the scale and low ones at the bottom rather than the reverse arrangement or a horizontal one. Similarly, high temperatures should be higher, and low temperatures lower. Displayed quantities, in short, should correspond with the operator's internal model of those quantities. Finally, compatibility may be violated by dissecting and displaying in separate parts a variable that is unitary. Grether (1949), for example, reports that operators have a more difficult time extracting altitude information from three concentric pointers of a rotating display (indicating units of 100, 1,000, and 10,000 feet) than from a single pointer.

Compatibility of display movement.   Roscoe (1968) and Roscoe et al. (1981) propose *the principle of the moving part* as a guideline dictating that the direction of movement of an indicator on a display should be compatible with the direction of movement of an operator's internal representation of the variable whose change is indicated. In the case of the household thermometer, this principle is typically adhered to, because a rise in the height of the mercury column indicates a rise in temperature. There are, however, circumstances in which the principle

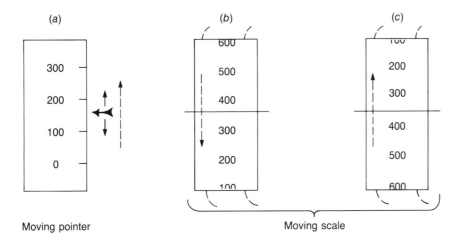

Figure 5.5 Display movement: *(a)* moving-pointer altimeter; *(b* and *c)* moving-scale or fixed-pointer altimeters. The dashed arrows show the direction of display movement to indicate an increase in altitude.

of the moving part and the compatibility of orientation operate in opposition to each other, and so one or the other must be violated. This occurs in so-called fixed-pointer/moving-scale indicators.

An example of this violation is the altimeter. Fixed-pointer altimeters, along with their opposite, the moving-pointer display, are shown in Figure 5.5. In the moving-pointer display (Figure 5.5*a*) both the principles of the moving part and compatibility of orientation are satisfied. High altitude is at the top, and an increase in altitude is indicated by an upward movement of the moving element on the display. Neither static nor motion compatibility is violated. However, in the moving-scale display, in order to print the numbers in such a way that high altitudes are at the top (Figure 5.5*b*), the scale must move *downward* to indicate an *increase* in altitude, in violation of the principle of the moving part. If the labeling is reversed to conform to the principle of the moving part (Figure 5.5*c*), this change will reverse the orientation and display high altitudes at the bottom. A final disadvantage with both the moving-scale displays is that, like a digital display, scale values become difficult to read when the variable is changing rapidly, since the digits themselves are moving.

The frequency-separated display.   An important design innovation, developed in an effort to adhere to both the static and motion-based aspects of compatibility, was described by Roscoe (1968, 1981; Roscoe et al., 1981) as a technique to display the bank angle, or attitude of an aircraft to pilots. Two conventional indicators of attitude are shown in Figure 5.6. The "outside-in" bird's-eye, or ground-referenced, display indicates the plane as it would appear to an outsider looking at

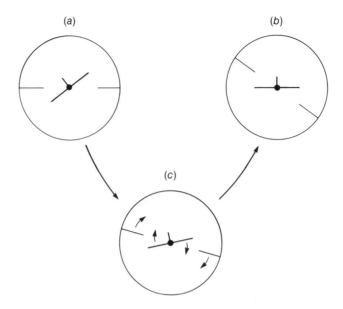

Figure 5.6 (a) Outside-in, (b) inside-out, and (c) frequency-separated display. All displays show an aircraft banking left. Low-frequency return to steady state is indicated by arrows.

the plane (Figure 5.6a). When the plane banks to the left, the display indicator also rotates left, and so the principle of the moving part is confirmed. On the other hand, the display intrinsically provides a frame of reference (a view from outside the aircraft) that is inconsistent with the pilot's actual frame of reference (inside the aircraft). Thus the static picture that the display provides, a horizontal horizon and tilted airplane, is incompatible with what the pilot actually sees out of the cockpit window: a tilted horizon and an aircraft that from the pilot's frame of reference is horizontal.

In contrast to the bird's-eye view which violates compatibility of orientation, the congruent correspondence of frames of reference is given by the "inside-out," moving-horizon, or pilot's-eye display shown in Figure 5.6b. However, this display now violates the principle of the moving part. For example, banking the aircraft to the left, which will generate a leftward rotation of the pilot's internal model, produces a rightward rotation of the moving element on the display (the horizon).

To achieve a compromise between the static and dynamic aspects of the display, Fogel (1959) proposed the concept of *frequency-separated display* (Johnson & Roscoe, 1972; Roscoe, 1968). In this display, rapid movement of the controls (and thus of the aircraft) will induce an outside-in (moving-aircraft) change on the display as in Figure 5.6a, so that display and aircraft *motion* are compatible when conditions are dynamic. When, however, the pilot enters into a gradual turn, the bank angle is held constant for some period of time, and perception is thereby static. Under these conditions the horizon and the plane both slowly rotate to an inside-out format, so that the frame of reference is now equated between the two. This progression is shown by the transition in Figure 5.6 from *a* to *c* to *b*. Thus, at high frequencies, when movement and motion perception are most dominant, the principle of the moving part is adhered to. At low frequencies of motion, to which human's velocity sensors are less sensitive, the static principle of compatibility of frame of reference is followed. An evaluation of this display by Roscoe and Williges (1975) indicated it to be more effective than either the inside-out or the outside-in display in terms of the accuracy of the pilot's control and the number of inadvertent control reversals that were made when flying.

In summary, the principle of compatibility, which will be discussed again in the context of responses in Chapter 9, is probably one of the most important guidelines in the engineering psychology of display design. Compatible displays are read more rapidly and accurately than incompatible ones under normal conditions. More important, their advantages increase under conditions of stress. Yet the nature of the internal mental models of complex systems such as computers or chemical processes which guide compatibility decisions must be better understood. Until these are carefully specified and the extent of individual differences in the format of representation become better known, it will often be difficult to translate the generic principles of display compatibility to the specific instances of display design.

SPACE PERCEPTION, MAPS, AND NAVIGATION

The use of maps in orientation and navigation integrates many of the different areas of knowledge discussed in this chapter. The information of concern (geographical) is clearly spatial, and the relevance of this information to the operator's behavior is normally defined in terms of spatial analog operations (movement and localization). Therefore, it is reasonable to assume that many aspects of the internal representation of geographical information are spatial and analog as well. In support of this assertion, Thorndyke and Stasz (1979) found that individuals with a higher spatial ability as measured on standardized paper and pencil tests performed significantly better on a map-learning task than did those with low spatial ability but an equivalent level of general intelligence.

The human factors problems associated with maps and navigation should be self-evident to anyone who has ever encountered the following circumstances: (1) driving through a complicated series of intersections, heading in a generally southerly direction while navigating from a conventional north-up road map; (2) following a list of instructions on how to get somewhere (". . . go south two blocks and then turn right . . .") and missing a turn; and (3) having carefully studied the map of a strange place, suddenly finding oneself unable to locate one's position within the region depicted by the map because the surrounding landmarks are all unfamiliar. We shall consider some of the particular characteristics of human spatial perception that lead to difficulties of these sorts.

Representation of Navigational Information

Learning.   Thorndyke (1980) proposes that as we become increasingly familiar with a geographical environment, such as a city to which we have just moved, the nature of our knowledge of that environment undergoes qualitative as well as quantitative changes. The qualitative changes are characterized by a progression through three levels of knowledge. Initially our representation is characterized by *landmark knowledge*. We orient ourselves exclusively by highly salient visual landmarks (statues, building, etc.), and so our knowledge may be characterized in large part by direct visual images of those features. The prominence of landmark knowledge at early stages of navigational understanding of an environment suggests the importance of including salient landmarks at intermittent locations in the design and planning of any city or neighborhood. These provide the skeletal frame of reference around which to build the two subsequent phases of learning: route knowledge and survey knowledge (Anderson, 1979).

Landmark knowledge is followed by *route knowledge*. Here, understanding is characterized by the ability to *navigate* from one spot to another, utilizing landmarks or other visual features to trigger the decisions to turn left, turn right, or continue straight at a given intersection. If we are able to recall these features, then we may offer verbal directions to someone else on how to navigate the route ("turn left when you get to the church"). If, however, we can only recognize them, then we may navigate ourselves but cannot give directions to others ("I can take you there, but I can't explain how to get there"). In this case, the landmarks must be perceived directly in order to trigger the action decisions. While route knowledge is based upon recognition of the visual features and therefore possesses spatial elements, it does contain a verbal component in the somewhat categorical statements of action (turn left, turn right). Also, route knowledge is knowledge from a uniquely ego-centered frame of reference.

Sufficient navigational experience eventually provides us with what Thorndyke refers to as *survey knowledge*. Here, our knowledge resides in the form of an internalized "cognitive map" (Tolman, 1948), the analog to the true physical map. Our ability to describe the relative location of two landmarks in a city, even though we may never have traveled a route that connects these landmarks, offers convincing evidence that survey knowledge can be acquired from navigation. Survey knowledge unlike route knowledge is based upon a world frame of reference.

Survey and route knowledge.   Because landmark knowledge by itself gives us little more than a primitive representation of the environment, our primary navigational abilities seem to reside in route and survey knowledge. The differences between these two have important implications for human performance and system design. Thorndyke's (1980) research has suggested that beyond describing two phases of navigational learning, landmark and survey knowledge forms may be contrasted in three other ways: (1) As we noted briefly, the two knowledges differ in their frame of reference. (2) Each is optimally suited for different kinds of geographical tasks. (3) Each may be acquired independently of the other through different training techniques, and possibly each is associated with individual differences in ability. The interactions between these contrasts will now be discussed in greater detail.

Route knowledge shares certain properties with the "inside-out" display described in the preceding section. The frame of reference in route knowledge is one's own position in the world and thus corresponds directly to what one sees as one follows a route. For example, if your route knowledge says "turn left" at a given intersection, you will always turn to *your* left, and the compass direction you may be facing makes no difference. On the other hand, survey knowledge is more like the outside-in display, since the frame of reference of knowledge (the internal map) is independent of the particular view that one has of the environment. As an example of the contrast between these frames of reference, if you are standing at an intersection, the instruction "turn right" (a command based on route knowledge) will lead to different actions from the perspective of an outside observer depending upon whether you are facing north or south. "Turn westward" (a command based on survey knowledge) will lead to the same ultimate action, independent of the initial orientation.

As Thorndyke (1980) has proposed, a logical consequence of this difference is that possession of route knowledge is optimal for judgments made from one's own frame of reference. These would include such tasks as pointing the direction to a given landmark not now visible (orientation) or judging the actual distance that must be traveled between two points and of course actually navigating that route. In contrast, individuals possessing survey knowledge should be relatively poor at these tasks but better at tasks requiring an independent, world frame of reference. These tasks, for example, might include estimating the relative direction between two different landmarks, estimating the euclidean distance (as opposed to the walking distance) that separates them, or identifying the absolute location of a single landmark on a map (object localization).

In terms of training, Thorndyke argues intuitively that route knowledge will be obtained best from direct navigation through an area, whereas survey knowledge will be most directly obtained by studying maps (as indicated above, however, survey knowledge will eventually be acquired from navigation as well). To test these two hypotheses, Thorndyke and Hayes-Roth (1978) compared performance between two groups of subjects on the orienting and distance-estimation tasks described above. One group had acquired knowledge of the geography of the Rand Corporation building through extensive navigation training (route knowledge). The other group acquired knowledge by map study (survey knowledge). Thorndyke and Hayes-Roth's results generally confirmed the predictions. At moderate levels of training, map-learning subjects showed better estimates of euclidian distance than of route distance and better judgments of object localization than of orientation. Subjects trained by navigation showed the reverse effect in both cases.

Further training of both groups, however, revealed an asymmetry of transfer. With more extensive learning, the subjects trained by navigation developed survey knowledge and eventually performed as well as or better than the map-learning subjects on all tasks. This finding is consistent with the order of the learning progression proposed by Thorndyke (1980). The map-learning subjects, on the other hand, failed to improve on the route-estimation and orientation tasks with more extensive map training. There is a certain analogy here to the discussion of schema formation earlier in the chapter. Acquiring survey knowledge through navigation is akin to learning category properties by repeated exposure to its exemplars. Acquiring survey knowledge through map study is more like what happens when one studies a list of the defining properties of the category.

The progression from route to survey knowledge with training suggests that our internal model slowly progresses from an inside-out, context-dependent representation to an outside-in, context-free representation. This progression is seemingly substantiated by the observation that experts in map reading tend to orient maps in a "north-up" direction (world frame of reference), while novices tend to rotate the map according to their own frame of reference (e.g., "up" in the direction they are heading) (Anderson, 1979).

These results have some important implications to training operators in tasks in which different kinds of spatial judgments must be made. Depending upon whether the task to be performed requires relative object localization from a neutral reference or navigation and orientation from one's own reference, different kinds of training should be employed. In particular, the results suggest that extensive map study may not be terribly effective in preparation for a task in which one must navigate through a strange environment. Following the analogy to schema learning, this is like having to recognize category instances after having been exposed only to the defining category rules. A more effective training procedure would be provided by an "inside-out" experience of the environment in which the operator actually "navigates" through videotapes or even views a highly abstracted movie that indicates the twists, turns, and landmarks to be encountered in navigation.

One relatively obvious advantage to survey knowledge (and map instructions rather than verbal-list instructions) concerns the consequence of errors. When a route list has guided navigation and one is found off course as the result of a wrong turn or an intentional side-track, the information provided by route knowledge, perfect while on course, suddenly becomes nearly meaningless. Survey knowledge on the other hand, may be used to guide one back either to the required route or to the desired goal by a different route. In a sense, the tradeoff between the two forms of knowledge seems to be one between automaticity and cognitive simplicity, on the one hand, favoring route lists, and flexibility and generality on the other, favoring survey knowledge and map displays. The flexibility provided by survey knowledge will assist the lost traveler. The automaticity of route knowledge will be useless. Where the likelihood or potential costs of errors are greater, survey knowledge becomes preferable.

Finally, some effects of individual differences in route versus survey knowledge may be suggested. As noted above, Thorndyke and Stasz (1979) found that high spatial ability facilitated map learning. One might anticipate, furthermore, that high verbal ability would be relatively more of a help (and low spatial ability less of a hindrance) in acquiring route knowledge from navigation. Given that individuals do clearly differ in both spatial and verbal ability, it would appear that redundant presentation of both kinds of information—for example, in maps of how to get from one place to another—would provide each group (high spatials and high verbals) with the kind of navigational information most appropriate to their preferred representation. The notion of redundancy in display format is an issue that will become relevant again in our discussion of spatial and verbal working memory as applied to human-computer interaction in Chapter 6.

Maps versus route lists.   Information presented to travelers on commercial transportation systems may be in the format of either a linear listing of the bus or subway stops (route list) or an actual map (sometimes highly schematized) showing the spatial relations between the stops. Bartram (1980) asked subjects to select the combination of bus routes that they would need to take in order to navigate between two selected locations on the London bus system. The information was presented either in a route list (the sequential listing of all bus stops on a route, showing stops that were common between lines) or in the form of a map (either schematic or accurate). Bartram found that decision times were considerably more rapid in the map condition and were uninfluenced by decision complexity. In contrast, the time required to make slower route-list decisions increased with complexity. These results are consistent with the conclusions drawn in the preceding discussion, since the judgment required in the task is one requiring survey knowledge (e.g., the chosen route is constant and independent of the subject's frame of reference). In this context-free situation the data of Thorndyke and Hayes-Roth suggest that the map display should be superior.

A second contrast between the two forms of geographical representation in terms of the predominant code of processing is illustrated in an experiment performed by Wetherell (1979) in which the two means of providing navigational instructions were compared. Wetherell had subjects learn the directions to navigate between two points in a driving "maze" using one of two techniques: either learning a linear list of turns (route knowledge) or studying a spatial map of how to get from *A* to *B* (map, or survey, knowledge). Subjects then actually drove the route (on an unfamiliar terrain). Subjects who were trained in the map condition made many more errors than those trained in the route-list condition. Through subsequent experimentation, Wetherell identified two contributing factors to the observed difference: (1) The spatial processing demands of actually driving, seeing, and orienting interfered with the spatial demands of maintaining the mental map in working memory in the survey-knowledge condition. This kind of interference between two spatial tasks will be discussed in greater detail in Chapters 6 and 8. (2) Subjects had great difficulty reorienting the map, which was learned in a north-up direction, to their own subjective frame of reference as they approached any intersection heading east, south, or west. Their survey knowledge was not what was required for the navigation task at hand.

The latter finding reiterates a point made when discussing the compatibility of frames of reference in visual displays. Route knowledge, because of its more verbal symbolic characteristic, is less sensitive to an incompatibility of orientation. The analog survey knowledge, at least as possessed by novices, is more adversely affected by the reversed frame of reference that can occur when navigating.

Frames of reference in flight navigation displays. The difference in frames of reference between earth-centered maps and ego-centered route lists has important consequences to aerial navigation aids or "horizontal situation displays" in aircraft (Roscoe, 1981). When a pilot views a map showing her aircraft position relative to a course to follow, should the map be in a fixed, north-up frame of reference with the aircraft rotating (earth frame of reference), or should the aircraft's heading always be up, with the map rotating underneath (ego frame of reference)? This contrast is quite analogous to the comparison of the outside-in, bird's-eye display with the inside-out, pilot's-eye display of aircraft-attitude information discussed earlier, where it was suggested that there were advantages to each and that the principle of frequency separation could provide them.

The situation with regard to horizontal situation displays is somewhat similar. Here too we find relative advantages to both frames of reference, some being better suited for certain parts of the flight task and others suited to others. The result is that there does not appear to be a great deal of difference between the two as measured by flight performance or by subjective preferences (Baty, 1976; Baty, Wempe, & Huff, 1974). As with attitude control, the principle of the moving part operates on heading control as well and suggests that the initial turn of the aircraft should be reflected by rotation of the aircraft symbol in the direction of the turn rather than rotation of the map in the opposite direction (Payne, 1952; Roscoe et al., 1981). Thus adherence to the principle of the moving part suggests a fixed-map display. The earth-referenced, fixed-map, outside-in display is also favored when pilots must solve navigational problems and plan alternate routes while interacting with air traffic control. Here the need for a common frame of reference between the two communicating parties is obvious, and only the earth frame of reference can provide this since the ego frame is always changing. Hart and Wempe (1979) report that pilots request that the fixed-map display be an available option when flying with cockpit displays of terminal traffic information (Hart & Loomis, 1980).

The primary difficulty with the north-up, earth-referenced, fixed-map display occurs when pilots fly in a generally southerly direction. Then there is a basic motion incompatibility between lateral movement of the flight controls and eventual movement of the aircraft (Baty et al., 1974). This disadvantage is sufficiently great that pilots who are given an option to select either of the two frames of reference will choose the ego-referenced, heading-up display a majority of the time (Hart & Wempe, 1979). The selective advantage of each reference frame to different aspects of the task seems to call for some combination of the following three options: (1) a frequency-separation principle in which immediate heading changes are indicated by aircraft symbol rotations and slower ones by map rotations, (2) a configuration such as that used by Hart and Wempe (1979) in which the pilot may select different frames of reference, and (3) an outside-in motion display in which the pilot can reorient the entire display with any direction facing up whenever required (Payne, 1952). By providing such flexibility the relative merits of each frame of reference can be realized when most needed.

Distortions of Spatial Cognition

Survey knowledge of an environment, derived either from maps or from navigational experience, is clearly spatial or analog in form. However, this knowledge, like the representation of visual images, is not a veridical photographic reproduction of the map but contains some interesting systematic distortions that are manifest as we try to use the information on the map. Two of these distortions, the *filled distance effect* and the effect of *rectilinear normalization*, have possible implications to map design.

Filled distance effect.   Maps or the mental representation of imagined maps must often be scanned. Kosslyn, Ball, and Reiser (1978) demonstrated that the time taken to scan a mental image is proportional to the distance of the scan. Thorndyke (1980) pursued this line of reasoning with the corollary logic that our estimation of size (area or distance) of an image is influenced by the time taken to scan its extent. If this premise is accepted, then anything that will slow down the scan time will increase the perceived size. Applying this reasoning to the estimation of map distances, Thorndyke performed the following experiment: Subjects first memorized simple schematic maps of roads and cities. Subsequently, when subjects were asked to estimate distances between target cities on the memorized maps (scanning their spatial image), those distances were systematically overestimated to the extent that more cities occurred between the target end points. The data were modeled by a counting-timer scheme, in which perceived distance is judged proportional to the elapsed time on a mental counter, from initiation to end point. This counter is slowed by intervening cities as the scan becomes "disrupted" by the added clutter. When subjects estimated distances on a map that was directly perceived instead of imagined, the filled distance effect was also observed but was of smaller magnitude.

The implications of this research to the development of computer-generated maps are demonstrated by considering the biases that may occur when unimportant information concerning cities or other geographical features is displayed. Such information might be made callable on request but should not necessarily be permanently displayed when rapid distance estimates are required.

Rectilinear normalization.   Our navigation and orientation judgments tend to be carried out in a normalized, right-angle-grid world, with orientation of most of our maps and cities emphasizing north, south, east, and west. (This is particularly true of those of us living in flat, midwestern cities.) Within this framework, there is some evidence that judgments of north and south (or up and down) are made with greater facility than those of east and west (Maki, Maki, & Marsh, 1977). Relative to other directions of the compass, however, all four of these primary directions exert a dominating influence. For example, Loftus (1978) found that both novice subjects and skilled pilots, when asked to indicate the direction of an auditorily presented compass heading ("230°"), showed the most rapid reactions to headings that corresponded with the four major compass directions. Furthermore, subjects showed relatively faster responses to directions close to the compass headings (e.g., 85°, 182°) than to those in the middle of a quadrant (45°, 150°). These data suggest that subjects employed the four directions as major orienting "landmarks." The heard direction is interpreted initially in terms of the nearest compass point and is then modified in terms of its degree of departure from that point. The closer the point is to the middle of the quadrant (and farther from the landmark), the longer will be the contribution of the second component to the estimation time. This result thus emphasizes both the categorical role of the N-S-E-W landmarks and the analog operations of subsequent orienting from these landmarks.

The strength of the predominant N-S-E-W grid we impose upon our internal representation of geography is sufficiently great that it induces some very systematic biases in our judgments of survey knowledge. Chase and Chi (1980) refer to these biases collectively as our tendency to engage in *rectilinear normalization*, demonstrated by an experiment of Stevens and Coupe (1978) in which subjects were asked to make judgments of the relative direction of one city to another—to decide, for example, the direction of Reno, Nevada, from San Diego, or of Santiago, Chile, from New York. In these examples, the tendency is to report that Reno, Nevada, is northeast of San Diego (it is, in fact, northwest) and that Santiago, Chile, is southwest of New York (it is southeast). The erroneous responses result from applying a grid or a set of propositional statements to reconstruct the geographical location in the following fashion: "Nevada is east of California; San Diego is in southern California; Reno is in northern Nevada; therefore, Reno is northeast of San Diego." Similarly, "New York is in eastern North America; Santiago is in western South America; South America is south of North America; therefore, Santiago is southwest of New York." The logic performed in making these judgments is in fact closer to a discrete propositional logic than to one based upon visual imagery and provides another example of the close cooperative interaction between verbal and spatial processes.

In a related example, Milgram and Jodelet (1976) note that Parisians tend to "straighten" their mental representation of the flow of the Seine River through Paris, forcing it into more of an east-west linear flow than it actually possesses. This cognitive distortion into the rectilinear grid thereby induces consequent errors and distortions in their survey knowledge of the area. It is important to note, however, that distortions of this kind will not generally disrupt navigation using route knowledge. Parisians can still easily navigate along familiar routes because the navigation does not require the precise analog judgments that are distorted in their survey knowledge. City navigation instead depends upon simple categorical decisions (left-right-straight) which are perfectly compatible with the distortions imposed by rectilinear normalization.

In yet another demonstration of rectilinear normalization, Chase and Chi (1980) asked subjects to draw a map of the Carnegie-Mellon University campus from memory. In their reconstruction, subjects tended to "force" nonrectilinear intersections into a rectilinear N-S-E-W grid (see Figure 5.7). Interestingly enough, only the architects in the sample failed to impose this distortion. Finally, Howard and Kerst (1981) found that when subjects were asked to recall maps of a campus environment, they tended to force their reconstruction into clusters of buildings oriented to the four primary directions.

The above examples illustrate the manner in which errors of misjudgment might occur when people attempt to navigate without maps in environments in which the normal N-S-E-W grid pattern is not imposed. One implication of this finding is that schematic maps, which themselves often systematically distort the analog environment and force it into the rectilinear grid, would serve to reinforce this tendency, with potentially unfortunate consequences where accurate survey knowledge is required.

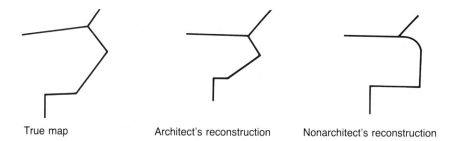

True map   Architect's reconstruction   Nonarchitect's reconstruction

Figure 5.7 Reconstruction of street map on Carnegie-Mellon University campus showing the bias due to rectilinear normalization. From *Cognitive Skill: Implications for Spatial Skill in Large-Scale Environments* (Tech. Rep. No. 1) by W. Chase and M. Chi, 1979, Pittsburgh, PA: University of Pittsburgh Learning and Development Center.

In summary, the discussion of maps and navigation reveals that our conception of the geography within which we negotiate involves a complex interaction between verbal and analog representations, between route and survey knowledge, and between inside-out and outside-in frames of reference. Which form of representation is "best" depends a good deal on what aspect of the geographic task is being performed. In general, outside-in, survey, analog conceptions are better suited for planning functions and inside-out, route conceptions better suited for actual travel. The discussion on the whole serves once again to emphasize that cooperation rather than competition seems to govern the interaction between our verbal and spatial processes.

Transition: Perception and Memory

Our discussion in the last two chapters has been presented under the categories of spatial and verbal processes in perception. Yet it is quite difficult to divorce these processes from those related to memory. There are three reasons for this close association:

1. Perceptual categorizations, as we saw, were guided by expectancy as manifest in top-down processing. Expectancy was based upon both recent experience—the active contents of working memory—and the contents of permanent or long-term memory. Indeed the rules for perceptual categorization themselves are formed only after repeated exposure to a stimulus. These exposures must be remembered in order to form the categories.

2. In many tasks when perception is not automatic, such as those related to navigation and comprehension as well as the processing of multiple cues in decision making, perceptual categorization must operate hand in hand with activities in working memory.

3. The dichotomy that distinguished codes of perceiving into spatial and verbal categories has a direct analog in terms of two codes of working memory.

It is with these associations in mind that we will turn in Chapter 6 to the topic of memory, its limitations, and its implications for engineering psychology.

References

Anderson, J. R. (1979). *Cognitive psychology.* New York: Academic Press.

Bamber, D. (1969). Reaction times and error rates for "same-different" judgments of multi-dimensional stimuli. *Perception & Psychophysics, 6,* 169–174.

Bartram, D. J. (1980). Comprehending spatial information: The relative efficiency of different methods of presenting information about bus routes. *Journal of Applied Psychology, 65,* 103–110.

Baty, D. L. (1976). Evaluating a CRT map predictor for airborne use. *IEEE Transactions on Systems, Man, & Cybernetics,* SMC-6, 209–215.

Baty, D. L., Wempe, T. E., & Huff, E. M. (1974). A study of aircraft map display location and orientation. *IEEE Transactions on Systems, Man, & Cybernetics,* SMC–4, 560–568.

Biederman, I., Mezzanotte, R. J., Rabinowitz, J. C., Francolin, C. M., & Plude, D. (1981). Detecting the unexpected in photo interpretation. *Human Factors, 23,* 153–163.

Broadbent, D., & Broadbent, M. H. (1977). General shapes and local detail in word perception. In S. Dornic (Ed.), *Attention & Performance VI.* Hillsdale, NJ: Erlbaum Associates.

Broadbent, D., & Broadbent, M. H. (1980). Priming and the passive/active model of word recognition. In R. Nickerson (Ed.), *Attention & Performance VIII.* New York: Academic Press.

Cantor, N., French, R., Smith, E. E., & Mezzich, J. (1980). Psychiatric diagnosis as prototype categorization. *Journal of Abnormal Psychology, 89* (2), 181–193.

Chase, W., & Chi, M. (1979). *Cognitive skill: Implications for spatial skill in large-scale environments* (Tech. Rep. No. 1). Pittsburgh, PA: University of Pittsburgh Learning and Development Center.

Cooper, L. (1976). Individual differences in visual comparison processes. *Perception & Psychophysics, 19,* 433–444.

Cooper, L. (1980). Visual information processing. In R. Nickerson (Ed.), *Attention and Performance VIII.* Hillsdale, NJ: Erlbaum Associates.

Dukes, W. F., & Bevon, W. (1967). Stimulus variation and repetition in the acquisition of naming responses. *Journal of Experimental Psychology, 74,* 178–181.

Dupree, D. A., & Wickens, C. D. (1982). Individual differences and stimulus discriminability in visual comparison reaction time. In R. E. Edwards (Ed.), *Proceedings of the 26th annual meeting of the Human Factors Society.* Santa Monica, CA: Human Factors.

Ells, J. G., & Dewar, R. E. (1978). Rapid comprehension of verbal and symbolic traffic sign messages. *Human Factors, 21,* 161–168.

Eriksen, C. W., and Hake, H. N. (1955). Absolute judgments as a function of stimulus range and number of stimulus and response categories. *Journal of Experimental Psychology, 49,* 323–332.

Fogel, L. J. (1959). A new concept: The kinalog display system. *Human Factors, 1,* 30–37.

Garner, W. R. (1974). *The processing of information and structure.* Hillsdale, NJ: Erlbaum Associates.

Garner, W. R., & Fefoldy, G. L. (1970). Integrality of stimulus dimensions in various types of information processing. *Cognitive Psychology, 1,* 225–241.

Gibson, J. J. (1950). *The ecological approach to visual perceptions.* Boston: Houghton Mifflin.

Ginsburg, A. P. (1971, March). Psychological correlates of a model of the human visual system. Masters thesis GE/EE/715-2, Air Force Institute of Technology, Wright Patterson Air Force Base, Dayton, Ohio.

Goodenough, D. R. (1976). A review of individual differences in field dependence as a factor in auto safety. *Human Factors, 18,* 53–62.

Green, P., & Pew, R. W. (1978). Evaluating pictographic symbols: An automotive application. *Human Factors, 20,* 103–114.

Grether, W. F. (1949). Instrument reading I: The design of long-scale indicators for speed and accuracy of quantitative readings. *Journal of Applied Psychology, 33*, 363–372.

Hahn, G., Morgan, C., & Lorensen, W. E. (1983, January). Colorface plots for displaying product performance. *Datamation*, pp. 23–29.

Hart, S. G., & Loomis, L. L. (1980). Evaluation of the potential format and content of a cockpit display of traffic information. *Human Factors, 22*, 591–604.

Hart, S. G., & Wempe, T. E. (1979, August). *Cockpit display of traffic information: Airline pilot's opinion about content, symbology, and format* (NASA Technical Memorandum No. 78601). Moffett Field, CA: NASA Ames Research Center.

Harvey, L. O., & Gervais, M. J. (1978). Visual texture perception and Fourier analysis. *Perception & Psychophysics, 24*, 534–542.

Howard, J. H., & Kerst, R. C. (1981). Memory and perception of cartographic information for familiar and unfamiliar environments. *Human Factors, 23*, 495–504.

Jacob, R. J. K., Egeth, H. E., & Bevon, W. (1976). The face as a data display. *Human Factors, 18*, 189–200.

Johnson, S. L., & Roscoe, S. N. (1972). What moves, the airplane or the world? *Human Factors, 14*, 107–129.

Kahneman, D., & Treisman, A. (in press). Changing views of attention and automaticity. In R. Parasuraman & R. Davies (Eds.), *Varieties of attention*. New York: Academic Press.

Kosslyn, S. M., Ball, T. M., & Reiser, B. J. (1978). Visual images preserve metric spatial information: Evidence from studies of image scanning. *Journal of Experimental Psychology: Human Perception and Performance, 4*, 47–60.

Lappin, J. S., Bell, H. H., Harm, O. O., & Kottas, B. (1975). On the relation between time and space in the visual discrimination of velocity. *Journal of Experimental Psychology: Human Perception and Performance, 1*, 383–394.

Levy, J. (1974). Psychobiological implications of bilateral asymmetry. In S. Dimond & G. Beaumont (Eds.), *Hemispheric function in the human brain*. London: Elek.

Lockhead, G. R. (1972). Processing dimensional stimuli. *Psychological Review, 79*, 410–419.

Lockhead, G. R. (1979). Holistic vs. analytic processing models: A reply. *Journal of Experimental Psychology: Human Perception and Performance, 5*, 740–755.

Loftus, G. R. (1978). Comprehending compass directions. *Memory & Cognition, 6*, 416–422.

Lohman, D. F. (1979). *Spatial ability: A review and reanalysis of the correlational literature* (Aptitude Research Project T.R. No. 8). Palo Alto, CA: Stanford University, School of Education.

Maki, R. H., Maki, W. S., & Marsh, L. G. (1977). Processing locational and orientational information. *Memory & Cognition, 5*, 602–612.

Medin, D. L., & Schaffer, M. M. (1978). Context theory of classification learning. *Psychological Review, 85*, 207–238.

Milgram, S., & Jodelet, D. (1976). Psychological maps of Paris. In H. M. Proshansky, W. H. Itelson, & L. G. Revlin, *Environmental psychology*. New York: Holt, Rinehart & Winston.

Miller, G., and Isard, S. (1963). Some perceptual consequences of linguistic rules. *Journal of Verbal Learning and Verbal Behavior, 2*, 217–228.

Moray, N. (1981). The role of attention in the detection of errors and the diagnosis of errors in man-machine systems. In J. Rasmussen and W. Rouse (Eds.), *Human Detection and Diagnosis of System Failures*. New York: Plenum Press.

Moriarity, S. (1979). Communicating financial information through multidimensional graphics. *Journal of Accounting Research, 17*, 205–224.

Navon, D. (1977). Forest before trees: The presence of global features in visual perception. *Cognitive Psychology, 9*, 353–383.

Palmer, S. E. (1975). The effects of contextual scenes on the identification of objects. *Memory & Cognition, 3*, 519–526.

Payne, T. A. (1952, November). *A study of the moving part, heading presentation, and map detail on pictorial air navigation displays* (Human Engineering Report SPEC-DEVCEN 71-16-10). Washington, DC: Office of Naval Research Special Devices Center.

Posner, M. I. (1969). Abstraction and the process of recognition. In G. H. Bower (Ed.), *The psychology of learning and motivation III*. New York: Academic Press.

Posner, M. I. (1973). *Cognition: An introduction*. Glenview, IL: Scott, Foresman.

Posner, M. I., & Keele, S. W. (1968). On the genesis of abstract ideas. *Journal of Experimental Psychology, 77*, 353–363.

Posner, M. I., & Keele, S. W. (1970). Retention of abstract ideas. *Journal of Experimental Psychology, 83*, 304–308.

Potter, M. C., & Faulconer, B. A. (1975). Time to understand pictures and words. *Nature, 253*, 437–438.

Potter, M. C., Kroll, J. F., & Harris, C. (1980). Comprehension and memory in rapid sequential reading. In R. S. Nickerson (Ed.), *Attention and Performance VIII*. New York: Academic Press.

Richards, W. (1978, January). *Experiments in Texture Perception* (Final Contract Report AFOSR Contract F44-62C-74-C-0076). Cambridge, MA: MIT Department of Psychology.

Rock, I. (1975). *An introduction to perception*. New York: Macmillan.

Roscoe, S. N. (1968). Airborne displays for flight and navigation. *Human Factors, 10*, 321–332.

Roscoe, S. N. (1981). *Aviation psychology*. Iowa City: University of Iowa Press.

Roscoe, S. N., Corl, L., & Jensen, R. S. (1981). Flight display dynamics revisited. *Human Factors, 23*, 341–353.

Roscoe, S. N., & Williges, R. C. (1975). Motion relationships and aircraft attitude guidance displays: A flight experiment. *Human Factors, 17*, 374–387.

Sekular, R. (1974). Spatial vision. *Annual Review of Psychology, 25*, 195–232.

Sekular, R., Pantle, A., & Levinson, E. (1978). Physiological basis of motion perception. In R. Held, H. Leibowitz, & H. Teuber (Eds.), *Handbook of sensory physiology*. Berlin: Springer.

Sergent, T. (1982). The cerebral balance of power: Confrontation or cooperation? *Journal of Experimental Psychology: Human Perception and Performance, 8* (2), 253–272.

Sheridan, T., & Ranadive, J. (1981). Video frame rate, resolution and gray scale trade-offs for undersea teleoperator control. In J. Lyman & T. Bejczy (Eds.), *Proceedings of the 17th Annual Conference on Manual Control* (JPL Publication 81–95). Pasadena, CA: Jet Propulsion Laboratory.

Snow, R. E. (1980, September). *Aptitudes and instructional methods* (Final Report, Aptitudes Research Project). Palo Alto, CA: Stanford University, School of Education.

Stevens, A., & Coupe, P. (1978). Distortions in judged spatial relations. *Cognitive Psychology, 10*, 422–437.

Thorndyke, P. W. (1980, December). *Performance models for spatial and locational cognition* (Technical Report R-2676-ONR). Washington, DC: Rand Corporation.

Thorndyke, P. W., & Hayes-Roth, B. (1978, November). *Spatial knowledge acquisition from maps and navigation*. Paper presented at the meetings of the Psychonomic Society, San Antonio, TX.

Thorndyke, P. W., & Stasz, C. (1979, January). *Individual differences in knowledge acquisition from maps and navigation*. (Technical Report R-2374-ONR). Washington, DC: Rand Corporation.

Tolman, E. C. (1948). Cognitive maps in rats and men. *Psychological Review, 55*, 189–208.

Treisman, A., & Gelade, G. (1980). A feature-integration theory of attention. *Cognitive Psychology, 12*, 97–136.

Tulving, E., Mandler, G., and Baumal, R. (1964). Interaction of two sources of information in tachistoscopic word recognition. *Canadian Journal of Psychology, 18*, 62–71.

Warren, R., & Owen, D. (1981, April). Functional optical invariants. In R. Jensen (Ed.), *First Symposium on Aviation Psychology* (ALP-1-81). Columbus: Ohio State University.

Welford, A. (1976). *Skilled performance*. Glenview, IL: Scott, Foresman.

Wetherell, A. (1979). Short-term memory for verbal and graphic route information. *Proceedings, 1979 meeting of the Human Factors Society*. Santa Monica, CA: Human Factors.

Whitaker, L. A., & Stacey, S. (1981). Response times to left and right directional signals. *Human Factors, 23*, 447–452.

Wickelgren, W. (1979). *Cognitive psychology*. Englewood Cliffs, NJ: Prentice-Hall.

Memory

OVERVIEW

The human memory is at once remarkable for its power and for its limitations. On the one hand, the vast store of information that we have in memory for word meaning, general knowledge, facts, and images is sufficient to put powerful computers to shame. On the other hand, the occasional constraints on memory are often severe enough to be major bottlenecks in human performance. It goes without saying that humans are often forgetful. We forget phone numbers entirely or recall them incorrectly. We make a wrong turn because our memory of a map fails, or we incorrectly remember the name of a person to whom we have just been introduced.

The limits of human memory are also demonstrated in more complex human-machine systems. For example, pilots may forget navigational instructions delivered to them by air traffic controllers, transposing digits (349 to 394) or confusing call letters; operators following a series of procedures may forget to perform critical operations; or an air traffic controller may forget to convey all pertinent information concerning the status of aircraft under supervision to his relief on the next shift, failing to warn of a pending crisis situation (Danaher, 1980). In Chapter 3 it was emphasized that the limitations of working memory were often major causes of departures from optimal decision making, restricting the hypotheses that decision makers were able to consider and limiting the decision maker's abilities to update old hypotheses on the basis of new data.

This chapter will discuss the mechanics of human memory, the nature of its limitations and constraints, and, as a direct implication, the factors that can alleviate some of the problems associated with these limitations. Two general themes or dimensions will be important in our treatment of memory. One, which has been discussed to some extent in the previous two chapters, emphasizes the nature of different *codes* or representations of memory. As the contrast between Chapters 4 and 5 suggests, not all information is stored in the same format in the brain. Major dichotomies can be drawn between verbal and spatial representation and between auditory and visual processes. Our concern in this chapter will be two characteristics of the codes of memory representation: (1) the interference or disruption of information in one code by another and (2) how the time course of enhancement, prolongation, and decay of those codes is affected by the nature of the code itself.

The second theme focuses specifically on the forgetting process and on those factors, other than the nature of the codes, that enhance or retard the loss of information in memory. These two themes are not independent of each other, and neither is entirely independent of the content of the chapters on perception and attention. The underlying theme, however, that characterizes and distinguishes this chapter from the others is the *loss* of availability of information either because of the passage of time or because of factors that operate over time. A final section of the chapter will deal with some of the properties of our permanent long-term memory.

CODES IN INFORMATION PROCESSING

As a stimulus is processed and transformations are performed upon it to generate a response, the stimulus may be represented by a number of different *codes*. For example, the letter *a* may be represented either by how it looks (its physical shape) or how it sounds. As we saw in Chapter 5, objects may be represented either by their meaning or by their visual appearance. The distinction between codes is a dimension that is independent of the stages of processing dimensions. Stages must operate at different points in time when processing a given stimulus. Codes can be alternative ways of representing a stimulus at a given point in time, although they may operate at different points in time as well.

We will distinguish between five codes of processing, depicted in Figure 6.1. Iconic and echoic codes (sensory codes) are raw "sensory-like" representations of visual and auditory stimuli, respectively, that must match the modality of input. These prolong the stimulus representation for a short time (less than a second for the iconic code, a few seconds for the echoic). However, the sensory codes do not require the operator's limited attentional resources to do so. Because the echoic

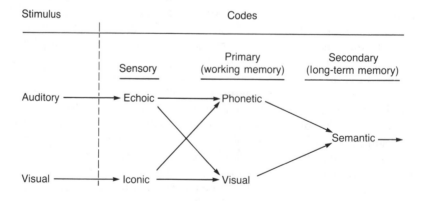

Figure 6.1 Relation between the five codes of memory.

and iconic sensory codes do not have any great implications for system design, they will be described in the chapter supplement.

The visual and phonetic primary codes are in many respects analogous to the visual and auditory stimulus modalities (and the iconic and echoic codes), respectively, but unlike the sensory codes they may be generated from stimuli of the opposite modality or from our own long-term memory. For example, we can rapidly form an "auditory image" (phonetic code) of a visually presented letter. In fact, we do so almost automatically. We can also form a visual image of an object that is neither seen or heard. The visual and phonetic primary codes, along with the semantic code, constitute the basis of our attention-demanding short-term memory, or *working memory*, necessary in so many skills yet limited in its capacity. The *semantic code*, the abstract representations in terms of the meaning rather than the sight or sound, may be generated from either visual or phonetic codes. The semantic code, while an important component of working memory, is even more important as the basis of information that is represented in the more permanent long-term memory.

Human experience is such that there appears to be a relatively natural progression—Posner (1973, 1978) refers to it as *abstraction*—in which codes that are closely related to the physical aspects of the stimulus produce codes that are progressively more symbolic in nature. For example, a visually presented word will produce an iconic representation followed by a phonetic code (the word sound) followed by the semantic code (the meaning of the word—and its implications for action). This general ordering in time does not imply that codes must follow each other in series. There is good evidence, for example, that all three nonsensory codes may exist at some level of strength at the same time, but that the time of maximum strength of each follows the progression outlined in Figure 6.2.

It is important also to emphasize that the "natural" order of appearance of codes is not invariant. We may, for example, "generate" either a phonetic or visual representation of an abstract idea or word. Posner (1973) uses the term *generation* to describe this reversal of the natural sequence. We are also able to sustain a code of representation for a duration longer than its natural "life-span"—for example, when we rehearse a name that we have just heard, thereby prolonging the natural decay of the phonetic code. Both generation and maintenance, however, work against the natural "gravity" of decay—the loss of code strength over time. As a consequence, they are achieved at a cost: the investment of cognitive effort or resources. We shall now consider the experimental evidence for the three primary codes of working memory and the usefulness of distinguishing them.

Primary Codes and Working Memory

Sensory codes are automatically and exclusively activated by input only from the corresponding sensory modality. The primary codes, on

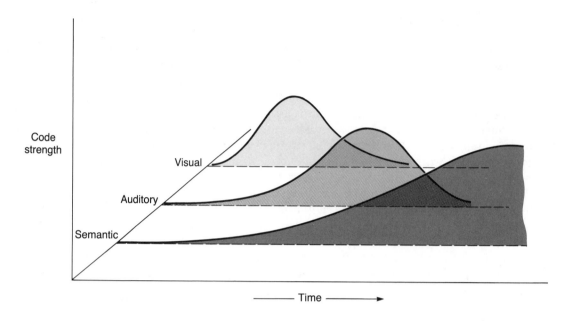

Figure 6.2   Time course of code development.

the other hand, are not. The phonetic and visual primary codes form the essence of what Baddeley and Hitch (1974) have defined as "working memory" and William James (1890) referred to as "primary memory." According to James's definition, primary memory represents "the width in time of the conscious present, and information within primary memory never leaves the conscious present" (1890, 646). Broadbent (1971) has ascribed to working memory the metaphor of a "desk top" upon which sit those pieces of information that are actively being considered in the present.

Working memory may receive input from sensory channels by way of sensory codes, as when we rehearse a phone number just read or the name of an individual just met. In this case, the functions of working memory become synonymous with the historical development of *short-term memory* (Brown, 1959; Melton, 1963; Peterson & Peterson, 1959). Alternatively, working memory may receive inputs from *long-term memory* or from itself: We recall the rules from long-term memory on how to execute a certain mathematical calculation and then maintain these rules in temporary activation while carrying out the calculation on a set of perceived values; in performing this calculation we may hold subsums or products in working memory. In decision making we recall weighting coefficients from long-term memory in order to integrate different attributes of information for different alternatives (see Chapter 3). Figure 6.3 represents the relations between the sources of input to working memory.

A common characteristic of working memory is that in the absence of attention—a continuous allocation of processing resources—the in-

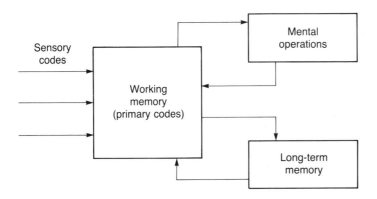

Figure 6.3   Working memory.

formation in working memory will degrade and any operations performed on this information will deteriorate accordingly. As one
example of the consequences of this resource demand, we saw in
Chapter 3 that people often employ *heuristics* that reduce the memory
load with the goal of economizing the resources demanded by working
memory operations. A second example is given by the process of
prediction. Prediction is quite useful in anticipating future states and
responding to them without lag (see Chapters 11 and 12). However,
prediction imposes a load on working memory because four kinds of
information must all be maintained or operated upon in working memory: present conditions, future conditions, the rules by which the
former generate the latter, and the actions that are appropriate to future
conditions. When heavy demands force the human's limited resources
to be allocated elsewhere, prediction will suffer, sometimes with unfortunate consequences. For example, the aircraft pilot who becomes
heavily stressed with momentary demands in the cockpit, diverting his
mental resources from working memory, will be less inclined to take
control actions based on future predicted events. As stress increases
our foresight declines.

The phonetic code.   The phonetic code has been a dominant concept for both historical and phenomenological reasons. The phonetic
code is analogous in many respects to the echoic, or "sounds-like,"
sensory code. When we rehearse a phone number, we "hear" the
sounds of the digits, although there may be no overt sound nor even
any covert activation of the articulatory muscles. The phonetic primary
code is automatically generated from an echoic sensory code, but it also
arises automatically from visual input in many circumstances. Conrad
(1964) demonstrated this effect by showing that when people were
presented with a visual sequence of letters to be recalled from working
memory, the errors of recall tended to reflect *acoustic* rather than visual
confusion factors. For example, if the letter *E* is presented in an original
list and is erroneously recalled, the error would be more likely to be a *D*

(which sounds like *E* but looks quite different) than an *F* (which looks like *E* but is quite different acoustically). Furthermore, the same acoustic confusion pattern is observed whether recall is vocal or written.

If forgetting results from a decay or degradation in the code of memory representation, so that letters sharing similar features in that code became progressively more confusable, then the existence of the "Conrad effect" strongly argues that the phonetic code replaces the iconic code of the visual input. If letters are represented in working memory by a phonetic code, then if a set of visually presented letters are quite similar acoustically (for example, *f, s, x, l* or *p, t, c, d*), they should be poorly recalled. This result was confirmed in an experiment by Conrad and Hull (1964).

While different letter sounds or phenomes (see Chapter 4) clearly have different representations in the phonetic code, Crowder (1971, 1978) has shown that only vowel differences are maintained in the sensory echoic code. Differences in consonant sounds do not "echo" or perseverate in the absence of attention. The fact that vowels but not consonants are preserved in echoic memory allows the principle of acoustic confusability demonstrated by Conrad and Hull (1964) to be refined. If a small set of letters are to be used in a coding task in which they might need to be stored in working memory, it is always desirable to minimize confusability by minimizing the similarity between letters. Therefore, initial effort should be devoted to avoiding confusable vowel sounds, since these will be a discriminating factor in both the echoic and phonetic codes. In fact, it is probably more important to avoid confusing vowel sounds with dissimilar consonants (*j* and *k*) than to avoid similar consonants with dissimilar vowels (*j* and *g*). If the number of letters used in the coding system is large enough so that more than two must share the same acoustic vowel sound (for example, *x* and *f*, or *b* and *v*), then emphasis must be placed on the phonetic discrimination of consonants.

Although the Conrad effect indicates that visual-verbal stimuli will normally produce a phonetic code, such stimuli may also give rise to a visual code as well. Conrad (1962) and Estes (1973) performed experiments in which subjects were forced to engage in irrelevant articulatory activity when each letter of a series was displayed. In both experiments the Conrad effect disappeared. These results suggest that when articulation of the displayed letters is prohibited by the competing task, the code of representation is something other than phonetic. A likely candidate is the *visual code*. When prevented from articulating, the subjects studied by Conrad and Estes maintained a visual image of the stimulus letter.

The primary visual code.   An important role of the primary visual code was discussed briefly in Chapter 5 in the context of navigation. There we saw that visual images were important for early phases of navigation, and we discussed the role of imagined maps in the "filled distance effect." Visual input does not automatically generate visual images. On the other hand, visual images can clearly be generated from nonvisual sources (Kosslyn, 1981). These effects may arise from long-term memory of various pictures, objects, and items that we have experienced. Insofar as verbal stimulus material (speech or print) is concerned, however, there is good evidence that the visual code does not naturally develop or persist in the same manner as the phonetic code.

Much of this evidence is provided by the work of Posner and his colleagues in the letter-name matching paradigm (see Posner, 1973, 1978 for a good review). In the basic variant of this paradigm (Posner & Mitchell, 1967), the subject is presented with two letters in sequence and is required to state as rapidly as possible whether they are of the same or different name (see Figure 6.4). Two critical experimental manipulations may be performed in this paradigm. These involve varying (1) the interval between the two stimuli (the interstimulus interval, or ISI) and (2) the relation between the two letters—that is, whether or not they are of the same name and/or the same case (capital or lower-case). Thus the pair may be physically identical (for example, AA, bb), may share the same name but be of different case (Aa, Gg), or may be of different name (AB, gF). In the first two cases a "yes" response is required. In the third case a "no" response is to be given (see Figure 6.4).

When the reaction times for "same" responses are examined, the results of the letter-name matching paradigm are instructive concerning the chronometric, or time-dependent, change in codes that occurs. Two major highlights of these results may be noted: First, when the ISI is very short (at an ISI of 0 seconds, the two letters are presented simultaneously side by side), physically identical stimuli are responded to more rapidly than stimuli sharing the same name but a different case (AA < Aa). This suggests that a visual-iconic code is immediately available, and if the stimuli are physically identical, this code can be used as the basis for a "yes" response. On the other hand, if the stimuli share the same name, but are of different case (Bb), a "yes" response must await the growth of the phonetic name code to establish

Stimulus 1	Stimulus 2	Response	
a	A	"Yes"	Name identity (NI)
A	A	"Yes"	Physical identity (PI)
B	d	"No"	
e	f	"No"	

Time

Figure 6.4 Posner and Mitchell's (1967) letter-name matching paradigm.

identity. This delay, or "waiting time," is responsible for the longer "yes" reaction times to physically different name-identical stimuli (Bb) in the name-match task.

Second, as the ISI is made progressively longer in the name-match task, the reaction-time advantage for physically identical stimuli declines so that with ISIs of greater than a second or so the advantage to physically matching stimuli is negligible (see Figure 6.5). This finding suggests that the strength of the iconic code of the first stimulus decays and/or the strength of the phonetic code grows as ISI is prolonged. While we know from our earlier discussion that the iconic code does decay in the absence of visual input, a study by Thorson, Hockhaus, and Stanners (1976) employing the name-match task seems to provide evidence that the phonetic code for this visual input grows over time.

Among the various combinations of first and second stimuli that were presented in their investigation, Thorsen et al. were particularly interested in those cases when the two stimuli were both of the same case but of a different name, thereby warranting a "no" response (for example, AF). The variable of particular interest was the visual and acoustic similarity between the two letters. If the similarity is great, there is a tendency to say "yes" which conflicts with and therefore delays the response to say "no." This effect may be labeled a *similarity decrement*. Thorsen et al. examined the similarity decrement as a joint function of the nature of the similarity (visual or acoustic) and the interstimulus interval. They observed that if the letters were visually similar (EF or QO), then the similarity decrement was greatest at short ISIs but declined to zero at longer ISIs. If the similarity was based upon acoustic features (fs or BT), then the opposite pattern was observed: the similarity decrement was negligible at ISI = 0 and increased to a steady-state value as the ISI increased. The former observation defines the decay of the iconic code, the latter the growth of the phonetic.

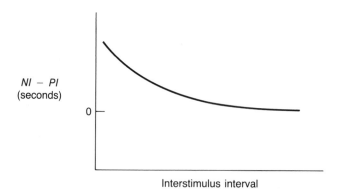

Figure 6.5   Effect of interstimulus interval on advantage of physical identity over name identity.

The preceding results have emphasized the highly transient nature of the primary visual or iconic code. There are nevertheless certain occasions when the visual code may be preserved for a relatively longer period of time or be generated from the echoic code. An example of visual code preservation is an experiment by Salzberg, Parks, Kroll, and Parkinson (1971) in which subjects were presented with two letters in sequence to make an identity-match judgment. While the first letter was retained but before the second was presented, subjects had to "shadow," or orally repeat, an auditory message. This task produced a heavy degree of phonetic interference. When the two letters were presented visually, the investigators observed that the advantage in response time for physically identical letters was retained even at intervals as long as 10–12 seconds, far longer than Posner had observed under conditions of no phonetic distraction. This finding suggests that when external conditions disrupt maintenance of a phonetic code, a visual code can be maintained.

The experiments described above indicate that the visual code does not appear to persist unless it is maintained by an effortful attention-demanding process. One characteristic of all the stimulus material employed in these demonstrations, however, is that they were primarily verbal (letters and words). Indeed, the prominence of verbal material in most memory research of the past two decades was undoubtedly responsible for the prevailing view that working memory is primarily phonetic. It would appear logical, however, that nonlinguistic material would be more naturally and compatibly stored by visual codes. Indeed, the discussion in the preceding chapter suggested that this is true with some navigational material.

An experiment by Bencomo and Daniel (1975) provides strong evidence for the persistence of secondary visual codes when material is intrinsically visual in nature. In a paradigm similar to Posner's letter-name matching task, subjects made a same-different judgment on two stimuli presented 30 seconds apart. The stimuli were either pictures or names of common objects. As in the experiment of Thorsen et al., Bencomo and Daniel were interested in the similarity decrement: the slowing of "different" responses that resulted when certain aspects of the two stimuli were made more similar, thereby inducing a perception of sameness. In this case similarity was visual, determined by the similarity of shape between the two objects. For example, a nail and a pencil are visually similar; a nail and a ball are not. The investigators observed that "different" reaction times were slowed by visual similarity when the two stimuli were pictures but not when they were words. Therefore, a visual code was assumed to persist with picture stimuli but not with verbal stimuli.

The semantic code.   While the discussion to this point has focused on the primary visual and phonetic codes, it is evident that material in working memory is also represented in terms of what it means—a semantic code. An experiment by Shulman (1972) demonstrated the importance of the semantic code in primary memory. He found that short-term memory for words suffered when the words were semantically similar, an analogous finding to the "Conrad effect" observed with acoustic similarity. Indeed, the role of the semantic code can be most clearly demonstrated phenomenologically by noting the disappearance of that code in the phenomenon of *semantic satiation.* Take a common word ("soap" is a personal favorite) and repeat it either aloud or covertly several times. Eventually, the word will lose its apparent meaning and become a mere "blob of sound." This demonstration reveals that the semantic code has somehow gone temporarily underground.

While the semantic code seems to be an equal partner to the visual and phonetic codes in working memory, it actually assumes dominance over these two in the storage and organization of material in the more permanent long-term or secondary memory. For this reason we shall defer further discussion of the semantic aspects of memory until the final sections of this chapter.

Practical Applications of Memory Codes

The practical implications of the distinction drawn between the different sensory and primary codes are provided primarily by four different phenomena: (1) The two kinds of working memory appear to be somewhat independent from each other and therefore are susceptible to interference from different sorts of concurrent activities (Baddeley, Grant, Wight, & Thompson, 1975). (2) As we have seen, the two codes are disrupted differentially by different kinds of similarity relations existing between stimulus material to be retained. Given that similarity induces error in retention, knowledge of how different codes change over time may assist in formulating stimulus materials that will be best retained for different intervals. (3) The relation of codes to sensory modalities has implications for the use of auditory versus visual displays and verbal versus spatial displays. (4) Visual and verbal codes may be differentially effective for different individuals, thereby providing a means for associating different operators with materials for which they are best suited. The following discussion will emphasize the first and third of these phenomena: code interference and display formatting.

Code interference.   The verbal-phonetic and visual-spatial codes of working memory appear to function more cooperatively than competitively. Posner (1978), for example, has argued that both may be activated in parallel by certain kinds of material (e.g., pictures of common objects). We also saw examples of the cooperation between codes in our discussion of navigation in Chapter 5. One implication of this cooperation, which will be considered more extensively in Chapter 8, is that the two codes do not compete for the same limited processing

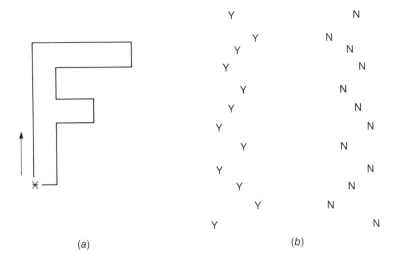

(a)   (b)

Figure 6.6   (a) Example of a stimulus in Brooks's study. (b) Display used in
pointing task. From "Spatial and Verbal Components in the Act of Recall,"
by L. R. Brooks, 1968, *Canadian Journal of Psychology, 22*, pp. 350–351.
Copyright 1968 by the Canadian Psychological Association. Reproduced
by permission.

resources or attention. That is, if two tasks employ different codes of
working memory, they will be time-shared more efficiently than if two
tasks share a common code. This is a theme that will be discussed in
some detail in Chapter 8. Here we restrict the discussion to time-
sharing with working memory.

The research of Baddeley and his colleagues (Baddeley, Grant,
Wight, & Thompson, 1975; Baddeley & Hitch, 1974; Baddeley & Lieberman, 1980) has contributed substantially to the understanding of this
dichotomy between the two forms of working memory, both in terms of
the kind of material that is manipulated within working memory (spa-
tial-visual or verbal-phonetic) and in terms of the separate processing
resources used by each. Brooks (1968) performed an important series of
experiments that stimulated a good deal of the subsequent research in
this area. In one experiment, Brooks required subjects to perform a
series of mental operations in spatial working memory.[1] Subjects imagined a capital letter, such as the letter *F*, as depicted in Figure 6.6. They
were then asked to "walk" around the perimeter of the letter indicating
in turn whether each corner was or was not in a designated orientation
(e.g., facing the lower right). The series of yes and no answers were
indicated either by vocal articulation (verbal response) or by pointing
to a column of Y's and N's (spatial response). Brooks found that per-

---

[1]The term *spatial* will be used in place of the term *visual*, since subsequent research
suggests the latter to be a more accurate description.

formance in the verbal condition was reliably better than in the spatial condition, suggesting that the verbal response code in the former case used different resources from those underlying the "walking" operations in spatial working memory. The greater resource competition with spatial working memory in the spatial response condition produced the greater degree of interference.

Of course, it is possible that the larger interference with the spatial-response mode occurred because this is a more difficult means of responding. However, the results of a second investigation by Brooks suggest that this interpretation is unlikely. In this study, the spatial working memory task was replaced with one relying upon verbal working memory. Here subjects imagined a familiar sentence (e.g., "The quick brown fox jumped over the lazy dog") and then "walked" through the sentence indicating in turn if each word was or was not a member of a particular grammatical category. Once again, responses were indicated either by speech or by pointing. Using the verbal task, Brooks now found that the reverse pattern of interference was observed. The task was performed better with the spatial-manual response than with the vocal-verbal one. Hence when different codes underlie working memory and response, performance will be improved.

Experiments conducted by Healy (1975) and by Baddeley and his colleagues offered further insight into the distinction between the two memory systems. Healy presented subjects with a brief series of 4 letters in sequence at different locations in a horizontal array of 4 windows, so that each letter followed another in turn (Figure 6.7). Subjects then had to recall the letters in two different conditions. In the temporal order condition, the 4 letters were to be recalled in the temporal order in which they occurred. In the spatial order condition, the 4 letters were to be recalled in a left-to-right fashion. Healy observed that a phonologically loading distractor task (repeating a sequence of heard digits) disrupted the temporal order recall but exerted almost no influence on spatial order recall. This contrast is important because the nature of the stimulus and response is identical in both conditions. The only difference between tasks concerns the attributes of information that need to be retained. Temporal-order information appears to be associated with verbal working memory, spatial-order information with visual memory (Crowder, 1978).

A study by Baddeley and Lieberman (1980) suggests that what we have referred to as visual working memory need not actually be visual at all but is primarily *spatial*. The term *visual* has often been applied simply because much visual information *is* spatial in nature and most auditory information (speech) is not. But this association is not mandatory. We read visual print (visual-verbal) and localize sounds in space (auditory-spatial). In Baddeley and Lieberman's investigation, subjects performed a spatial working memory task developed by Brooks (1967) in which a 4 × 4 matrix of blank squares was imagined and then "filled in" on the basis of a series of verbal commands of the following form:

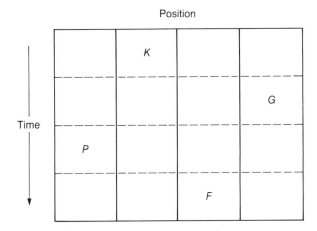

Position

Time

Spatial report: *P K F G*

Temporal report: *K G P F*

Figure 6.7 Spatial and temporal display of 4 letters. From "Coding of Temporal-Spatial Patterns in Short-Term Memory" by A. F. Healy, 1975, *Journal of Verbal Learning and Verbal Behavior, 14,* p. 492. Copyright 1975 by Academic Press. Adapted by permission of the author.

"In the starting square place a 1; in the next square *up* place a 2; in the square to the *left* place a 3"; and so forth. Subjects then recalled the relative locations of the digits by associating each digit with the directional indicator (e.g., up, 2). The task was performed alone, and then concurrently with an *auditory spatial* task, and then with a *visual nonspatial* task. The auditory spatial task required subjects to "track" or point to a continuously moving sound source while blindfolded. The visual task required detection of threshold visual signals. Baddeley and Lieberman observed that greater interference with the spatial memory task was produced by the concurrent task that was also spatial but not visual (auditory tracking) than by the task that was visual but not spatial (signal detection).

The relevance of these and related experimental findings may be summarized as follows: We seem to have two forms of working memory. Each is used to process or retain qualitatively different kinds of information (spatial and visual versus temporal, verbal, and phonetic), and each can be disrupted by different concurrent activities. Tasks should be designed therefore in such a manner that this disruption does not occur. Tasks that impose high loads on spatial working memory (e.g., an air traffic controller's requirement to maintain a mental model of the aircraft in his purview) should not be performed concurrently with tasks that will also use the visual-manual-spatial system.

Spatial tasks will be less disrupted by employment of an auditory-phonetic-vocal loop to handle subsidiary information-processing tasks.

This guideline does not necessarily suggest that the primary task display should be in a different format from the code of working memory used in the task. In the case of the air traffic controller, this is probably unwise. A visual-spatial display will be optimally compatible with the spatial representation in visual working memory. However, there are other circumstances in which it would be best to display information relevant to the primary task by a code different from that used in processing the primary task information. An example was suggested by the investigation of navigation conducted by Wetherell (1979), discussed in Chapter 5. Wetherell concluded that navigational information was better represented in phonetic route-list format than in map format, in part because the latter interfered more with the spatial characteristics of driving.

The implications of code-specific interference have not been extensively followed up with investigations in more applied settings. Hoffman and MacDonald (1980) studied how the retention of information on highway traffic signs displayed verbally ("no left turn") or spatially (by a ⊘) was influenced by concurrent verbal or spatial activity during the retention interval. When the subjects' task was to recognize the sign that had been originally presented, their results provide some evidence for code-specific interference. However, when the subjects' task actually involved implementing a response on the basis of the information presented on the sign, Hoffman and MacDonald found that the spatial task consistently produced greater interference independent of the code of the original sign. This suggests that whether the original sign was verbal or spatial, the implications for action were directly coded in a spatial format, more disrupted by the spatial task. These results suggest in turn that much of driving involves spatial processing (as suggested as well by Wetherell's 1979 results). Since sign-format difference had little influence on interference during retention in Hoffman and MacDonald's experiment, it appears that the guidelines for sign formatting can best follow the principle of display compatibility described in Chapter 5; that is, because the implications for action are spatial, the optimum display code will be spatial as well.

Two analogous investigations by Weingartner (1982) and by C. D. Wickens, Sandry, and Vidulich (1983) did find evidence for code-specific interference in working memory. Weingartner's subjects engaged in a simulated process-control monitoring task (see Chapter 12) in which several cross-coupled slowly changing variables were monitored to detect periodic "failures," changes in the dynamic equations governing their relations. Concurrently, subjects performed either Brooks's (1967) spatial-memory matrix task or a verbal task requiring the memory of abstract words. Weingartner found that this visual/spatial monitoring task was disrupted more by the spatial than by the verbal matrix task. Since input for both matrix tasks was always auditory, the source of interference was necessarily the code of working memory. Wickens, Sandry, and Vidulich examined how a pilot's task when flying an aircraft simulator was disrupted by either a task requiring memory of verbal navigational information or a task requiring the localization of a target in space. Even when both tasks were presented visually, they found that the localization task, depending upon transformations in spatial working memory, disrupted flying more than did the verbal-memory task.

Display modality and working memory code.   Figure 6.8 shows four different possible formats of information display defined by the auditory and visual input modalities and the two primary perceptual and memory codes. These can be potentially associated with either of the two different codes of working memory, depending upon which is required by the task at hand. Experimental data suggest that the assignment of formats to memory codes should not be arbitrary. C. D. Wickens et al. (1983) have described the principle of stimulus/central-processing/response compatibility that prescribes the best association of display formats to codes of working memory. In this principle S refers to the four stimulus display formats, C to the two possible central processing codes, and R to the two possible response modalities (manual and speech). The following discussion concerns the optimum matching between stimulus and working memory codes (S–C compatibility). The compatibility of response modalities will be dealt with in Chapter 9.

In Figure 6.8 the shaded cells indicate the optimum format (combination of mode and code) for each memory task. Thus, although it is

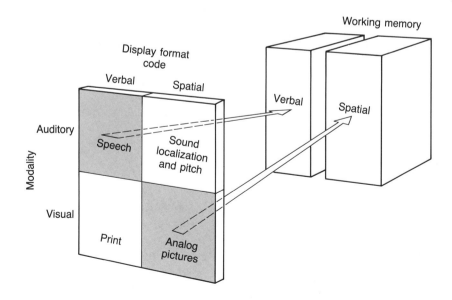

Figure 6.8   Optimum assignment of display format to working memory code.

possible to employ auditory-spatial displays in the service of tasks that demand spatial working memory, such a format does not generally provide an effective display because the auditory modality is less attuned to processing spatial information than is the visual. For example, Vidulich and Wickens (1982) employed an auditory display of spatial material in which the location and velocity of an aircraft on a two-dimensional grid were indicated by tones varying in pitch (corresponding to the vertical dimension) and apparent location (corresponding to the horizontal). Performance with this display was far less proficient than with a visual spatial counterpart, even in the presence of a concurrent visual task. Further examples of using the auditory modality in the service of spatial tasks will be discussed in the treatment of auditory tracking in Chapter 11.

In contrast to spatial memory, tasks that demand verbal working memory can be more readily served by either speech or print. However, the fact that echoic memory has a slower decay than iconic and that speech displays are more directly compatible with the vocalization we use in rehearsal suggests that speech should be employed as a display for verbal tasks. This is particularly true if the verbal material can only be displayed for a transient interval (C. D. Wickens et al., 1983). This guideline is supported by a series of laboratory studies that find verbal material is better retained for short periods when presented by auditory rather than visual means (Murdock, 1968; Nilsson, Ohlsson, & Ronnberg, 1977).

This observation, of course, has considerable practical implications when verbal material is to be presented for temporary storage (for example, navigational entries presented to the aircraft pilot or the outcome of diagnostic tests presented to the physician or process control monitor by automatic means). Such information will be less susceptible to short-term loss when presented by way of auditory channels (either spoken or through speech synthesis). C. D. Wickens et al. (1983), for example, found that pilots could retain navigational information better from auditory than visual channels, and this advantage was enhanced under conditions of high work load. Of course, it is possible to compensate for the more rapid decay of the visual presentation by displaying the visual information for a longer duration. This option may not be available, however, if the information must be presented at a rapid rate or if display space is at a premium.

Spatial-verbal displays in instructions. The relative merits of spatial and verbal displays are directly relevant to the issue of how to format procedures and instructions. Whether pictorial or verbal instructions are superior depends in part upon whether the nature of transformations and operations to be performed on that material is spatial or verbal. In criticizing the use of purely verbal instructions, Kammann (1975) has articulated the so-called two-thirds rule. He summarized a number of investigations of instruction formats and concluded that printed instructions are understood on the average only two-thirds of the time. Kammann then compared comprehension of a set of dialing instructions for telephone switchboard operators when these were presented entirely in verbal text format and when they were presented in a "spatial" flow-chart format that emphasized the decision-tree aspect of the task. Kammann found that the flow charts were better understood than the verbal instructions. Furthermore, the superiority in comprehension was maintained in a field trial 1–2 months after users were initially introduced to the system by the various formats.

Thomas and Gould (1975) observed a similar advantage for spatial over verbal formatting in the domain of human-computer interaction. They compared two interactive computer systems whereby users query a data base to obtain particular aspects of information in computer memory (e.g., information about trails at different ski areas or demographic information about employers in a company). One system, called *sequel*, is essentially verbal in format and is based upon key words used to retrieve the desired information. The other, *query-by-example*, is more spatial in that it presents a matrix of attributes within which the user specifies the attribute combination desired by designating the appropriate columns and rows (Schneiderman, 1980). Thomas and Gould found that naive users were far more successful in retrieving data with the latter system, a conclusion supported by a subsequent study by Greenblatt and Waxman (1978).

The advantage of spatial data displays for comprehension is not, however, universal. In another computer-related study of the relative merits of spatial flowcharts and verbal lists, Wright and Reid (1973) observed that flow charts were better for immediate comprehension, but noted also that information presented in flow-chart form showed a greater loss of retention over time. Brooke and Duncan (1980) investigated subjects' abilities to locate faults in a program designed to run a fuel distribution and pricing system. Subjects were given statements about the nature of the error and could then use either a verbal statement listing or a flow chart to assist in their diagnosis. Brooke and Duncan found little difference in performance between the two groups of subjects who had the two formats of information available. A study conducted by Schneiderman, Mayer, McKay, and Heller (1977) also failed to observe any difference between the two alternative display formats in terms of their assistance in computer programming, computer program comprehension, or program debugging.

Ramsey, Atwood, and Van Doren (1978) compared the performance of computer programmers who used flow-charting techniques with those who used a *program design language* (PDL)—a simplified set of verbal statements used to help design program formats. Like Schneiderman et al. (1977), they noted no difference in comprehension or recall between the two modes but found that use of the program design language mode generated better quality programs. Ramsey et al. concluded that the reason for this difference in performance was based upon the nature of the information emphasized by the two. Flow charts emphasize the flow of information, while a PDL emphasizes its hierarchical structure. To the extent that efficient programming requires use of hierarchical relations in working memory, it would seem that the PDL provides a representation more compatible with this structure. It should be noted, however, that the absence of hierarchical structure is not an inevitable characteristic of all spatial displays. Ramsey et al. did not evaluate a format in which hierarchical structure was displayed spatially, although this alternative is clearly feasible.

Individual differences and redundancy.   It is apparent from the preceding discussion that the experimental data do not clearly point to one mode as superior to the other. Some of the ambivalence may result because different modes are preferred by different categories of users. Different users may have different "internal models" of the system. The advantages of flow-charting found by Kammann, for example, were observed with nonprogrammers, while the verbal advantage obtained by Ramsey et al. used experienced computer programmers as subjects. Schneiderman (1980) reports a study in which users of high verbal ability were more facile with a key-word data-base query system, whereas those with high mathematical aptitude were superior with the spatially oriented query-by-example format. Schneiderman suggests, in general, that programmers may prefer verbal listings while managers prefer flow-charting.

A second related reason for the lack of pronounced differences between verbal and spatial formats concerns the tremendous sources of variance, often confounding, that enter any experimental comparison of different programming aids (Brooke & Duncan, 1980). Different investigations employ quite different tasks and measures, some employing debugging, some programming quality, some programming speed. Performance criteria are often difficult to specify. If problems are formulated in available computer-programming languages, then subjects must be trained programmers familiar with those languages. These programmers in turn often bring certain biases and differences in familiarity into the laboratory and these biases may confound any comparisons.

A third reason for the ambivalence of results is that different display formats emphasize different properties of the information to be represented in working and in long-term memory. Depending upon the task, either the spatial, causal flow of information favoring flow charts or the hierarchical semantic network relationships emphasized by the verbal statements may be more important (Fitter & Green, 1979). The joint consequences of individual differences and of the different kinds of information emphasized in the two formats suggest a fairly obvious yet important principle: Redundant information format provides the flexibility for different users to capitalize on the format they process best and for the information desired to be represented in whichever mode is most salient.

Three investigations reinforce this conclusion, even as they emphasize the relative strengths of different formats. Booher (1975) evaluated
subjects who were mastering a series of procedures required to turn on
a piece of equipment. Six different instructional formats were compared: one purely verbal, one purely pictorial, and four combinations
of the two codes. Two of these combinations were *redundant*: one code
was emphasized and the other provided supplementary cues. Two
others were *related*: the nonemphasized mode gave related but not
redundant information to the emphasized mode. Booher found the
worst performance with the printed instructions and best with the
pictorial-emphasis/redundant-print format. While the picture was of
primary benefit in this condition, the redundant print clearly provided
useful information that was not available in the picture-only condition.

In the second study, Stone and Gluck (1980) compared subjects'
performance in assembling a model using pictorial instruction, text, or
a completely redundant presentation of both. Like Booher, Stone and
Gluck found the best performance in the redundant condition. The
investigators also monitored eye fixation in the redundant condition
and found that five times as much time was spent fixating the text as
the picture. This finding is consistent with a conclusion drawn by both
Booher and by Stone and Gluck: The picture provides an overall context or "frame" within which the words can be used to fill in the details
of the procedures or instructions. The importance of context was, of
course, emphasized in Chapter 4.

The question of individual differences in spatial and verbal ability
and their interaction with display format was examined in an investigation by Yallow (1980). Subjects with high and low ability in both verbal
and spatial aptitude were presented material on the topic of economics,
in a form that emphasized either a verbal or a spatial (i.e., graphic)
representation of the material. Not surprisingly, Yallow observed that
immediate retention of the material was helped by formats that capitalize on one's strengths (e.g., graphic displays for subjects of high spatial
ability). However, she also found that this advantage was markedly
short-lived and dissipated when long-term retention was assessed, a
finding similar to that of Wright and Reid (1973) with regard to flow-
charting.

It is apparent from this discussion that the comparative evaluations
of pictorial/symbolic versus verbal displays do not provide the clear-
cut results upon which firm design guidelines could be based. This
state of affairs is sadly noted by Moran (1981) with regard to human-
computer interfacing. A number of factors cause this ambivalence,
including the variance in individual abilities, the variance in the information required to perform tasks involving working memory, variability of concurrent task demands, and the degree to which immediate
use versus long-term retention is required. If any conclusion emerges
from these factors (besides the obvious one that "more good research is
needed"), it is that (a) the contribution of the four factors described

above can be evaluated to provide some guidelines of what format may be best in what circumstances, (b) whichever format is chosen, basic principles of good display should be adhered to (both well described in Bailey, 1982), and (c) where possible, a redundant verbal-spatial format is probably best.

TIME-DEPENDENT PROCESSES IN WORKING MEMORY

The rapid rate of decay or loss of availability is one of the greatest limitations of working memory. When verbal information is presented auditorily, the decay may be slightly postponed because of the transient benefits of the echoic code. When information is presented visually, the decay will be more rapid. The consequence of decay before material is used is the increased likelihood of error. The pilot may forget navigational instructions delivered by the air traffic controller before they are implemented. The computer programmer may lose his or her place in an interactive data program after having briefly diverted attention to another task. Or the process control monitor may forget which systems on an operator-controlled display have been recently checked in the sequence of routine monitoring operations. In fact, Moray (1980) concludes that "the task of monitoring a large display with many instruments is one for which human memory is ill-suited, especially when it is necessary to combine information from different parts of the display and the information is dynamic" (p. 33).

An apparent solution to the problem of such memory failures is to augment the initial transient stimulus (whether visual or auditory) with a more tonic visual display—a visual "echo" of the message a pilot receives from air traffic control, for example, or a continuous record of one's location in a hierarchical computer program. However, a memory-aiding alternative such as this is not free. Instead, it represents one of the many tradeoffs in human engineering design, because display augmentation is generally achieved only at the cost of added display clutter. This tradeoff is manifest in a number of environments. For example, in an air traffic controller's display, computer-generated tags placed next to each aircraft symbol help to off-load working memory by depicting such information as flight number and altitude. But it is likely that much more information presented about each flight might be excessive, so that the added visual clutter would disrupt the controller's overall image of the flight space.

An explicit example of the tradeoff between memory aiding and display clutter is reflected in deciding whether to equip commercial aircraft cockpits with cockpit displays of traffic information, or CDTI's (Hart & Loomis, 1980; Palmer, Jago, Baty, & O'Connor, 1980). Such a display, while off-loading the pilot's working memory of the surrounding airspace, will also increase the visual workload imposed. Is the increase worth the benefits? Clearly there is no direct answer to the question. The relative benefits of the CDTI will obviously depend upon the degree of crowding of the airspace, the pilot's spatial abilities, and a host of other factors. The important point, however, is that simple memory aids that replace memory store with physical store are not an unmixed blessing. At best they will provide a valuable service if they are needed, and they can be perceptually filtered (or physically turned off) by the user if they are not. However, at worst, memory aids may prove unnecessary in most circumstances and either distract or disrupt the perception of more recent events.

Fortunately, there exist other techniques besides the simple prolongation of visual information that can be used to detour around the bottlenecks imposed by the frailties of working memory. To understand these techniques, we shall describe a number of basic time-dependent properties of working memory and consider each one's design implications. More specifically, we will describe variables that either enhance or lessen the critical decay of material in working memory. Many of these variables interact with each other, and all are manifest outside as well as inside the laboratory. Equipped with knowledge of these variables, the system designer can proceed to reduce the effects of those that are harmful and exploit those that are helpful.

Time and Attention Demands

In the late 1950s experiments conducted by Brown (1959) and Peterson and Peterson (1959) used similar techniques to demonstrate two of the most fundamental properties of working memory: its dependence on time and on attention. Using what has subsequently been labeled the Brown-Peterson paradigm, Peterson and Peterson (1959) asked subjects to retain a simple sequence of three random letters in memory for short intervals. In order to prevent subjects from rehearsing the digits, they were asked to count backwards aloud by threes from a designated number, presented just after the item to be remembered. Upon hearing a recall cue, the subject stopped the count and attempted to retrieve the appropriate item. Both Brown and the Petersons found that retention dropped to nearly zero after only 20 seconds when rehearsal was prevented in this manner. This decay function is shown schematically by curve *b* of Figure 6.9.

This highly transient characteristic of short-term memory has been demonstrated repeatedly in numerous variations of the Brown-Peterson paradigm. The various estimates generally suggest that in the absence of attention devoted to continuous rehearsal, little information is retained beyond 10–15 seconds. For example, Loftus, Dark, and Williams (1979) obtained decay functions very similar to those in Figure 6.9 for subjects attempting to remember navigational information similar to that which would be delivered to pilots from air traffic controllers. Moray and Richards (Moray, 1980) replicated the same

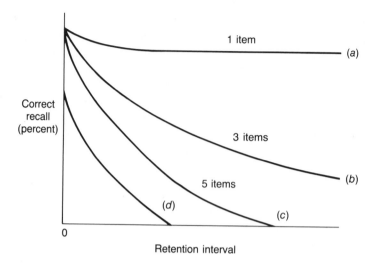

Figure 6.9   Effect of retention interval on recall from working memory with rehearsal prevented. From "Implications of Short-Term Memory for a General Theory of Memory" by A. W. Melton, 1963, *Journal of Verbal Learning and Verbal Behavior, 2*, p. 9. Copyright 1963 by Academic Press. Reprinted by permission.

general decay trend of Figure 6.9 for radar controllers attempting to recall displayed information on a radar scope. Their findings suggest that the transience is equally applicable to spatial and to verbal working memory.

Time and Capacity

Working memory is quite limited in its capacity, or the amount of information that it can hold, and this limit directly interacts with time. Curves *a* and *c* of Figure 6.9 represent decay functions in a Brown-Peterson paradigm that would be generated by 1- and by 5-letter items, respectively (Melton, 1963). Not surprisingly, faster decay is observed when more items are held in working memory. A major reason for this difference is that rehearsal itself (covert speech) is not instantaneous. With more items to be rehearsed in working memory, there will be a longer delay between successive rehearsals of each item. This delay increases the chance that a given item will have decayed below some minimum retrieval threshold before it is next encountered in the rehearsal sequence.

The limiting case occurs when a number of items cannot be successfully recalled even immediately after their presentation and with full attention allocated to their rehearsal (Curve *d* in Figure 6.9). This limiting number is sometimes referred to as the *memory span*. In a classic paper discussed previously in Chapter 2 in the context of absolute judgment, George Miller (1956) identified the limit of memory span as "the magical number seven plus or minus two" (the title of the paper). Thus somewhere between five and nine items defines the maximum capacity of working memory when full attention is deployed.

To this point, we have spoken somewhat generically of an "item" in working memory, defining it explicitly as a letter in the Brown-Peterson paradigm but not being terribly specific in other cases. Can an "item" (of which 7 $\pm$ 2 defines the working memory capacity) also be a number, or a word even if that word consists of 3 letters? Miller (1956) answered this question by proposing the concept of a "chunk." The capacity of working memory he asserted is 7 $\pm$ 2 *chunks* of information. A chunk can be a letter, a digit, a word, or some other unit. The defining properties of a chunk are a set of adjacent stimulus units which also are represented together as a single unit or node in the subject's long-term memory. Thus seven 3-letter words will define the capacity of working memory, even though this represents 21 letters, because the letter trigrams (cat, dog, etc.) are each familiar sequences to the subject—stored together in long-term memory—and thereby define seven chunks. Furthermore, if the seven words are combined in a familiar sequence so that the rules that combine the units are also stored in long-term memory ("London is the largest city in England"), then the entire string consists only of a single chunk. Thus, the family of decay curves shown in Figure 6.9 would describe equally well a string of 1, 3, 5, or 8 unrelated letters, 1, 3, 5, or 8 unrelated words, or 1, 3, 5, or 8 unrelated but familiar idioms. In each case, the items constituting each chunk within the string are bound together by the "glue" of associations in long-term memory.

It is of course, important to acknowledge not only the mean (7) but also the variability ($\pm$ 2) of Miller's limit. For example, some investigators (Conrad, 1957; Jacobs, 1887) have found that unrelated digits have a slightly longer span than unrelated letters, perhaps because we are more familiar with random digit strings or because there are fewer digits and therefore less acoustic confusability between them. Correspondingly, as chunks grow excessively long, the rules for sequencing their elements will be less solid, and so capacity will be reduced somewhat.

The 7 $\pm$ 2 limit is a critically important one in system design. When either presenting information that will not persist or presenting more persistent information that must be integrated in working memory, tasks that encroach upon the limits of 5–9 chunks should be avoided. In the former category, we might consider the length of strings of navigational information that might be issued to a pilot. For example, the message, "Change heading to 179 and speed to 240 knots when you reach latitude 47° 21', longitude 15° 30'," clearly exceeds the limits. As an example of an overload of information that must be integrated, consider the number of choices or options to be selected from a menu of a computer options on an interactive display terminal. If all alternatives must be compared simultaneously with each other to select the optimum, the choice will be easier if their number does not exceed working memory limits (Schneiderman, 1980). As another example, the discussion in Chapter 3 emphasized how the limiting characteristics of working memory capacity reduced the efficiency of good decision making by limiting the number of hypotheses that could be entertained or cues that could be processed.

The Exploitation of Chunking

The data just described suggest that one of the optimum ways to avoid or at least minimize the capacity and decay limitations of working memory is to facilitate "chunking" of material whenever possible. There is indeed substantial experimental support for this assertion. For example, while normal memory span for binomial digits (0's and 1's) is around 7 ± 2, Miller (1956) reports a subject who could remember well over 20 such digits. His technique was to "chunk" triads of digits into an octal code, familiar to many computer users. In this code, for example, 010 becomes the octal digit 2, and 111 becomes the digit 7. Thus, the 6-item string 010111 would be encoded into the 2-chunk string 27. Hence 20 items is no more than 7 chunks.

A still more dramatic example of chunking is provided by Chase and Ericsson (1981). These investigators employed a subject who initially had a normal memory span and trained him to retain and correctly recall strings as long as 82 digits! A second subject was trained to spans of up to 68. Both of these subjects adopted a number of interesting storage and retrieval strategies in the course of the several months of training, and foremost among these was chunking. In particular both subjects, avid runners, learned to encode many of the digit strings as running times for races of a certain distance, and these times became chunks. Thus they might have coded 353431653 as a mile run in 3 minutes, 53.4 seconds followed by a marathon run in 3 hours, 16 minutes, 53 seconds. Where race times were not appropriate they chunked in terms of familiar dates or ages. Furthermore, in order to accommodate the longer lists that were finally recalled, groups of chunks were combined hierarchically to form "super chunks." Chase and Ericsson argue that subjects do not in fact retain the entire list in working memory. Rather, they use working memory and their chunking strategy to set up a highly efficient storage and retrieval system.

In general, the procedure followed in chunking is either to find or to create a meaningful sequence of stimuli within the total string that, because of its meaningfulness, has an integral representation already stored in long-term memory. If the physical sequence is close but not identical to the sequence in memory, then it may be stored as the memory sequence plus an exception (Chase & Ericsson, 1981). For example, 1676 might be stored as "100 years before Independence Day." As a consequence of this representation, each original string of 2–5 items that formed the sequence is "condensed" for storage purpose to a single node. This new set of higher-order nodes is then stored (although these also may be chunked in turn into a second-order node if possible). Upon recall, only the node sequence must be recalled.

When each node is activated in its turn, its contents—the original items—are simply "unpacked" and recalled in their stored order. The issue of how the nodes are recalled in proper sequence will be considered later in this chapter.

Chunking in fact has two correlated benefits: (a) Because it effectively reduces the number of items in working memory, it thereby increases the apparent capacity of this limited store. (b) Because of the reduced load on working memory, material contained therein is more easily rehearsed and is thereby more easily transferred to permanent long-term memory. This transfer is facilitated further because, by definition, the associations between the items that constitute a chunk already exist in long-term memory. There appear to be three important characteristics of chunking that may be utilized to advantage in system design. These relate to teaching strategies, to capitalizing on familiarity, and to the physical spacing of stimuli.

Teaching chunking.   Chunking can be thought of in part as a strategy or mnemonic device that may be taught. The subjects described by Miller and by Chase and Ericsson learned to apply grouping principles to facilitate storage. In the same manner, we can learn to search for meaningful subgroups of digits or letters in license plate numbers or other codes that must be learned. In fact, an emerging conclusion from several studies of "expert behavior" in a variety of disciplines such as computer programming (Norcio, 1981a, 1981b), chess (Chase & Simon, 1973; deGroot, 1965), or decision making (Payne, 1980) is that the expert is able to perceive and store the relevant stimulus material in working memory in terms of its chunks rather than its lowest-level units (Anderson, 1979, 1981).

A dramatic demonstration of this phenomenon was provided by Chase and Simon (1973) comparing recall of a chessboard layout by chess masters and by novices. If the board position was taken from the progression of a logical game, experts recalled far more accurately than did novices. However, if the board position was created by a random placement of pieces, no difference between the two groups was evident. Only in the first case could the expert "chunk" the various pieces into familiar subpatterns—stored in long-term memory in terms of the evolution of different game strategies. When confronted with the random board that had no correspondence in their long-term memory with what a chessboard should look like, chunks were unavailable and the expert's recall regressed to the level of the novice. In a related fashion, our discussion in Chapter 3 of expert decision making emphasized the use of patterns of correlated attributes, or "syndromes," that behaved essentially as chunks and allowed the decision maker to encode more levels on more different attributes than the limits of working memory would predict (Phelps & Shanteau, 1978).

Capitalizing on familiarity.   While chunking is a strategy that is a property of the memorizer, it may also be hindered or helped by properties of the stimulus material that is to be memorized. Obviously, the difference between 21 random letters that exceed working memory capacity and the 21 letters of seven 3-letter words that do not is that in the second case the sequencing of letters has been constrained to form chunks. Exploiting this difference, system designers should formulate codes in such a way as to facilitate chunking. California license plates contain a large number of words—a strategy that takes advantage of this principle. Commercial phone numbers often try to accomplish this by using familiar alphabetic strings in place of digits ("dial 263 HELP"). In fact, the abandonment of the alphabetic prefix (AM 3–7539) for the numeric one (263–7539) in phone numbers seemed to represent a move away from effective chunking. In general, letters induce better chunking than digits because of their greater number of sequential associations.

The importance of chunking to telephoning is also illustrated by considering the order in which a telephone number is dialed. Since the prefix of a phone number (263) is generally more familiar and more likely to be stored in long-term memory as a chunk than is the suffix (7539), less of a demand is placed on memory if the more vulnerable suffix is held in memory for a shorter period of time. This goal would be accomplished if the suffix were dialed first rather than the other way around. Memory would be unburdened of the more fragile information as rapidly as possible. (This procedure has not been implemented presumably because of the fairly complex changes in switching logic of the switchboard circuiting that would be required. Also, like the redesign of the typewriter keyboard, to be discussed in Chapter 10, any major redesign of a familiar household system such as the telephone would be exceedingly difficult to implement in practice).

Parsing.   Chunking may be facilitated by parsing or placing physical discontinuities between subsets that are likely to reflect chunks. Thus the digit sequence 4149283141865 is probably less easily encoded than 4 1492 8 314 1865, which is parsed in such a way as to emphasize five chunks ("for Columbus ate pie at Appomattox"). For the sufficiently imaginative reader these five chunks in turn may be chunked hierarchically as a single visual image. When Loftus et al. (1979) investigated subjects' memory of air traffic control information, they observed that four-digit codes were better retained when parsed into two chunks ("twenty-seven eighty-four") than when presented as four digits (two seven eight four). Bower and Springston (1970) presented subjects with sequences of letters that contained familiar acronyms and found that memory was better if pauses separated the acronyms (FBI . . . JFK . . . TV) than if they did not (FB . . . IJF . . . KTV).

When long stimulus sequences are parsed in a way that will facilitate chunking, it is important to consider the optimum size of a chunk for storage. When stimulus-controlled chunks (through parsing) become too long, the likelihood that the units within a chunk will constitute a familiar sequence in long-term memory diminishes, to the extent that the parsed unit itself may have to be subdivided by the viewer into two or more chunks. On the other hand, decreasing the size of the parsed unit will, of course, increase the total number of chunks. An experiment by Wickelgren (1964) seems to suggest that the optimum chunk size is 3–4. Subjects were presented strings of 6–10 digits and asked to rehearse them in groups of 1, 2, 3, 4 or 5. Recall was best when three-digit rehearsal groups were employed, and nearly as good with four-digit groups. This conclusion agrees with the recommendation made by Bailey (1982) with regard to the optimum size of grouping for any arbitrary alphanumeric strings used in codes.

Recently the concept of chunking has been applied to the memory and comprehension of computer programs. Norcio (1981a) has examined the kinds of errors that programmers make when they must fill in a missing line of program code. This technique represents an indirect yet objective way of assessing program comprehension. His analysis of these data suggests that replacement is more difficult for lines at the beginning of an algorithmic or logical statement than in the middle. This finding indicates that the logical statement itself represents a chunk. When the first line is present, the following lines are more automatically activated or "unpacked" from long-term memory and may be easily retrieved, if they are physically missing. On the other hand, the first line, or "chunk label," is not so easily accessed when it is missing. A second study by Norcio (1981b) suggests that *indentation* of program statements, a form of parsing, can facilitate the chunking process if the indentation is imposed at natural boundaries between chunks (i.e., between logical statements).

In contrast to the benefits of indentation seen by Norcio, there is also experimental evidence that in some circumstances indentation may be harmful to program comprehension and recall (Schneiderman, 1980). Similar harmful effects are caused by program documentation statements that are inserted within the program (in contrast to the benefits of the initial, context-setting documentation described in Chapter 4). These negative effects may also be attributed to chunking: (a) to the extent that the program aids (indentation or documentation) do not correspond with natural chunks in the programmer's memory or (b) to the extent that either indentation or documentation so expands the physical size of the program that parsimonious groupings can no longer be easily visualized. In the case of indentation, the harm will occur if the shortened lines imposed by indentation require that statements be divided between lines (Schneiderman, 1980). The moral is straightforward. Physical parsing will assist chunking if (a) it conforms to rather than violates subjective chunking boundaries and (b) it does not greatly increase the physical size of the material.

Forgetting

When we lose material from working memory because we engage in a competing activity, two major causes seem to contribute to the disruption of the memory trace: (1) The memory "decays" and grows less salient over time—like the passive decay of the sensory code. (2) The competing activity disrupts the trace through an active process of *interference* (Underwood, 1957). In the Brown-Peterson paradigm described on page 217, the relative contribution of the two effects cannot easily be distinguished since the activity that diverted attention from rehearsal (counting backwards), thereby allowing passive decay to operate, might also have disrupted the phonetic code of the letters because of the great degree of similarity between these two activities. The separate roles of the decay and interference factors in working memory were demonstrated by Reitman (1974). She found that when a distractor task that was quite different from the verbal-memory material was employed during the retention interval (auditory signal detection of faint tones), there was still forgetting—a "decay" without interference. When a highly similar activity was employed (counting backwards), the rate of forgetting increased. Thus time and similarity both operate in conjunction to attenuate the strength of the memory trace (Klatzky, 1980).

Similarity.   It is instructive to consider in more detail what is meant by "the effect of similarity on interference." We shall do so by analyzing the two terms *similarity* and *interference* in greater detail. Similarity may be defined at two different levels. One level is defined in terms of codes of processing. The greater interference found by Reitman (1974) between memory of verbal information and the verbal counting task presumably reflects the common competition for verbal or phonetic processes in working memory. Similar examples of interference between spatial tasks and spatial working memory were described earlier in this chapter. At a finer level of detail, similarity differences *within* codes may affect interference as well. The "Conrad effect," described on page 201, in which greater interference within the phonetic code is observed when acoustically similar letters are employed, provides such an example. Within the semantic code, Shulman (1972) obtained an analogous effect of similarity, finding that a list of semantically similar items were forgotten more rapidly than a list of semantically different items.

Interference.   While similarity may be defined at two levels—similarity of codes and similarity within a code—interference may result from one of three sources: within-list, retroactive, and proactive. *Within-list interference*, exemplified by the "Conrad effect" with acoustically similar items, describes the forgetting that results from similarity between the items within a group. For example, when an air traffic controller must deal with a number of aircraft from one fleet, all possessing similar identifier codes (AI3404, AI3402, AI3401), the interference due to the similarity between items makes it difficult for the controller to maintain their separate identity in working memory (Fowler, 1980). Interestingly, while this kind of similarity will disrupt the ordered recall of the items it may actually enhance their *free* *recall*—that is, the ability to recall all items independent of their spatial or temporal order (Crowder, 1978). Unfortunately for the air traffic controller in the above situation, "free recall" will not be sufficient. The controller must also maintain in working memory the identity of the separate aircraft along some ordered continuum (e.g., projected time of arrival, position in airspace). It is knowing which item goes where that is devastated by interference from within-list similarity.

Retroactive interference is interference due to any activity that takes place between the time that material is encoded into memory and the time that it is retrieved for later use (Underwood, 1957). This prominent source of forgetting would occur, for example, when a process control monitor forgets the value of a diagnostic test before it can be written down because she must perform an additional numerical calculation during the interim. The greater the similarity between the item to be retained and the intervening activity, the greater the likelihood of interference. In Peterson and Peterson's experiment, the similarity between counting backwards and the verbal items was high and so the retroactive interference was great.

Proactive interference is the consequence of activity that precedes the encoding of the item to be remembered. A particularly graphic demonstration of proactive interference was observed by Keppel and Underwood (1962) when they repeated the Brown-Peterson experiment. On almost all trials they obtained similar retention functions to those depicted for 3 and 5 items in Figure 6.9. However, on the very first trial of a series, the functions were flat, showing perfect recall at all retention intervals like the 1-item curve of Figure 6.9. Keppel and Underwood interpreted their finding by assuming that on most trials, forgetting was enhanced by interference from the items previously learned (i.e., proactive interference). On the very first trial, however, there were no preceding items and so forgetting was nil.

Assuming that the phenomenon observed by Keppel and Underwood is indeed interference, and knowing that interference is enhanced by similarity, then we should observe a reduction in proactive interference, or a "release from PI" if the delay between successive items to be learned is increased or if the kind of materials used between successive memory trials of the Brown-Peterson paradigm is suddenly changed. Immediately after the change there is less similarity of the new items with the preceding items. D. Wickens, Born, and Allen (1963) observed a release from PI when trigrams of consonant letters (*KJF*) were suddenly switched to trigrams of digits (284). Similar effects were observed when other changes in item similarity were used. For example, when word lists are used, altering the taxonomic category or the general meaning of the words to be stored will cause a temporary release from PI (D. Wickens, 1970).

A manifestation of the buildup and release of proactive interference was observed by Loftus et al. (1979) in their study of communications between air traffic control and pilots. The investigators observed that recall on a given trial was significantly disrupted if it followed the preceding trial by less than 10 seconds. Intervals of greater length were necessary for material in memory from the earlier trial to dissipate so that its subsequent interference effect on recall for the later trial would be minimized. Loftus et al. did not change the material from trial to trial. However, the release from PI phenomena suggests that when an operator must encode and then retrieve repeated series of heterogeneous material, retrieval will be more accurate if the items are presented in varying order, so that qualitatively different kinds of information are stored in sequence. For example, suppose a supervisor must read and then enter the status of several systems along several attributes (i.e., several turbines each with a set of different instrument readings). Assuming that attributes were either objectively or subjectively coded differently, then performance would be superior if the supervisor dealt in turn with all attributes of one system before progressing to the next system, rather than dealing with all systems on one attribute before progressing to the next attribute. This manner of dealing with the multiple attributes of several items has further benefits in performance related to chunking. These benefits will be discussed later in the chapter in the context of running memory.

Intentional forgetting.   The experiment of Loftus et al. demonstrated that the buildup of proactive interference from prior learning (or encoding in memory) can disrupt later memory. Since much of the information that we retain briefly in working memory is information that we will probably never need or want again, it would seem useful to be able to "purge" our memory of this material. Hopkin (1980) suggests that a major potential source of errors for the air traffic controller results when "unwanted baggage" of previously used information disrupts the retention of later material. In order to attenuate this undesirable effect, besides spacing memory trials as Loftus et al.'s data suggest and varying the nature of material to be held in memory as the phenomenon of PI release would indicate, it is also possible to purposely "purge" memory of unneeded items after their recall. Research by Bjork (1972) suggests that such purposeful forgetting can be easily trained; furthermore, it achieves the desired goal of reducing the harmful influence of PI upon the working memory of future items. Martin and Kelly (1974) found that when subjects are asked to purge memory of unwanted items, they have more attention available for dealing with a concurrent task. It would seem then that this technique, like chunking, is a potentially valuable strategy that can be learned and subsequently employed for more efficient storage and retrieval of subsequent memory items.

Running Memory

Our discussion so far has identified the typical estimate of working memory capacity as 7 ± 2 chunks. In many real-world environments, however, this is probably an optimistic figure (Moray, 1980). In the *running memory* task typical of much human-machine interaction, the human's capacity is greatly attenuated. In the running memory task, a continuous sequence of stimuli is presented to the operator, who neither knows its length nor is expected to retain the entire string. Instead, a different response must be made to each stimulus or series of stimuli at some lag after they occur. For example, a series of aircraft might be "handled" by the air traffic controller, inspected items might be categorized by the quality control inspector, or check readings on instruments of a dynamic system might be stored in order to update an internal model of the system. In this paradigm, if the operator is interrupted and asked to recall the last few items, the memory span will be considerably reduced from the 7 ± 2 value. More difficult still is the situation depicted in Figure 6.10 in which the operator must continuously retrieve information after a lag of a few items. More formally, she must hold a certain number K of the most recent items in working memory and, as item n is presented, respond with the identity of item $n - K$. In this case, typical of what happens if an operator falls behind in processing stimuli that are transient in nature, performance falls off rapidly for values of K greater than 2 (Moray, 1980).

Fortunately, in a running memory task when stimuli are complex and multidimensional, some corrective solutions are available to reduce the effect of the severe limitations. Yntema (1963) has demonstrated some of these solutions in a task analogous to that confronting an air traffic controller. In Yntema's task the subject must keep track of a large number of *objects* (e.g., identified aircraft), each of which varies along a number of *attributes* (e.g., altitude, airspeed, location), which in turn can take on a number of specific *values* (e.g., 6,000 ft, 400 mph, 60°N). The value of a particular attribute for one object is periodically updated and the subject must revise memory of the current conditions. At random times the subject's memory is probed concerning the nature of one of the objects, and the fidelity of running memory is reflected in the accuracy of report. Yntema investigated different conditions in which the number of objects, the number of attributes per object, and

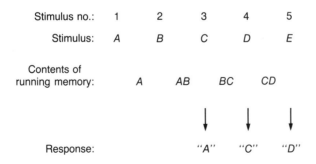

Figure 6.10 Running memory task with $K = 2$.

the number of values per attribute were varied. The values were chosen so that each of these manipulations produced formally equivalent changes in the amount of information load.

Three important conclusions were obtained from Yntema's data: First, subjects performed much better with a few objects that varied on a number of attributes than with many objects that varied on a few attributes, when the total number of variables was the same in both cases. Presumably, the object acts as a sort of integrating "chunk" that readily incorporates its various attributes (Kahneman & Treisman, in press). This finding suggests that if responsibility in such a monitoring situation is to be divided between operators, then it is advisable (from the standpoint of memory limits) to assign each operator to monitor all attributes of a few objects rather than a particular set of attributes for all objects. Fortunately, this finding is concordant with the recommendation made in the previous discussion concerning release from proactive interference. We suggested there that it is better for the operator to deal sequentially with all attributes of a single object than to do the reverse. Second, the number of values per attribute has little influence on accuracy of memory as long as changes in condition take place at a relatively constant frequency. Thus accuracy will not be greatly sacrificed by increasing the precision with which values are presented and must be retained, despite the fact that more information in the formal sense is required per item. (We should note that Yntema's values were digital readouts and therefore were not susceptible to the absolute judgment limitations described in Chapter 2.)

Third, performance is much better if each attribute has its own unique scale, discriminable from the others. This conclusion is consistent with the detrimental effect of item similarity on ordered recall described above. For example, this guideline would dictate that height and speed be coded in different scale units—height perhaps in 1,000-ft units and speed in miles per hour. Thus the values 15.2 and 400 would represent an altitude of 15,200 ft and a speed of 400 mph. The discriminability could be enhanced further by presenting the information for each attribute in print of different size, case, or color.

In Yntema's experiment, subjects did not have the information continuously available. After an attribute was updated, it was hidden and its value was *required* to be maintained in memory. This state of affairs might correspond to the process monitor reviewing the status of several plant variables on a selective centralized display. On the other hand, this restricted availability is not always typical. For example, the air traffic controller normally has the status of all relevant aircraft continuously visible and so is able to respond on the basis of perceptual rather than memory data. However, even in systems in which there *is* a continuous display of the updated items, the principles described should still apply. As discussed in Chapter 4, an efficiently updated memory will ease the process of perception through top-down processing and will unburden the operator when perception may be directed away from the display. Furthermore, there is always the possibility of a system failure, in which the perceptual information will no longer exist, not a trivial occurrence with air-traffic-control computers. In this case, an efficient memory becomes *essential* and not just useful.

Retrieval from Working Memory: The Memory-Search Task

The majority of research on working memory has addressed the process of memory storage and has employed a *recall* measure of performance (i.e., the subject is asked to repeat or write down the material that remains available). Sternberg (1966, 1969, 1975) has focused attention more explicitly on the search for and retrieval of information from working memory in a *recognition* task. In the classic memory-search paradigm developed by Sternberg, the subject is given a small set of characters (e.g., 1–6 letters) to hold in working memory. These letters are referred to as the *positive set*. Then a probe letter is presented and the subject is required to decide as rapidly as possible if the probe was or was not a member of the previously memorized positive set—that is, to search through working memory. When the reaction time to make responses is plotted as a function of the memory-set size from which the probe was selected, the data typically appear as shown in Figure 6.11. Positive responses are shown in the lower function, negative responses are in the upper. Numerous variations of the task have been implemented and some will be considered in later chapters, but the

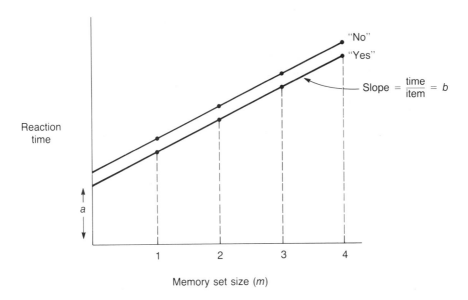

Figure 6.11 Hypothetical data from the Sternberg memory-search task.

aspect of the data most important to the current discussion concerns the linear effect of positive set size on the latency of response shown in Figure 6.11.

The data depicted in Figure 6.11 may be interpreted as follows: (a) A constant time is required to encode the probe stimulus and to execute the response after a decision has been made whether or not it matches one of the positive set. This constant is independent of the size of the memory set and is reflected by the intercept a of Figure 6.11. (b) The actual comparison search time varies linearly with the size of the positive set and is independent of the time to encode and respond. Since the function in Figure 6.11 is linear (each character added to the positive memory set adds a constant time to the search), the search is argued to be a serial one; the encoded probe is compared in turn to each item in working memory, and each comparison requires a constant time. As more items are present, proportionately more time is taken until the search process will encounter a match. The search time per item is given by the slope of the function b. (c) The fact that the slope of the function is found to be as steep for "yes" as for "no" responses suggests that the search is not only serial but *exhaustive*. That is, even after a match is found for "yes" responses the serial comparison process is carried through until all of the memory items have been compared with the probe. Failure to find a match only adds the same constant time to all responses, independent of the set size. The logic behind this reasoning is as follows. If the search were *self-terminating* (stopping upon reaching a match), then when a positive probe is presented, sometimes a match will be encountered on the first comparison, sometimes not until the last, but on the average, matches should be made after the set is only half searched. For mismatches, the search must always be carried to completion. This difference between positive and negative searches caused by a self-terminating search predicts that adding members to the memory set should increase "no" response latencies twice as much as "yes" latencies, thereby producing a steeper slope for noes than for yesses. The fact that the slopes are parallel suggests that the search is not self-terminating but is exhaustive.

It is useful to compare these data with the data obtained by Neisser and his colleagues in the *perceptual* search task discussed in Chapter 4 and shown in Figure 4.2 (Neisser, 1967). Here a very similar linear function was obtained from a perceptual search that was also serial. The more items that have to be searched (on the display, not in memory) until a target is encountered, the longer will be response time. However, in the perceptual search task the search is clearly self-terminating. The subject stops when a target is encountered and does not search the rest of the list. Hence the reaction-time function of list length (the perceptual equivalent to memory-set size) when a target is not encountered would be twice as steep as that when a target is encountered.

While Sternberg's memory-search paradigm and the resulting model of working memory retrieval does not have direct implications for system design innovations to unburden memory, it does have a number of other uses that are of importance to engineering psychology. In this regard, a notable aspect of the paradigm is the consistency of its results over a wide variety of different experiments and conditions. The intercept parameter a may vary with changes in encoding and response difficulty (see Chapter 9), but using alphanumeric stimuli the value of the slope parameter b is remarkably stable at 38 msec/character.[2] This stability suggests that the paradigm taps quite directly a consistent, measurable, and fundamental parameter of human processing efficiency that seems to be little affected by strategies and practice. A further elaboration by Cavanaugh (1972) reveals that the basic search rate parameter b changes systematically with different types of material (e.g., words, nonsense syllables) in a function that is inversely proportional to the capacity of short-term memory for the material in question. For example, nonsense syllables for which working memory capacity is low (around 5 items) show a proportionately slower search rate in the paradigm than do letters with the higher capacity of around 7–9 items.

Considering these two characteristics together, we note that the Sternberg task provides both a good predictor of differences in working memory capacity and a measure of this capacity that is reliable and stable and not influenced by differences in chunking strategy. As a consequence, the Sternberg task can be employed as a highly sensitive index to reveal changes in memory function induced by such variables as age (Salthouse & Sonberg, 1982), Mental Retardation (Harris & Fleer, 1974), task loading (Crosby & Parkinson, 1979; C. D. Wickens & Derrick, 1981; Wetherell, 1979), environmental stress, or toxic chemicals (Smith & Langolf, 1981). Given the pronounced individual differences that result from the use of chunking when conventional measures of memory capacity are employed, the Sternberg search parameter appears to represent a more consistent and therefore more powerful technique for estimating changes in memory capacity (Smith & Langolf, 1981).

LONG-TERM MEMORY

Transfer to Long-Term Memory

A detailed treatment of the characteristics of the more permanent store of information that we describe as long-term memory (LTM) is beyond the scope of this book, and the reader is referred to treatments such as those of Anderson and Bower (1973), Adams (1980), or Klatzky (1980) for deeper coverage. Similarly, the issue of training—the acquisition of

---

[2]The slope is constant, however, only so long as the positive memory set to be searched is randomly varied between trials or blocks of trials. If it is consistent (e.g., the subject always searches for letters J, F, and W), then this subset of letters will eventually benefit from an automatic processing response and the slope will drop to near zero (Schneider & Shiffrin, 1977; see also Chapter 4).

more permanent information—will not be considered. Stammers and Patrick (1975) and Anderson (1981) provide good treatments. Instead, the final pages of this chapter will touch briefly on three characteristics of long-term memory that relate to the previous discussion and have important implications for system design: the role of working memory in the transfer of information to long-term memory, the role of mnemonics, and the importance of organization.

Earlier in this chapter we suggested that chunking provided benefits to two distinct processes: (1) It helps to maintain information in working memory for a few seconds until it can be "dumped" or "offloaded." (2) It helps to transfer material to a more permanent long-term memory store. A typical task of the first category is performed by the air traffic controller who must "handle" the data concerned with many aircraft, but after each leaves her field of responsibility, these data are no longer relevant and may be purged from memory. In the second category, an example is the industrial process monitor or military watch officer who assumes the watch and is presented with a series of current status reports and instrument checks that must be integrated and stored in long-term memory in order to create an overall mental representation of the state of the system. Furthermore, this representation must be updated as new information arrives. Thus all samples should be transferred to a more permanent representation. By "permanent" we do not mean that the information will be stored forever but that it must be available for subsequent decisions and actions, even after it may have left the direct focus of consciousness (i.e., after it has left working memory; James, 1890).

The distinction between these two characteristics of working memory—maintenance and transfer—is important from both a practical and a theoretical perspective. At a practical level, we noted that after information has been used, it should be purged from our memory system in order to reduce the harmful effects of proactive interference on subsequent material. This goal is precisely the opposite of the optimal strategy for transfer to LTM, which emphasizes retention.

At a theoretical level Craik and his colleagues (Craik and Lockhart, 1972; Craik & Watkins, 1973; Watkins, Watkins, Craik, & Mazuryk, 1973) have articulated the distinction between the two qualitatively different rehearsal strategies involved in maintenance and transfer. The two strategies are similar in that both involve working memory and compete for processing resources with concurrent tasks. However, they are distinct in that *maintenance rehearsal* tends to emphasize the phonetic aspects of the stimulus for immediate retrieval, while *elaborative rehearsal* emphasizes the semantic attributes of the material and their association with information that is already stored in LTM.

Different experimental instructions can be employed to induce predominantly one rehearsal strategy or the other. When this is done, several investigators find that the amount of time spent in elaborative rehearsal increases the likelihood of later recall from LTM (i.e., transfer to LTM). On the other hand, recall from LTM is unrelated to the amount of time spent in maintenance rehearsal (Craik & Watkins, 1973; Glenberg, Smith, & Green, 1977). Since transfer to LTM will increase the likelihood of later interference, these data suggest that proactive interference may be reduced by employing phonetic maintenance rather than semantic elaboration when information is only to be retained for a short time. This strategy could then be coupled with the purging strategy investigated by Bjork (1972) and discussed above.

The two types of rehearsal also differ in their dependence upon information content. As we noted, the capacity of working memory for maintaining information seems to be relatively independent of the information content of the material. Chunks may contain either a lot of information or a little as formally measured, depending upon the degree of associations existing between material existing in long-term memory. For example, seven digits contain 3.3 bits/chunk; seven letters contain 4.7 bits; seven three-letter words contain formally 14.8 bits/chunk. In a related way, as Yntema demonstrated, the capacity of running memory in terms of formal information may be either high or low, depending upon whether several attributes of a few objects or a few attributes of several objects are monitored. Performance also was affected little by the number of possible levels of the variables retained, even though varying the number of possible levels changes the formal information content.

On the other hand, there is good evidence that the rate of transferring material into long-term memory is an inverse function of its information content (Adelson, Muckler, & Williams, 1955; Keele, 1973). This observation is not surprising in light of the fact that the key to efficient storage and retrieval of information from LTM is good organization (Bower, 1970; Bower et al., 1969). As described in Chapter 2, organization is inversely related to information content. Therefore, material that is poorly organized (possessing a high information content) will require longer for the organization to be imposed so that efficient learning can take place.

Mnemonics

The previous discussion has suggested that both attention and organization were critical factors for the transfer of information to long-term memory. *Mnemonic strategies* refer to a series of techniques that explicitly detail these organizational rules and demonstrate their utility in encoding or storing new information (Bower & Reitman, 1972; Harris, 1980; Luria, 1968; Yates, 1966). Chunking is one such technique that we have described already, used by Chase and Ericsson's subjects to greatly increase the apparent memory span for digits. We shall not review the discussion of chunking here except to reemphasize one of its fundamental characteristics, the importance of hierarchical organization. Any list of items that is to be permanently stored in memory will be better memorized if these can somehow be grouped into hierarchical clusters (Bower, Clark, Lesgold, & Winzenz, 1969). The advantage is particularly evident in recall when members of each cluster can be exhaustively listed as a relatively small group. Therefore, the hierarchical relations existing in any list of items that must be memorized should be stressed.

While chunking and hierarchical organization may help the retention of a large number of items, they do not by themselves provide a means of recalling the *order* in which the chunks are to occur. Yet often the order of recall is critical, as when a set of procedures must be executed in fixed sequence. To aid the storage and retrieval of ordered information, variations of the mnemonic *method of loci* have been proposed (Yates, 1966).

The method of loci works by providing a visual memory framework of a fixed sequence order. Each item to be memorized can be associated through visual imagery with a point in the ordered framework. As originally performed by Roman orators to memorize the order of points delivered in orations, the sequence was a series of "loci," or familiar landmarks, that would be encountered as the orator took a walk along a certain familiar route (i.e., traversing the inside of a house or following a familiar path from one point to the next). Each landmark on the route is well known and therefore is associated with a visual image in LTM. Then, as the new chunks or points are to be stored in order, visual images of each one are associated with the image of a landmark. At the time of recall, the route can be mentally "traversed," each landmark "examined," and the stored chunk "found" and recalled.

More recently, Bower and Reitman (1972) investigated a variant of the method of loci in which the ordered image sequence was created by memorizing a rhyming list: "one is a bun, two is a shoe, three is a tree. . . ." In this case, the order stored by the method of loci is replaced by the order of the numerical scale, and the landmarks of the loci are replaced by visual images of the associated "peg words" (i.e., bun, shoe, tree). If bizarre images are formed relating each item on the list to be remembered to the peg words, then recall is greatly facilitated. Of course, mnemonic techniques such as the method of loci require some initial investment to learn the strategy. However, once learned, the benefits of the techniques are well validated.

Congruence of Organization in Long-Term Memory

The discussion in Chapter 4 emphasized that the value of context lies in the organizing framework it provides for the material. This framework then makes subsequent encoding more efficient—like a filing system that is constructed prior to the deposit of papers within the system. It is essential to realize, however, that not all organizational systems are equally efficient for storing all kinds of material. An alphabetical filing system may be appropriate for names but not for storing the results of experiments. In an analogous manner, it is important for the organizational format of data with which an operator must interact to be compatible or congruent with the organizational form of storage in the operator's memory.

A data base may be organized in many formats—a linear list ordered along some dimension (e.g., an alphabetic name list), a hierarchical organization with dominant categories subsuming lower-level categories (e.g., a company's organizational structure), a network with no particular dominance ordering but with connecting relations between items that are related (e.g., a network describing the frequency of communications between employees of a company or mechanical elements of a system), or a table defining separate attributes of the items in rows and columns (e.g., a table presenting the frequency of different military job classifications and ranks). Human operators must often interact with such data bases, retrieving pertinent pieces of information or determining the relation between items in different locations. It is important therefore that a compatibility of organizational structure exist between long-term memory and the data base with which an operator must interact.

The importance of this compatibility was demonstrated by Durding, Becker, and Gould (1977), who presented subjects with various lists of 15 words to be memorized. The words in each list could in fact be appropriately described by one of the four organizational schemes described above: lists, hierarchies, networks, or tables. Prototypical examples of these schemes are shown in Figure 6.12. However, the subjects only saw the words presented in the same linear format. When subjects were asked to place the word list in an organizational scheme, the scheme that they imposed on paper corresponded quite closely with the scheme inherent in the data. When asked, however, to fit another set of word lists into schemes different from those implicit in the words (e.g., fitting the hierarchical words of Figure 6.12a into a network format), subjects encountered considerable difficulties.

Durding et al. emphasized that their data have implications for the design of computer-based data entry and retrieval systems. Such systems are typical, for example, of those used in business to maintain the records of employees or in health information systems in which a data base concerning various disease prevalence rates and symptoms is stored. The user will probably have some implicit assumption of the organization of the data structure—a mental model of the relationships existing between the elements of that structure. It is important then that the data base be organized in a compatible and congruent format so that the user can interact with the data through the terminal by using commands and language that are appropriate to his or her internal data structure. Durding et al. suggested that those commands be supplemented with visual spatial displays of skeletal organizational structures that depict the format in which the data are actually represented.

The research of Durding et al. identifies one of the most challenging issues in human factors design emerging in this era in which computers are increasingly used not only by "expert" programmers, but by consumers, business people, and clerical workers as well. How does the

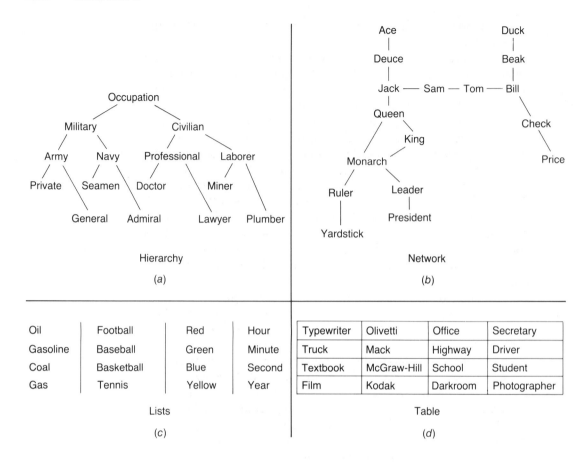

Figure 6.12 Examples of word sets for each of the four major types of organizations. From "Data Organization" by B. M. Durding, C. A. Becker, and J. D. Gould, 1977, *Human Factors, 19*, p. 4. Copyright 1977 by the Human Factors Society, Inc. Reproduced by permission.

computer model or understand the user's conception of the data and logic within the computer itself? Clearly the computer should organize data in a form compatible with the user's mental model. But what if different individuals possess different styles of organization? Are different organizational formats appropriate for spatial versus verbal modes of thinking, as suggested by Schneiderman (1980)? A related question concerns the assumptions that the computer should make about the level of knowledge of the user. For the same program, a computer's interaction with a novice should probably be different from its interaction with an expert user. A novice, for example, would benefit from a menu selection program in which all options are offered, since many of them are not likely to be stored in long-term memory. For the expert, this format will probably give unnecessary clutter, since the alternatives are stored and available in LTM in any case. An intriguing question from the viewpoint of systems design is how the computer can either explicitly assess or implicitly deduce the level of knowledge or the format of organization employed by the user. The challenge offered to research in long-term memory is to provide models of the organization of human memory that can be effectively used by the computer system designer in order to optimize this interaction.

Transition: Memory and Attention

Our survey of memory and its implications to system design has focused more on the transient characteristics of working memory than on the more permanent representation in long-term memory. One reason for this bias is that the discussions of long-term memory are closely linked to issues of training and skill acquisition. These issues, as we have noted, are beyond the scope of the current text. A second reason, however, is that most aspects of long-term memory storage and retrieval operate relatively automatically; as a result, they do not represent bottlenecks in human performance. In contrast, working memory, as we have repeatedly emphasized, is heavily dependent upon the human's limited processing resources. As a consequence, operations in working memory are normally disrupted when attention is diverted to other tasks, thereby representing a substantial bottleneck in complex task performance. It is appropriate therefore that we turn now to a detailed discussion of the nature of the limits of attention and their relevance to engineering psychology. These issues will be the focus of Chapters 7 and 8.

SUPPLEMENT C

Sensory Codes in Memory

THE ICONIC CODE
(SHORT-TERM VISUAL STORE)

The classic experimental demonstration of the briefly prolonged visual image—referred to as the iconic code—is provided by the *partial report paradigm* developed by Sperling (1960) and represented schematically in Figure 6.13. In this paradigm, a 4 × 3 array of letters is briefly presented (i.e., 100–200 msec). While the subject feels phenomenologically that all letters are visible, he typically can only report four or five of them before the others are "lost" from the decaying visual image. To validate that all stimuli are indeed present in iconic memory just after stimulus termination, Sperling *cued* one of the three rows by presenting a low, medium, or high tone. This tone was the cue for the subjects to give a report of letters from the bottom, middle, or top row, respectively. The term "partial report" refers to the fact that only part of the full array is thereby reported. In the example of Figure 6.13,

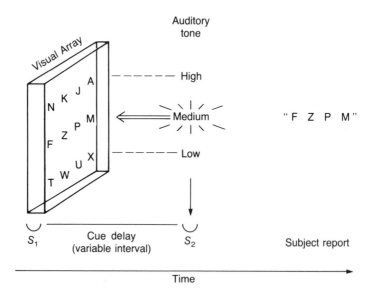

Figure 6.13 Sperling's (1960) partial report paradigm. In this example a medium-tone poststimulus cue is provided, cuing a report of the middle row.

the middle row is cued. Therefore, only the four letters in the cued row are to be reported.

While the cue is only presented *after* stimulus termination, the partial report is normally found to be perfect (⁴⁄₄) if the cue delay is short. This finding suggests that the full array is indeed present in the visual system after the stimulus has terminated, and the subject can use the poststimulus information to read from iconic memory the contents of the appropriate row from the array. Since all information is initially present after stimulus termination, the recall is perfect. As the cue is progressively delayed, however, recall drops off to a steady-state level at delays of around 0.5 second. This duration defines the life-span of iconic memory. To be more precise, the time at which asymptotic performance is reached will depend somewhat on the luminance of the poststimulus background. If it is very low, the icon will last longer, if it is bright, the icon will be shorter lived.

Sperling's results clearly demonstrate that the iconic code fades or decays over time as its information is recoded to a more abstract level. Numerous other experiments have shown how the iconic code can also be "masked," terminated, or overwritten by a following stimulus, thereby truncating the availability of the information (e.g., Averbach & Coriell, 1961; Neisser, 1967; Turvey, 1973; see Coltheart, 1980, for a summary). This masking process appears to result from an *integration* or summation over time of the two stimuli which results from the limited temporal resolving power of the visual system (Eriksen & Schultz, 1978). Visual events that occur together within a sufficiently short temporal window will integrate with each other, and the information content in one stimulus will disrupt and obliterate that in the other.

There are some circumstances, however, in which the integration may enhance recognition of the stimuli if the information content in the two stimuli reinforce each other, so that the integrated montage generates a more meaningful stimulus. This phenomenon was demonstrated by Eriksen and Collins (1971) who presented subjects with sequences of stimuli such as those shown in Figure 6.14. When superimposed, the two random-appearing dot patterns form an integrated pattern (the letter N). At 0 msec interstimulus interval (ISI), the pattern at the right of Figure 6.14 was directly perceived. As the interstimulus interval was increased, the letter was perceived with progressively less accuracy out to ISIs of 500 msec At this delay, the icon of the first stimulus had entirely decayed before the second arrived and so perception was poor.

Coltheart (1980) has presented a comprehensive review of research and theory related to iconic memory. Some of this research has shown convincingly that the iconic representation is not just a peripheral retinal phenomenon but that the persisting code exists within the central nervous system. This evidence has been based primarily upon experiments in which the stimulus and a mask are presented to different eyes (e.g., Turvey, 1973). If the iconic image of a stimulus were

Figure 6.14 Example of stimuli used by Eriksen and Collins to demonstrate integration of two "random" stimuli to create a meaningful pattern. From "Some Temperal Characteristics of Visual Pattern" by C. W. Eriksen and J. R. Collins, 1967, *Journal of Experimental Psychology, 74*, p. 478. Copyright 1967 by the American Psychological Association. Reprinted by permission of the authors.

preserved only at a retinal level, then when the stimulus is presented to one eye and the mask to another, the mask would not be expected to obliterate the stimulus. It does, however, and thereby indicates that the level of processing at which iconic memory resides is higher in the nervous system than the level at which the visual pathways from the two eyes combine.

THE ECHOIC CODE
(SHORT-TERM AUDITORY STORE)

The echoic code, the auditory counterpart of the iconic code, represents the phenomonological "echo" of a sound that persists after the physical stimulus has terminated. A number of different research paradigms have indicated that the duration of the echoic code is considerably longer than that of the iconic, although the estimates of echoic duration have also been somewhat more variable. An important characteristic of the echoic code is that it is *preattentive*—that is, the echoic code will retain information from an auditory source whether that source is attended to or not. As such, it provides the explanation for the "Whadyasay? Oh yeah" effect. This effect occurs when someone addresses you while your attention is directed elsewhere. Your immediate impression, since you were not attending to the spoken message, is that you did not extract its meaning, hence the "Whadyasay." However, focusing your attention on what remains of the trace in echoic memory, you can often "perceive" the echo and interpret its semantic content, hence the "Oh yeah." Three experiments will be described that nicely demonstrate the phenomenon of echoic memory.

Darwin, Turvey, and Crowder (1972) presented subjects with an auditory analog of Sperling's partial report paradigm. Subjects heard three sequences of three alphanumeric characters presented through headphones to three apparent spatial locations: the left ear, the midplane of the head, and the right ear. This was accomplished by modulating the relative intensity of the message delivered to the two headphones. When intensity is equal, the message is localized in the middle of the head. All nine items were presented in one second, so that one-third of a second separated each item triad. In the *whole-report* condition, subjects had to repeat as many letters and digits as they could recall in the correct order and location. Performance in this condition was relatively poor, with less than half of the items correctly recalled. In the *partial report* condition, one of three visual-spatial cues was presented shortly after the last stimulus. These cues, which were lights to the left, front, and right of the subject, indicated which of the three spatial channels to report. When the cue occurred immediately after stimulus offset, a much greater proportion of the cued channel could be recalled than in the whole-report condition. As the cue was progressively delayed, the proportion of a location correctly reported deteriorated, to approach a level equal to the whole-report condition when the delay reached 4 seconds. The data of Darwin et al. were thus analogous to those reported by Sperling but indicated a longer life-span of echoic memory. From these data, Darwin et al. inferred that the trace from which the report was made followed a decay function, reaching a steady asymptotic level by 4–5 seconds.

In the second study, Wickelgren (1969) examined the way in which echoic memory could assist subjects in determining if two successive tones were of the same or different pitch. When the tones were presented in close temporal proximity, subjects naturally had an easy task. When one was presented long after the other, performance was much lower since a comparison was required between the perception of the second pitch and the fallible memory of the first. The important characteristic of Wickelgren's data is reflected in the function relating percent correct to temporal interval as this interval is varied from short to long. This function showed a decay that reached an asymptotic level at around 8 seconds. The inference can therefore be made that for delays as long as 8 seconds, the subjects possessed an echoic trace of the first stimulus to assist in their judgment. After this time the echo had vanished and judgment was based upon a longer-term store.

The discrepancy between Wickelgren's estimate and that of Darwin et al. (8 versus 4 seconds) may be easily resolved by considering the difference in difficulty of the two tasks. Wickelgren's task was fairly simple, involving simply a 1-bit discrimination along a single dimension (pitch). Therefore, even a very faint trace could help the comparison. In contrast, the task used by Darwin, Turvey, and Crowder was much more complex, requiring *recall* of three 5.2-bit items of information. The usable information in the trace would therefore be far more disrupted by the decay function, and so the trace would cease to provide assistance after a much shorter delay.

A third manifestation of echoic memory is demonstrated by the phenomena of *recency* and the *suffix* effect (Crowder, 1978; Crowder & Morton, 1969). In the experimental paradigm, subjects hear a list of 7 or

8 letters that they must subsequently recall. In a control condition, the recall cue is simply a short silent interval. Under these conditions, subjects tend to show better recall for letters at the beginning of the list (called a primacy effect) and those at the end of the list (called a recency effect) (see Figure 6.15). Crowder and Morton interpret the recency effect to reflect the advantage to the last few letters that remain in echoic memory and are little disrupted by subsequent sound. This advantage is not provided to the earlier letters in the list. To demonstrate this difference, recall in the control condition is contrasted with recall in the *suffix* condition. Here the recall cue is given by an articulated sound such as the word "go" heard immediately after the last item. The subject knows that this cue is not part of the list and doesn't have to be stored in memory; yet this stimulus suffix is sufficient to abolish the recency portion of the curve (Figure 6.15).

Crowder and Morton also found that the suffix effect is *not* observed when a visual suffix is presented. This implies that it is the auditory quality of the suffix and not its information content that disrupts the recency portion of recall. The uniquely auditory nature of the recency effect, upon which the suffix effect is dependent, is demonstrated when a visual analog of the same procedure is used. In this case, primacy of recall is again observed, but there is no recency effect, even in the absence of a suffix. When the letters are presented visually the decay of iconic memory is too rapid to benefit recall for the final stimuli in the list.

An obvious implication of the suffix effect to system design principles is that when an auditory message is presented that may have to be retained in memory, it should where possible be preceded and not followed by any supplementary auditory information. Sometimes information must follow the message. For example, in presenting an auditory message, some terminal stimulus is often required to indicate that the message has indeed ended. In this case, the signal should be of a dissimilar acoustic nature to the actual speech stimulus (e.g., a tone rather than a word) to reduce the interference of the suffix effect.

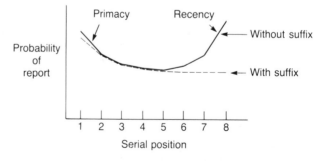

Figure 6.15   The suffix effect, demonstrating the loss of recency.

ICONIC VERSUS ECHOIC MEMORY

The longer delay of echoic than iconic memory may be explained in terms of the former's adaptive function for perception. Auditory stimuli are intrinsically more transient than visual, since a major component of the auditory stimulus information is in the temporal domain, and temporal information is by definition transient. Much of the content of visual information is spatial, which does not need to be transient. Therefore, the greater "persistence" of the auditory store seems adaptively suited to accommodate the greater transience of auditory stimuli. The slower decay time of echoic memory appears to be responsible for the fairly consistent observation in memory research that working memory for verbal material is enhanced when its presentation is auditory rather than visual (Murdock, 1968; Nilsson, Ohlsson, & Ronnberg, 1977; Watkins, 1972). A logical explanation for this finding is that the more persistent availability of the auditory, as opposed to the visual, sensory code allows information to be translated to the phonetic code of working memory with greater fidelity.

ECHOIC MEMORY VERSUS
THE PHONETIC CODE

It is both tempting and parsimonious to assume that the phonetic and echoic codes are one and the same phenomenon, yet Crowder (1971, 1978) has presented convincing evidence that they are not. This evidence is provided by his demonstration that the recency effect and the suffix effect, both phenomena of the echoic code, operate on vowels but not on consonants. In this demonstration, subjects are asked to recall a string of stimuli that are differentiated only in terms of their consonant characters (*ba, da, ga, fa, ta, ma*). The recency advantage to the last few characters disappears, indicating that there is no "echo" of the discriminating consonants persisting to aid their recall. Consonants can be rehearsed and are a part of our phonetic experience, and this fact indicates that they clearly are represented in phonetic working memory. Yet they do not exist within echoic memory. Such a conclusion conforms with our intuitions that vowels echo but consonants do not. It also confirms the dichotomy between echoic and phonetic codes: The former only maintains transient sounds of vowels. The latter maintains the phonetic representations of both vowels and consonants.

References

Adams, J. (1980). *Human memory* (2nd ed.). New York: McGraw-Hill.

Adelson, M., Muckler, F. A., & Williams, A. C. (1955). Verbal learning and message variables related to amount of information. In H. Quastler (Ed.), *Information theory in psychology*. Glencoe, IL: Free Press.

Anderson, J. R. (1979). *Cognitive psychology*. New York: Academic Press.

Anderson, J. R. (1981). *Cognitive skills and their acquisition*. Hillsdale, NJ: Erlbaum Associates.

Anderson, J. R., & Bower, G. H. (1973). *Human associative memory*. Washington, DC: Winston.

Averbach, I., & Coriell, A. S. (1961). Short-term memory in vision. *Bell System Technical Journal, 40*, 309–328.

Baddeley, A. D., Grant, S., Wight, E., & Thompson, N. (1975). Imagery and visual working memory. In P. M. Rabbitt & S. Dornic (Eds.), *Attention and Performance V.* New York: Academic Press.

Baddeley, A. D., & Hitch, G. (1974). Working memory. In G. Bower (Ed.), *Recent advances in learning and motivation* (vol. 8). New York: Academic Press.

Baddeley, A. D., & Lieberman, K. (1980). Spatial working memory. In R. S. Nickerson (Ed.), *Attention and Performance VIII.* Hillsdale, NJ: Erlbaum Associates.

Bailey, R. W. (1982). *Human performance engineering: A guide for system designers*. Englewood Cliffs, NJ: Prentice-Hall.

Bencomo, A. A., & Daniel, T. C. (1975). Recognition latency for pictures and words as a function of encoded-feature similarity. *Journal of Experimental Psychology: Human Learning and Memory, 1*, 119–125.

Bjork, R. A. (1972). Theoretical implications of directed forgetting. In A. W. Melton & E. Martin (Eds.), *Coding processes in human memory*. Washington, DC: Winston.

Boies, S. (1971). *Memory codes in a speeded classification task*. Unpublished Ph.D. dissertation, University of Oregon, Eugene, OR.

Booher, H. R. (1975). Relative comprehensibility of pictorial information and printed words in proceduralized instructions. *Human Factors, 17*, 266–277.

Bower, G. H. (1970). Organizational factors in memory. *Cognitive Psychology, 1*, 18–46.

Bower, G. H., Clark, M. C., Lesgold, A. M., & Winzenz, D. (1969). Hierarchical retrieval schemes in recall of categorized word lists. *Journal of Verbal Learning and Verbal Behavior, 8*, 323–343.

Bower, G. H., & Reitman, J. S. (1972). Mnemonic elaboration in multilist learning. *Journal of Verbal Learning and Verbal Behavior, 253*, 478–485.

Bower, G. H. & Springston, F. (1970). Pauses as recoding points in letter series. *Journal of Experimental Psychology, 83*, 421–430.

Broadbent, D. E. (1971). *Decision and stress*. New York: Academic Press.

Brooke, J. B., & Duncan, K. D. (1980). Flowcharts versus lists as aides in program debugging. *Ergonomics, 23*, 387–399.

Brooks, L. R. (1967). The suppression of visualization in reading. *Quarterly Journal of Experimental Psychology, 19*, 289–299.

Brooks, L. R. (1968). Spatial and verbal components in the act of recall. *Canadian Journal of Psychology, 22*, 349–368.

Brown, J. (1959). Some tests of the decay theory of immediate memory. *Quarterly Journal of Experimental Psychology, 10*, 12–21.

Cavanaugh, J. P. (1972). Relation between immediate memory span and the memory search rate. *Psychological Review, 79*, 525–530.

Chase, W. G., & Ericsson, A. (1981). Skilled memory. In S. A. Anderson (Ed.), *Cognitive skills and their acquisition*. Hillsdale, NJ: Erlbaum Associates.

Chase, W. G., & Simon, H. A. (1973). The mind's eye in chess. In W. G. Chase (Ed.), *Visual information processing*. New York: Academic Press.

Coltheart, M. (1980). Iconic memory and visible persistence. *Perception & Psychophysics, 27*(3), 183–228.

Conrad, R. (1957). Accuracy of recall using keyset and telephone dial and the effect of a prefix digit. *Journal of Applied Psychology, 42*, 285–288.

Conrad, R. (1962). An association between memory errors and errors due to acoustic masking of speed. *Nature, 193*, 1314–1315.

Conrad, R. (1964). Acoustic comparisons in immediate memory. *British Journal of Psychology, 55*, 75–84.

Conrad, R., & Hull, A. J. (1964). Information, acoustic confusions, and memory span. *British Journal of Psychology, 55*, 429–432.

Craik, F. I. M., & Lockhart, R. S. (1972). Levels of processing: A framework for memory research. *Journal of Verbal Learning and Verbal Behavior, 11*, 671–684.

Craik, F. I. M., & Watkins, M. J. (1973). The role of rehearsal in short-term memory. *Journal of Verbal Learning and Verbal Behavior, 12,* 599–607.

Crosby, J. V., & Parkinson, S. (1979). A dual-task investigation of pilot's skill level. *Ergonomics, 22,* 1301–1313.

Crowder, R. G. (1971). The sound of vowels and consonants in immediate memory. *Journal of Verbal Learning and Verbal Behavior, 10,* 587–596.

Crowder, R. (1978). Audition and speech coding in short-term memory. In J. Requin (Ed.), *Attention and Performance VII.* Hillsdale, NJ: Erlbaum Associates.

Crowder, R. G., & Morton, J. (1969). Precategorical acoustic storage (PAS). *Perception & Psychophysics, 5,* 365–371.

Danaher, J. W. (1980). Human error in ATC system. *Human Factors, 22,* 535–546.

Darwin, C. J., Turvey, M. T., & Crowder, R. G. (1972). An auditory analogue of the Sperling partial report procedure. *Cognitive Psychology, 3,* 255–267.

deRoot, A. D. (1965). *Thought and choice in chess.* The Hague, Netherlands: Mouton.

Durding, B. M., Becker, C. A., & Gould, J. D. (1977). Data organization. *Human Factors, 19,* 1–14.

Ells, J. G., & Dewar, R. E. (1979). Rapid comprehension of verbal and symbolic traffic sign messages. *Human Factors, 21,* 161–168.

Eriksen, C. W., & Collins, J. F. (1971). Sensory traces versus the psychological moment in temporal organization of form. *Journal of Experimental Psychology, 89,* 306–313.

Eriksen, C. W., & Schulz, D. (1978). A continuous flow model of human information processing. In J. Requin (Ed.), *Attention and Performance VII.* Hillsdale, NJ: Erlbaum Associates.

Estes, W. K. (1973). Phonemic encoding and rehearsal in short-term memory for letter strings. *Journal of Verbal Learning and Verbal Behavior, 12,* 360–372.

Fitter, M., & Green, T. R. (1979). When do diagrams make good computer languages? *International Journal of Man-Machine Studies, 11,* 235–261.

Fowler, F. D. (1980). Air traffic control problem: A pilot's view. *Human Factors, 22,* 645–654.

Glenberg, A., Smith, S. M., & Green, C. (1977). Type 1 rehearsal: Maintenance and more. *Journal of Verbal Learning and Verbal Behavior, 16,* 339–352.

Greenblatt, D., & Waxman, J. (1978). A study of three data base query languages. In B. Schneiderman (Ed.), *Databases: Improving usability and responsiveness.* New York: Academic Press.

Harris, C., & Fleer, R. (1974). High-speed memory scanning in mental retardates: Evidence for central processing deficits. *Journal of Experimental Child Psychology, 17,* 452–459.

Harris, J. E. (1980). Memory aids people use: Two interview studies. *Memory & Cognition, 253,* 31–38.

Hart, S. G., and Loomis, L. L. (1980). Evaluation of the potential format and content of a cockpit display of traffic information. *Human Factors, 22,* 591–604.

Healy, A. F. (1975). Temporal-spatial patterns in short-term memory. *Journal of Verbal Learning and Verbal Behavior, 14,* 481–495.

Hoffman, E. R., & MacDonald, W. (1980). Short-term retention of traffic turn restriction signs. *Human Factors, 22,* 241–252.

Hopkin, V. S. (1980). The measurement of the air traffic controller. *Human Factors, 22,* 347–360.

Jacobs, J. (1887). Experiments in prehension. *Mind, 126,* 75–79.

James, W. (1890, reprinted 1950). *The principles of psychology.* New York: Dover.

Kahneman, D., & Treisman, A. (in press). Changing views of attention and automaticity. In R. Parasuraman & R. Davies (Eds.), *Varieties of attention.* New York: Academic Press.

Kammann, R. (1975). The comprehensibility of printed instructions and the flow chart alternative. *Human Factors, 17,* 183–191.

Keele, S. W. (1973). *Attention and human performance.* Pacific Palisades, CA: Goodyear.

Keppel, G., & Underwood, B. J. (1962). Proactive inhibition in short-term retention of single items. *Journal of Verbal Learning and Verbal Behavior, 1,* 153–161.

Klatzky, R. L. (1980). *Human memory: Structures and processes.* San Francisco: Freeman.

Kosslyn, S. (1981). The medium and the message in mental imagery: A theory. *Psychological Review, 88,* 46–66.

Loftus, G. R., Dark, V. J., & Williams, D. (1979). Short-term memory factors in ground controller/pilot communications. *Human Factors, 21,* 169–181.

Luria, A. R. (1968). *The mind of a mnemonist.* New York: Basic Books.

Martin, P. W., & Kelly, R. T. (1974). Secondary task performance during directed forgetting. *Journal of Experimental Psychology, 103,* 1074–1079.

Melton, A. W. (1963). Implications of short-term memory for a general theory of memory. *Journal of Verbal Learning and Verbal Behavior, 2,* 1–21.

Miller, G. A. (1956). The magical number seven plus or minus two: Some limits on our capacity for processing information. *Psychological Review, 63,* 81–97.

Moran, T. (1981). An applied psychology of the user. *Computing Surveys, 13,* 1–11.

Moray, N. (1980, May). *Human information processing and supervisory control* (Technical Report). Cambridge, MA: Massachusetts Institute of Technology, Man-machine System Laboratory.

Murdock, B. B. (1968). Modality effects in short-term memory: Storage or retrieval? *Journal of Experimental Psychology, 77,* 79–86.

Neisser, U. (1967). *Cognitive psychology.* New York: Appleton-Century-Crofts.

Nilsson, L. G., Ohlsson, K., & Ronnberg, J. (1977). Capacity differences in processing and storage of auditory and visual input. In S. Dornick (Ed.), *Attention and Performance VI.* Hillsdale, NJ: Erlbaum Associates.

Norcio, A. F. (1981a). *Comprehension aids for computer programs* (Technical Report AS–1–81). Annapolis, MD: U.S. Naval Academy, Applied Sciences Department.

Norcio, A. F. (1981b). *Human memory processes for comprehending computing programs* (Technical Report AS–2–81). U.S. Naval Academy, Applied Sciences Department.

Palmer, E., Jago, S., Baty, D., & O'Connor, S. (1980). Perception of horizontal aircraft separation in a cockpit display of traffic information. *Human Factors, 22,* 605–620.

Payne, J. W. (1980). Information processing theory: Some concepts and methods applied to decision research. In T. S. Wallsten (Ed.), *Cognitive processes in choice and decision behavior.* Hillsdale, NJ: Erlbaum Associates.

Peterson, L. R., Peterson, M. J. (1959). Short-term retention of individual verbal items. *Journal of Experimental Psychology, 58,* 193–198.

Phelps, R. H., & Shanteau, J. (1978). Livestock judges: How much information can an expert use? *Organizational Behavior and Human Performance, 21,* 209–219.

Posner, M. I. (1973). *Cognition: An introduction.* Glenview, IL: Scott, Foresman.

Posner, M. I. (1978). *Chronometric explorations of the mind.* Hillsdale, NJ: Erlbaum Associates.

Posner, M. I., & Mitchell, R. F. (1967). Chronometric analysis of classification. *Psychological Review, 74,* 392–409.

Ramsey, H. R., Atwood, M. E., & Van Doren, J. R. (1978). Flow charts vs. program design languages. In E. Baise & S. Mitter (Eds.), *Proceedings, 22nd annual meeting of the Human Factors Society.* Santa Monica, CA: Human Factors.

Reitman, J. S. (1974). Without surreptitious rehearsal: Information and short-term memory decays. *Journal of Verbal Learning and Verbal Behavior, 13,* 365–377.

Salthouse, T. A. and Sonberg, B. L. (1982). Isolating the age deficit of speeded performance. *Journal of Gerontology, 37,* 59–63.

Salzberg, P. M., Parks, T. E., Kroll, N. E. A., & Parkinson, S. R. (1971). Retroactive effects of phonemic similarity on short-term recall of visual and auditory stimuli. *Journal of Experimental Psychology, 91,* 43–46.

Schneider, W., & Shiffrin, R. M. (1977). Controlled and automatic human information processing I: Detection, search, and attention. *Psychological Review, 84,* 1–66.

Schneiderman, B. (1980). *Software psychology,* Cambridge, MA: Winthrop.

Schneiderman, B. L., Mayer, R., McKay, D., & Heller, P. (1977). Experimental investigations of the utility of detailed flowcharts in programming. *Communications of the Association for Computing Machinery, 20,* 373–381.

Shulman, H. G. (1972). Semantic confusion errors in short-term memory. *Journal of Verbal Learning and Verbal Behavior, 11,* 221–227.

Smith, P., & Langolf, G. D. (1981). Sternberg's memory scanning paradigm in assessing effects of chemical exposure. *Human Factors, 23,* 701–708.

Sperling, G. (1960). The information available in brief visual presentations. *Psychological Monographs, 74* (Whole No. 11).

Stammers, R., & Patrick, J. (1975). *The psychology of training.* London: Methuen.

Sternberg, S. (1966). High-speed scanning in human memory. *Science, 153,* 652–654.

Sternberg, S. (1969). The discovery of processing stages: Extensions of Donders' method. *Acta Psychologica, 30,* 276–315.

Sternberg, S. (1975). Memory scanning: New findings and current controversies. *Quarterly Journal of Experimental Psychology, 27,* 1–32.

Stone, D. E., & Gluck, M. D. (1980). *How do young adults read directions with and without pictures?* (Technical Report). Ithaca, NY: Cornell University, Department of Education.

Thomas, J. C., & Gould, J. D. (1975). A psychological study of query by example. In *Proceedings of the National Computer Conference* (pp. 439–445). Arlington, VA: AFIPS Press.

Thorson, G., Hockhaus, L., & Stanners, R. F. (1976). Temporal changes in visual and acoustic codes in a lettermatching task. *Perception & Psychophysics, 19,* 346–348.

Turvey, M. T. (1973). On peripheral and central processes in vision: Inferences from information processing analysis of masking with patterned stimuli. *Psychological Review, 80,* 1–52.

Underwood, B. J. (1957). Interference and forgetting. *Psychological Review, 64,* 49–60.

Vidulich, M. D., & Wickens, C. D. (1982). The influence of S-C-R compatibility and resource conflict on performance of threat evaluation and fault diagnosis. In R. E. Edwards (Ed.), *Proceedings, 27th annual conference of the Human Factors Society.* Santa Monica, CA: Human Factors.

Watkins, M. J., Watkins, O. C., Craik, F. I., & Mazuryk, G. (1973). Effect of nonverbal distraction on short-term storage. *Journal of Experimental Psychology, 101,* 296–300.

Weingartner, A. M. (1982). *The internal model of dynamic systems: An investigation of its mode of representation.* Unpublished undergraduate honors thesis, University of Illinois, Champaign, IL.

Wetherell, A. (1979). Short-term memory for verbal and graphic route information. *Proceedings, 1979 meeting of the Human Factors Society.* Santa Monica, CA: Human Factors.

Wickelgren, W. A. (1964). Size of rehearsal group in short-term memory. *Journal of Experimental Psychology, 68,* 413–419.

Wickelgren, W. A. (1969). Associative strength theory of memory for pitch. *Journal of Mathematical Psychology, 6,* 13–61.

Wickens, C. D., & Derrick, W. (1981). *The processing demands of higher order manual control: Application of additive factors methodology* (Technical Report EPL–81–/ONR–81–1). Champaign, IL: University of Illinois, Engineering Psychology Research Laboratory.

Wickens, C. D., Sandry, D., & Vidulich, M. (1983). Compatibility and resource competition between modalities of input, central processing, and output: Testing a model of complex task performance. *Human Factors, 25,* 227–248.

Wickens, D. D. (1970). Encoding categories of words: An empirical approach to meaning. *Psychological Review, 77,* 1–15.

Wickens, D. D., Born, D. G., & Allen, C. K. (1963). Proactive inhibition and item similarity in short-term memory. *Journal of Verbal Learning and Verbal Behavior, 2,* 440–445.

Wright, P., & Reid, F. (1973). Written information: Some alternatives to prose for expressing the outcomes of complex contingencies. *Journal of Applied Psychology, 57,* 160–166.

Yallow, E. (1980). *Individual differences in learning from verbal and figural materials* (Aptitude Research Project T. R. No. 12). Palo Alto, CA: Stanford University, School of Education.

Yates, F. H. (1966). *The art of memory.* Chicago: University of Chicago Press.

Yntema, D. (1963). Keeping track of several things at once. *Human Factors, 6,* 7–17.

Attention and Perception

OVERVIEW

The limitations of human attention, like those of working memory, constitute one of the most formidable bottlenecks in human information processing. We can all easily relate instances when we failed to notice or attend to the words of a speaker because we were distracted or our attention was directed elsewhere or occasions on which we had so many tasks to perform that some were neglected. These intuitive examples of failures of attention may be described more formally in terms of three categories:

1. *The limits of selective attention.* In some instances we select inappropriate aspects of the environment to process. For example, as discussed in Chapter 3, decision makers sometimes select salient rather than diagnostic cues. A more dramatic example is provided by behavior of the flight crew of the Eastern Airlines L–1011 flight that crashed in the Everglades in 1972. Because of their preoccupation with a malfunction elsewhere in the cockpit, none of the personnel on the flight deck attended to the critical altimeter reading and to subsequent warnings that showed that the plane was entering gradual descent to ground level (Wiener, 1977; see also Chapter 12).

2. *The limits of focused attention.* Occasionally we are unable to concentrate upon one source of information in the environment—that is, we have a tendency to be distracted. The clerical worker transcribing a tape in a room filled with extraneous conversation encounters such a problem. So also does the translator who must ignore the feedback provided by her own voice in order to concentrate solely on the incoming message. Another example is the pilot attempting to locate rapidly a critical item of information in the midst of a "busy" display consisting of a multitude of other changing variables. The difference between failures of selective attention and failures of focused attention is that in the former case there is an intentional but unwise choice to process nonoptimal environmental sources. In the latter case this processing of nonoptimal sources is "driven" by external environmental events despite the operator's efforts to shut them out.

3. *The limits of divided attention.* On the other side of the coin from the limits of focused attention are the limits of divided attention. When problems of focused attention are encountered, our attention is inadvertently directed to stimuli we do not wish to process. When problems of divided attention are encountered, we are unable to divide our attention between stimuli or tasks all of which we wish to process. Here

we may consider the automobile driver who must scan the highway for road signs while maintaining control of the vehicle or the examples encountered in Chapter 3 of the fault diagnostician who must maintain several hypotheses in working memory while scanning the environment for diagnostic information and also entering this information into a recording device. Thus the limits of divided attention are often considered to be synonymous with our limited ability to *time-share* performance of two or more tasks.

Attention in perception may be described by the metaphor of a searchlight (Wachtel, 1967). The momentary direction of our attention is a searchlight. The focus of this searchlight falls on that which is in momentary consciousness. Everything within the beam of the light is processed whether wanted (successful focusing) or not (failure to focus). Within the context of this metaphor, three properties of the searchlight are relevant to phenomena of human experience: (1) The "breadth" of the beam and the distinction, if any, between the desired focus, or "penumbra" (that which we want to process), and the "umbra" (that which we must process but do not want to). These represent the issues of divided and focused attention, respectively. (2) The hand that guides the searchlight: How rapidly can it shift from one location to another? This represents the hardware "switching" aspects of selective attention. (3) The brain controlling the hand that guides the searchlight. This represents the software, or "executive," properties of selective attention. Each of these will be considered in some detail below, as we consider examples of how operators search the complex stimulus world for critical pieces of information and how that information is processed once it has been found.

The searchlight metaphor conveniently describes the various characteristics of attention and perception, the topic of this chapter. Yet the concept of attention is relevant to a range of activities that are not only perceptual. We can speak of dividing attention between two tasks, no matter what stage of processing they require. The broader issue of divided attention as it relates to the time-sharing of activities will be the concern of the next chapter.

The following chapter will present a basic overview of the experimental findings of selective and focused attention, with details of the laboratory experiments upon which these findings are based presented in the chapter supplement. Then two practical implications of this research will be addressed in some detail: the implications to display design and those to visual search and target acquisition.

SELECTIVE ATTENTION

Optimality of Selection

In situations such as the aircraft cockpit or the process control console, there are many sources of information in the environment which must be sampled periodically. In these situations, engineering psychologists have studied how optimal the performance of subjects is when selecting the relevant stimuli to attend at the appropriate times. As in our discussions of signal detection theory, *optimal* is defined in terms of a behavior that will maximize an expected value or minimize an expected cost. For example, an aircraft pilot who continuously fixates (samples) the airspeed indicator but ignores the altimeter is clearly nonoptimal. One who samples both of these but never checks outside the aircraft is doing better but is still not optimal, for he is incurring the expected costs of missing important stimuli (e.g., other aircraft) that can only be seen by looking out the window.

Engineering psychologists who have worked in this area divide the environment into *channels*, along which critical *events* may periodically occur. They have assumed that environmental sampling is guided by the expected cost that results when an event is missed. The *expected* cost of missing an event is equal to the true cost of missing it multiplied by the probability that the event will be missed. The probability of missing an event in turn is directly related to event frequency. Those events that occur often are more likely to be missed if the channels along which they occur are not sampled. Also in most real-world tasks, the probability of missing an event on a channel increases with the amount of time since the channel has last been sampled. For example, the probability of missing an aircraft outside the pilot's window increases with the time that has passed since he has last looked out.

When optimum sampling is examined in the laboratory the subject has typically been presented with two or more potential channels of stimulus information, along which events may arrive at semipredictable rates. For example, a channel might be an instrument dial, with an "event" defined as the needle moving into a danger zone. The channels may be visual locations in space (e.g., Senders, 1964; Sheridan, 1972) or spatial locations where auditory stimuli may occur (Moray, Fitter, Ostry, Favreau, & Nagy, 1976). The general conclusions of these studies, which are well summarized by Moray (1978, 1981), suggest the four conclusions discussed in the following paragraphs.

First, sampling is guided by an "internal model" of the statistical properties of the environment. The frequency of events on different channels and the correlation of channels with each other determine where and when the operator will sample. Because sampling strategies provide estimates of the operator's internal model of a system under supervision, the patterns of fixations will also be of help to the system designer in locating information displays more optimally. Dating from the pioneering work of Fitts, Jones, and Milton (1950), aviation psychologists have employed scanning data to configure displays according to two principles: Frequently sampled displays should be placed centrally. Pairs of displays that are often sampled sequentially should be located close together (Frost, 1972).

Second, people learn to sample channels with higher event rates more frequently and lower rates less frequently. However, sampling rate is not quite adjusted upward or downward with event frequency as much as optimal. This is similar to the "sluggish beta" phenomenon discussed in Chapter 2, describing people's reluctance to adjust the response criterion in signal detection as far as optimal.

Third, human memory is imperfect and sampling reflects this fact. Hence people tend to sample information sources more often than they would need to if they had perfect memory about the status of an information source when it was last sampled. Also because of limits of memory people may forget to sample a particular display source if there are a multitude of possible sources. This, for example, might well be the case for the monitor of a nuclear process control console. These limitations in memory suggest the usefulness of computer aiding to present "sampling reminders" (Moray, 1981).

Fourth, when people are presented with a preview of scheduled events that are likely to occur in the future, then sampling and switching become somewhat more optimal. Now subjects' sampling can be guided by an "external model," the display of the previewed events. However, even here there are departures from optimal behavior if the number of channels is large. In this case, the "planning horizon"—the use of the future to guide in sampling—declines (Tulga & Sheridan, 1980). All in all, then, sampling may be described as reasonably optimal but not perfect, subject to certain limitations of human memory. The experiments upon which these conclusions are based will be described in greater detail in the chapter supplement.

Stress and Arousal

The laboratory research on optimal sampling has usually taken place in a fairly relaxed experimental environment. However, there is some anecdotal evidence that under conditions of high stress, sampling departs from optimal in a systematic fashion: fewer environmental cues are sampled. For example, the process control monitor when confronted with a sudden emergency—several indicators flashing—may observe progressively fewer symptoms in diagnosing the failure as his overall level of stress and arousal is increased by the pending emergency. (See Chapters 3 and 12 for more discussion of this issue.)

The research that has supported this conclusion is again based primarily upon laboratory investigations (see Broadbent, 1971; Kahneman, 1973). Subjects are given a set of tasks to perform—for example, tracking a moving target and detecting visual events in the periphery. Then arousal is manipulated by various means; stressors, incentives, shocks, or drugs have been employed. Two general conclusions emerge from these experiments: (1) Under high arousal the focus of attention is more restricted: fewer cues are sampled. (2) Fortunately and adaptively, the channels that are sampled generally tend to be those that are perceived by the subject as being more *important*. Importance here is determined by the frequency of events—their costs or values and their salience.

It is this third factor—salience—that presents potential problems. As discussed in Chapter 3, salient cues (those in central vision or those that are intense) may not necessarily be the best ones to sample. An important objective of engineering psychology then should be to establish ways to ensure that *all* important channels and not just the salient ones will be sampled in times of high stress.

Sampling Between Modalities

Often the operator may be in an environment in which there are both auditory and visual events. In the airborne environment, for example, most information is delivered through visual instruments, but the ear receives both communications from ground control and auditory warning signals. Research on selective attention between the eye and the ear seems to indicate two general principles: (1) It takes about 100 msec longer to switch attention between modalities than between two sources of information within a modality. (2) Humans have a general bias to process information within the visual modality. This bias is referred to as *visual dominance.*

The experiments on attention switching typically employ a manipulation of *expectancy* (LaBerge, Van Gelder, & Yellott, 1971). Stimuli may be delivered on either auditory or visual channels, but on a given trial the subject is led to expect that the stimulus will occur on one channel. For example, a visual stimulus may be expected. When the time to react to that expected visual stimulus is compared with the time to react to the same visual stimulus when an auditory one is expected, reaction time is about 100 msec longer in the latter case. Here it is assumed that attention was "aligned" to the expected auditory channel and needed to be "switched" to the visual modality before processing of the visual stimulus could begin.

The difference in switching time between and within modalities probably has few major implications outside the laboratory. It may suggest that information from two tasks can be dealt with better if information from both is presented visually as long as both are presented within foveal vision and information concerning the two is presented sequentially. If either of these conditions is not met, then further difficulties of dividing attention will arise. These will be discussed later in this chapter and again in Chapter 8.

The phenomenon of visual dominance is exhibited in a wide number of different situations summarized by Posner, Nissen, and Klein (1976). Typical is the result of an experiment by Colovita (1971). Subjects in a reaction-time experiment were presented with either an auditory or visual stimulus to respond to. On a very few occasions both stimuli were presented at once. During these "conflict trials" Colovita found that subjects almost always made only the response appropriate to the visual stimulus as if that stimulus suppressed or dominated the processing of the auditory.

The occurrence of visual dominance might at first glance lead one to conclude that critical information such as warning signals should be presented visually. However, the theory of visual dominance proposed by Posner et al. (1976) in fact states just the opposite. It is precisely *because* auditory (and kinesthetic) stimuli are intrinsically more intrusive upon consciousness that we adopt a more general bias in favor of the less intrusive visual modality. Hence stimuli that are unexpected and infrequent (as warnings typically are) will be better processed if they are auditory.

The visual dominance phenomenon seems to operate instead in situations where streams of auditory and visual information may be delivered to the operator at roughly the same rate. Under these circumstances, Posner et al. (1976) conclude that humans will tend to "shut out" the more intruding auditory (or kinesthetic) stimuli in favor of vision. Therefore the auditory information will be less likely to be processed.

FOCUSED AND DIVIDED ATTENTION

The discussion of selective attention has generally assumed that the spotlight of attention is unitary. A channel is either attended or not, and the research concerns the factors that drive the spotlight to one source or another. However, by now a considerable amount of experimental evidence indicates that attention is not strictly serial in this fashion. Several "things" may be processed by the brain "in parallel." Stated in other terms, we can divide attention between the processing of different kinds of information. Because divided attention leads to the processing of more information and because there are several applied situations in which the operator is overloaded with information (the nuclear control room after a malfunction is a case in point), it seems important to understand how and when we can successfully divide attention.

Defining a Channel

It is useful to think about divided and focused attention in terms of a channel model (Treisman, 1969). Our attention can be focused on a channel of information. All events within the channel can roughly be processed in parallel. Furthermore, simultaneous events within a channel *must* be processed in parallel even if we don't want to. Thus one event may intrude upon the ability to focus attention exclusively on the other. On the other hand, information on separate channels must be processed serially so that even if we wish, we cannot process in parallel. But at the same time, if desired, we can focus attention on information in one channel while excluding that from another.

The most obvious physical property that defines a channel is space. A good deal of research indicates that it is difficult to process information in parallel when it comes from different locations in the visual field. Similarly, it is hard to *focus* attention on information coming from one location of space, whether these are sights or sounds, if there is a second kind of information (i.e., change in stimulus properties) delivered at that same location in space. Here again the spotlight metaphor is directly applicable.

While different locations in two-dimensional space (vertical and horizontal) seem to be the most effective way of defining channels of focused attention, there are other dimensions that can be used as well. In vision, for example, there is some evidence that distance can also define a channel. Neisser and Beckman (1975) had subjects view video displays of two different animated games projected at different apparent distances from the observer. A hand-clapping game displayed the hands up close; a ball-tossing game showed figures in the distance. They found that subjects showed only a very limited ability to divide attention between the two channels defined by the two games. This experiment has direct relevance to recent technological innovations in aviation. The "heads up display," or HUD, is a means of displaying some parts of the pilot's instrument panel directly on the windscreen, ostensibly so that the pilot may monitor the instruments while viewing in parallel the outside pattern of visual landing cues. Yet Neisser and Beckman's findings suggest that parallel processing of the two distances may not necessarily take place. In fact, it may be relatively easy for the pilot to focus on the near instruments while ignoring potentially critical information outside the aircraft (Fischer, Haines, & Price, 1980).

In audition, too, there are dimensions other than space that may show the properties of channels (how hard it is to divide between or focus within them). For example, Treisman (1964b) found that pitch can define a channel. It is more difficult to focus attention on one of two voices of the same sex (and therefore pitch range) than on one of two voices of different sex. Treisman also found that the semantic content of a message can define a channel. She found that subjects had greater difficulty excluding an unwanted message from the focus of attention when the message dealt with the same material as the attended message.

Parallel Processing Within a Channel

The channel model asserts that two sources of information within a channel—here close together in space—will be processed in parallel. This turns out to be a mixed blessing because parallel processing can have either of two helpful or either of two harmful effects. Which effect prevails depends upon the nature of the tasks confronting the operator and the implications of the displayed material for action. Each of these effects will be considered in turn.

When the two sources have independent implications for action—that is, when the operator is performing two tasks—then parallel processing is generally helpful. For example, the pilot controlling both pitch and roll of an aircraft (two relatively independent dimensions) will benefit from parallel processing when these are displayed close together (see Chapter 11). Also, as discussed in Chapter 2, when an operator is reading coded symbols it is useful to convey two different kinds of information as two dimensions of a single symbol.

Sometimes two different sources of information imply the same common action. Here again, close proximity in space is helpful. This mutual assistance is called a *redundancy gain*, also discussed in Chapter 2. For example, using color to represent redundantly a dimension that is also coded by shape will speed up the processing of that information (Kopala, 1979).

The harmful effects of close spatial proximity may be perceptual. Any time that similar aspects of two different stimuli must be processed in parallel, there will be some competition for the *perceptual analyzers* that are used (Treisman, 1969). For example, if you must read two words close together in space there will be competition for the perceptual analyzers involved in reading (see Chapter 4). This phenomenon, called *resource competition*, will be discussed further in Chapter 8. It may occur even if your attention is focused on only one of the stimuli.

When two stimulus aspects are close in space and you only wish to process one, performance will be disrupted if the unwanted aspect has implications for action that are similar to but *incompatible with* the wanted aspect. For example, suppose an operator is to press a lever down if a symbol occurs low on the display and up if it occurs high. The speed with which he does so will be slowed if the symbol is an arrow which points in the opposite direction to the commanded movement (Clark & Brownell, 1975).

Variables that Enhance Parallel Processing

The previous discussion suggests that two sources of information delivered within a channel may have any of four effects depending upon the nature of that information and its implications for action. While our earlier discussion of channels described the dimensions that could be used to define a channel quite precisely, it was less precise in defining the point when two channels become one. How close together in space, for example, do two sources of visual information have to be before they belong to one channel? Eriksen and Eriksen (1974) examined this question. They had subjects make either a right or left lever movement to one of two centrally displayed letter stimuli. Each central stimulus could be surrounded by flanking letters. These could present either compatible or incompatible movement information. The first condition, when the flanking letter indicated movement in the same direction as the central letter, created redundancy gain (but some perceptual conflict); the second condition produced response conflict. As the flanking letters were moved progressively closer to the central letter the magnitude of both redundancy gain and response conflict increased. It appears that information that is within 1° of visual angle of a focused target will be likely to receive some processing in parallel with the focused stimulus (Broadbent, 1982).

The extreme case of close spatial proximity occurs when the two sources of information represent different dimensions of a single object. In this case, an exceptional increase in parallel processing occurs, in part because there is no longer any perceptual competition, since different dimensions of an object (i.e., shape and color) are processed by different perceptual analyzers (Treisman, 1969). However, the use of different perceptual analyzers is apparently not the only reason for good parallel processing of the dimensions of an object because two different attributes of different stimuli do not show as much parallel processing as the attributes of a single stimulus. This suggests a certain intrinsic benefit to the fact that attributes belong to the same object. (See Chapter 5.) Separate dimensions of an object "want" to be processed together (Kahneman and Treisman, in press).

A nice example of parallel processing of dimensions of an object is provided by the Stroop effect (Keele, 1972; Stroop, 1935). Subjects are asked to attend to and report the color of ink in which each word of a series is printed. When each word spells a color name that is different from its ink color there is considerable response conflict. The semantic dimension of the stimulus cannot be filtered out. On the other hand, if the ink color and color name are the same, there is redundancy gain.

The previous discussion relates to earlier topics covered in both Chapters 2 and 5. Stimuli that have the same implication for action show a redundancy gain. This gain is increased when the stimuli are attributes of a single object. Often stimuli will have not identical but similar implications for action: In this case, they will normally be highly correlated. For example, in an energy process, high temperature and high pressure will often co-occur, and the implications for action for both are similar. The data described above indicate that when values are correlated, they will be represented best in terms of an integrated "object display," a concept covered in Chapter 5. The research of Garner and Fefoldy (1970) discussed in Chapter 2 supports this claim.

The research that has been conducted on selected and divided attention has a number of important practical applications to the design of complex multielement displays. Many of the issues here are also directly relevant to the task of *visual search*. This task is required when an operator must scan a visual array, such as a piece of sheet metal, for a target or flaw. These categories of application will be described in the following section.

PRACTICAL IMPLICATIONS OF RESEARCH ON FOCUSED AND DIVIDED ATTENTION

Implications for Display Design

The implications of the basic research on focused and divided attention to human factors relate primarily to issues in display design and display formatting. Clearly there are times when it is important that information concerning several different attributes (either different objects or different dimensions of a single object) be made available to the operator's working memory or decision-making system "in parallel." For example, the air traffic controller might need to know information about the heading of two aircraft at once or about the heading and velocity of a single aircraft. Likewise, when the operator must monitor a number of information channels, the probability of detecting an event on any channel will be enhanced if all are processed in parallel. Under these circumstances, displays should be formatted to emphasize conditions of parallel processing, and designers should consider the role of the four variables that influence this processing.

In many complex systems a large amount of visual information must be compressed into a relatively small area of the visual field. This compression will have the advantage of avoiding excessive scanning. However, the small separation of elements within the display will influence the extent to which problems of focal attention disrupt the processing of information. Furthermore, the research on response conflict indicates that more than the mere presence of "display clutter" should be taken into account. Closely adjacent stimuli that have opposite or incompatible implications for action will be disruptive (e.g., an upward moving indicator adjacent to a downward indicator that requires a downward movement of a control). In a similar manner, when multiple attributes of a single display element are used to provide efficient multidimensional coding (see Chapter 2), caution should be exercised to ensure that the various attributes rarely generate incompatible response tendencies that may in turn produce a "Stroop effect." Eriksen and O'Hara (1982) have shown that these tendencies will be manifest not only for physical dimensions like "left" and "right," but for conceptual dimensions like "same" and "different" as well. Of course, the other side of the coin is that processing will be enhanced by redundancy gain if the implications for action of the two attributes are the same.

It may, of course, be important to design displays that *will* be processed in series. Here the goal is to build separate channels. With visual displays this can easily be done by using widely separated spatial locations so that unwanted information occurs outside of foveal vision. In the auditory domain, the means of formatting separate channels are less clear. Yet it may be desirable to do so since the auditory modality is often underused in systems such as aircraft or process control stations. As with vision, the spatial domain can again be employed. The experiment by Darwin, Turvey, and Crowder (1972) reported in the previous chapter suggests that three "spatial" channels may be processed without distraction if one is presented to each ear and a third is presented with equal intensity to both ears, thereby appearing to originate from the midplane of the head.

In this manner, the pilot of an aircraft might have available three distinct audio channels: one for messages from the copilot, one for messages from air traffic control, and a third reserved for messages from other aircraft or for synthesized voice warnings from his own aircraft. He could not process the three in parallel since all would call for common semantic analysis, but he could at least focus on one without the unwanted intrusion of information from the others. The definition of channels in terms of the pitch dimension suggests that additional separation might be obtained by distinguishing the three spatial channels, redundantly, through variation in pitch quality. Thus the center message that is most likely to be confused with the other two could be presented at a substantially different pitch (or with a different speaker's voice) than the others.

Visual Search and Target Acquisition

An operator's ability to locate targets in a complex background is an important domain of practical application that combines many of the characteristics of both selective and divided attention in perception. This task is relevant, for example, to the quality control inspector locating faults or scratches in sheet metal (Drury, 1975), to the pilot locating targets on the ground from an aircraft (Scanlan, 1977), or to the supervisor locating coded symbols on a complex video display (Teichner & Mocharnuk, 1979). The applied literature on visual search is an extension of the letter-search paradigm of Neisser, Novick, and Lazar (1964) described in Chapter 4 combined with the Sternberg (1975) memory-search-task paradigm described in Chapter 6. However, a major difference from Neisser's paradigm is that the target elements in most applied search paradigms are not presented in an ordered array as shown in Figure 4.2. Instead, the target may appear anywhere in a random field. As a consequence, the searcher can apply neither a linear search procedure, as Neisser's subjects were able to do, nor a search guided by an internal model which generates expected locations, as described earlier in this chapter. Within the visual-search paradigms, interest has focused on two general issues: developing models of search time and identifying factors that influence search speed.

As an example of the first of these issues, Drury (1975; 1982) has derived a model which predicts the probability of detecting a flaw in an inspected industrial commodity as a function of the time allowed for search. A two-stage process is modeled, the first stage reflecting the scanning and searching behavior and the second the sort of detection decision discussed in Chapter 2. Spitz and Drury (1978) conclude that these two phases are relatively independent of each other. Drury identified factors such as batch failure rate that affect the decision component of the second phase, and other factors such as fault conspicuity and the amount of surface area to be inspected that influence the probability of locating a flaw in the first phase. These factors are incorporated into a quantitative model.

The most important general characteristic of this model is that the probability of detecting a target by the time T follows a negatively accelerating function of T. This function in turn dictates that there is an optimum time during which any item should be inspected. Longer times will produce diminishing gains on accuracy. Drury (1975; 1982) discusses the manner in which this optimum time should be established, given such factors as the specification by the industrial manager of the desired rate at which products should be inspected, the probability of fault occurrence, and the desired overall level of inspection accuracy. Then industrial material to be inspected can be presented at a rate determined by this optimal time.

The second class of research has focused directly on the factors that influence the speed of target detection and localization (Drury & Clement, 1978; Mocharnuk, 1978; Scanlan, 1977; Teichner & Mocharnuk, 1979). Four fairly general conclusions have been drawn from this research.

First, the dominant effect on search time is the number of elements to be searched (Drury & Clement, 1978; Mocharnuk, 1978). It matters relatively little if the elements are closely spaced, requiring little scanning, or are widely dispersed. The increased scanning that is required with wide dispersal does increase search time slightly. However, the high density of nontarget elements when the items are closely spaced also has a small retarding influence on search. Thus the two factors, scanning and visual clutter, essentially trade off with each other as target dispersion is varied.

Second, on the basis of a summary of a large number of experimental results, Teichner and Mocharnuk (1979) conclude that overall search *rate* (speed of inspecting an item or a dimension of each item) increases as the total amount of information in the display increases. Information is increased by increasing the number of items to be inspected, the number of relevant dimensions of variation, or the number of possible targets. However, the increase in search rate (items/unit time) with more items is not sufficient to compensate for the increased number of items that have to be inspected, so that the total search time is prolonged either by including more items or a larger number of relevant dimensions.

Third, searching for one of several targets is generally slower than searching for only one. An exception to this conclusion occurs when the multiple targets in the searched-for class can all be defined in terms of one unique, salient feature—for example, if all targets and only targets are red.

The conclusion that multiple targets slow the rate of search, supported by the findings of most applied search results (Craig, 1981;

Geyer, Patel, & Perry, 1979; Monk, 1976; Sheehan & Drury, 1967), is somewhat at odds with Neisser, Novick, and Lazar's (1964) original finding that subjects could search for 10 target letters in a linear display as fast as one (see Chapter 4). These differences in results may seemingly be reconciled by considering the role of practice. Neisser et al. found that multiple letter search was as rapid as single letter search only after several days of practice, more time than was provided in many of the inspection studies. The implications would seem to be that, since the inspector normally is highly practiced in any case, search time would not be greatly hampered by multiple search. However, this is an area that needs considerably more research. Fisk and Schneider (1982), for example, have found that the critical factor determining the efficiency of search is the number of times a target is actually detected and not how often it is *searched for*. Thus to be efficient parallel search must also be practiced with detection of the multiple targets.

Fourth, the different dimensions that can be used to define a target differentially affect search rate. Here the conclusions offered follow relatively directly from those discussed in Chapter 2: (1) Search for targets defined by one dimension in an array that varies only in that one dimension is more efficient than search for targets in a multidimensional array whether the targets are defined by one dimension (i.e., blue) or two (i.e., blue squares). (2) An exception to the above principle occurs when targets are defined redundantly. There is often a redundancy gain when two features uniquely define the target (i.e., the target is a blue square, and no nontargets are ever blue or square). (3) Color appears to be a particularly salient dimension in defining targets for search, being more proficient than shape, size, or alphanumeric characters in defining targets (Christ, 1975). As a consequence, color can be very useful as a redundant code for other dimensions (Kopala, 1979), but it will interfere with perception of those other dimensions when it varies independently of them (Christ, 1975). The advantages of color coding appears to decline however as more than five or six codes are employed (Carter & Cahill, 1979).

In conclusion, the research on visual search—the allocation and processing of visual attention across space—has produced relatively accurate models of human performance that can be applied to real-world behavior to predict performance in complex environments. As we shall see in Chapter 8, this precision is greatly reduced when the domain expands to the broader area of modeling the dividing of attention between tasks that may require activity at all stages of processing.

SUPPLEMENT D

Attention and Perception: Theories and Experiments

SELECTIVE ATTENTION

In many situations we select, choose, or process one source of environmental information before moving to the next, just as the searchlight moves serially from one object to the next (Moray, 1978, 1981). We shall consider in later sections the way in which people can process different aspects of information "in parallel." However, in this section our primary concern is with two questions: How rapidly can we or do we "switch" attention from one source of information to another? How optimally do we select (attend to) those sources of environmental information that are important to us and ignore those that are trivial? These questions portray the selective aspects of attention in the framework of a skill (Moray, 1978; Moray & Fitter, 1973). The operator must develop an internal model of the environment and its sources of information. He must know when these sources become available and what the costs and benefits of attending to or ignoring them might be. This "internal model" thus becomes a program to guide the searchlight beam of attention. Consequently, by monitoring the focus of attention in a complex environment—that is, by determining how frequently the operator looks at or listens to those aspects of the environment that convey important information—the investigator can make inferences about the fidelity of the internal model.

Optimality of Selection

Information sources. Senders (1964) examined the degree to which operators select information optimally from several sources. In his experiment, subjects monitored deviations of indicators in four separately spaced dials. The focus of attention was defined as the direction of visual fixation as this was monitored with a movie camera. The subjects' task was to detect if any of the dials entered a danger zone. Each meter was "driven" by a random noise function with a different maximum frequency of variation. This method varied the probability of signals appearing in the different channels. Senders applied information theory to determine precisely how often the operator *should* sample each signal in order to detect accurately the excursions when they occurred. When compared with this optimal specification, subjects were observed to be relatively close. Sources of greater information content were fixated proportionately more frequently. However, a slight departure from optimality was observed as subjects sampled rare (high information) sources more often than optimal and frequent sources less often. This phenomenon is reminiscent both of the sluggish beta described in Chapter 2 and of the way subjects misestimate event proportions discussed in Chapter 3.

Using auditory stimuli, Moray et al. (1976) have also observed that sampling becomes more optimal as the operator forms a model of the statistical properties of the environment. In a two-channel signal detection experiment, subjects monitored a series of tones appearing in both ears. Occasionally, one tone was more intense, a signal to be detected. Signal contingencies were adjusted so that, while the overall probability of a signal was 0.10, the occurrence of a signal tone on one ear increased the probability of a signal arriving at the same time on the other to 0.50. This is a contingency that often occurs in the real world. Events do not always occur randomly, but signals at one place may correlate with signals at other places.

Moray et al. observed that with practice subjects' sampling of the environment as reflected by their response criterion (willingness to report a tone) reflected this statistical structure. When the response criterion beta was examined for one channel, as a function of whether a detection response (a hit or false alarm) was made on a given trial on the other channel, beta was reduced to a level that was near to the optimal levels for a trial in which $P(Sn) = 0.5$. On the other hand, beta was increased to near the optimal levels for $P(Sn) = 0.10$ on the trials when a correct rejection or miss occurred on the other channel—that is, when the subject decided that a signal was not present. This experiment, like Senders' investigation, demonstrates that attentional sampling is sensitive to event frequency since a "sample" may be equated with a momentary lowering of the response criterion. As with Senders' finding, Moray et al. found that the adjustment of sampling rules is almost, but not quite, as great as that dictated by the optimal model.

Costs and payoffs. Senders' (1964) research took place in a "free" environment, in which no penalties or rewards were imposed. Usually, however, this is not the case. The operator outside the laboratory will probably be penalized for missing a critical event (e.g., attending somewhere else at the time the event occurs). Furthermore, costs are sometimes associated with switching attention. For example, it is often effortful to scan too much and too rapidly, and the cost of switching away from one source might well be the failure to observe something that later occurs at that source. Sometimes sampling information imposes a cost in terms of mechanical implementation. In a wartime submarine environment, sampling information might involve turning on a sonar. The cost occurs because the sonar signal will possibly reveal the submarine's location. Sheridan (1972) has developed an optimum model dictating how often one should sample from a given information source, given that there are costs associated with sampling it but also costs associated with leaving it unsampled. The latter costs are reflected by the fact that leaving an information source unsampled increases the uncertainty about its state since it was last sampled.

An example of such sampling strategies is provided by the aircraft pilot on an approach to landing. She is concerned with how frequently she needs to observe either the primary flight display depicting the aircraft's attitude in airspace or the cloud-obscured world outside the cockpit. A more detailed hypothetical example is provided by a father who is watching a baby (who crawls but does not walk) beside a swimming pool while reading an important contract. There is clearly a *cost* associated with *not* attending the baby (she will plop into the pool if she crawls far enough). There is, however, a cost associated with sampling or attending the baby: Comprehension of the language of the contract will be disrupted. So, optimally, how often should the father observe the baby? If he continually watches the infant, the latter will never fall into the water (no cost there), but he also will never master the contract (high cost). Obviously if he never looks up from his contract, on the other hand, the potential cost of losing the infant is too high. So where in between is the optimum frequency of sampling in the tradeoff? This is the issue addressed by Sheridan's model.

This optimum can be determined mathematically if we assume that the cost of not reading the contract is a mathematically increasing function of how often the father samples the baby, shown by the heavy dashed line of Figure 7.1. The expected cost of losing the baby is a mathematically decreasing function of how often he observes the baby. This function results because the expected cost is equal to the actual cost multiplied by the probability that the baby will fall in the water, as shown by the solid line of Figure 7.1. This probability is a direct function of the uncertainty of the baby's current location. If the father has looked once, and the baby is far from the pool, he can be confident that another sample will not be required until some time has passed. With progressively longer passage of time, his uncertainty of the baby's whereabouts grows. Therefore to maintain the greatest certainty he must sample often. Slower sampling increases uncertainty and therefore increases the expected cost.

Optimal sampling is therefore dictated by two opposing cost functions: one of these increases (based on the cost of not completing the contract) with sampling frequency, and one (based on the "cost" of the infant falling into the water multiplied by the expectancy that she will do so) decreases with sampling frequency. How often the father should optimally look is the point at which the summed value of the two curves, shown by the dotted line, is least (point A of Figure 7.1). The location of this optimum point will also be influenced by the speed with which the baby moves. Formally we can describe this speed as the information content provided by her motion, or her "bandwidth" (see Chapter 2). The father will lose certainty of the infant's location more

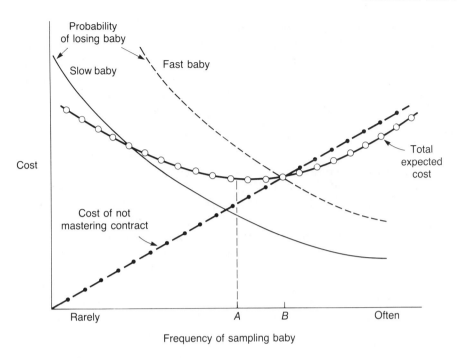

Figure 7.1 Optimum sampling. The optimum sampling frequency occurs
where the total expected cost is least. From "On How Often a Supervisor
Should Sample" by T. Sheridan, 1972, *IEEE Transactions on Systems,
Sciences, and Cybernetics, SSC–6*, p. 141. Copyright 1972 by the Institute
of Electrical and Electronics Engineers, Inc. Adapted by permission.

rapidly if the baby moves faster, thus driving the optimum sampling
frequency upward. The uncertainty and expected cost of a "high bandwidth baby" is shown by the light dashed line in Figure 7.1.

Using an experimental simulation, Sheridan and Rouse (1971) found
that people in fact sample an uncertain process more often than they
should (point *B* of Figure 7.1). They argue that this shift results from
failures of human memory regarding the level of uncertainty of the
sampled information (e.g., the father forgets what he should have remembered about the infant's previous location). In this sense the increased sampling frequency of point *B* may actually be considered as
closer to optimal, *given* the constraints of memory. Moray (1981) observes that this increase in sampling frequency resulting from memory
loss is enhanced for targets of greater importance (i.e., stimuli which if
missed will have more adverse consequences).

These observations provide another example of a source of nonoptimal behavior that potentially results from limitations of human memory. Indeed similar memory limits may also readily explain the finding
by Senders (1964) and by Moray et al. (1976) that low-probability

events were sampled more frequently than optimal. As Moray (1981) has pointed out, the occurrence of this memory limitation in sampling provides an important lesson to system designers. When operators must monitor systems with multiple sources of information, computer assistance should be given to inform them how often and what sources should be sampled, when this assistance can be based upon the computer's accurate representation of the statistical properties of the system.

Correlation between variables. In an article that attempts to bridge the gap between the laboratory work on attention and the applied world of real systems, Moray (1981) emphasizes that in most complex systems with many variables to be monitored, the variables are not independent of each other as in Senders' paradigm, but are correlated. The bank angle of an aircraft at one moment will be closely related to its heading the next. The temperature of a gas at a relief valve will probably correlate either with its pressure, with the valve position, or both. Of course, this pattern of correlation is in part dictated by the dynamics of the system and therefore may well be reflected in the operator's "internal model" or understanding of the system (see Chapters 3 and 5).

Moray argues that this pattern of correlation should influence the sampling strategy in the following way: In order for the operator to obtain as much information as possible about the state of a normally functioning system as rapidly as possible, variables or instruments that are closely connected or highly correlated with each other (e.g., the temperature in two adjoining pipes) should not be sampled consecutively. This is because the operator will gain very little added information from the second variable, having known the value of the first. Therefore, an optimum sampling strategy dictates that sequential samples should be taken from loosely coupled or poorly correlated variables as the supervisor monitors the normally operating system. The operator should sequence between "prototypical variables" of different major independent clusters in turn and avoid sequential sampling of the correlated variables within a cluster. Moray (personal communication) cites data of Allen, Clement, and Jex (1970) to suggest that well-trained operators do tend to follow the appropriate strategy under conditions of normal operation. An investigation by Cuqlock (1982) also supports this conclusion. She found that monitors of a simulated energy conversion process became less likely to sample two correlated variables as the value of the correlations between them increased.

However, once a failure or a malfunction is suspected and a given variable is perceived to be in an abnormal region, Moray proposes that the optimum sampling strategy should change dramatically. The operator should now examine variables that are normally highly correlated with the initial abnormal reading to assess whether they represent the source of abnormality. One direct implication of Moray's principle is that it is important to teach operators about the fundamental causal relations that exist within a complex system, so that the appropriate scanning strategies may be carried out. A similar argument was made in Chapter 3, where it was pointed out that the "internal model" could assist operators in failure detection and fault diagnosis.

Moray's proposal is particularly relevant to the development of automated centralized monitoring stations of complex processes, where the operator does not physically look at different dials on a wide-panel physical display but instead must request the display of different quantities on a centralized video terminal by keyboard entry. Under these circumstances it is easy to think of ways in which a smart computer system can make recommendations for optimal sampling strategies.

Planning and preview.   Our discussion of information sampling in dynamic processes suggests that the internal model of the process or of the statistical properties of the environment may serve as a guide for expecting when and where events are likely to occur and therefore when appropriate channels should be sampled. Tulga and Sheridan (1980) examined the situation in which forthcoming events on different channels could be directly *previewed* rather than simply predicted statistically. An analogy to their paradigm is the task of the chef who has several dishes at various stages of cooking, each of which must be "attended" (removed from the flame or stirred) at critical times that can be predicted in the future. Each of these events has a *cost* of missing (e.g., a burnt quiche). However, dealing with one dish increases the likelihood of missing another. Events also vary in their importance. The cost of a burnt quiche is far greater than the cost of a burnt piece of toast. The central characteristic of this paradigm is that time relentlessly moves forward. The chef cannot easily call all other processes to a halt while he deals with one. Another example of the paradigm is the pilot on an approach to landing who has a number of operations that must be performed in sequence, all of which are "driven" by the continuous approach to the ultimate touchdown on the runway. (In this case, it may sometimes be possible to halt the passage of time by aborting the landing and circling for another attempt. In principle, however, this is not a desirable procedure.)

In Tulga and Sheridan's paradigm, represented in Figure 7.2, the subject views a series of time lines (each corresponding to an information source), with blocks (corresponding to events that must be dealt with) that have varying heights (corresponding to their value). All blocks slide continuously toward the "present" time "slice" at a constant rate. For a block to be "attended" the subject must touch a light pen or attention pointer to that block's time line. When the block reaches the present, the subject earns points proportional to the block's height (its value) and how long it is attended (its width). However, as a consequence of dealing with one block the subject must ignore any other blocks that may simultaneously be sliding through the present.

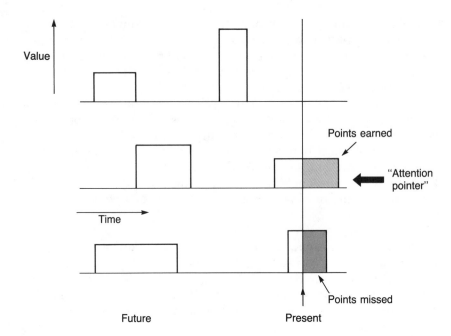

Figure 7.2 Multitask supervisory control and sampling paradigm. From "Dynamic decisions and work load in multitask Supervisory Control" by M. K. Tulga and T. B. Sheridan, 1980, *IEEE Transactions on Systems, Sciences, and Cybernetics, SMC–10*, p. 218. Copyright 1980 by the Institute of Electrical and Electronics Engineers, Inc. Reprinted by permission.

(This feature of the paradigm models the human as a single channel processor, an issue that will be discussed later in the chapter.) There is also a further penalty imposed for rapid switching of attention. After leaving one channel the subject can only begin to earn value from another channel after a minimum time has passed. Therefore, rapidly moving the pen back and forth between two lines will fail to earn any points.

Tulga and Sheridan found that subjects are reasonably optimal in their sampling relative to a complex "optimal" solution so long as the number of channels to be dealt with—possible sources of events—is reasonably small. Under these circumstances optimal performance is maintained even if the event rate is increased. However, when the number of channels increases, performance deteriorates rapidly from the optimal solution. The cause of this deterioration is that operator behavior at the present time is less influenced by anticipation of important events to be dealt with in the future. That is, the "planning horizon" is reduced. Once again, a recurring theme appears. Increased load imposes more restrictions upon working memory. These restrictions in turn reduce the ability to anticipate, which in the present case disrupts the efficient and optimal allocation of attention. As noted in our discussion of decision making, the operator performs reasonably optimally only as long as the structure of the problem is simple. When it is not, then working memory imposes a bottleneck.

Arousal and Selective Attention

In Chapter 3 we described how operators under stress often show what Sheridan (1981) identifies as "cognitive tunnel vision." In the decision-making task discussed there, this phenomenon of narrowing occurs at two levels: The number of hypotheses considered and the number of perceptual cues processed are both reduced. Over the past few decades there appears to have developed a reasonable body of experimental evidence indicating that the "breadth" of the selective attention beam and the number of sources of information it processes both decline under conditions of stress that increase the operator's *level of arousal* (Easterbrook, 1959; Kahneman, 1973). Furthermore, the qualitative nature of this perceptual narrowing is such that it is specifically cues of greater importance and greater centrality that are processed, at the expense of peripheral cues of lesser importance.

Across these studies, the concept of "level of arousal" is employed as a hypothetical construct or intervening variable that is associated with a number of experimental manipulations and measurable characteristics. These include stressors such as noise (which increases arousal) and sleep loss (which decreases it), incentives (increase), emotional states (generally increase), drugs such as caffeine and amphetamines (increase) or atropine (decrease), personality variables, and physiological responses that indicate arousal level such as the galvanic skin response or the electroencephalogram. A detailed listing of all of these variables, their effects on arousal, and the arousal concept in general is beyond the scope of this book (see Broadbent, 1971, and Hamilton & Warburton, 1979, for a more detailed discussion). However, there does appear to be enough consistency in experimental results to suggest that the arousal concept is a meaningful one which does influence the degree of perceptual narrowing.

Thus, for example, Hockey (1970) compared the changes in performance on a central tracking and a peripheral monitoring task induced by noise stress. He found that increasing noise facilitated tracking but disrupted peripheral monitoring. Calloway (1959; Calloway & Stone, 1960) concluded that amphetamines facilitated central monitoring performance more than peripheral, while atropine, reducing arousal, exerted the opposite effect. Anderson and Revelle (1982) investigated the effects of arousal (as manipulated by personality type and caffeine intake) on proofreading errors. The authors reasoned that a more diverse number of cues must be sampled to detect interword errors, such as grammatical or semantic mistakes, than to detect intraword errors such as misspellings. Consistent with the narrowing hypothesis, they found that manipulations that increased arousal produced a greater disruption on the detection of interword errors than on the detection of spelling errors. Although arousal increases lowered the accuracy of detecting both kinds of errors, the speed of performance for spelling errors was increased, while that for interword errors, using more cues, was decreased.

In many of the experiments that have supported the concept of attentional narrowing, the importance of cues and their position on a display are confounded. That is, display variables that are centrally located are also either explicitly or implicitly assumed to be more important because, for example, they are continuous, while peripheral tasks are intermittent. Broadbent argued that when these two variables are unconfounded, it is the cues of greater importance rather than those of greater centrality that command the focus of attention under conditions of high arousal (Bacon, 1974; Broadbent, 1971; Cornsweet, 1969). An experiment by Bacon (1974) demonstrates this effect. Subjects performed a central tracking task and a peripheral auditory detection task while arousal was manipulated by imposing shocks. When the tracking task was designated of greatest importance, Bacon obtained the expected effects of arousal: auditory detection was more disrupted than the continuous tracking task. However, he found that when payoffs and task structure were altered so as to place greater attentional focus on detection, the arousal effect was still present but was considerably reduced in magnitude.

The phenomenon of perceptual narrowing with arousal increase has not apparently been experimentally demonstrated in more applied multicue situations such as the aircraft cockpit or the industrial monitoring station, although anecdotal reports indicate that it is present there as well (Sheridan, 1981). An important point, however, is that the phenomenon represents a mixture of optimal and nonoptimal behavior. Arousal produces a nonoptimal response by limiting the breadth of attention. But subject to this limit the human appears to respond optimally by focusing the restricted searchlight on those environmental sources that are judged to be important.

Stimulus Influences in Selection

The "searchlight" model of attention assumes that an operator's internal model of the environment is primarily responsible for directing the allocation and selection of attention. There are, however, two other factors that may sometimes override this purposefully driven switching process. These concern the physical properties or salience of the stimuli themselves, and the phenomenon of visual dominance. Stimuli that are central, loud, bright, novel, dynamic, or otherwise salient will have a tendency to direct attention away from a previously focused channel. These are of course the principles upon which effective alerting and warning signals are based. However, as noted in Chapters 4 and 11, mere stimulus intensity may create an annoyance that is counterproductive as a warning signal, because the operator might have a tendency to deactivate the signal. I have often wished to disconnect an alerting buzzer in my car that informs me that my seatbelt has not been fastened. The second factor, visual dominance, will be considered in greater detail.

Visual dominance.   The phenomenon of visual dominance opposes the instinctive tendency that humans have to switch attention to stimuli in the auditory and tactile modalities. These stimuli are intrusive and the peripheral receptors have no natural way to shut out auditory or tactile information. We cannot "close our earlids" in the same way we close our eyelids. Nor can we "shift our ears" in the same manner that we shift our gaze. As a consequence, auditory devices are generally preferred to visual signals as warning indicators (Cooper, 1977; Deatherage, 1972; Simpson & Williams, 1980).

The superior ability of the auditory modality to alert was demonstrated by Nissen (1974). In her experiment subjects were led to expect a stimulus on a particular sensory channel by increasing the probability that it would occur on that channel. Nissen found that if subjects expected an auditory stimulus to occur and instead received a low-probability visual stimulus, then reaction time to the visual stimulus was quite delayed relative to reaction time to an expected visual stimulus. On the other hand, when subjects expected a visual stimulus and received a low-probability auditory stimulus, there was little delay. The unexpected auditory stimulus in the second condition automatically called attention to itself. The visual stimulus in the first condition did not. In a related study, Posner et al. (1976) found that an auditory warning stimulus will speed the response to both a subsequent auditory and visual stimulus; however, a visual warning will not speed the response to the auditory stimulus. Posner et al. have argued that in order for humans to counteract these automatic alerting tendencies of the auditory modality, they possess a bias to process information in the visual channels—a bias that is intended to overcome this alerting handicap of the visual modality. This bias is known as visual dominance.

Examples of visual dominance over auditory or proprioceptive stimuli are abundant. For example, Colovita (1971) required subjects to respond as fast as possible to either a light (with one hand) or a tone (with the other hand). On very infrequent occasions, both stimuli were presented simultaneously. When this occurred, subjects invariably made *only* the response with the hand appropriate to the light. In fact, subjects were even unaware that a tone had been presented on these trials! Experiments by Jordan (1972) and by Klein and Posner (1974) supported the dominance of vision over proprioception. These investigators found that reaction time to a stimulus that was a compound of a light and a proprioceptive displacement of a limb was slower than reaction time to the proprioceptive stimulus alone. This result suggests that the light "captured" attention and slowed down the more rapid processing of the proprioceptive information. Different examples of visual dominance are observed in experiments in which the sense of vision and proprioception are placed in conflict through prismatically distorted lenses (Rock, 1975). Behavior in these situations suggests that the subject responds appropriately to the visual information and disregards that provided by other modalities.

Some dual task time-sharing situations described in Chapter 8 also appear to manifest a form of visual dominance when an auditory task is performed concurrently with a visual one. In these circumstances, the auditory task tends to be hurt more by the division of attention than the visual task (Isreal, 1980; Massaro & Warner, 1977; Treisman & Davies, 1973).

While visual dominance may be viewed as an adaptive and therefore useful adjustment of information processing made in response to anatomical differences between sensory channels, there are of course circumstances in which it can lead to nonadaptive behavior. Illusions of movement provide one such example. When the visual system gives ambiguous cues concerning the state of motion, the correct information provided by proprioceptive, vestibular, or "seat of the pants" cues is often misinterpreted and distorted. For example, while sitting in a train at a station with another train just outside the window, the passenger may experience the illusion that his train is moving forward, while in fact his train is stationary and the adjacent train is moving backwards. The passenger's model of the world has "discounted" the proprioceptive evidence from the seat of the pants that informs him that no inertial forces are operating.

In summary, when an abrupt auditory stimulus may intrude upon a background of ongoing visual activity, it will probably alert the operator and call attention to itself. However, if visual stimuli are appearing at the same frequency and providing information of the same general type or importance as auditory or proprioceptive stimuli, then biases toward the visual source at the expense of the other two must be expected.

Attention switching.   Our previous discussion has considered the factors that lead the human operator to switch attention from one source of information to another but has not described the dynamics of the switch process itself. While a large number of ingenious paradigms have been employed (e.g., Broadbent, 1958; Kristofferson, 1967; LaBerge et al. 1971; Moray, 1969). The "expectancy reaction time" paradigm employed by LaBerge et al. (1971) provides a set of typical results. In this experiment, subjects performed a choice reaction-time task in which the alternatives were either pressing a left key to a left stimulus or a right key to a right stimulus. In addition to the uncertainty of stimulus (and response) *location*, however, the stimulus *modality* was also uncertain. The stimulus could be auditory or visual with equal probability. The response location was independent of stimulus modality. Thus the subject might receive a left or right stimulus in either the auditory or visual modality.

Prior to stimulus presentation, a precue was presented concerning which stimulus was about to occur. The cue stated that a given one of the four stimuli would occur with a 90% probability. The experiment-ers assumed therefore that the subject's attention "switch" would align itself to the expected channel. LaBerge et al. were interested in deter-mining how much of an added delay to reaction time there would be when attention was aligned to the wrong sensory channel (i.e., when the unexpected modality occurred with the expected response) and a switch was required. This delay was found to be approximately 100 msec. Other research summarized by Moray and Fitter (1973) has indicated that this 100-msec. delay is not really "dead time" during which attention is actually "moving" (like the saccadic movement of the eyeball) but rather is the time required to initiate the switching command away from the expected modality.

FOCUSED ATTENTION

In the previous discussion of selection, attention was represented es-sentially as a dichotomous variable. A stimulus or source of informa-tion was either assumed to be fully attended, or it was ignored. While this dichotomization is a useful simplification for describing selection, it fails to account for the continuous gradations in the degree to which stimuli may be processed. In terms of the searchlight analogy, certain stimuli may be in the "penumbra" of attentional focus, others may be out of the spotlight altogether, and still others may be in the "umbra," receiving some but not full processing. These partially processed stim-uli may represent sources of distraction, as when a background conver-sation disrupts comprehension of a lecture. Alternatively, they may be a source of assistance, as when information from peripheral vision is found to assist such tasks as flying (Bermudez, Harris, & Schwank, 1979; Brown, Holmquist, & Woodhouse, 1961), driving (Reid & Sol-owka, 1981), photo interpretation (Leatchenauer, 1978), or reading (McConkie & Raynor, 1974). In this section we consider the properties of stimuli that increase the likelihood that two or more sources will be processed in parallel (i.e., will receive access to attention together) and therefore cause one to disrupt focused attention on the other.

The most important general property of a pair of stimuli that leads to parallel processing is their similarity or "proximity." The term "sim-ilarity," however, is a highly general one. In the following section we will consider more precisely how similarity affects parallel processing and so disrupts focal attention in both the auditory and visual modalities.

The Auditory Modality

Two sources of information must differ along one or more dimensions that can be perceived by the operator in order to be discriminated so that one may be processed and the other rejected. The particular *level* along a given dimension at which a given information source appears is defined as a *channel*. For example, in the dimension defined by spatial location, a voice that is heard in the right ear is presenting information along the "right-ear channel" of the spatial dimension. In the dimension of pitch, a voice that chants a message on the note of F sharp is presenting information on the "F-sharp channel" of the pitch dimension. Increasing degrees of separation of channels of the two messages along these dimensions will increase the ability to select one and ignore the other.

Spatial location is perhaps the most salient physical dimension of selectivity. When attempting to attend to one auditory message and ignore the contents of another, our ability to do so declines as the two messages become closer in space (Egan, Carterette, & Thwing, 1954; Spieth, Curtis, & Webester, 1954; Treisman, 1964b). An extreme example of this difference in separation is provided by the distinction between monaural and dichotic listening. In monaural listening, two messages are presented with equal relative intensity to both ears, a stimulus that would be experienced when listening to two speakers at the same physical location. In dichotic listening the two messages are delivered to opposite ears by headphones, so that no overlap is possible. Egan et al. (1954) found that there are large benefits of dichotic over monaural listening in terms of the operator's ability to filter out distraction from an unwanted channel.

We are, of course, also able to attend selectively to auditory messages even from highly similar locations. The classic "cocktail party effect" describes this ability to attend to a speaker at a noisy party and selectively filter out (with varying degrees of success) other conversations coming from the same spatial location. In this case, we must be able to use dimensions of similarity other than location to focus attention selectively. One such dimension for selection is defined by pitch. It is easier to selectively process a male or female voice in the presence of a second voice of the opposite sex (and thereby different pitch) than to selectively attend when confronted with two voices of the same sex (Treisman, 1964b). Interestingly, this pitch discrimination can be made even though there is a large overlap between the frequency content or spectrum of male and female voices (see Figure 4.12).

Intensity may also serve as a dimension of selection. It is clearly easy to attend to a loud message and "tune out" a soft one. However, Egan et al. (1954) found that it is also possible to selectively attend the softer channel and ignore a louder one as long as the differences in intensity are less than 10 decibels. When the difference is greater than this, however, the advantage to the selective difference is more than outweighed by the fact that the louder channel physically masks the softer one.

We usually think about channels of selective attention defined in terms of physical properties of the stimulus. However, Treisman (1964a) has observed that semantic content, a cognitive dimension, can readily serve as a "cue" for selection. Bilingual subjects were asked to attend to an auditory message played to one ear as a second message to be ignored was played in the other. Treisman found that subjects' ability to process the attended message was disrupted if the message delivered to the opposite ear and spoken in a different language was of a similar semantic content. The fact that a different language was employed ensured that the failure of selection could not have resulted from any *physical* similarity between messages. Although these results show that similar semantic content impairs the ability to focus attention, they do not imply that differences in semantic content provides a perfect cue for selection. Two messages from identical locations and spoken with identical voices will still create considerable problems in selection, even if each is quite different in meaning.

We might summarize these results by noting that auditory messages differ from each other along a wide variety of different dimensions such as pitch, location, loudness, and semantic content. The greater the difference between two messages along a given dimension and the greater the number of dimensions of difference, the easier it will be to focus on one message and ignore the other. There is not, however, a great deal of evidence pertaining to the degree to which one dimension is more important than another. For example, if two messages differ from each other in pitch but are identical in spatial location, will focused attention be easier or more difficult than if they differ in location but are of equivalent pitch? Electrophysiological data obtained by Hansen and Hillyard (1983) suggest that there may be little pronounced difference in selection ability along the two dimensions when they are equated for discriminability.

The Visual Modality

Many of the properties of focused attention in the auditory modality have counterparts in vision as well. It is considerably easier to "tune" the eyes than the ears to fixed locations in space. However, there is ample evidence as well that irrelevant visual stimulation may intrude upon the focus of visual attention. As with audition, the degree of intrusion seems to depend upon the degree of similarity or proximity. An experiment by Eriksen and Eriksen (1974) demonstrated the role of spatial proximity in focused attention. Subjects were required to respond to one of four letters (e.g., H, S, F, R) with either a leftward or rightward deflection of a stick. Two of the letters (H and S) received a left response, and the other two (F and R) a right response. The stimulus letter was presented either by itself or flanked by surrounding letters. The flanking letters, in turn, could either be of a compatible class, indicating the same direction of motion, or an incompatible response class with the central stimulus letter (see Figure 7.3).

Eriksen and Eriksen observed that reaction time was generally slowed by the presence of flanking letters relative to a single-letter control condition. However, the magnitude of this effect was influenced by two important factors: (1) The delay was enhanced when the flanking letters called for a response that was in the opposite direction to that demanded by the center letters (e.g., an H surrounded by F and

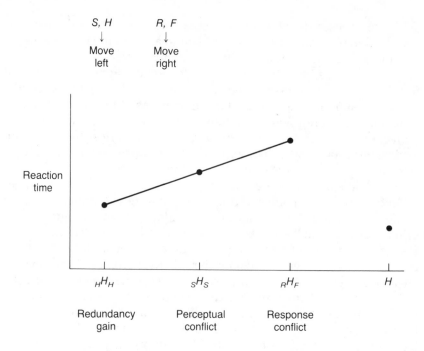

Figure 7.3   Data from focused attention study. From "Effects of Noise Letters upon the Identification of a Target Letter in a Nonsearch Task" by B. A. Eriksen and C. W. Eriksen, 1974, *Perception and Psychophysics, 16,* p. 146. Copyright 1974 by the Psychonomics Society, Inc. Reproduced by permission of the authors.

R. (2) The effect was greatly reduced if the flanking letters were of a compatible response class (e.g., an *H* flanked by *S*'s, when both are stimuli signaling a left movement). In fact, when the stimuli are identical (an *H* flanked by *H*'s) the response is typically as fast as and may even be faster than in the single letter case (Morton, 1969).

These findings suggest that two phenomena operate to delay the response. On the one hand, *perceptual competition* disrupts processing of one stimulus because perceptual analysis of closely adjacent stimuli is required. This would occur in all flanking conditions when letters were different from the central letter. On the other hand, *response conflict* operates when two stimuli, both within the focus of the searchlight of attention, elicit incompatible responses. It is response conflict that must account for the difference in processing speed between the two conditions described above. In both conditions there is additional perceptual information that is different from the stimulus of importance, leading to perceptual competition. When that information generates an incompatible response (e.g., *H* flanked by *R*), the automatic processing of the flanking stimuli generates response tendencies that are incompatible with those indicated by the central stimulus. The need to suppress these tendencies slows the response.

Response conflict and redundancy.   The response-compatible condition of Eriksen and Eriksen's (1974) data shown on the left of Figure 7.3 suggests that there may be a benefit as well as a cost to the close proximity of information within a spatial channel. This benefit will be realized if this flanking information has implications for action that are compatible with those of the central stimulus. Even when the information is physically different (an *H* flanked by *S*'s), the congruence of action may be great enough to outweigh any slowing due to perceptual competition. If the perceptual information is identical (*H* flanked by *H*'s), then a *redundancy gain* is enjoyed: The processing of the *H* is speeded by the presence of adjacent *H*'s.

Response conflict and redundancy gain are thus opposite sides of the same conceptual coin. If two perceptual channels are proximate they will both be processed, even if only one is desired. This processing will inevitably lead to some competition (intrusion, distraction) at a perceptual level. If they have common implications for action, the perceptual competition will be balanced by the fact that both channels activate the same response. If, however, their implications for action are incompatible, then the amount of competition is magnified.

In Eriksen and Eriksen's study, the similarity between stimuli was defined by spatial location. When flanking letters were moved closer to the central letter, the similarity effects were enhanced. We might expect that the observed effects of response conflict and redundancy gain would be amplified even further if the different sources of information represented different attributes of a *single* stimulus object at one spatial location. Indeed several studies have shown that this is the case. Many of these investigations have employed some variation of the Stroop task (Stroop, 1935), in which the subject is asked to report the color of a series of colored stimuli as rapidly as possible. In a typical control condition (e.g., Hintzman et al., 1972; Keele, 1972), the stimuli consist of colored symbols—for example, a row of four *X*'s in the same color. In the critical *conflict* condition, the stimuli are color names which do not match the ink color in which they are printed (e.g., the word "red" printed in blue ink). Here the results are dramatic. The reporting of ink color is painfully slow and fumbling when compared to the control condition, as the semantic attribute of the stimulus ("red") activates a response that is fundamentally incompatible with information that the subject is instructed to process (the color blue). The mouth cannot articulate the words "red" and "blue" at the same time, yet both are called for by different attributes of the single stimulus.

In light of the strong Stroop effects resulting from response conflict, it is not surprising that *cooperation* through redundancy gain has also been observed when the color of the ink matches the semantic content of the word (Hintzman et al., 1972; Keele, 1972, 1973). In fact, several examples of both cooperation and competition resulting from the relation between two attributes of a stimulus have been reported. Clark and Brownell (1975) observed that judgments of an arrow's direction (up or down) were influenced by the arrow's location within the display. "Up" judgments were made faster when the arrow was higher in the display. "Down" judgments were made faster when it was low. Similarly, Rogers (1979) found that the time that it took to decide if a word was "left" or "right" was influenced by whether the word was to the right or left of the display.

In a more complex extention of the conflict/cooperation paradigm, Laxar and Olsen (1978) assessed the speed with which naval submarine officers could process relative velocity vectors to determine whether one ship would pass to the left or right of the other. The top row of Figure 7.4 portrays two such stimuli typical of those used by Laxar and Olsen. The direction of each arrow represents the compass heading of the ship, and the length is proportional to the ship's velocity. The left column portrays a situation in which your ship is heading northeast at a speed roughly three times that which the other ship is heading southeast. Therefore, the other ship will pass to the left of yours. Laxar and Olsen found that the subjects' decision times to judge "left" or

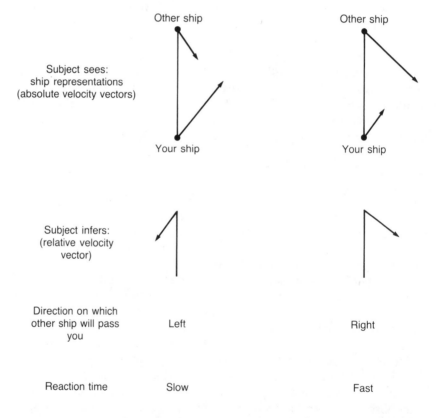

Figure 7.4   The stimuli used and representative results in the conflict/cooperation paradigm. From "Human Information Processing in Navigational Displays" by K. Laxar and G. M. Olsen, 1978, *Journal of Applied Psychology,* *63,* p. 736. Copyright 1978 by the American Psychological Association. Adapted by permission of the authors.

"right" were influenced in a cooperative/competitive fashion by the compatibility of the final judgment ("left" or "right") with the direction of heading (left or right) portrayed in the display of movement of the two ships. Thus, as shown in the left-hand column of Figure 7.4, the most difficult judgments were those in which the absolute velocity vector of both ships was to the right (left), yet the relative velocity vector of the other ship compared to your own ship was to the left (right). In the right-hand column, on the other hand, the response is fast, because both the absolute and relative velocity vectors are in the same direction.

The issue of redundancy gain has been addressed in a much broader context when the effectiveness of redundant presentation of instructional material is considered. A general conclusion is that redundancy will be particularly helpful when the formats of information of the two sources are sufficiently different that they will not compete for resources. This principle was discussed in Chapter 6 when the argument for the redundant use of verbal and spatial formats was presented. Baggett and Ehrenfeucht (1981) examined whether redundant presentation of auditory and visual information assisted the comprehension of material about carnivorous plants. They found that redundant presentation in which sound track and film were presented simultaneously yielded far better comprehension than did a sequential presentation in which one modality followed the other. This advantage was observed despite the fact that in the redundant condition only half as much study time was afforded and there was possible competition for attention between the two information sources. As we will discuss in Chapter 8, however, presenting two information sources in different modalities reduces the extent of competition between processing channels.

DIVIDED ATTENTION IN PERCEPTION

It is apparent that some of the factors that lead to difficulties of focused attention may provide advantages for tasks requiring divided attention. If two sources of information both *must* gain access to the common "searchlight" of attention when we wish to process only one, then when in fact we wish to process both, our performance should benefit from the very same conditions that led to a failure of focal attention. For example, while an operator may experience difficulty in ignoring a distracting tone played in the same ear as a vocal message, the fact that the two are delivered to a common channel will be beneficial if she is trying to determine the relevance and meaning of both signals at once.

Much of the research on divided attention in perception was stimulated by the question, raised in the 1950s, of whether people could determine the meaning of two independent messages delivered simultaneously. Broadbent (1958), in his book *Perception and Communication*, concluded on the basis of data then available that people could not. Based upon these data, Broadbent proposed a single-channel or "filter" model of perception. This model asserted that information was processed in parallel only as far as the stage of short-term sensory store. Beyond this level, only a single message could be selected at a time for the further processing involved in perception, determination of meaning, decision, and response selection.

While Broadbent's strict single-channel or "filter" model of attention has some useful features, as described in the first section of this chapter, experimental evidence collected since the late 1950s suggests that the assumption that humans are strict serial processors of information is untenable. Instead, a considerable amount of environmental information is processed in parallel. We shall consider below the nature of this evidence.

Evidence for Parallel Processing

Some of the initial evidence pertaining to Broadbent's single-channel model of attention was provided by a series of *dichotic listening* studies in which subjects listened to a different stream of verbal information in each ear (Cherry, 1953). Typically, subjects would be asked to "shadow" one stream—that is, to repeat the message out loud and thereby focus attention directly on its contents. The investigators argued that evidence against a serial model and for parallel processing would be provided if it could be shown that certain semantic (meaningful) aspects of the nonattended channel were processed.

The initial research in this paradigm by Cherry (1953) at first suggested a serial model. In Cherry's experiment, after shadowing one attended message (A), subjects were unable to recall anything about the meaning of the message played to the nonattended (NA) ear. The conclusion offered by Cherry was that nonattended material is not processed at a semantic level (its meaning is not determined by the brain). Soon, however, evidence began to appear to the contrary. Four investigations are prototypical of a larger number of studies that refuted the strict filter model of attention.

In 1959, Moray showed that while subjects shadowed one message, they were often aware of the presence of their own name when it occurred in the other. Therefore, Moray concluded that the nonattended channel was continuously monitored for semantically defined target information such as that inherent in the name. Treisman (1960) demonstrated that if the message in the nonattended ear was made semantically related to, or in context with, information in the attended ear, subjects might inadvertently switch and shadow the nonattended (NA) channel. Her experimental result is shown in Figure 7.5. The subject is instructed to attend to the upper message. Yet when the contextually relevant message ("for all good men") continues on the bottom channel, the subject occasionally cannot help uttering a few words of the message to be ignored. This inadvertent switch is shown by the arrow. The fact that subjects did this implies that the brain must be determining the meaning of the NA channel all along, in order to

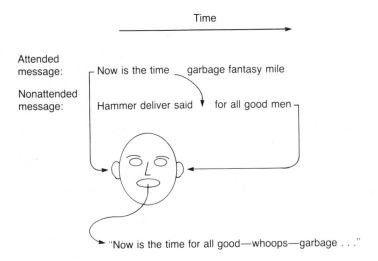

Time

Attended
message: Now is the time   garbage fantasy mile

Nonattended
message: Hammer deliver said   for all good men

"Now is the time for all good—whoops—garbage . . ."

Figure 7.5 Contextual effects on the nonattended message cause inadvertent switch in the experiment by Treisman (1960).

establish its pertinence when it is in context with prior words on the attended channel.

Another study by Treisman (1964a), described in the previous section, also provided evidence for semantic processing of nonattended information. She observed that bilinguals shadowing a passage in one language delivered to an attended ear were disrupted if the same passage in a different language was presented to the opposite ear. This disruption of focused attention would not have occurred if the brain did not determine the meaning of the nonattended material. Finally, Lewis (1970) found that if subjects shadowed a random string of words, the vocal reaction time to repeat a given word on the attended channel was slowed down if a simultaneous word appearing on the nonattended channel was semantically related. Here again, semantic processing of nonattended as well as attended material is implicated.

These four studies, along with several others (see Kahneman, 1973), suggested a model of selective attention in which there is a certain amount of "free" semantic processing of nonattended material. It is "free" in the sense that the brain carries it out even when the conscious limited supply of attention is focused elsewhere. It is important in this context to make the distinction between attention and processing. In dichotic listening studies, for example, attention is operationally defined in terms of the task that the subject is instructed to perform, and therefore the allocation of attention is inferred from his shadowing performance. Attention is to be focused on one channel. This is not synonymous with processing, since the data suggest that processing can take place on one channel as the focus of attention is allocated elsewhere. In this sense, attention is clearly defined only in terms of a conscious intention. Processing encompasses a broader domain.

Once the strict filter model of attention was discredited, experimental interests then shifted to determining exactly how much nonattended material benefited from some sort of free preattentive processing. Alternatively, what were the added benefits of allocating attention to a channel? Research employing nonsemantic material suggested that a considerable quantity of processing was indeed free. Thus, investigators such as Lappin (1967) showed that subjects could report three dimensions (size, shape, and color) of a tachistoscopically presented stimulus as accurately as they could report a single dimension alone. Attention could be successfully divided between these three attributes. A series of experiments conducted by Schneider and Shiffrin (1977) indicated that when subjects consistently categorize a set of letters into one response class over a number of trials (see Chapter 4), they can search for several letters in an array in parallel, as if the perceptual system is simultaneously scanning all locations in that array. In the auditory domain, Moore and Massaro (1973) reported a similar success of divided attention. Their subjects were able to judge the quality and pitch of a tone simultaneously as well as either of the two attributes could be judged in isolation.

Early Versus Late Selection of Attention

Although early investigations conclusively showed that nonattended material was processed to some degree, it also became apparent that it is not processed to the same extent as material that we are consciously choosing to attend. Experimenters therefore began to examine where in the sequence of information-processing stages the benefit of focused attention was realized. One view asserted that this benefit is in the form of enhanced signal quality or increased strength of representation at the earlier stages of processing. This was the view proposed by Treisman (1969) and referred to as "early selection." An alternative view suggested that all material, whether attended or not, activates its representation in memory with equal strength, and the effect of attention then is one of a "response bias." Thus perceptual processing of attended and unattended material is similar, but the subject is more likely to select attended material for response, rehearsal, storage in memory, or further elaboration. The latter view, proposed by Deutsch and Deutsch (1963) and elaborated by Norman (1968) and Keele (1973), is referred to as a "late-selection" theory.

As so often happens in psychology, there appears to be some truth in both theoretical positions (see Broadbent, 1982, for a good integrative summary). There does indeed seem to be fairly conclusive evidence that early stages of processing benefit from focused attention, as Treisman proposes (see also Kahneman & Treisman, in press). Some of this support is provided by the research of Hillyard and his co-workers employing the evoked brain potential (e.g., Hansen & Hillyard, 1983;

Hillyard, Picton, & Regan, 1978; Hink & Hillyard, 1976). The evoked potential is the brain's electrical response to discrete environmental events. This electrical potential can be recorded from the scalp by electrodes. In Hillyard's investigations, a series of tones is played to two ears, and the subject is instructed to attend to tones in one ear and ignore those in the other. Evoked potentials elicited by the tones in both the attended and nonattended ears are recorded. Hillyard finds that the evoked potential to tones in the attended ear is enhanced in its negative-going voltage at a very early latency—only around 100 msec after the stimulus. This enhancement suggests that attended and nonattended materials are treated differently by the brain after only 100 msec of processing. This is a sufficiently short latency to argue that the difference in processing is occurring prior to the selection of a response.

A further test of the difference between the early- and late-selection theories has employed signal detection theory, discussed in Chapter 2 (Moray & O'Brien, 1967; Moray et al., 1976; Treisman & Riley, 1969). According to this logic, detection of a single sort of target is compared between trials when that target arrives along a channel that is attended (A) and trials when it arrives along one which is not (NA). Treisman's early-selection view predicts that the major effect of focal attention will be to enhance the *sensitivity* of target detection along the A as compared to the NA channel. Late-selection theories, on the other hand, predict that there will be no difference in sensitivity but only a criterion bias in favor of information on the A channel. That is, there will be more detections of the relevant stimulus on the attended channel but also proportionately more false alarms.

In fact, the data collected by Treisman and Riley and by Moray and his co-workers seem to confirm that there is a loss in *sensitivity* that results from not attending a channel, a result that supports the early-selection theory. This sensitivity loss may be observed either when attended and nonattended channels are contrasted in a conventional shadowing study (Treisman & Riley, 1969) or when the sensitivity to detect physical or semantic targets in one ear is compared with the ability to divide attention between two ears (e.g., Moray & O'Brien, 1967; Moray et al., 1976; Ninio & Kahneman, 1973). In this second class of experiments, when detection is compared between two and one channels, Moray and O'Brien (1967) and Moray et al. (1976) noted that the greatest loss of sensitivity occurred when targets on the two monitored channels appeared simultaneously. This finding has significance later.

While the experimental results thus argue that there exists a benefit for focused attention at earlier processing stages, the data are also consistent with a late-selection benefit to attention as proposed by Norman (1968) and by Deutsch and Deutsch (1963). Moray et al. (1976), for example, found that there are criterion shifts in addition to sensitivity changes associated with selective attention. When subjects know there is a greater likelihood of a target appearing on a given channel, they are more willing to respond "yes," thereby increasing both hit and false-alarm rate. Furthermore, a late-selection mechanism would appear necessary to account for the manner in which contextually relevant or pertinent material on a nonattended channel can intrude into the focus of attention. The conclusion that emerges then is one that seems quite reasonable. Focused attention helps information processing at all information stages beyond short-term sensory store (Kahneman & Treisman, in press).

Parallel Versus Serial Processing

In concluding the discussion of divided attention, it is useful to identify explicitly the four experimental variables that will enhance or attenuate the degree of "free" parallel processing of unattended relative to attended material. These variables are of considerable importance because enhanced free processing will increase the possibility of response conflict and thus be costly if a competing response is triggered, but it may be beneficial if the two information sources feed into a common compatible response or into different tasks.

Separate dimensions of a single stimulus. The demonstrations of the Stroop effect and, in the auditory modality, the experiment of Moore and Massaro (1973) are typical of the general finding that different attributes of a single stimulus may be processed in parallel. This produces a redundancy gain, successful divided attention, or response conflict depending upon whether the responses which the various attributes of the stimuli activate are identical, independent, or incompatible, respectively. In her model of selective attention, Treisman (1969) proposed that perceptual analyzers, each responsible for processing a given dimension of the stimulus, may operate in parallel. When, however, a single perceptual analyzer must process information from two stimuli, then problems are encountered.

This principle is relevant to material covered in both Chapters 2 and 5. In Chapter 5 the concept of the object display was introduced. Part of the advantage of an object display was achieved because of the apparent benefit received by the parallel processing of the several dimensions of a single object (Kahneman & Treisman, in press). The discussion in Chapter 2 of Garner's research on stimulus dimensions suggests that the amount of parallel processing of two dimensions of an object may furthermore depend upon whether those dimensions are integral or separable. A pair of integral dimensions are those in which it is impossible to specify the level on one dimension without specifying the level on the other. Garner and Fefoldy's (1970) research demonstrated that integral dimensions, such as the hue and brightness of a single object, showed both an increased redundancy gain when levels on the dimensions are correlated and response conflict (a failure to focus) when the dimensions are independent.

Close proximity in space. A logical extention of the single-stimulus principle is that if the two aspects of information are pulled apart from the single stimulus source and gradually separated along a given stimulus dimension, redundancy gains will be reduced, focused attention will be easier, and divided attention will be more difficult. Eriksen and Eriksen (1974) demonstrated this effect on focal attention by varying the spatial separation of the central and peripheral stimuli. Treisman (1964a) showed that successful focused attention increased with increased spatial separation of sound sources. Lappin (1967) demonstrated that it was more difficult to judge the size of one object, the shape of another, and the color of a third than the size, shape, and color of a single object. Finally, as discussed in Chapter 2, Garner and Fefoldy (1970) observed that when two dimensions were made separable, so that the level on one could be specified without needing to specify the other, there was neither a redundancy gain nor response conflict when stimuli were to be sorted on correlated and uncorrelated dimensions, respectively.

One apparent exception to this principle of close proximity concerns the processing of bimodal information. Presenting one stimulus visually and another one auditorily may be thought of as an extreme level of separation. Hence, according to the principle, divided attention should be difficult. On the contrary, however, experimental data indicate that attention divided between the eye and ear is quite effective (Treisman & Davies, 1973). Thus the principle of proximity appears to hold only within a sensory modality. The issue of divided attention between modalities will be considered in Chapter 8.

The previous discussion has defined separation in terms of visual or retinal space. An experiment by Neisser and Beckman (1975) indicates that separation may be defined by other dimensions at a given spatial location. Their subjects watched a display on which two video games were presented simultaneously, one superimposed over the other. One showed distant figures tossing a ball, the other two pairs of hands slapping each other. One game was designated as "relevant," and critical elements were to be monitored and detected. Neisser and Beckman found that while monitoring one game, subjects failed to see events on the other game, even when these were unusual or novel (e.g., the ball tossers paused to shake hands). These results suggest that "separation" may be defined not only in terms of differences in visual or retinal location, but also in terms of the nature of the perceived activity or possibly the inferred distance from the observer.

Neisser and Beckman's finding has a direct counterpart in aviation, where pilots may be presented with a "heads-up display" in which critical instrument readings are displayed on the glass windscreen. While this procedure was meant to ensure that information inside and outside the cockpit could be processed simultaneously without visual scanning, the conclusions of Neisser and Beckman's study indicate that there is no guarantee that this will happen. A pilot may treat the two distances as different attentional channels and become engrossed in processing instrument information while ignoring critical cues from outside the aircraft (Fischer, Haines, & Price, 1980).

Automaticity.   The automaticity of processing defines the characteristics of processing stimuli that have been frequently encountered and consistently assigned to a given response category; hence these are stimuli for which there is a well-formed representation in memory (Schneider & Shiffrin, 1977). Therefore, the representation of such stimuli in memory becomes activated automatically even if attention is directed elsewhere. Initial perceptual processing of automated stimuli is "free." This property of perceptual processing was discussed in some detail in Chapter 4. The costs of response conflict will be observed if the automatically processed stimulus calls for an incompatible response to one that is desired. The benefits of parallel processing (successful division of attention) will be observed if the responses are not conflicting. Both benefits and costs will be reduced if the stimuli are not well learned or if their processing is made difficult by other means (e.g., degrading stimulus quality). Under these circumstances, parallel processing of two relevant stimuli will be less likely to occur, while an unwanted stimulus will be less likely to intrude on the focus of attention. Unfamiliar (or degraded) stimuli not relevant to a task will be unlikely to induce response conflict.

Simultaneous decisions.   Parallel processing either stops or is greatly attenuated at the point in the processing sequence at which selection of a discrete action is called for. As Moray and O'Brien (1967) and Moray et al. (1976) observed, humans can monitor for targets in two ears at one time as well as in one. But when simultaneous targets in both ears appear, calling for a detection response of both at once, one of the targets will almost always be detected, but the other will often be missed. This finding suggests that a breakdown in parallel processing occurs at decision making. Duncan (1980) observed that this failure to detect simultaneous visual targets occurred even when the paradigm required that only a single response be made to both targets if they appeared together. Thus it was not the selection of two overt responses but the detection of two independent targets that led to the disruption of efficient parallel processing. This restricting "bottleneck" of decision making and the reduced ability to engage in parallel processing at later information-processing stages will be covered in more detail in Chapter 8.

References

Allen, R. W., Clement, W. F., & Jex, H. R. (1970, July). *Research on display scanning, sampling, and reconstruction using separate main and secondary tracking tasks* (NASA CR–1569). Washington, DC: NASA.

Anderson, K. J. & Revelle, W. (1982). Impulsivity, caffeine, and proofreading: A test of the Esterbrook hypothesis. *Journal of Experimental Psychology: Human Perception and Performance, 8,* 614–624.

Bacon, S. J. (1974). Arousal and the range of cue utilization. *Journal of Experimental Psychology, 102,* 81–87.

Baggett, P., & Ehrenfeucht, A. (1981, September). *Encoding and retaining information in the visuals and verbals of an educational movie* (Technical Report 108 ONR). Boulder, CO: University of Colorado, Institute of Cognitive Science.

Bermudez, J. M., Harris, D. A., & Schwank, T. C. H. (1979). Peripheral vision and tracking performance under stress. In C. Bensel (Ed.), *Proceedings, 23rd annual meeting of the Human Factors Society.* Santa Monica, CA: Human Factors.

Broadbent, D. E. (1958). *Perception and communications.* London: Pergamon Press.

Broadbent, D. E. (1971). *Decision and stress,* New York: Academic Press.

Broadbent, D. E. (1982). Task combination and selective intake of information. *Acta Psychologica, 50,* 253–290.

Brooks, L. (1968). Spatial and verbal components of the act of recall. *Canadian Journal of Psychology, 22,* 349–368.

Brown, I. D., Holmquist, S. D., & Woodhouse, M. C. (1961). A laboratory comparison of tracking with four flight-director displays. *Ergonomics, 4,* 229–251.

Calloway, E. (1959). The influence of amobarbital and methamphetamine on the focus of attention. *Journal of Medical Science, 105,* 382–392.

Calloway, E. & Stone, G. (1960). Re-evaluating the focus of attention. In L. Uhr & S. Miller (Eds.), *Drugs and Behavior.* New York: Wiley.

Carter, R. C., & Cahill, M. C. (1979). Regression models of search time for color-coded information displays. *Human Factors, 21,* 293–302.

Cherry, C. (1953). Some experiments on the reception of speech with one and with two ears. *Journal of the Acoustical Society of America, 25,* 975–979.

Christ, R. E. (1975). Review and analysis of color coding research for visual displays. *Human Factors, 17,* 542–570.

Clark, H. H., & Brownell, H. H. (1975). Judging up and down. *Journal of Experimental Psychology: Human Perception and Performance, 1,* 339–352.

Colovita, F. B. (1971). Human sensory dominance. *Perception & Psychophysics, 16,* 409–412.

Cooper, G. E. (1977, June). *A summary of the status and philosophy relating to cockpit warning system* (NASA Contractor Report NAS 2-9117). Washington, DC: NASA.

Cornsweet, D. J. (1969). Use of cues in the periphery under conditions of arousal. *Journal of Experimental Psychology, 80,* 14–18.

Craig, A. (1981). Monitoring for one kind of signal in the presence of another. *Human Factors, 23,* 191–198.

Cuqlock, V. B. (1982). *A behavioral assessment of the weights applied to redundant cues.* Unpublished doctoral thesis, University of Illinois, Urbana-Champaign.

Darwin, C., Turvey, M. T., & Crowder, R. G. (1972). An analog of the Sperling partial report procedure. *Cognitive Psychology, 3,* 255–267.

Deatherage, B. H. (1972). Auditory and other sensory forms of information presentation. In H. P. VanCott & G. R. Kinkade, *Human Engineering Guide to Equipment Design.* Washington, DC: U.S. Government Printing Office.

Deutsch, J., & Deutsch, D. (1963). Attention: Some theoretical considerations. *Psychological Review, 70,* 80–90.

Drury, C. (1975). Inspection of sheet metal: Model and data. *Human Factors, 17,* 257–265.

Drury, C. (1982). Improving inspection performance. In G. Salvendy (Ed.), *Handbook of Industrial Engineering.* New York: Wiley.

Drury, C. G., & Clement, M. R. (1978). The effect of area, density, and number of background characters on visual search. *Human Factors, 20,* 597–602.

Duncan, J. (1980). The locus of interference in the perception of simultaneous stimuli. *Psychological Review, 87,* 272–300.

Easterbrook, J. A. (1959). The effect of emotion on cue utilization and the organization of behavior. *Psychological Review, 66,* 183–207.

Egan, J., Carterette, E., & Thwing, E. (1954). Some factors affecting multichannel listening. *Journal of the Acoustical Society of America, 26,* 774–782.

Eriksen, B. A., & Eriksen, C. W. (1974). Effects of noise letters upon the identification of a target letter in a non-search task. *Perception & Psychophysics, 16,* 143–149.

Eriksen, C. W., & O'Hara, W. P. (1982). Are nominal same-different matches slower due to differences in level of processing or to response competition? *Perception & Psychophysics, 32,* 335–344.

Fischer, E., Haines, R., & Price, T. (1980, December). *Cognitive issues in head-up displays* (NASA Technical Paper 1711). Washington, DC: NASA.

Fisk, A. D., & Schneider, W. (1982, March). *Task versus component consistency in the development of automatic processes: Consistent attending versus consistent responding* (Technical Report 8106). University of Illinois at Urbana-Champaign, Human Attention Research Laboratory.

Fitts, P., Jones, R. E., & Milton, E. (1950). Eye movements of aircraft pilots during instrument landing approaches. *Aeronautical Engineering Review, 9,* 24–29.

Frost, G. (1972). Man-machine system dynamics. In H. P. VanCott & G. R. Kinkade (Eds.), *Human engineering guide to systems design.* Washington, DC: U.S. Government Printing Office.

Garner, W. R., & Fefoldy, G. L. (1970). Integrality of stimulus dimensions in various types of information processing. *Cognitive Psychology, 1,* 225–241.

Geyer, L. H., Patel, S., & Perry, R. F. (1979). Detectability of multiple flaws. *Human Factors, 21,* 7–12.

Hamilton, V., & Warburton, D. M. (Eds.). (1979). *Human Stress and Cognition: An Information Processing Approach.* Chichester, England: Wiley.

Hansen, J. C., & Hillyard, S. A. (1983). Selective attention to multidimensional stimuli. *Journal of Experimental Psychology: Human Perception and Performance, 9,* 1–19.

Hillyard, S. A., Picton, R. W., & Regan, D. (1978). Sensation, perception, and attention: Analysis using ERPs. In E. Calloway, P. Teuting, & S. H. Koslow (Eds.), *Evoked Potentials.* New York: Academic Press.

Hink, R. G., & Hillyard, S. A. (1976). Auditory evoked potentials during selective attention to dichotic speech messages. *Perception & Psychophysics, 20,* 236–242.

Hintzman, D. L., Carre, F. A., Eskridge, V. L., Ownes, A. M., Shaff, S. S., & Sparks, M. E. (1972). "Stroop" effect: Input or output phenomenon? *Journal of Experimental Psychology, 95,* 458–459.

Hockey, G. R. (1970). Signal probability and spatial location as possible bases for increased selectivity in noise. *Quarterly Journal of Experimental Psychology, 22,* 37–42.

Isreal, J. (1980). *Structural interference in dual task performance: Behavioral and electrophysiological data.* Unpublished Ph.D. dissertation, University of Illinois, Champaign, IL.

Jordan, T. C. (1972). Characteristics of visual and proprioceptive response times in the learning of a motor skill. *Quarterly Journal of Experimental Psychology, 24,* 536–543.

Kahneman, D. (1973). *Attention and effort.* Englewood Cliffs, NJ: Prentice-Hall.

Kahneman, D., & Treisman, A. (in press). Changing views of attention and automaticity. In R. Parasuraman & R. Davies (Eds.), *Varieties of attention.* New York: Academic Press.

Keele, S. W. (1972). Attention demands of memory retrieval. *Journal of Experimental Psychology, 93,* 245–248.

Keele, S. W. (1973). *Attention and human performance.* Pacific Palisades, CA: Goodyear.

Klein, R. M., & Posner, M. I. (1974). Attention to visual and kinesthetic components of skills. *Brain Research, 71,* 401–411.

Kopala, C. (1979). The use of color-coded symbols in a highly dense situation display. In C. Bensel (Ed.), *Proceedings of the 23rd annual meeting of the Human Factors Society.* Santa Monica, CA: Human Factors.

Kristofferson, J. (1967). Attention and psychophysical time. *Acta Psychologica, 27,* 93–100.

LaBerge, D., VanGelder, P., & Yellott, S. (1971). A cueing technique in choice reaction time. *Journal of Experimental Psychology, 87,* 225–228.

Lappin, J. S. (1967). Attention in the identification of stimuli in complex visual displays. *Journal of Experimental Psychology, 75,* 321–328.

Laxar, K., & Olsen, G. M. (1978). Human information processing in navigational displays. *Journal of Applied Psychology, 63,* 734–740.

Leatchenauer, J. C. (1978). Peripheral acuity and photo-interpretation performance. *Human Factors, 20,* 537–552.

Lewis, J. L. (1970). Semantic processing of unattended messages using dichotic listening. *Journal of Experimental Psychology, 85,* 225–228.

Massaro, D. W., & Warner, D. S. (1977). Dividing attention between auditory and visual perception. *Perception & Psychophysics, 21,* 569–574.

McConkie, G. W., & Raynor, K. (1974). *Identifying the span of the effective stimulus in reading* (Final Report OEG 2-71-0537). Washington, DC: U.S. Office of Education.

Mocharnuk, J. B. (1978). Visual target acquisition and ocular scanning performance. *Human Factors, 20,* 611–632.

Monk, T. H. (1976). Target uncertainty in applied visual search. *Human Factors, 18,* 607–612.

Moore, J. J., & Massaro, D. W. (1973). Attention and processing capacity in auditory recognition. *Journal of Experimental Psychology, 99,* 49–54.

Moray, N. (1959). Attention in dichotic listening. *Quarterly Journal of Experimental Psychology, 11,* 56–60.

Moray, N. (1969). *Listening and attention.* Baltimore: Penguin.

Moray, N. (1978). Strategic control of information processing. In G. Underwood (Ed.), *Human information processing.* New York: Academic Press.

Moray, N. (1981). The role of attention in the detection of errors and the diagnosis of errors in man-machine systems. In J. Rasmussen & W. Rouse (Eds.), *Human detection and diagnosis of system failures.* New York: Plenum Press.

Moray, N., & Fitter, M. (1973). A theory and measurement of attention. In S. Kornblum (Ed.), *Attention and performance IV.* New York: Academic Press.

Moray, N., Fitter, M., Ostry, D., Favreau, D., & Nagy, V. (1976). Attention to pure tones. *Quarterly Journal of Experimental Psychology, 28,* 271–285.

Moray, N., & O'Brien, T. (1967). Signal detection theory applied to selective listening. *Journal of the Acoustical Society of America, 42,* 765–772.

Morton, J. (1969). The use of correlated stimulus information in card sorting. *Perception & Psychophysics, 5,* 374–376.

Neisser, U., & Beckman, R. (1975). Selective looking: Attention to visually specified events. *Cognitive Psychology, 7,* 480–494.

Neisser, U., Novick, R., & Lazar, R. (1964). Searching for novel targets. *Perceptual and Motor Skills, 19,* 427–432.

Ninio, A., & Kahneman, D. (1973). Reaction time in focused and divided attention. *Journal of Experimental Psychology, 103,* 394–399.

Nissen, M. J. M. (1974). *Facilitation and selection: Two modes of sensory interaction.* Unpublished master's thesis, University of Oregon, Eugene, OR.

Norman, D. (1968). Toward a theory of memory and attention. *Psychological Review, 75,* 522–536.

Posner, M. I., Nissen J. M., & Klein, R. (1976). Visual dominance: An information processing account of its origins and significance. *Psychological Review, 83,* 157–171.

Reid, L. D., & Solowka, E. N. (1981). A systematic study of driver steering behavior. *Ergonomics, 24,* 447–462.

Rock, I. (1975). *An introduction to perception.* New York: Macmillan.

Rogers, S. P. (1979). Stimulus-response incompatibility: Extra processing stages versus response competition. In C. Bensel (Ed.), *Proceedings, 23rd annual meeting of the Human Factors Society.* Santa Monica, CA: Human Factors.

Scanlan, L. A. (1977). Target acquisition in realistic terrain. In A. S. Neal & R. Palasek (Eds.), *Proceedings, 21st annual meeting of the Human Factors Society.* Santa Monica, CA: Human Factors.

Schneider, W., & Fisk, A. D. (1982). Concurrent automatic and controlled visual search: Can processing occur without cost? *Journal of Experimental Psychology: Learning, Memory, & Cognition, 8,* 261–278.

Schneider, W., & Shiffrin, R. (1977). Controlled and automatic human information processing. *Psychological Review, 84,* 1–66.

Senders, J. (1964). The human operator as a monitor and controller of multidegree of freedom systems. *IEEE Transactions on Human Factors in Electronics, HFE-5,* 2–6.

Sheehan, J. J., & Drury, C. G. (1967). Ergonomic and economic factors in an industrial inspection task. *International Journal of Production Research, 7,* 333–341.

Sheridan, T. (1972). On how often the supervisor should sample. *IEEE Transactions on Systems, Sciences, and Cybernetics, SSC-6*, 140–145.

Sheridan, T. (1981). Understanding human error and aiding human diagnostic behavior in nuclear power plants. In J. Rasmussen & W. B. Rouse (Eds.), *Human detection and diagnoses of system failures*. New York: Plenum Press.

Sheridan, T., & Rouse, W. B. (1971). Supervisory sampling and control: Sources of suboptimality. *Proceedings of the 7th Annual Conference on Manual Control* (NASA SP-281). Washington, DC: U.S. Government Printing Office.

Simpson, C., & Williams, D. H. (1980). Response time effects of alerting tone and semantic context for synthesized voice cockpit warnings. *Human Factors, 22*, 319–330.

Spieth, W., Curtis, J., & Webester, J. (1954). Responding to one of two simultaneous messages. *Journal of the Acoustical Society of America, 26*, 391–396.

Spitz, G., & Drury, C. C. (1978). Inspection of sheet materials test of model predictions. *Human Factors, 20*, 521–528.

Sternberg, S. (1975). Memory scanning: New findings and current controversies. *Quarterly Journal of Experimental Psychology, 27*, 1–32.

Stroop, J. R. (1935). Studies of interference in serial verbal reactions. *Journal of Experimental Psychology, 18*, 643–662.

Teichner, W. H., & Mocharnuk, J. B. (1979). Visual search for complex targets. *Human Factors, 21*, 259–276.

Treisman, A. (1960). Contextual cues in selective listening. *Quarterly Journal of Experimental Psychology, 12*, 242–248.

Treisman, A. (1964a). Verbal cues, language, and meaning in attention. *American Journal of Psychology, 77*, 206–214.

Treisman, A. (1964b). The effect of irrelevant material on the efficiency of selective listening. *American Journal of Psychology, 77*, 533–546.

Treisman, A. (1969). Strategies and models of selective attention. *Psychological Review, 76*, 282–299.

Treisman, A., & Davies, A. (1973). Divided attention to eye and ear. In S. Kornblum (Ed.), *Attention and Performance IV*. New York: Academic Press.

Treisman, A., & Riley, R. G. (1969). Is selective attention selective perception or selective response? A further test. *Journal of Experimental Psychology, 79*, 27–34.

Tulga, M. K., & Sheridan, T. B. (1980). Dynamic decisions and workload in multitask supervisory control. *IEEE Transactions on Systems, Man, and Cybernetics, SMC-10*, 217–232.

Wachtel, P. L. (1967). Conceptions of broad and narrow attention. *Psychological Bulletin, 68*, 417–419.

Wiener, E. L. (1977). Controlled flight into terrain accidents: System-induced errors. *Human Factors, 19*, 171.

Attention, Time-Sharing, and Workload

8

OVERVIEW

Chapter 7 described the searchlight metaphor of attention. As the searchlight illuminates the visual world, so attention guides our perception of the environment. The concept of attention, however, is relevant to a much broader range of human performance than merely perception. We speak, for example, of dividing attention between tasks that may not be perceptual. To account for the role of attention in time-sharing between tasks, in this chapter we adopt a different metaphor, that of the resource (Norman & Bobrow, 1975).

While the searchlight metaphor emphasizes the unity of attention, the *resource metaphor* emphasizes its divisibility. When performing any task, different mental operations must be carried out (responding, rehearsing, perceiving, etc.). Performance of each of these activities requires some degree of the operator's limited processing resources. Since these resources are limited, the metaphor readily accounts for our failures of time-sharing. Two activities will demand more resources than a single activity, and so there will be a greater deficiency between supply and demand in the former case. The resource metaphor also proposes that some operations may require resources that are different from others (just as some furnaces utilize oil and others natural gas or coal). As a consequence, there is less competition between these processes for their enabling resources, and time-sharing between them may be more successful. An example of separate resources was provided in Chapter 6 in terms of the distinction between verbal and spatial working memory (Baddeley & Hitch, 1974). Different resources underlie the two memory codes, and so there is relatively effective time-sharing between two tasks using the two types of memory. The resource metaphor is particularly important because it specifies that resources are not synonymous with a unitary "consciousness" but instead represent a hypothetical construct that enables performance to proceed whether conscious or not (Navon & Gopher, 1979).

TIME-SHARING

Theories of divided attention in perception, such as those proposed by Treisman (1969), Deutsch and Deutsch (1963), and Norman (1968), stimulated research and theory into the more general limitations of attention in dual-task performance. This research was concerned not only with the limits of dividing attention between two perceptual channels, but between two activities of any sort, whether they involved perception, working memory, decision, or response. The theoretical treatments of time-sharing that were proposed in the 1960s and 1970s identified two major influences on the efficiency with which two tasks could be performed concurrently.

The first of these influences relates to the *structural similarity* of the tasks. Thus we find it more difficult to read two messages at once than to read one message and listen to another. The greater the similarity, the more likely it is that two tasks will compete for common processing mechanisms (or for processing mechanisms that both demand the same higher-level processor to function). This competition produces greater interference. A prototypical example of a theory of task interference based upon structural aspects of processing is that proposed by Kerr (1973). Kerr summarized the data from a number of dual-task studies and integrated these with a model that accounts for interference between tasks in terms of a taxonomy of seven basic processing operations demanded by the tasks. These operations are encoding of single inputs, multiple-input encoding, transformations, rehearsal, response selection, response execution, and response execution to a physical stop. All operations but the first and last are assumed to demand a *limited-capacity central processor*. Two tasks will interfere if both require operations that depend upon the limited-capacity processor.

Kerr's theory was parsimonious and did account for many of the differences in the success with which two tasks could be carried out concurrently. However, it did not incorporate a second influence on time-sharing efficiency that relates to the *difficulty* of the tasks. We can time-share driving and conversation on an open freeway, but on a crowded freeway our conversation will deteriorate because driving in heavy traffic is more difficult. Thus a second approach to time-sharing emphasizes the quantitative, intensive aspects of attention demands. This approach is compatible with the resource metaphor of attention: The demand for resources is determined by the difficulty of the task confronting the operator. The discussion of time-sharing that follows presents first a resource model of time-sharing that does not deal with the structural aspects of the tasks. This model will then be elaborated to account for the structural characteristics of both the task and of the human operator (Wickens, in press).

Resource Theory

Emphasis on the quantitative attributes of attention owes much to a seminal paper by Moray (1967), who proposed that attention was like the limited processing capacity of a general-purpose computer. This capacity could be allocated in graded amounts to various activities performed, depending upon their difficulty or demand for that capacity. The capacity concept emphasizes both the flexible and the sharable nature of attention or processing resources. The "executive" is not considered to be a dedicated processor that handles all of one mental operation or another as these are "demanded" by the task. Instead, the executive is a resource manager. Tasks demand more of these hypothetical resources (attention or "mental effort") as they become more difficult or their desired level of performance increases. With fewer resources available for other tasks, performance on them will deteriorate.

The concept of attention as a flexible, sharable, processing resource of limited availability was elaborated by Kahneman (1973), Norman and Bobrow (1975), and Navon and Gopher (1979). The major contributions of these investigators in refining the theoretical representation of processing resources will be described first, before further data on dual-task performance is considered.

Kahneman (1973) was the first who stated some of the predictions of the resource or capacity concept of human attention. In the early chapters of his comprehensive book, he proposes that there is a single undifferentiated pool of such resources, available to all tasks and mental activities, as shown in Figure 8.1*a*. (This is a position that he qualifies later in the book.) As task demands increase either by making a given task more difficult or by imposing additional tasks, physiological arousal mechanisms produce an increase in the supply of resources. However, this increase is insufficient to compensate entirely for the increased resource demands (Figure 8.1*b*). Thus performance falls off as the supply-demand shortfall increases, while physiological manifestations of increased arousal such as heart rate and pupil diameter are evident as indices of resource mobilization. These physiological indicators of resource demand or *mental workload* will be discussed in more detail later in the chapter.

Norman and Bobrow's (1975) theoretical paper introduced the important concept of the *performance-resource function*. If two tasks do in fact interfere with each other (are performed less well) because they are sharing resources to which each previously had exclusive access, then there must be some underlying function that relates the quality of performance to the quantity of resources invested in a task. This hypothetical function is the performance-resource function, or PRF, an example of which is shown in Figure 8.2. Single-task performance occurs when all resources are invested in the task (point *A*) and is the best that can be obtained. Diverting a large amount of resources away from the task to a concurrent one will depress performance accordingly as indicated by point *B*. As more resources are then reinvested back into the task, performance will improve up to the point (*C*) at which no further change in performance is possible. To the right of this point, the task is said to be *data-limited* (limited by the quality of data, not by the resources invested). Data limits may occur at any level of performance. Remembering a 2-digit number is data-limited, since perfect performance can be obtained with few resources and further

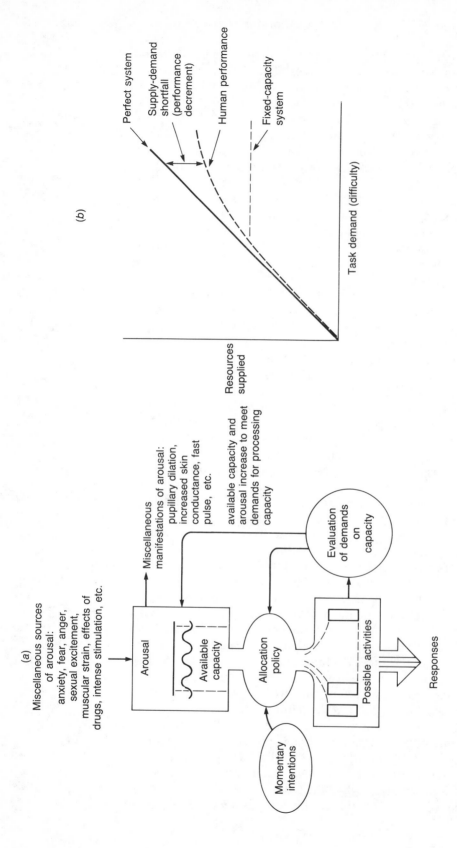

Figure 8.1 (a) Relation between capacity demands, arousal, and performance. This conception is a foundation for the "undifferentiated" capacity view of resources. (b) Hypothetical relation between resources demanded with increasing task difficulty and resources supplied. From *Attention and Effort* (pp. 10, 15) by D. Kahneman, 1973, Englewood Cliffs, NJ: Prentice-Hall. Copyright 1973 by Prentice-Hall. Reprinted by permission.

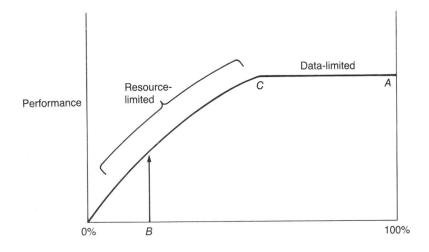

Figure 8.2 A hypothetical performance-resource function. Points A, B, and C are described in the text. From "On Data-Limited and Resource-Limited Processing" by D. Norman and D. Bobrow, 1975, *Journal of Cognitive Psychology, 7,* p. 49. Copyright 1975 by the American Psychological Association. Adapted by permission of the authors.

effort will lead to no improvement. As a second example of data limits, you also may not be able to understand a faint conversation no matter how hard you "strain your ears." In this case, the data limit is at less-than-perfect performance. When performance does change with added or depleted resources, the task is said to be *resource-limited,* the region to the left of point C in Figure 8.2.

Time-sharing and the PRF. If all resources are presumed to come from the same "reservoir" of capacity, then the amount of interference between two tasks is determined by the form of the two PRFs. In Figure 8.3a the PRF for the bottom task (B) is plotted backwards so that a given vertical slice through both functions will represent a single policy of allocating X% resources to task A and (100 − X)% to task B. Maximum single-task performance on each task is given by 100% resource allocation. If resources are divided between tasks, performance must drop off on one or both unless both are data-limited. An example in which both tasks are data-limited is shown in Figure 8.3b. In this case the allocation of 40% resources to task A and 60% to task B will provide perfect time-sharing. That is, both tasks performed concurrently can be done as well as either task performed alone.

Allocation of resources. In the discussion of Figure 8.3 we have assumed that subjects can *allocate* their resources flexibly between tasks in any proportion desired. That is, they need not adopt a "50/50 split" but can choose any other proportional allocation. This ability has been well documented in a number of investigations (e.g., Gopher &

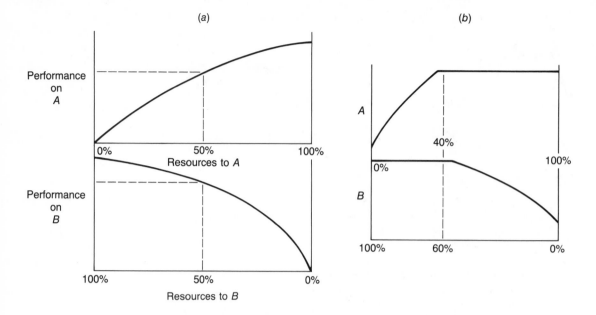

Figure 8.3   Performance-resource functions of two time-shared tasks. (a) Two resource-limited tasks with 50/50 allocation policy. (b) Two data-limited tasks with 40/60 allocation policy demonstrating perfect time-sharing.

Navon, 1980; Gopher, Brickner, & Navon, 1982; Schneider & Fisk, 1982; Sperling & Melchner, 1978; Wickens & Gopher, 1977). In fact, when two tasks are time-shared and the subject is asked to adopt different strategies of allocation on successive trials, it is possible to cross-plot performance of the two tasks on a single graph. This curve is called a *performance operating characteristic*, or *POC* (Norman & Bobrow, 1975). A POC constructed from the two PRFs of Figure 8.3a is shown in Figure 8.4a. The POC has proven to be a useful way of summarizing a number of characteristics of two time-shared tasks. Hence it is important to describe certain important "landmarks" or characteristics of the POC.

1.   *Single-task performance* is shown by points on the two axes (*A* and *B*) that have a hypothetical intersection in the POC space at *P*. This point represents perfect time-sharing. As shown in Figure 8.4, the single-task points may not be continuous with the extention of the POC to the single-task axes. If, as is shown in Figure 8.4, these single-task points are higher (better performance), then there is, in the words of Gopher and Navon, a "cost of concurrence." The act of time-sharing itself pulls resources away from both tasks above and beyond the resources that each task demands by itself. Thus, time-sharing, even with no resources allocated to the other task, produces worse performance than the single-task condition. This cost of concurrence will be imposed, for example, if the two tasks are displayed at different locations in the visual field. Thus both cannot be fixated simultaneously. The cost of concurrence may also reflect the resource demands of an executive time-sharing mechanism that is responsible for coordinating responses, sampling display locations, and deciding how to allocate resources. Such a mechanism will be called into play only in time-sharing conditions (Allport, 1980; Hunt & Lansman, 1981; McLeod, 1977).

2. The *time-sharing efficiency* of the two tasks is indicated by the average distance of the curve from the origin (O); obviously the farther from the origin, the more nearly dual-task performance is close to the single-task performance point P. This is efficient time-sharing.

3. The *linearity* or "smoothness" of the function indicates the extent of shared or exchangeable resources between the tasks. A curve such as shown in Figure 8.4a is smooth, and so indicates that a given number of hypothetical units of resources removed from task A (thereby decreasing its performance) can be transferred to and efficiently utilized to improve the performance of task B. A discontinuous or "boxlike" POC (Figure 8.4b) suggests that resources are not as interchangeable. When resources are withdrawn from one task they cannot be used to improve performance of the other one. This may occur when there are data limits in the tasks. The POC shown in Figure 8.4b would result from the

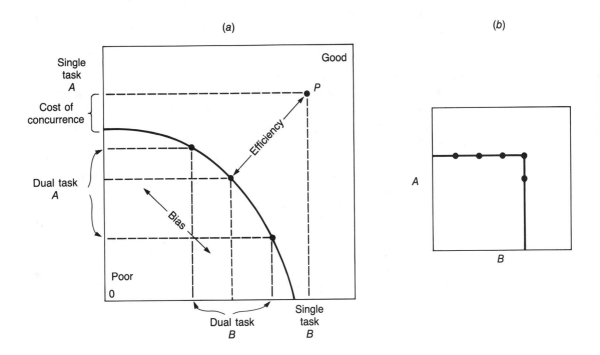

Figure 8.4   The performance operating characteristic (POC). (*a*) Two resource-limited tasks shown in Figure 8.3*a* with three allocation policies. (*b*) Two data-limited tasks shown in Figure 8.3*b* with five allocation policies.

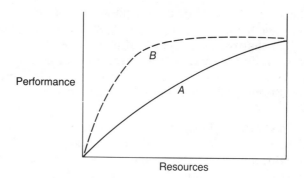

Figure 8.5   The performance resource function and practice. Task *B*: practiced or easy; task *A*: unpracticed or difficult.

shift in resource allocation across the PRFs of the two data-limited tasks shown in Figure 8.3*b*. A second cause of the boxlike POC is the possibility that different resources may be used in the two tasks. This possibility will be considered in some detail in the later sections of this chapter.

4.   Since the POC is actually a series of points, each one collected in a different time-sharing trial, the *allocation bias* of a given condition is indicated by the proximity of a given point on the POC to one axis over the other. A point on the positive diagonal indicates an "equal allocation" of resources between tasks.[1]

   Automation and difficulty.   The effects of both practice and task difficulty can be easily represented by the resource metaphor. Figure 8.5 shows the PRF represented by two tasks, *A* and *B*. Task *B* demands fewer resources to reach performance levels that are equivalent to *A*. Task *B* also contains a greater "data-limited" region. Task *B* then differs from *A* by being of lesser difficulty or having received more practice (being more automatic). Note that task *B* may not necessarily be performed better than task *A* if full resources are invested into *A* but will simply be performed at that level with more "spare capacity." Hence task *B* will be less disrupted by diverting resources from its performance than will task *A*. Within this framework, the "automaticity" of perceiving words or letters discussed in Chapter 4 need not be viewed as a qualitatively different phenomenon from attention-demanding perceptual activities (e.g., Kerr, 1973). These tasks merely differ on the

---

[1]When the two tasks in a POC are the same, with the same performance measure, there is no problem with the assumption that equal allocation of resources lies on the positive diagonal. However, when the tasks are different, with different dependent variables (e.g., tracking and reaction time), a question arises as to how to scale the two axes into common units so that spatial relations along the two axes may be meaningfully compared. One possible solution is to convert both to common dimensionless units such as standardized scores (Wickens, Mountford, & Schreiner, 1981).

basis of a quantitative change in the data-limited regions of the PRF. The experiment performed by LaBerge (1973), described in Chapter 4, on automatic processing of letters and nonletters demonstrated the way in which the resource-limited curve *A* changes to the data-limited curve *B* when subjects receive extensive practice in categorizing letterlike symbols.

The implication of the difference between PRFs such as those shown in Figure 8.5 is that the extent of differences between two tasks or two situations may not be appreciated by examining primary-task performance alone. Only when the primary task is time-shared with a concurrent task will the differences be realized. Thus, for example, Bahrick, Noble, and Fitts (1954) employed the secondary task to index differences in learning of a perceptual-motor primary task that were not revealed by primary-task performance. Dornic (1980) compared comprehension of first and second languages by bilingual speakers and observed differences in the secondary task while none were found in the primary. The use of secondary tasks as one means to assess mental workload will be considered in greater detail later in the chapter.

When task difficulty is represented in the POC, the easier or more practiced version of a task yields a POC that is farther from the origin. This contrast is shown in Figure 8.6. An easy and a difficult version of task *A* are time-shared with task *B*. Curve I is that underlying time-

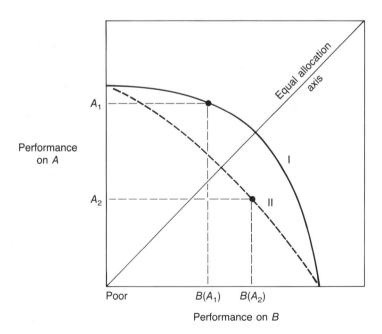

Figure 8.6   POC representation of an easy (A_1) and a difficult (A_2) primary task time-shared with a secondary task (*B*). The figure shows how difficulty and allocation bias can be confounded if only the secondary task is examined.

sharing performance with the easier version of task A, while curve II underlies the more difficult version.

The representation shown in Figure 8.6 conveys an important message to investigators who employ performance on a secondary task to make inferences concerning primary-task resource demands (Rolfe, 1973). Suppose that only two measures were taken of secondary-task performance B, one with each of the two versions of the primary task (I and II) whose resource demands are to be compared. These measures are shown as $B(A_1)$ and $B(A_2)$ in Figure 8.6. On the basis of these data, since $B(A_2)$ is performed better than $B(A_1)$, the investigator would conclude that A_2 must be an easier version of the primary task than A_1. Yet as we look at the shape of the underlying POC, we see this is not the case. Curve II runs closer to the origin. It is only because our subjects allocated resources in favor of task B when paired with A_2 to a greater extent than when paired with A_1 that this spurious result was observed. When comparing resource demands between tasks, it is important then to represent dual-task results in a POC space and not just in terms of "secondary task-performance decrements." When data are presented in the POC space, investigators have a better idea how the subject is controlling allocation (Navon & Gopher, 1979; Wickens, in press).

Limitation of the "undifferentiated" aspect of capacity theory.   A major limitation with single-resource theory is that it cannot account for several aspects of the data from dual-task interference studies (Wickens, 1980, in press). The concept of a resource translates closely to that of "difficulty." Tasks of greater difficulty performed at the same level of performance demand more resources. Yet examples abound in which interference between tasks is predicted not by their difficulty but by their structure (e.g., the stages, codes, and modalities of processing required). For example, Wickens (1976) found that performance on a manual tracking task (a task similar to flying an aircraft, discussed in detail in Chapter 11) was more disrupted by a concurrent task requiring a pure response (maintaining constant pressure on a stick) than by auditory signal detection, even though the latter was judged by subjects to be the more difficult task and therefore presumably demanded more resources.

Other examples may be cited in which increases in the difficulty of one task, which should presumably consume more resources (as allocation is held constant), fail to influence the performance of a second task. An example of such "difficulty insensitivity" is provided in a study by North (1977). Subjects time-shared a tracking task with a discrete task in which mental operations of varying degrees of complexity were performed on visually displayed digits. Responses were indicated with a manual key press. In the simplest condition subjects merely pressed the key corresponding to the displayed digit (digit canceling). A condition of intermediate demand required the subject to indicate the digit immediately preceding the displayed digit. In the most demanding condition, subjects were required to perform a classification operation on a pair of displayed digits. These three operations apparently imposed different demands, as indicated by their single-task performance and their interference with simple digit canceling. However, when they were performed concurrently with the tracking task, all three disrupted tracking performance equally.

A third class of results that is not readily explained by an undifferentiated capacity viewpoint are those in which two tasks that are both clearly attention demanding can be time-shared perfectly. This success is possible because the two tasks have separate structures. Allport, Antonis, and Reynolds (1972), for example, showed that skilled pianists could sight-read music and engage in verbal shadowing with no disruption of either task by the other. Shaffer (1975) reports an experiment in which a skilled typist can transcribe a written message and engage in verbal shadowing, again with perfect time-sharing efficiency.

An explanation for some of these phenomena is that the characteristic that allows for efficient time-sharing and difficulty insensitivity is simply the shape of the performance-resource functions. As shown in Figure 8.3*b*, PRFs with large data limits for one or both tasks should allow either of the above phenomena to occur. Indeed, it is true that much of the variation in the efficiency with which any two tasks may be time-shared is due to the automation of one or both of the component tasks (Hunt & Lansman, 1981; Schneider & Fisk, 1982; Schneider & Shiffrin, 1977). Yet this argument cannot explain differences in the amount of interference when the structure or processing mechanism of one of the paired tasks is altered. These changes are referred to as *structural alteration effects* (Wickens, 1980). McLeod (1977), for example, observed a change in the interference between tracking and tone discrimination when the discrimination response was made vocally rather than manually. Wickens, Sandry, and Vidulich (1983) observed a reduction in interference between tracking and memory search when the input of the latter task was changed from visual to auditory and when the response was changed from manual to vocal. Many other examples of structural alteration effects are identified by Wickens (1980).

Kahneman's (1973) theory has accounted for these limitations of single-resource theory by proposing that task interference results from competition both for an undifferentiated capacity for which all tasks compete and for a series of satellite *structures* (eyes, ears, hands, voice). Competition for common structures between two tasks will produce *structural interference*, which can be distinguished from *capacity interference* (demand for resources from the central pool). An alternative to Kahneman's model is the multiple-resource theory, which will be discussed in the next section.

Multiple-Resource Theory

The multiple-resource view argues that instead of one central "pool" of resources with satellite structures, humans possess several different capacities with resource properties. Tasks will interfere more if more resources are shared. This position has been proposed by Kantowitz and Knight (1976), McLeod (1977), Roediger, Knight, and Kantowitz (1977), Allport (1980), and Kinsbourne and Hicks (1978) but has received the most explicit theoretical development within the framework of the POC by Navon and Gopher (1979). Wickens (1980, in press), drawing upon the results from a large number of dual-task studies, has argued that resources may be defined by three relatively simple dichotomous dimensions. There are two stage-defined resources (early versus late processes), two modality-defined resources (auditory versus visual encoding), and two resources defined by processing codes (spatial versus verbal).

Figure 8.7 presents the dimensional representation of multiple resources. To the extent that any two tasks demand separate rather than common resources on any of the three dimensions, three phenomena will occur: (1) Time-sharing will be more efficient. (2) Changes in the difficulty of one task will be less likely to influence performance of the other; that is, difficulty insensitivity will be observed. (3) The POC constructed between the tasks will be more of the "boxlike" form of Figure 8.4*b*, because resources withdrawn from one task cannot be used to advantage by the other, since it is dependent upon different resources (Gopher & Navon, 1980; Wickens, in press). The nature of the three dimensions that constitute the multiple-resource model will now be considered in more detail.

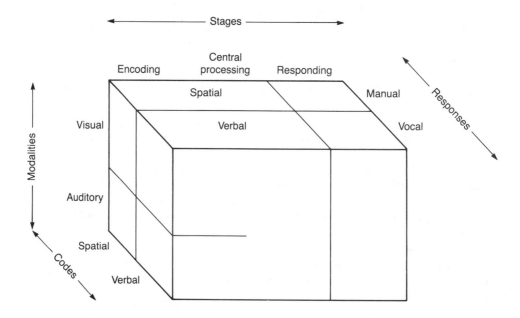

Figure 8.7 The proposed structure of processing resources. From "Processing Resources in Attention" by C. D. Wickens, in press, in R. Parasuraman and R. Davies (Eds.), *Varieties of Attention*, New York: Academic Press. Reproduced by permission.

Stages.   The resources used for perceptual and central-processing activities appear to be the same and functionally separate from those underlying the selection and execution of responses. Evidence for this dichotomy is provided when the difficulty of responding in a task is varied and this manipulation does not affect performance of a concurrent task whose demands are more perceptual in nature. A series of experiments by Isreal, Chesney, Wickens, and Donchin (1980) and Isreal, Wickens, Chesney, and Donchin (1980) provide examples of this evidence. In these experiments the amplitude of an evoked brain potential (see Chapter 7) elicited by a series of counted tones is assumed to reflect the investment of perceptual and central processing resources, since the evoked potential can be elicited without requiring any responses. The experiments find that the evoked potential is insensitive to response-related manipulations of tracking difficulty but is influenced by manipulations of display load. Wickens and Kessel (1980) observed that a tracking task, demanding response resources, was disrupted by a concurrent tracking task but not by a mental arithmetic task demanding central-processing activities in working memory.

Modalities.   It is apparent that we can sometimes divide attention between the eye and ear better than between two auditory channels or two visual channels. That is, bimodal time-sharing is better than intramodal. It is obvious that poor time-sharing with intramodal displays will be found if the two visual sources are spatially separated so that both cannot access foveal vision simultaneously or if the two auditory sources mask each other. However, some studies have carefully controlled for this form of peripheral interference (Isreal, 1980; Rollins & Hendricks, 1980; Treisman & Davies, 1973), and these investigations suggest that there is still an advantage to cross-modal displays even when scanning is not required or masking does not occur.

Treisman and Davies (1973) asked subjects to detect simultaneous pairs of targets when these were both delivered visually, both auditorily, or cross-modally. They found that both spatial-temporal patterns and semantic targets were detected better in the cross-modal condition than in either of the intramodal conditions. Rollins and Hendricks (1980) replicated this result even when the depth of semantic processing of the auditory stimuli was systematically varied (see Chapter 6). Isreal (1980) also obtained greater intra- than cross-modality interference between tracking and a reaction time task, when the modality of both tasks was manipulated. That is, whether the reaction-time stimuli were auditory or visual or the tracking error was displayed auditorily or visually, time-sharing was consistently better in the cross-modal conditions.

Processing codes.   The role that the spatial and verbal codes play in defining separate processing resources has been discussed in some detail already in previous chapters. In Chapter 4 spatial and verbal perceptual processes were contrasted, and their possible association with cerebral hemispheres was discussed. In Chapter 6 the degree of interference within and between spatial and verbal working memory was considered. In those chapters we described the research of Baddeley and his colleagues, of Brooks, and of Kinsbourne and Hicks and concluded that spatial and verbal processes, whether functioning in perception, working memory, or response, depend upon separate resources. The conclusion offered by Kinsbourne and Hicks (1978) that performance of a tracking task with the right hand disrupts a verbal task more than performing it with the left hand suggests furthermore that the code-related resource dimension may well be defined by the two cerebral hemispheres. Resources underlying spatial processing and left-hand control resides predominantly in the right hemisphere. Resources underlying verbal processing, speech responses, and right-hand control reside in the left.

The separation of spatial and verbal resources seemingly accounts for the high degree of efficiency with which manual and vocal outputs can be time-shared, assuming that manual responses are usually spatial in nature and vocal ones are verbal. In this regard investigations by McLeod (1977), Wickens (1980), and Wickens et al. (1983) have shown that tracking and a discrete verbal task are time-shared more efficiently when the verbal task employs vocal as opposed to manual response mechanisms. Discrete manual responses using the nontracking hand appear to interrupt the continuous flow of the tracking response. Discrete vocal responses leave this flow untouched.

Finally, consider the near perfect time-sharing efficiency with which shadowing and visual-manual transcription tasks (typing and sight-reading) can be carried out, as demonstrated by Shaffer (1975) and by Allport et al. (1972), respectively. This success is clearly due in large part to the separation of codes of processing and modalities of processing of the two tasks. If the auditory shadowing and piano sight-reading task pair investigated by Allport, Antonis, and Reynolds (1972) are examined within the framework of Figure 8.7, we see that auditory shadowing is clearly auditory, verbal and vocal. Piano sight-reading is visual and manual. If the further assumption is made that music involves right-hemispheric processing (Nebes, 1977), then the two tasks may be considered to require entirely separate resources.

Limits of the multiple-resource model.   The three dimensions of the multiple-resource model do not intend to account for all structural influences on dual-task performance and time-sharing efficiency. Instead, they indicate three major dichotomies that can account for a reasonably large portion of these influences and can be readily used by the system designer. In this sense, the model is an effort to gain usability and parsimony by sacrificing some degree of precision (see Chapter 1).

It is important to realize, however, that there are many ways in which two tasks can be similar that influence the efficiency of their time-sharing but are not accounted for by the three dimensions portrayed in Figure 8.7. These include several factors: (1) Two tasks may have different timing requirements. Tasks with different rhythmic requirements are hard to time-share (Peters, 1977). (2) Manual control tasks with different control dynamics reduce time-sharing efficiency (Chernikoff, Duey, & Taylor, 1960; Moray, 1982; see Chapter 11). (3) Two tasks may have similar processing elements. Two tasks that use both digits and letters will be more easily time-shared than two tasks using the same material. Two tracking tasks that use horizontal and vertical axes will be better time-shared than two tasks using the common direction. This similarity factor was identified in Chapter 6 as a major source of interference in memory. (4) The astute reader will also note that data reported earlier in Chapter 7 suggested that it is faster to switch attention to sources within a stimulus modality than between modalities. This effect is in direct contrast with the superior cross-modal time-sharing predicted by the multiple-resource model.

The previous discussion has suggested that there are two competing factors that influence time-sharing efficiency. On the one hand, greater *similarity* between the overall control laws, rhythm, or executive timing parameters of two tasks seems to improve time-sharing efficiency. (This assumes that presenting stimuli from both tasks within a modality, by eliminating the need to switch, simplifies the timing requirements.) On the other hand, greater *differences* in the structural elements that are employed by the tasks (whether these are resources within the model or elements within a resource) also improve time-sharing efficiency. This contrast between the two aspects of similarity creates difficulty in making a general statement concerning the effect of similarity on time-sharing. When two factors oppose each other in this fashion, then it is difficult to formulate quantitative predictive models of time-sharing performance.

Finally, a literal interpretation of Figure 8.7 suggests that any two tasks demanding separate "cells" within the model should yield perfect time-sharing. The tasks of Allport et al. (1972) and Shaffer (1975) were interpreted in this light. Yet in fact some have argued that the time-sharing in these conditions is very good, but not perfect (Broadbent, 1982). There may well be a layer of "general capacity" that is added, like frosting on a cake, to the top and front of the separate resources of Figure 8.7. These resources, competed for by all tasks, would then prevent perfect time-sharing of all but heavily data-limited tasks.

Individual Differences and Learning in Time-Sharing

When observing the expert perform a complex task, whether juggling, performing secretarial work, flying an aircraft, or engaging in industrial inspection and assembly, the novice is often overwhelmed at the ease with which the expert can effectively time-share a number of separate activities. Indeed, across people and across age groups there is a considerable diversity in the efficiency with which tasks may be time-shared. Some of these differences may result from differences in practice and others from differences in a basic time-sharing ability. We shall consider below some of the evidence for the role of practice and of ability differences in time-sharing efficiency.

Attention as a skill.   Continued practice in a dual-task environment will soon lead to improved dual-task performance. Naturally some component of this gain could result simply from an improvement in the single-task component skills. As the two skills demand fewer resources and become more data-limited through practice, their combined resource demand will be diminished, and their dual task efficiency will improve accordingly. It should also be recalled that the change in single-task performance we call *automation* need not entail an increased level of single-task performance, but only an increase in the data-limited region of a PRF that asymptotes at the same level (see Figure 8.5).

This form of improvement does not actually result from acquisition of an attentional or time-sharing "skill" but from a reduced resource demand of single-task performance. It may be acquired, in short, from extensive single-task practice. Such a mechanism explains the "automatic" processing of familiar perceptual stimuli such as letters (LaBerge, 1973), of consistently assigned targets (Schneider & Fisk, 1982; Schneider & Shiffrin, 1977), and of repeated sequences of stimuli (Bahrick, Noble, & Fitts, 1954; Bahrick & Shelley, 1958), as well as the automatic performance of habitual motor acts such as signing one's own name. When describing the response end of processing, this automaticity of performance is referred as the *motor program* (Keele, 1973; Summers, 1981); it will be discussed more in Chapter 11.

Distinct from automation is the true skill in time-sharing that results explicitly and exclusively from multiple-task practice. In order to show that improvements in time-sharing efficiency are due to the development of a time-sharing skill and not simply to increased automation of the component task, Damos and Wickens (1980) suggest that one of three procedures must be employed: (1) Show that time-sharing performance for a given task pair develops more rapidly with dual- than with single-task performance. (2) Show that a time-sharing "skill" developed with one task combination transfers to a qualitatively different time-sharing task combination. (3) Demonstrate, through a microscopic analysis of the timing of responses in dual-task performance, that changes in strategy develop which directly reflect differences in the manner in which the tasks are interwoven. As Moray and Fitter (1973) argue, such changes would indicate that time-sharing, switching, or sampling behavior is learned, similar to the optimal sampling prescriptions discussed at the beginning of Chapter 7.

Damos and Wickens (1980) conducted an investigation of time-sharing skill development that emphasized the three procedures described above. They asked subjects to time-share two speeded tasks: a digit running-memory task (see Chapter 6) and a digit categorization task (digit pairs were judged on their similarity of value and physical size). Performance on both tasks was assessed when they were performed together and on periodic single-task trials, spaced throughout roughly 25 trials of dual-task training. When dual-task training was terminated, the subjects then transferred to a dual-axis tracking task, and the same training procedure that was employed on the two discrete tasks was repeated.

Three kinds of analyses suggested that a unique time-sharing skill was developing: (1) The degree to which time-sharing on the dual-axis tracking task benefited from the prior dual-task exposure was evaluated in a transfer of training design by comparing the performance of these subjects with that of a control group. The control group had experienced corresponding single-task practice on the two discrete tasks but had never performed them together. The transfer was positive, suggesting that the earlier dual-task training of the transfer group did, in fact, develop a "generalizable" dual-task skill. (2) Microscopic analysis of dual-task performance of both the discrete task pair and the tracking pair revealed that practiced subjects engaged in more parallel processing of the stimuli. (3) Individual differences between subjects revealed some who alternated rapidly between the discrete tasks and others who processed both in parallel. The latter group consistently performed better.

The learning of perceptual sampling strategies related to time-sharing was discussed in Chapter 7 in regard to the experiments performed by Senders (1964), Moray et al. (1976), and Sheridan (1972). The notion that strategies in allocating and switching attention contribute to improved time-sharing performance has received further support from three more recent experiments. Gopher and Brickner (1980) observed that subjects who were trained in a time-sharing regime which successively emphasized different resource allocation policies became more efficient time-sharers in general than did a group trained only with equal priorities. The former group was also better able to adjust performance in response to changes in dual-task difficulty.

Schneider and Fisk (1982) found that subjects could time-share an automated and a resource-demanding letter-detection task with perfect efficiency if they received training to allocate their attention *away from* the automatic task. In the absence of this training, subjects allocated resources in a nonoptimal fashion by providing more resources to the automated task than it needed, at the expense of the resource-limited task. In the example presented in Figure 8.3b, this was as if subjects initially allocated 60% of the resources to A and 40% to B and needed to be trained to adopt the more optimal 40/60 split. This paradigm is interesting in that it showed the contributions of both single-task automation (for the automated detection task) *and* dual-task resource allocation training to overall time-sharing efficiency.

A related finding was obtained by Wickens et al. (1982). Their subjects performed a letter-recognition task concurrently with tracking.

The two tasks were widely spaced on the display with the tracking symbol to the left and letters to the right. When adopting their own strategy, subjects time-shared the two tasks reasonably well. However, their time-sharing performance on both tasks improved markedly when they were instructed to fixate the verbal display. Under these conditions the spatial tracking display fell in left peripheral vision, which, although not of high enough acuity for seeing letters, is adequate for processing the position and velocity information used in tracking. Furthermore, the fact that tracking information was to the left provided it direct access to the "spatial" processing hemisphere, as discussed in Chapter 4. The role of hemispheric laterality was suggested here because a similar benefit of fixating the verbal display was not found when the tracking task was displayed to the right. Under this configuration, such a strategy would direct spatial information to the right visual field, that is, the "verbal" hemisphere.

Finally, two studies by Neisser and his collaborators (Neisser, Hirst, Spelke, Reaves, & Coharack, 1980; Spelke, Hirst, & Neisser, 1976) have made strong assertions concerning the development of divided attention skills. With extensive training, their subjects were able to time-share perfectly the tasks of reading one message for comprehension and writing a separate message from oral dictation. In subsequent experiments, the investigators established that the learning was not simply achieved through the automation of single-task skills. In fact, their subjects showed a reasonably good level of semantic comprehension both of the passage read and of the relation between the orally dictated words. This level of comprehension would not have been obtained had either or both tasks been processed at an "automated" level.

It appears safe to conclude, therefore, that the very efficient time-sharing performance of the expert results not only from the more automated performance of component tasks, but also from a true skill in time-sharing: knowing when to sample what from the display, when to make which response, and how to better integrate the flow of information in the two tasks. To what extent the time-sharing skill acquired in one environment is generalizable to others is not well established. Damos and Wickens (1980), as noted, did find some transfer. The amount of transfer, however, was not large relative to the amount of skill learning demonstrated by both groups on the new task. It seems then that most time-sharing skills that are learned are probably fairly specific to a given task combination and are not of the generic kind.

Attention as an ability.   Differences between individuals in time-sharing efficiency may to a large extent be related either to differences in automation of single-task skills or to the practice-related acquisition of time-sharing skills described in the previous section. Thus the observation made by Damos (1978) that flight instructors have greater reserve capacity than novices, as measured by performance of a secondary task, was undoubtedly related to the greater degree of automation of the flight task for the instructors. On the other hand, Hunt and

Lansman (1981), using an information-theory approach, provided evidence that some portion of the individual differences in dual-task performance is related directly to the amount of resources available and not to the level of skill on the component tasks. A similar conclusion was offered by Fogarty and Stankov (1982) on the basis of factor analysis of single- and dual-task scores. Gopher and Kahneman (1971), Kahneman, Ben-Ishai, and Lotan (1973), and Gopher (1982) report that measures of the flexibility of attention switching in a dichotic listening task predict both the success in flight training and the frequency of accidents encountered by bus drivers. Keele, Neill, and DeLemos (1977) report tentative evidence for a fairly general skill of attention switching that correlates across a number of different paradigms.

Given that there are individual differences in time-sharing skill, to what extent are these general—applicable across a wide variety of different dual-task combinations? Phrased in other terms, if there is such a "general" time-sharing ability, would it suggest that individuals who perform well on one dual-task combination will be more likely to perform better on a second combination involving entirely different component tasks?

The statistical and analytical problems involved in answering this question are numerous (Ackerman & Wickens, 1982). However, except for the attention-switching ability described by Keele et al. (1977) and more tentative data offered by Fogarty and Stankov (1982), there is not much evidence at this point for a general time-sharing ability. Wickens, Mountford, and Schreiner (1981), for example, examined differences in time-sharing ability of 40 subjects performing four different tasks in nine different pairwise combinations. Although they observed substantial individual differences in the efficiency of time-sharing a given task pair, these differences did not correlate highly across the different task combinations. These findings, along with similar results obtained by Jennings and Chiles (1977) and Sverko (1977) suggest that what accounts for differences between individuals in time-sharing ability is related either to differences in automation of the component tasks or to differences in the ability to time-share a specific task pair. As an example of the second kind of ability, Damos, Smist, and Bittner (1983) have identified a fairly stable difference in the ability with which people can process information in parallel between two discrete tasks. Some can easily process stimuli and responses of the two at the same time (but with some slowing), while others must deal with epochs of one task followed by epochs of the other. Once again, however, this dichotomy is probably somewhat task specific. It is not likely that it also accounts for differences in the ability of people to perform dual-axis tracking or to the ability to read one message and shadow another.

Practical Implications

The practical implications of research and theory on attention and time-sharing are as numerous as the cases in which a human operator is called upon to perform two activities concurrently, and his or her limitations in doing so represent a bottleneck in performance. These instances include the pilot of the high-performance aircraft, who may have a variety of component tasks simultaneously imposed; the process control or nuclear power plant monitor who is trying to diagnose a fault and is simultaneously deciding, remembering, and scanning to acquire new information; the musical performer who is attending to notes, rhythm, her accompanist, and the quality of her own performance; the translator who must concurrently listen, translate, and speak; and the learner of any skill who must concurrently perceive different stimuli associated with a task, make responses, and process feedback.

In all of these examples, it is important to know the limits of attention and to understand of how these limits might be overcome through system redesign (e.g., use of mixed auditory and visual displays instead of all visual), through training of time-sharing skills, and through the selection of operators with high time-sharing ability (Wickens et al., 1981). Research suggesting that there is increased time-sharing efficiency between processing modalities, for example, is helping to pave the way for implementing voice recognition and synthesis technology into the cockpit of high-performance aircraft (Harris, Owens, & North, 1978; Wickens et al., 1983). Evidence for generalizable individual differences in time-sharing performance could greatly increase the efficiency with which operators are selected for tasks in which such time-sharing may be heavily demanded (Damos, 1978; Gopher, 1982; Hunt & Lansman, 1981).

A somewhat different application has related to the use of secondary and loading task techniques to provide a sensitive methodology for assessing mental workload. We have argued above that differences between conditions may be revealed better by relying on secondary tasks than upon primary-task performance alone. Yet theories of resources are necessary in order to understand when certain tasks will and will not compete for resources. The topic of mental-workload assessment and the use of secondary tasks in that assessment will be the focus of interest of the rest of this chapter.

MENTAL WORKLOAD

Within the last few years, the applied community has demonstrated considerable interest in the concept of *mental workload*: How busy is the operator? How complex are the tasks that confront him? Can he handle any additional tasks above and beyond those that are already performed? Will he be able to respond to uncertain stimuli? How does the operator feel about the tasks being performed? The growing number of articles, books, and symposia in the field (Leplat & Welford, 1978; Moray, 1979; Roscoe, 1978; Smith, G., 1979; Williges & Wierwille, 1979; Wierwille & Williges, 1978, 1980) is testimony to the fact that system designers realize that workload is an important concern that should be studied. In fact, a major issue between management and labor in the airline industry has concerned the concept of workload. The Air Line Pilots Association has argued that the workload demanded at peak times in the class of narrow-body air transports such as the DC–9 or Boeing 737 is excessive for a two-man crew on the flight deck. They assert that a three-man complement is required. Correspondingly, the airlines industry has argued that the workload can be adequately handled by the two-man crew (Lerner, 1983). The Federal Aviation Administration will soon require certification of aircraft in terms of workload metric, and the Air Force also imposes workload criteria on newly designed systems. All of these concerns lead to the very relevant question: What is mental workload and how is it measured?

Importance of Workload

Both the designers and the operators of systems realize that performance is not all that matters in the design of a good system. It is just as important to consider what demand a task imposes on the operator's limited resources. This may or may not correspond with performance. More specifically, the importance of research on mental workload may be viewed in three different contexts. These relate to workload prediction, to the assessment of workload imposed *by* equipment, and to the assessment of workload experienced by the human operator. The difference between the second and third is their implications for action. When the workload of systems is assessed or compared, the purpose of such a comparison is to optimize the system. When the workload experienced by an operator is assessed, it is for the purpose of choosing between operators or providing an operator with further training. Workload in all three contexts may be initially represented in terms of a simplified "undifferentiated capacity" model of human processing resources. (We shall consider the added complexities of multiple-resource theory later.) Figure 8.8 shows the relations between the important variables in this model. The resources demanded by a task are shown on the horizontal axis. The resources supplied (or needed for adequate performance) are shown on the vertical axis. If adequate performance of a task demands more resources from the operator than are available, performance will break down. If, on the other hand, the supply exceeds the demand, the amount of this excess expresses the amount of "residual capacity."

We make the assumption in this chapter that the concept of workload is fundamentally defined in terms of this relation between resource supply and task demand. In the region to the left of the "break point" of Figure 8.8, workload is inversely related to reserve capacity. In the region to the right, it is inversely related to the level of task performance. These two quantities are primarily what system designers should wish to predict. Note that changes in "workload" according to this conception may result either from fluctuations of *operator* capacity

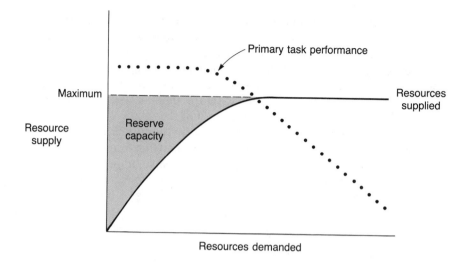

Figure 8.8 Schematic relation between primary-task resource demand, resources supplied, and performance.

or from changes in *task* resource demands. The sections below will first consider the uses that are made of workload measures and then present a detailed description of the various indices proposed to measure workload, highlighting the costs and benefits of each.

Workload prediction. As discussed in Chapter 1, it is often desirable to know whether a system will be satisfactory before it enters final production. It should be able to perform its mission adequately, without placing excessive demands upon the human operator. In order to make such a determination, it is essential to have some predictive model that will map system demands onto operator capacities and determine the extent to which the demands do or do not exceed the capacities. If they do, then breakdowns in performance will occur. Even if demands do not exceed supply, it is important to ensure that the system is designed with a sufficient margin of residual capacity or resources so that unexpected failures or environmental events may be handled satisfactorily. Thus predictive models of workload are needed at this stage.

Equipment assessment. While workload prediction before a system is designed is desirable, it is often essential to assess workload of a system already existing at some stage of production. This assessment may be done for the purpose of identifying those "bottlenecks" in system or mission performance in which resource demands momentarily exceed supply and performance breaks down. Alternatively, workload may be assessed to compare two alternate pieces of equipment that may achieve similar performance but differ in their resource demands because they possess differently shaped performance-resource functions (see for example Figure 8.5). Sometimes the criterion of workload may offer the only satisfactory means of choosing between alternatives. An even greater challenge to workload-assessment techniques is posed by the requirement to determine if the *absolute level* of workload imposed by a system is above or below a given absolute criterion level. The goal of developing workload-certification criteria for complex systems has spawned the need for such absolute scales of workload.

Assessing operator differences.   Workload measures may also assess differences in the residual capacity available to a given operator and not that imposed by a given system. This may be done in one of two contexts: (1) The level of skill or automation achieved by different operators who may be equivalent in terms of their primary-task performance may be compared. For an example, Damos (1978) and Crosby and Parkinson (1979) showed that flight instructors differed from student pilots in their level of residual attention. Damos furthermore found that this measure applied to students was a good predictor of success in pilot training. (2) Operators may be monitored on line in real-task performance. In this case, intelligent computer-based systems could decide to assume responsibility for performance of certain tasks from the human operator when momentary demands were measured to exceed capacity (Enstrom & Rouse, 1977; Wickens, 1979; Wickens & Gopher, 1977), although this form of on-line human-computer interaction requires a certain level of cooperation from the human operator (see Chapter 12).

Workload Prediction and Assessment Techniques

Criteria for workload indices.   Sheridan and Stassen (1979) have proposed a number of criteria that should ideally be met by any technique to assess workload. Of course it is true that some of these criteria may "trade off" with each other (Wickens, 1979, 1981), and so rarely if ever will one technique be found that satisfies all criteria. The following list of five criteria of a workload index is similar to the list proposed by Sheridan and Stassen.

Sensitivity.   The index should be sensitive to changes in task difficulty or resource demand.

Diagnosticity.   An index should not only identify when workload varies, but also indicate the cause of such variation. In terms of multiple-resource theory, it should indicate *which* of the capacities or resources are varied by demand changes in the system. This information makes it possible to implement better solutions.

Selectivity.   The index should be selectively sensitive only to differences in capacity demand and not reflect changes in such factors as physical load or emotional stress that may be unrelated to mental workload or information-processing ability.

Obtrusiveness.   The index should not interfere with, contaminate, or disrupt performance of the primary task whose workload is being assessed.

Bandwidth and reliability.   As with any measure of behavior, a workload index should be reliable. However, if workload is assessed in a time-varying environment (e.g., if it is necessary to "track" workload changes over the course of a mission), it is important that the index offer a reliable estimate of workload rapidly enough so that the transient changes may be estimated.

A myriad of workload prediction and assessment techniques have been proposed, some meeting many of the criteria described above, but few satisfying all of them. These may be classified into five broad categories. The first is the technique of workload prediction based on time-line analysis. The remaining four are categories of assessment (Wierwille & Williges, 1978) related to primary-task measures, secondary-task measures of spare capacity, physiological measures, and subjective rating techniques.

Time-line analysis.   The technique of time-line analysis is designed primarily for workload prediction and will enable the system designer to "profile" the workload of operators encountered during a typical mission such as landing an aircraft or starting up a power-generating plant. In a simplified but readily usable version, it assumes that workload is proportional to the ratio of the time occupied performing tasks to total time available. If one is busy with some measurable task for 100% of a duration of time, then workload is 100% during that interval. Thus, for example, the workload of a mission would be computed by drawing lines representing different activities, of lengths proportional to their duration. The total length of the lines would be summed and then divided by the total time (Parks, 1979), as shown in Figure 8.9. In such a way the workload encountered by different members of a team (e.g., pilot, copilot, and flight engineer) may be compared and tasks reallocated if there exists a great imbalance. Furthermore, epochs of peak workload or work overload in which "load" is calculated as greater than 100% can be identified as potential bottlenecks.

Time-line analysis, although a valuable predictive tool, has some important limitations. It normally assumes a simplified "single-channel" model of human attention, in which any observable activity demands full attention (independent of its actual resource demand or the particular resources involved), whereas nonobservable activities (decision making, problem solving), because they are not associated with observable motor acts, may be assigned zero values. Time-sharing of any two activities will, according to this technique, render 200% workload. Yet we know in certain cases that time-sharing may indeed proceed very efficiently if resource demands are low or if separate resources are deployed.

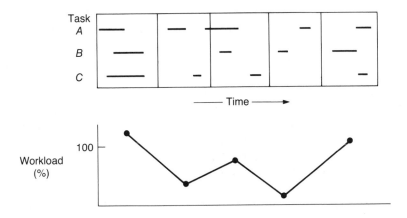

Figure 8.9   Time-line analysis.

Another limitation of time-line analysis is that it is most applicable in a relatively "force-paced" schedule in which activities cannot be shifted about in time. The original Mercury flights of the space program provided examples, since here the astronaut's every action was tightly constrained in time. On the other hand, as Gottsdanker and Senders (1980) point out, in a number of environments (typical, for example, of air transport) the operator has a considerable degree of self-paced flexibility to perform required tasks at alternative times (e.g., calibrating a dial, testing a reading, checking the weather). The operator is thereby able to minimize overlapping demands by efficient scheduling. The more loosely coupled or self-paced task demands become, the more difficult it is to rigorously apply time-line analysis in order to predict epochs of peak overload.

Primary-task measures.   Primary-task measures describe performance characteristics of the task whose workload is measured. Increases in the difficulty of a task should normally be expected to increase its resource demands or workload. At first glance then, task difficulty parameters expressed by such measures as interstimulus interval, information input bandwidth, number of information sources, control complexity, or compatibility should offer good "predictive" measures of task workload. There are, however, two potential shortcomings to the use of primary task parameters to predict and assess workload. First it is often hard to cross-calibrate such diverse measures between tasks. For example: How many keys must be added to a complex keyboard transcription task? Or, how much control instability must be added to a tracking task in order to produce a workload increase that is equivalent to increasing the information transmitted from 2 to 3 bits in a choice-reaction-time task or increasing the load on working memory from 5 to 7 chunks?

Second, it is impossible to say what *task difficulty* (manipulated by the experimenter) does to workload (experienced by the subject) unless primary task performance is carefully controlled and measured. In terms of a performance-resource function, we can contrast the two cases shown in Figure 8.10. Each graph shows three different tasks performed and experienced by the operator. In Figure 8.10a the increase in task difficulty clearly increases the workload (resources demanded), because in every case the operator is allocating sufficient resources to obtain maximum performance. As difficulty increases, this amount increases accordingly. In Figure 8.10b, however, the constant resource-allocation policy that the operator adopts will lead to progressively greater performance decrements as difficulty is increased. It does not then seem appropriate to conclude that workload has been varied in this case since the same amount of resources is invested in all conditions. In short, one must be careful of assuming *a priori* that a manipulation of task demands has increased workload. This assumption can only be made if it is known that the subject is investing a sufficient amount of resources to produce the best possible performance in every condition. The subject in Figure 8.10a was doing this. The subject in Figure 8.10b was not.

The primary-task workload margin. If the intent of a workload assessment or prediction technique is to determine the level of residual capacity for a particular task performed at a given level, one means of achieving this goal is by assessing the *primary-task workload* margin. To derive the workload margin, the following steps are taken: First, a criterion level at which a task is to be performed is specified. In applied contexts, this criterion is often supplied by a systems engineer—for example, the maximum allowable deviation from a flight path in an

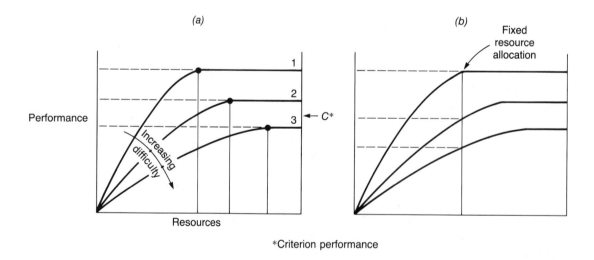

Figure 8.10 Relations between the performance-resource function, resource allocation, and primary-task difficulty, indicating the problem with primary-task workload measures.

approach to landing an aircraft or the allowable error rate and typing speed for a clerk-typist. Then a primary task parameter is chosen that will affect the demand for resources of a particular nature. Finally, this parameter is manipulated until it reaches a level such that performance falls below the criterion. In Figure 8.10a we can imagine increasing the difficulty until maximum obtainable performance falls below Criterion C. The workload margin is thus defined as a difficulty level that would fall between Curves 2 and 3. In the aviation example described above, a small dynamic instability in the actual flight-control surface could be gradually increased until flight error is sufficiently deviant to exceed a maximum allowable limit (Jex & Clement, 1979). The magnitude of the instability parameter manipulation is defined as the workload margin. This provides an index of how much additional demand from the initial task conditions the resource in question can bear before performance becomes unsatisfactory.

Within the context of multiple-resource theory, the workload margin is a "vector" measure, since ideally one such dimension should be supplied for each resource that is postulated. For example, a task that is heavily loaded on perceptual/central processing (e.g., signal detection or verbal comprehension) will be disrupted more by increases in its encoding demands than by increases in response complexity. Thus we would anticipate a smaller workload margin measured by the former technique than by the latter. On the other hand, a task with greater motor demands such as tracking would be relatively more sensitive to an increase in the difficulty of its response characteristics than to a decrease in display quality.

The secondary-task technique.   Imposing a secondary task as a measure of residual resources or capacity not utilized in the primary task (Ogden, Levine, & Eisner, 1979; Rolfe, 1973) is a technique that is in some aspects closely related to the primary-task workload margin. In the workload margin, the reserve capacity is "absorbed" by increasing the difficulty of the original activity. In the secondary-task technique, residual resources are absorbed by introducing a new activity, the secondary task. As described earlier in this chapter, secondary-task performance is thus, ideally, inversely proportional to the primary-task resource demands. In this way, secondary tasks may reflect differences in task resource demand, automation, or practice that are not reflected in primary-task performance. As examples of the use of this technique, Garvey and Taylor (1959) found differences in tracking controls to be reflected only with the addition of a secondary-task measure. Bahrick and Shelly (1958) found that the secondary task was sensitive to differences in automation. Performance of their subjects on a serial-reaction-time task did not differ between a random and a predictable sequence of stimuli. However, performance on a secondary task discriminated these two; with practice the repeated sequence required fewer resources. Papers by Rolfe (1973) and Ogden et al. (1979) provide numerous other examples of the use of secondary-task techniques.

A variant of the secondary-task technique is the use of a *loading task* (Ogden et al., 1979; Rolfe, 1973). When using the secondary-task technique, the investigator is interested in variation in the secondary-task decrement (from a single secondary-task control condition) to infer differences in primary-task demand. The primary task is thus both the task of interest and the task whose priority is emphasized. In the loading-task technique, on the other hand, different allocation instructions are provided. The subject is asked to devote all necessary resources to the loading task, and the degree of intrusion of this task on performance of the "primary" task is examined to compare differences between primary tasks. In the context of our previous discussion of resource theory, it is apparent that the secondary-task and loading-task techniques merely direct the focus of experimental interest to different areas of the performance operating characteristic.

Figure 8.11 shows a POC constructed with two versions of a primary task of different difficulty, performed concurrently with a second task. When the secondary-task technique is employed (closed points), the primary task is emphasized and performance variance is cast onto the secondary task. When the loading-task technique is used (open points), variance is now cast into the primary task. One may see from this hypothetical representation that the secondary-task method will be likely to reflect greater difference than the loading-task method (compare $S_E - S_D$ with $L_E - L_D$).

A multitude of secondary tasks have been proposed and employed at one time or another to assess the residual capacity of primary tasks.

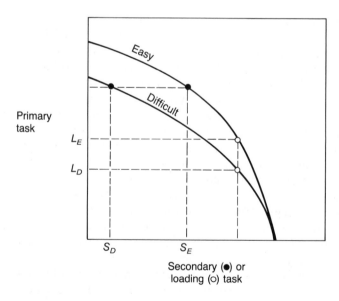

Figure 8.11 Relation between secondary task, loading tasks, and the performance operating characteristic (POC).

While the reader is referred to reviews by Wierwille and Williges (1978, 1980) and Ogden et al. (1979) for a more exhaustive listing of these tasks, a few prominent candidates will be described here.

Rhythmic-tapping or interval-production task. Michon (1966) has proposed a secondary task in which the subject is required to maintain a constant rate of finger tapping. Michon assumes that primary-task demands will disrupt the ability of a central response-selection/decision-making stage to initiate the finger taps at the precisely timed intervals. Primary-task workload will thereby be reflected by increased *variability* of the intertap interval. Michon and Van Doorne (1967) have developed a semiportable apparatus that allows the measurements of tapping variability to be taken in the field. The main limitation of this technique is its obtrusiveness. One hand of the operator must be continuously available to make the taps and so cannot readily be employed for any primary-task activity.

Probe reaction time. Posner and his colleagues have employed the probe-reaction-time technique with considerable success in exploring the resource demands associated with a number of perceptual/motor and cognitive tasks (e.g., Ells, 1973; Posner & Boies, 1971; Posner & Keele, 1969). An unpredictable stimulus is presented and a rapid response is required. As the primary task demands more resources, slower processing is offered to the probe. Unfortunately, because of its minimal central processing demands, the probe-reaction-time technique appears to be heavily "perceptual-motor" and therefore particularly sensitive to competition for input and output modalities. As a consequence, it can provide quite different results as a function of whether within-modality or cross-modality probes are used (McLeod, 1978; Wicken et al., 1983).

The Sternberg memory-search task. Closely related to the probe task is the Sternberg memory-search task, described in Chapter 6. Like the probe task, reaction time (RT) to an unexpected stimulus is again measured, but the response is now a yes-no decision whether the probe is or is not a member of a previously memorized "positive set" of characters held in working memory. When the size of this memory set is varied (i.e., from 2 to 4 characters), then reaction time is prolonged as more items must be searched in working memory (see Chapter 6). Because RT is typically plotted as a function of memory load on the abscissa, this increase in RT is referred to as the "slope," shown by the solid line in Figure 8.12.

When measuring workload, reaction time is measured under low and high memory load both alone and concurrently with the primary task, the dashed line of Figure 8.12. Then a comparison of the two functions allows the investigator to make inferences about the *locus* of resource demands of the primary task. If the slope is steeper in dual- than in single-task conditions (Figure 8.12a), then the primary task slows down the rate of memory search. Since memory search is a central-processing activity, this indicates that the primary task has a central-processing load. If, as shown in Figure 8.12b, only the "inter-

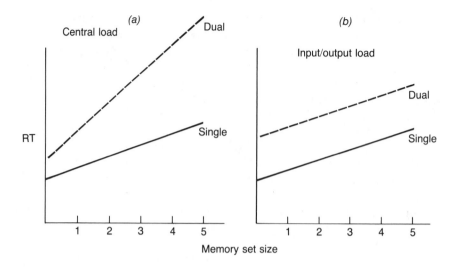

Figure 8.12 Sternberg memory-search task as an RT workload index (with and without primary task). (*a*) A primary task demanding both central processing and input/output load; (*b*) a primary task demanding only input/output load.

cept" is raised by the primary task but the slope is unchanged, then the main source of resource demands of the primary task is assumed to be perceptual/motor. Of course both slope and intercept may be affected, as in the left panel.

The technique may be equally well employed with an "easy" and "difficult" version of the primary task to assess the locus of the resource demands imposed by increasing difficulty. This ability to discriminate load effects provides the technique with some diagnosticity while not sacrificing sensitivity. The technique has been well validated both in the laboratory (Briggs, Peters, & Fisher, 1972; Logan, 1979; Wickens & Derrick, 1981) and in the field as a measure of pilot workload (Crosby & Parkinson, 1979; Schiflett, 1980; Spicuzza, Pincus, & O'Donnell, 1974; see Micalizzi & Wickens, 1980, for a survey). The Sternberg task has been packaged as a commercially available instrument called the *workload assessment device*.

Time estimation and production. Intuitively, we realize that the subjective passage of time is in some way affected by how busy we are. This phenomenon has been translated into two different techniques of workload measurement: time estimation and time production. In fact, two different models underlie the two techniques. In the *production* task, the subject is asked to produce constant intervals of time (e.g., tap a finger every 10 seconds). Hart (1975) makes the assumption that an internal mental counter will be employed to produce these intervals. As primary-task load increases, the counter will be disrupted or slowed, and intervals will be systematically *overestimated*.

In the retrospective *estimation* task, the subject is asked after a task is performed to estimate the amount of time that has passed. Hicks, Miller, and Gaies (1977) observed that time durations are underestimated with high workload. Thus they found that performance of a more difficult choice-reaction-time task led subjects to underestimate the time they spent performing the task. In a second study, they observed that time was underestimated while subjects performed a memory task, as compared to a control condition during which they did nothing. The effects obtained by Hicks et al. are thus explained by an active counter model or by the intuitive explanation that "time flies when you're busy and drags when you're bored." On the other hand, using the retrospective estimation procedure with aviation tasks, Hart (1975) observed opposite effects. Greater difficulty of a primary task caused longer estimates of time. Her data were accounted for by an "adaption level" model in which the subject is presumed to base his assessment of time on a "typical" level of activity. If a period of time is empty, less activity has occurred than normal and the interval will be underestimated. If the period is quite full (high workload), duration will be overestimated ("In order to accomplish that much, I must have been working longer than X seconds").

The main problem then with time estimation as a workload measure is the inconsistency of possible processes and results. Its major benefits are, of course, its low degree of obtrusiveness. Time productions need be made only infrequently during a primary task, and retrospective estimations, of course, need be made only after a primary task is completed.

Random-number generation. Baddeley (1966) has investigated a secondary task in which subjects are required to generate a stream of random digits. In calibrating his task, Baddeley observes that the degree of randomness (as measured by information theory) declines monotonically with increasing primary-task load. As load increases, subjects begin to generate repeating sequences such as "123 123 123." In terms of information theory, less information is provided at higher workload. This, like Michon's interval-production task, is also somewhat intrusive since a continuous stream of responses must be generated.

The critical-instability-tracking task. Jex (1967, 1979) has developed a tracking task that is the computerized analog of the task of balancing a dowel rod on the end of one's finger. The dowel rod can be shortened or lengthened, thereby either decreasing or increasing the stability of the task, respectively, and so altering the frequency with which stabilizing control is required. This task is described in more detail in Chapter 11. Jex has found that performance on the critical task is quite effective in measuring differences in the workload of other tracking tasks. Furthermore, the critical task is quite useful as a loading task since a given level of the difficulty parameter lambda (the computer analog of the reciprocal of dowel length) essentially forces the subject to allocate a minimum amount of resources to the task in order that control is not lost completely (i.e., the stick is dropped). As lambda is increased (the dowel length decreased), this minimum requirement increases. Like the Sternberg task and the interval-production task, the critical task has also been packaged as a commercially available tool (Jex, 1979). The critical task, like random-number generation, has the disadvantage that it requires a continuous generation of output and therefore is likely to be intrusive.

Benefits and costs of the secondary-task technique. The secondary task technique possesses two very distinct benefits. First, it has a high degree of face validity. To the extent that a workload measure is designed to predict the amount of residual attention that an operator will have available in case an unexpected failure or environmental event occurs, then this is exactly what the secondary task measures. This validity places the technique in contrast with the physiological and subjective measures described below. Of course, the primary-task workload margin also possesses the same face validity. However, primary-task measures of workload margin may be more difficult to compare across different primary tasks because the investigator is constrained to manipulate a dimension of difficulty that is part of the primary task. The same dimension of difficulty may not be present in two primary tasks which are to be compared. In contrast, a given secondary task can, in principle, be applied to any primary task. We see below, however, that this assumption encounters some theoretical difficulties.

One cost associated with this technique is that, like the workload margin, the secondary-task technique as a vector quantity must also account for the fact that there are different kinds of resources. Workload differences that result from manipulating a primary-task variable can be greatly underestimated if the resource demands of the primary-task manipulation do not match those of most importance in the secondary task. Wickens and Kessel (1980), for example, found that the response-loading critical-tracking task developed by Jex (1967) was not appropriate as a loading task when employed with a perceptual monitoring task. Another example of such a mismatch would occur if an auditory-word-comprehension or mental-arithmetic task (auditory, verbal, perceptual/central demands) were used to assess the workload attributable to manipulations of tracking-response load (visual, spatial, response demands). Our previous discussions of time-sharing would suggest that these two tasks use very different resources in the model of Figure 8.7. Hence the resource demands of tracking would be greatly underestimated.

A second problem often encountered with the secondary-task technique is that it may interfere with and disrupt performance of the primary task. On the one hand, this may be inconvenient or even dangerous if the primary task is one like flying; a diversion of resources at the wrong time could lead to an accident. On the other hand, disruption of the primary task could present problems of interpretation. These problems were discussed earlier in this chapter in the context of Figure 8.6. When two primary tasks are compared by a secondary task and there is a negative correlation between primary and secondary task decrements across the two comparisons, then it may be difficult to interpret which primary task really does demand more resources.

It is interesting that one of the solutions offered to these problems is to choose secondary tasks that are highly dissimilar from the primary task (Rolfe, 1973). The preceding discussion suggests that this remedy may be employed only with a potential cost—a reduced sensitivity to resource-specific attributes of primary-task workload. In fact, the best answer to both of these problems is through instructions. Emphasize consistently that the secondary task is secondary and that the primary-task decrements should be equal across all primary tasks and as small as possible. Referring to Figure 8.10, we see that these instructions would lead the operator to allocate resources to the secondary task in proportion to the areas to the right of the vertical lines intersecting Curves 1, 2, and 3.

Kahneman (1973) has proposed that the ideal secondary-task technique is one that employs a battery of secondary-task measures sensitive to different resources in the system. Brown (1968) proposes that Baddeley's number-generation task be associated with perceptual/cognitive load and Michon's tapping task (or Jex's critical task) with response load. In cases where it is clear that one level of a dimension does not contribute to primary-task performance, the dimensionality of the battery may be reduced accordingly. For example, a verbal processing task with no spatial components need not be assessed using a spatial secondary task. However, in cases in which an activity is performed that potentially engages all "cells" of processing resources as depicted in Figure 8.7, a secure workload measure should involve a battery that also incorporates those cells or at least taps early and late processing of a verbal and spatial nature (Wickens, in press).

The various secondary-task workload measures may, of course, be evaluated in terms of their likelihood of intrusion. This factor, incidentally, is negatively correlated with operator acceptance—a critical criterion for consideration if the method is to be used for on-line workload monitoring. In this regard, methods such as the critical-tracking task, the tapping task, or random-number generation that require a fairly continuous motor output will be highly intrusive. Time estimation, on the other hand, is ideal since responses are few. The probe or Sternberg RT tasks are of intermediate disruption if the probe frequency is relatively low (4 to 7 seconds). The problem with reducing stimulus rate, of course, is that when responses are made less frequent in order to avoid intrusion, then the data by definition become less reliable (or require larger epochs to obtain equivalent reliability). As a result, the ability of the measure to provide information concerning transient changes in demands deteriorates accordingly. If it requires a minute's worth of data recording to assess workload reliably, then it will be impossible to pick up changes in workload in the order of seconds.

Physiological measures.   One solution to problems of the performance intrusiveness encountered by secondary tasks is to record, unobtrusively, the manifestations of workload or increased resource mobilization through appropriately chosen physiological measures of autonomic or central nervous system activity (see Figure 8.1).

From the standpoint of multiple-resource theory, physiological measures are generally less precise than are secondary tasks. The secondary tasks can be associated reasonably well with demands imposed upon the different resources of Figure 8.7. However, there are not yet enough data to establish conclusively whether changes in a particular physiological index reflect changes in the demands on certain specific resources, in which case the measure is *diagnostic*, or changes in any and all resources, in which case its diagnosticity is sacrificed for greater total *sensitivity*. Three prototypical measures will be briefly described and contrasted in terms of two criteria, diagnosticity and sensitivity.

The evoked brain potential. When the evoked brain potential described in Chapter 7 is used to assess workload, many of the same assumptions are made as those underlying secondary-task measures. When the evoked potential (EP) is employed as a secondary task (Isreal, Chesney, Wickens, & Donchin, 1980; Isreal, Wickens, Chesney, & Donchin, 1980; Natani & Gomer, 1981), the subject sees or hears a Bernoulli series of stimuli (AABABABBB . . .) and is asked to count covertly one of the two classes of stimuli. The processing of the stimuli elicits a prominent late-positive or P300 component in the wave form of the evoked potential recorded from the scalp. Isreal and his colleagues found that introducing a concurrent primary task of a perceptual/ cognitive nature, typical of the air traffic controller's task, will attenuate P300 amplitude. Increasing the difficulty of the task by requiring more display elements to be monitored reduces P300 amplitude still further (see Figure 8.13). The EP measure is somewhat *diagnostic*, in that it reflects perceptual/cognitive load but is relatively insensitive to variations in response load. Therefore, its diagnosticity is obtained at the expense of *sensitivity*. The EP measure has two particular advantages. In contrast to the other physiological measures described below, it provides a graded measure of direct cognitive activity rather than an indirect measure of autonomic activity. In contrast to most secondary tasks, on the other hand, it does not require overt responses and therefore is less likely to be intrusive.

Pupil diameter. Several investigators have observed that the diameter of the pupil correlates quite closely and accurately with the resource demands of a large number of diverse cognitive activities (Beatty, 1982). These include mental arithmetic (Kahneman, Beatty, & Pollack, 1967), short-term memory load (Beatty & Kahneman, 1966; Peavler, 1974), and logical problem solving (Bradshaw, 1968; see Beatty, 1982, for an integrative summary). This diversity of responsiveness suggests that the pupilometric measure may be a highly sensitive one, although as a result it is undiagnostic. It will reflect demands imposed anywhere within the system. Its disadvantage, of course, is that relevant pupil

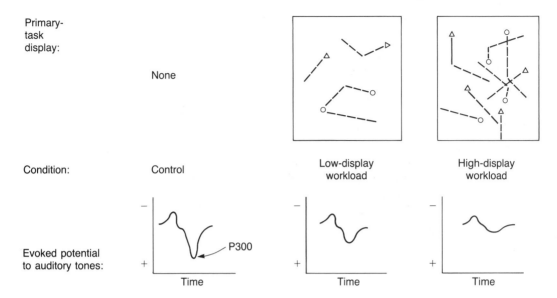

Figure 8.13   The evoked potential (EP) as a workload index. The amplitude of
the P300 component of the EP to counted auditory tones declines systematically as the complexity of the simulated air traffic control display increases. Subjects must detect course changes of each circular element on
the display. The dashed lines indicating the trajectory are not visible on
the subjects' display. From "The Event-Related Brain Potential as an Index
of Display-Monitoring Workload" by J. B. Isreal, C. D. Wickens, G. L.
Chesney, and E. Donchin, 1980, *Human Factors, 22,* pp. 214 and 217.
Copyright 1980 by The Human Factors Society, Inc. Reproduced by permission.

changes are in the order of tenths of a millimeter. This means that
accurate measurement requires considerable head constraint and precise measuring equipment. Also, changes in ambient illumination must
be monitored, since these also affect the pupil. Because of its association with the autonomic nervous system, the measure will also be
susceptible to variations in emotional arousal.

Heart-rate variability. A number of investigators have examined
different measures associated with the variability or regularity of heart
rate as a measure of mental load. Variability is generally found to
decrease as load increases (Mulder & Mulder, 1981). When this variability is associated specifically with the periodicities resulting from
respiration, the measure is termed sinus arhythmia (Kalsbeek & Sykes,
1967; Mulder & Mulder, 1981). Like pupil diameter, the sinus
arhythmia measure is sensitive to a number of different difficulty
manipulations and therefore appears to be more *sensitive* than *diagnostic.* Wickens and Derrick (1981) investigated this measure with four
quite different tasks performed in different combinations within the
framework of the multiple-resource model. They concluded that the

variability measure reflected the total demand imposed on all resources within the processing system rather than the amount of resource competition (and therefore dual-task decrement) between tasks.

Costs and benefits. Physiological indices have two great advantages: (1) They provide a relatively continuous record of data over time. (2) They are not disruptive of primary-task performance. On the other hand, they do often require that electrodes be attached (EPs and heart measures) or some degree of physical constraints be imposed (pupilometric measures), and therefore they are not really unobtrusive in a physical sense. These constraints will influence user acceptance. They have a further potential cost in that they are, generally, one conceptual step removed from the inference that the system designers would like to make. That is, workload differences measured by physiological means must be used to *infer* that performance breakdowns would result or to *infer* how the operator would feel about the task. Secondary- or primary-task measures assess the former directly, while subjective measures, which we will now consider, assess the latter.

Subjective measures.   Subjective ratings of task difficulty represent perhaps the most acceptable measure of workload from the standpoint of the actual system user, who feels quite comfortable in simply stating, or ranking, the subjective feelings of "effort" or attention demands encountered in performing a given task or set of tasks (Eggemeier, 1981; Moray, 1982; Reid, Shingledecker, & Eggemeier, 1981). Some have argued (Sheridan, 1980) that these measures come nearest to tapping the essence of mental workload. Indeed, there may be circumstances in which a system designer would feel more comfortable using data concerning how an operator feels about a task than concerning how the task is performed. It is important, however, to know how accurately an operator can assess the demands imposed upon his limited resources, what the dimensions underlying his ratings might be, and how these are scaled.

Cooper-Harper scale. Perhaps the oldest and best-validated subjective measure of workload is the Cooper-Harper rating scale of aircraft handling qualities, a decision-tree procedure that rates handling qualities on a 10-point scale (Cooper & Harper, 1969). Within the relatively restricted domain of the tracking or manual-control task (and the system-dynamics dimension of task difficulty, see Chapter 11), the Cooper-Harper ratings provide a reliable and acceptable measure. In fact, Jex and Clement (1979) found that this measure correlated quite highly ($r = 0.96$) with a measure of residual capacity as assessed by the critical tracking task. On the other hand, the scale is tailored specifically to the flight task and thereby is less applicable to tasks of a different nature (e.g., communications, monitoring).

Sheridan's dimensional scale. Sheridan (1980; Sheridan & Simpson, 1979) has proposed that three dimensions define the subjective experience of mental workload. These are related to the proportion of time busy or information-processing load, the mental effort invested in a task or its complexity, and the "emotional stress" of the task. Reid et al. (1981) find that subjects are readily able to rate tasks along these three dimensions, and there appears to be a fair degree of agreement between operators on the relative ratings of different tasks along the three dimensions. What is not known, however, is whether these dimensions are truly independent, whether they reflect all of the variance in subjective workload, or how they relate to the processing-resource dimensions that underlie task performance.

Multidimensional scaling. An alternative approach to providing operators with a multidimensional scale and having them rate tasks is to provide them with a unidimensional scale, collect rating data on a wide variety of tasks, and employ multidimensional scaling techniques (Kruskall & Wish, 1978) to assess how many dimensions actually underlie variability in subjective judgments of task workload. Derrick (1981) used this procedure to evaluate the differences that operators perceived in subjective workload between four tasks, performed alone, and in all pairwise combinations. Applying multidimensional scaling techniques to the subjective data, Derrick concluded that three primary dimensions seemed to underlie differences in subjective ratings of workload. These related to (1) resource competition, (2) adequacy of feedback, and (3) the heart-rate-variability measure described in the previous section. It is possible, therefore, that the third dimension of subjective measure correlated explicitly with the total resource demand within the system, since this dimension was reflected by heart-rate variability.

Costs and benefits. The benefits of subjective techniques are apparent. They do not disrupt primary-task performance, and they are relatively easy to derive. Their costs relate to the uncertainty with which an operator's verbal statement truly reflects the availability of or demand for processing resources. These problems will be considered below as we consider some of the relations between the different measures of workload.

Relation Between Workload Measures

If all measures of workload demonstrated high correlation with each other and the residual variance was due to random error, there would be little need for further validation research in the area. The practitioner could adopt whichever technique is methodologically simplest and most reliable for the workload measurement problem at hand. Generally, high correlations between measures will be found if these measures are assessed across tasks of similar structure and widely varying degrees of difficulty. An example is Jex and Clement's (1979) finding of the high correlation between subjective and secondary task measures of flight-control difficulty. However, the correlations may not be high and may even be negative when quite different tasks are contrasted. An example is an experiment conducted by Herron (1980) in which an innovation designed to assist in a target-aiming task was subjectively preferred by users over the original prototype but generated reliably poorer performance than the original. Similar dissociations have been observed by Childress, Hart & Bortalussi (1982), measuring pilot workload associated with cockpit-display innovations. It is important then to determine which characteristics of a task most strongly influence a given workload measure while leaving another measure unaffected. For example, Wickens and Derrick (1981) and Wickens and Yeh (1983) concluded that subjective measures were relatively more sensitive to the number of competing activities an operator must perform, while primary task performance measures reflected to a greater extent the difficulty of a given *single* task activity. Primary task performance is also relatively more sensitive to the competition between tasks for resources than are subjective measures.

When such dissociations between measures appear, the question of which is the "best" measure clearly depends upon the use to be derived from that information. If workload is intended to predict performance margins or the "residual attention" available to cope with failures in critical operational environments, it seems wiser to use secondary- or primary-task measures to guide the choice of systems, despite the fact that the chosen system may demonstrate higher subjective or physiological ratings of difficulty. If, on the other hand, the issue is one of consumer usability, of setting of work-rest schedules, or of job satisfaction and variations in performance are relatively less critical, then greater weight should be provided to the subjective measures (and perhaps to their physiological counterparts).

Strategies and Workload

The strategies an operator may select in performing a task are relevant to the topic of workload because the operator may choose to perform a task in a qualitatively different manner as task difficulty increases. This change could occur in such a way as to maintain certain aspects of performance constant while sacrificing others (Welford, 1978). The ability to do so, of course, is a natural consequence of the fact that performance may be varied along different dimensions (hits and false alarms, speed and accuracy, allocation of resources to task A and task B). The operator is thereby capable of altering performance in various ways as the situation (resource demands) requires. As a consequence, all aspects of performance must be carefully measured to ensure a true understanding of the effects of workload variables.

A hypothetical example is provided by operator who chooses to sacrifice latency, as resource demands increase at low levels, and accuracy, as the demands are imposed at higher levels. We have described in both Chapters 3 and 7 the manner in which operator's information-seeking strategies and attention allocation alter under conditions of high workload, focusing relatively more on fewer channels of higher information content. Rasmussen's (1981) distinction between symptomatic and topographic troubleshooting strategies described in

Chapter 3 provided a further example of how strategies shift from the former to the latter with increasing load. As a final example, Sperandio (1978) investigated the workload experienced by air traffic controllers as the number of planes under their purview increased. He found that quite different chunking strategies, applied to the grouping of aircraft by attributes, were employed as workload was increased from a low (fewer than 6 aircraft) to a moderate (6 to 10) to a high (more than 10) level.

The previous discussion emphasized the difficulties which the flexibility of human behavior imposes on accurate engineering measurement. A further intervening variable must often be considered between task demands and performance. In a nonstrategic, nonflexible system, increasing task demands will invariably sacrifice either performance or residual capacity by a constant amount. In a strategically adapting system such as the human operator, increasing demands may lead to altered (and perhaps more efficient) strategies. These strategies consume fewer resources to obtain the same level of performance along the dimension of greatest importance or highest utility to the operator (while sacrificing performance along another). Models of workload must ultimately take this flexibility into account.

Transition

The concept of performance has been a critical element in our discussion. Performance on a task was assessed to rise or fall as a function of concurrent task conditions. It is now important that we consider precisely what is meant by performance in terms of the two dimensions most often used to describe it: speed and accuracy. Also, as the earlier chapters of the book have dealt primarily with the perceptual and cognitive processes that preceded action, it is now appropriate that we consider the selection of actions themselves. Chapter 9 will consider the two variables of speed and accuracy (with particular emphasis on speed) as they are employed to measure and describe the actions that subjects make in response to environmental stimuli. We shall consider how these variables are used as a tool to infer the nature of not only the actions themselves, but also the preceding perceptual and cognitive processes.

References

Ackerman, P., & Wickens, C. D. (1982). Task methodology and the use of dual and complex task paradigm in human factors. In R. Edwards (Ed.), *Proceedings, 26th annual meeting of the Human Factors Society*. Santa Monica, CA: Human Factors.

Allen, R. W., Clement, W. F., & Jex, H. R. (1970, July). *Research on display scanning, sampling, and reconstruction using separate main and secondary tracking tasks* (NASA CR-1569). Washington, DC: NASA.

Allport, D. A. (1980). Attention. In G. L. Claxton (Ed.), *New directions in cognitive psychology*. London: Routledge & Kegan Paul.

Allport, D. A., Antonis, B., & Reynolds, P. (1972). On the division of attention: A disproof of the single channel hypothesis. *Quarterly Journal of Experimental Psychology, 24,* 255–265.

Baddeley, A. (1966). The capacity for generating information by randomization. *Quarterly Journal of Experimental Psychology, 18*, 119–130.

Baddeley, A. D., & Hitch, G. (1974). Working memory. In G. Bower (Ed.), *Recent advances in learning and motivation* (Vol. 8). New York: Academic Press.

Bahrick, H. P., Noble, M., & Fitts, P. M. (1954). Extra task performance as a measure of learning a primary task. *Journal of Experimental Psychology, 48*, 298–302.

Bahrick, H. P., & Shelly, C. (1958). Time-sharing as an index of automatization. *Journal of Experimental Psychology, 56*, 288–293.

Beatty, J. (1982). Task-evoked pupillary responses, processing load, and the structure of processing resources. *Psychological Bulletin, 91*, 276–292.

Beatty, J., & Kahneman, D. (1966). Pupillary changes in two memory tasks. *Psychonomic Science, 5*, 371–372.

Bradshaw, J. L. (1968). Pupil size and problem-solving. *Quarterly Journal of Experimental Psychology, 20*, 116–122.

Briggs, G., Peters, G. C., & Fisher, R. P. (1972). On the locus of the divided attention effect. *Perception & Psychophysics, 11*, 315–320.

Broadbent, D. E. (1982). Task combination and selective intake of information. *Acta Psychologica, 50*, 253–290.

Brooks, L. (1968). Spatial and verbal components of the act of recall. *Canadian Journal of Psychology, 22*, 349–368.

Brown, I. D. (1968). Criticism of time-sharing techniques for the measurement of perceptual-motor difficulty. *XVI International Congress of Applied Psychology*, Amsterdam: Swets & Zeitlinger.

Chernikoff, R., Duey, J. W., & Taylor, F. V. (1960). Effect of various display-control configurations on tracking with identical and different coordinate dynamics. *Journal of Experimental Psychology, 60*, 318–322.

Childress, M. E., Hart, S. G., & Bortalussi, M. R. (1982). The reliability and validity of flight task workload ratings. In R. Edwards (Ed.), *Proceedings, 26th annual meeting of the Human Factors Society*. Santa Monica, CA: Human Factors.

Cooper, G. E., & Harper, R. P. (1969, April). *The use of pilot ratings in the evaluation of aircraft handling qualities* (NASA Ames Technical Report NASA TN-D-5153). Moffett Field, CA: NASA Ames Research Center.

Crosby, J. V., & Parkinson, S. (1979). A dual task investigation of pilot's skill level. *Ergonomics, 22*, 1301–1313.

Damos, D. (1978). Residual attention as a predictor of pilot performance. *Human Factors, 20*, 435–440.

Damos, D., Smist, T., & Bittner, A. C. (1983). Individual differences in multiple task performance as a function of response strategies. *Human Factors, 25*, 215–226.

Damos, D., & Wickens, C. D. (1980). The acquisition and transfer of time-sharing skills. *Acta Psychologica, 6*, 569–577.

Derrick, W. L. (1981). The relationship between processing resources and subjective dimensions of operator workload. In R. Sugarman (Ed.), *Proceedings, 25th annual meeting of the Human Factors Society*. Santa Monica, CA: Human Factors.

Deutsch, J., & Deutsch, D. (1963). Attention: Some theoretical considerations. *Psychological Review, 70*, 80–90.

Dornic, S. S. (1980). Language dominance, spare capacity, and perceived effort in bilinguals. *Ergonomics, 23*, 369–378.

Eggemeier, T. F. (1981). Current issues in subjective assessment of workload. In R. Sugarman (Ed.), *Proceedings, 25th annual meeting of the Human Factors Society*, Santa Monica, CA: Human Factors.

Ells, J. G. (1973). Analysis of temporal and attentional aspects of movement control. *Journal of Experimental Psychology, 99*, 10–21.

Enstrom, K. D., & Rouse, W. B. (1977). Real-time determination of how a human has allocated his attention between control and monitoring tasks. *IEEE Transactions on Systems, Man, and Cybernetics, SMC-7*, 153–161.

Fogarty, G., & Stankov, L. (1982). Competing tasks as an index of intelligence. *Personality and Individual Differences, 3*, 407–422.

Garvey, W. D., & Taylor, F. V. (1959). Interactions among operator variables, system dynamics, and task-induced stress. *Journal of Applied Psychology, 43*, 79–85.

Gopher, D. (1982). A selective attention test as a prediction of success in flight training. *Human Factors, 24*, 173–184.

Gopher, D., & Brickner, M. (1980). On the training of time-sharing skills: An attention viewpoint. In G. Corrick, M. Hazeltine, & R. Durst (Eds.), *Proceedings, 24th annual meeting of the Human Factors Society*, Santa Monica, CA: Human Factors.

Gopher, D., Brickner, M., & Navon, D. (1982). Different difficulty manipulations interact differently with task emphasis: Evidence for multiple resources. *Journal of Experimental Psychology: Human Perception and Performance, 8*, 146–158.

Gopher, D., & Kahneman, D. (1971). Individual differences in attention and the prediction of flight criteria. *Perception and Motor Skills, 33*, 1335–1342.

Gopher, D., & Navon, D. (1980). How is performance limited: Testing the notion of central capacity. *Acta Psychologica, 46*, 161–180.

Gottsdanker, R. M. & Senders, J. W. (1980, November). *On the estimation of mental load.* (Tech. Rep. No. AFOSR–79–0122). Santa Barbara: University of California, Psychology Department.

Harris, S., Owens, J., & North, R. A. (1978). A system for the assessment of human performance in concurrent verbal and manual control tasks. *Behavior Research Methods and Instrumentation, 10*, 329–333.

Hart, S. G. (1975, May). Time estimation as a secondary task to measure workload. *Proceedings, 11th Annual Conference on Manual Control* (NASA TMX–62, N75–33679, 53), pp. 64–77. Washington, DC: U.S. Government Printing Office.

Herron, S. (1980). A case for early objective evaluation of candidate displays. In G. Corrick, M. Hazeltine, & R. Durst (Eds.), *Proceedings, 24th annual meeting of the Human Factors Society*. Santa Monica, CA: Human Factors.

Hicks, R. E., Miller, G. W., & Gaies, G. (1977). Concurrent processing demands and the experience of time in passing. *American Journal of Psychology, 90*, 431–446.

Hunt, E., & Lansman, M. (1981). Individual differences in attention. In R. Sternberg (Ed.), *Advances in the psychology of intelligence* (Vol. 1). Hillsdale, NJ: Erlbaum Associates.

Isreal, J. (1980). *Structural interference in dual task performance: Behavioral and electrophysiological data.* Unpublished Ph.D. dissertation, University of Illinois, Champaign.

Isreal, J. B., Chesney, G. L., Wickens, C. D., & Donchin, E. (1980). P300 and tracking difficulty: Evidence for multiple resources in dual-task performance. *Psychophysiology, 17*, 259–273.

Isreal, J. B., Wickens, C. D., Chesney, G. L., & Donchin, E. (1980). The event-related brain potential as an index of display-monitoring workload. *Human Factors, 22*, 211–224.

Jennings, A. E., & Chiles, W. D. (1977). An investigation of time-sharing ability as a factor in complex performance. *Human Factors, 19*, 535–547.

Jex, H. R. (1967). Two applications of the critical instability task to secondary task workload research. *IEEE Transactions on Human Factors in Electronics, HFE-8*, 279–282.

Jex, H. R. (1979). A proposed set of standardized subcritical tasks for tracking workload calibration. In N. Moray (Ed.), *Mental Workload.* New York: Plenum Press.

Jex, H. R., & Clement, W. F. (1979). Defining and measuring perceptual-motor workload in manual control tasks. In N. Moray (Ed.), *Mental workload: Its theory and measurement.* New York: Plenum Press.

Kahneman, D. (1973). *Attention and Effort.* Englewood Cliffs, NJ: Prentice-Hall.

Kahneman, D., Beatty, J., & Pollack, I. (1967). Perceptual deficits during a mental task. *Science, 157*, 218–219.

Kahneman, D., Ben-Ishai, R., & Lotan, M. (1973). Relation of a test of attention to road accidents. *Journal of Applied Psychology, 58*, 113–115.

Kalsbeek, J. W., & Sykes, R. W. (1967). Objective measurement of mental load. *Acta Psychologica, 27*, 253–261.

Kantowitz, B. H., & Knight, J. L. (1976). Testing tapping time-sharing I: Auditory secondary task. *Acta Psychologica, 40*, 343–362.

Keele, S. W. (1973). *Attention and human performance.* Pacific Palisades, CA: Goodyear.

Keele, S. W., Neill, W. T., & DeLemos, S. M. (1977). *Individual differences in attentional flexibility* (Technical Report). Eugene, OR: University of Oregon, Center for Cognitive and Perceptual Research.

Kerr, B. (1973). Processing demands during mental operations. *Memory & Cognition, 1*, 401–412.

Kinsbourne, M., & Hicks, R. (1978). Functional cerebral space. In J. Requin (Ed.), *Attention and Performance VII.* Hillsdale, NJ: Erlbaum Associates.

Kruskall, J., & Wish, M. (1978). *Multidimensional scaling.* Beverly Hills, CA: Sage.

LaBerge, D. (1973). Attention and the measurement of perceptual learning. *Memory & Cognition, 1,* 268–276.

Leplat, J., & Welford, A. T. (Eds.). (1978). *Ergonomics,* 21(3).

Lerner, E. J. (1983). The automated cockpit. *IEEE Spectrum, 20,* 57–62.

Logan, G. D. (1979). On the use of a concurrent memory load to measure attention and automaticity. *Journal of Experimental Psychology: Human Perception and Performance, 5,* 198–207.

McLeod, P. (1977). A dual task response modality effect: Support for multiprocessor models of attention. *Quarterly Journal of Experimental Psychology, 29,* 651–667.

McLeod, P. (1978). Does probe RT measure central processing demand? *Quarterly Journal of Experimental Psychology, 30,* 83–90.

Micalizzi, J., & Wickens, C. D. (1980, December). *The applications of factor methodology to workload assessment in a dynamic system monitoring task* (Engineering Psychology Technical Report EPL–80–2/ONR–80–2). Champaign: University of Illinois.

Michon, J. A. (1966). Tapping regularity as a measure of perceptual motor load. *Ergonomics, 9,* 401–412.

Michon, J. A. & Van Doorne, H. (1967). A semi-portable apparatus for measuring perceptual motor load. *Ergonomics, 10,* 67–72.

Moray, N. (1967). Where is attention limited? A survey and a model. *Acta Psychologica, 27,* 84–92.

Moray, N. (Ed.). (1979). *Mental workload: Its theory and measurement.* New York: Plenum Press.

Moray, N. (1982). Subjective mental load. *Human Factors, 23,* 25–40.

Moray, N., & Fitter, M. (1973). A theory and measurement of attention. In S. Kornblum (Ed.), *Attention and Performance IV.* New York: Academic Press.

Moray, N., Fitter, M., Ostry, D., Favreau, D., & Nagy, V. (1976). Attention to pure tones. *Quarterly Journal of Experimental Psychology, 28,* 271–285.

Mulder, G., & Mulder, L. J. (1981). Information processing and cardiovascular control. *Psychophysiology, 18,* 392–401.

Natani, K., & Gomer, F. E. (1981). *Electro-cortical activity and operator workload: A comparison of changes in the EEG and in event-related potentials* (Technical Report MDC E2427). St. Louis: McDonnell Douglas Corporation.

Navon, D., & Gopher, D. (1979). On the economy of the human processing system. *Psychological Review, 86,* 254–255.

Nebes, R. D. (1977). Man's so-called minor hemisphere. In M. C. Wittrock (Ed.), *The Human Brain,* Englewood Cliffs, NJ: Prentice-Hall, 1977.

Neisser, U., Hirst, W., Spelker, E. S., Reaves, C. C., & Coharack, G. (1980). Dividing attention without alternation or automaticity. *Journal of Experimental Psychology: General, 109,* 98–117.

Norman, D. (1968). Toward a theory of memory and attention. *Psychological Review, 75,* 522–536.

Norman, D., & Bobrow, D. (1975). On data-limited and resource-limited processing. *Journal of Cognitive Psychology, 7,* 44–60.

North, R. A. (1977, October). *Task functional demands as factors in dual-task performance.* Paper presented at the 25th annual meeting of the Human Factors Society, San Francisco.

Ogden, G. D., Levine, J. M., & Eisner, E. J. (1979). Measurement of workload by secondary tasks. *Human Factors, 21,* 529–548.

Parks, D. (1979). Current workload methods and emerging challenges. In N. Moray (Ed.), *Mental workload: Its theory and measurement.* New York: Plenum Press.

Peavler, W. S. (1974). Individual differences in pupil size and performance. In M. Janisse (Ed.), *Pupillary dynamics and behavior.* New York: Plenum Press.

Peters, M. (1977). Simultaneous performance of two motor activities: The factor of timing. *Neuropsychologia, 15,* 461–465.

Posner, M. I., & Boies, S. J. (1971). Components of attention. *Psychological Review 78,* 391–408.

Posner, M. I., & Keele, S. W. (1969). Attention demands of movements. *Proceedings of the 17th Annual Congress of Applied Psychology.* Amsterdam: Zeitlinger.

Rasmussen, J. (1981). Models of mental strategies in process control. In J. Rasmussen & W. Rouse (Eds.), *Human detection and diagnosis of system failures*. New York: Plenum Press.

Reid, G. B., Shingledecker, C., & Eggemeier, T. (1981). Application of conjoint measurement to workload scale development. In R. Sugarman (Ed.), *Proceedings, 25th annual meeting of the Human Factors Society*. Santa Monica, CA: Human Factors.

Roedinger, H. L., Knight, J. L., & Kantowitz, B. H. (1977). Inferring decay in short-term memory: The issue of capacity. *Memory & Cognition, 5*, 156–176.

Rolfe, J. M. (1973). The secondary task as a measure of mental load. In W. T. Singleton, J. G. Fox, & D. Whitfield (Eds.), *Measurement of man at work* (pp. 135–148). London: Taylor & Francis.

Rollins, R. A., & Hendricks, R. (1980). Processing of words presented simultaneously to eye and ear. *Journal of Experimental Psychology: Human Perception and Performance, 6*, 99–109.

Roscoe, A. H. (Ed.). (1978, February). *Assessing pilot workload* (AGARD–AG–233, AD A051 587). Paris: NATO.

Schiflett, S. G. (1980). *Evaluation of a pilot workload assessment device to test alternate display formats and control handling qualities*. (Technical Report SY–33R–80). Patuxent River, MD: Naval Air Test Center.

Schneider, W., & Fisk, A. D. (1982). Concurrent automatic and controlled visual search: Can processing occur without cost? *Journal of Experimental Psychology: Learning, Memory, and Cognition, 8*, 261–278.

Schneider, W., & Shiffrin, R. (1977). Controlled and automatic human information processing. *Psychological Review, 84*, 1–66.

Senders, J. (1964). The human operator as a monitor and controller of multidegree freedom systems. *IEEE Transactions on Human Factors in Electronics, HFE–5*, 2–6.

Shaffer, L. H. (1975). Multiple attention in continuous verbal tasks. In S. Dornic (Ed.), *Attention and Performance V*. New York: Academic Press.

Sheridan, T. (1972). On how often the supervisor should sample. *IEEE Transactions on Systems, Science, and Cybernetics, SSC–6*, 140–145.

Sheridan, T. (1980). Mental workload: What is it? Why bother with it? *Human Factors Society Bulletin, 23*, 1–2.

Sheridan, T. B., & Simpson, R. W. (1979, January). *Toward the definition and measurement of the mental workload of transport pilots*. (FTL Report R 79–4). Cambridge, MA: Massachusetts Institute of Technology, Flight Transportation Laboratory.

Sheridan, T., & Stassen, H. (1979). Definitions, models and measures of human workload. In N. Moray (Ed.), *Mental workload: Its theory and measurement*. New York: Plenum Press.

Smith G. (Ed.). (1979). *Human Factors, 21*(5).

Spelke, E., Hirst, W., & Neisser, U. (1976). Skills of divided attention. *Cognition, 4*, 215–230.

Sperandio, J. C. (1978). The regulation of working methods as a function of workload among air traffic controllers. *Ergonomics, 21*, 193–202.

Sperling, G., & Melchner, M. (1978). Visual search, visual attention, and the attention operating characteristic. In J. Requin (Ed.), *Attention and Performance VIII*. Hillsdale, NJ: Erlbaum Associates.

Spicuzza, R., Pincus, A., & O'Donnell, R. D. (1974, August). *Development of performance assessment methodology for the digital avionics information system*. Dayton, OH: Systems Research Laboratories, Inc.

Summers, J. J. (1981). Motor programs. In D. H. Holding (Ed.), *Human skills*. New York: Wiley.

Sverko, B. (1977). Individual differences in time-sharing performance. *Acta Instituti Psychologici, 79*, 17–30.

Treisman, A. (1969). Strategies and models of selective attention. *Psychological Review, 76*, 282–299.

Treisman, A., & Davies, A. (1973). Divided attention to eye and ear. In S. Kornblum (Ed.), *Attention and Performance IV*. New York: Academic Press.

Vidulich, M. & Wickens, C. D. (1982). The influence of S–C–R compatibility and resource competition on performance of threat evaluation and fault diagnosis. In R. Edwards (Ed.), *Proceedings, 26th annual meeting of the Human Factors Society*. Santa Monica, CA: Human Factors.

Welford, A. T. (1978). Mental workload as a function of demand, capacity, strategy, and skill. *Ergonomics, 21,* 151–167.

Wickens, C. D. (1976). The effects of divided attention in information processing in tracking. *Journal of Experimental Psychology: Human Perception and Performance, 2,* 1–13.

Wickens, C. D. (1979). Measures of workload, stress, and secondary tasks. In N. Moray (Ed.), *Mental workload: Its theory and measurement* (pp. 79–99). New York: Plenum Press.

Wickens, C. D. (1980). The structure of attentional resources. In R. Nickerson and R. Pew (Eds.), *Attention and Performance VIII.* Hillsdale, NJ: Erlbaum Associates.

Wickens, C. D. (1981). Workload: In defense of the secondary task. *Personnel Training and Selection Bulletin, 2,* 119–123.

Wickens, C. D. (in press). Processing resources in attention. In R. Parasuraman & R. Davies (Eds.), *Varieties of attention.* New York: Academic Press.

Wickens, C. D., & Benel, D. (1981). The development of time-sharing skills. In J. A. S. Kelso & J. Clark (Eds.), *Motor development.* New York: Wiley.

Wickens, C. D., & Derrick, W. (1981). Workload measurement and multiple resources. *Proceedings, 1981 IEEE Conference on Cybernetics and Society.* New York: Institute of Electrical and Electronics Engineers, Inc.

Wickens, C. D., & Gopher, D. (1977). Control theory measures of tracking as indices of attention allocation strategies. *Human Factors, 19,* 349–365.

Wickens, C. D., & Kessel, C. (1980). The processing resource demands of failure detection in dynamic systems. *Journal of Experimental Psychology: Human Perception and Performance, 6,* 564–577.

Wickens, C. D., Mountford, S. J., & Schreiner, W. (1981). Multiple resources, task hemispheric integrity, and individual differences in time-sharing. *Human Factors, 23,* 211–229.

Wickens, C. D., Sandry, D., & Hightower, R. (1982, October). *Display location of verbal and spatial material: The joint effects of task hemispheric integrity and processing strategy* (Technical Report EPL–82–2/ONR–82–2). Champaign, IL: University of Illinois, Engineering Psychology Research Laboratory.

Wickens, C. D., Sandry, D., & Vidulich, M. (1983). Compatibility and resource competition between modalities of input, central processing, and output: Testing a model of complex task performance. *Human Factors, 25,* 227–248.

Wickens, C. D., & Yeh, Y. Y. (1983). The dissociation of subjective ratings and performance: A multiple resources approach. In L. Haugh & A. Pope (Eds.), *Proceedings of the 27th Annual Conference of the Human Factors Society.* Santa Monica, CA: Human Factors Press.

Wierwille, W. W., & Williges, R. C. (1978, September). *Survey and analysis of operator workload assessment techniques.* (Report No. S–78–101). Blacksburg, VA: Systemetrics, Inc.

Wierwille, W. W., & Williges, B. H. (1980, March). *An annotated bibliography on operator mental workload assessment.* (Report SY–27R–80). Patuxent River, MD: Naval Air Test Center.

Williges, R. C., & Wierwille, W. W. (1979). Behavioral measures of aircrew mental workload. *Human Factors, 21,* 549–574.

Selection of Action

OVERVIEW

In most systems the human operator must translate the information that is perceived about the environment into an action. Sometimes the action is an immediate response to a perceived stimulus: We slam on the brake when a car unexpectedly pulls into the intersection ahead; the pilot makes a rapid correction in the face of an oncoming target; or the monitor in an intensive care ward must make an immediate corrective action to a loss of blood pressure observed in a patient. At other times, the action is based more upon a thorough, time-consuming evaluation of the current state of the world, integrating information from a large number of sources over a longer period of time. Many examples of this latter process were considered in Chapter 3, including the medical diagnosis and selection of treatment performed by a physician.

The two types of selection of action represent end points on a continuum related to the degree of automaticity with which the action is chosen. This automaticity in turn is determined by the amount of practice that operators have had in applying the rules of action selection. In the context of the nuclear power monitor, Rasmussen (1980) has distinguished three distinct levels on this continuum: skill-based, rule-based, and knowledge-based behavior. At the most automated level, *skill-based* behavior assigns stimuli to responses in a rapid automatic mode with a minimum investment of resources (see Chapter 8). Applying the brake on a car in response to the appearance of a red light is skill-based behavior. The stimuli for skill-based behavior need not necessarily be simple. For example, certain combinations of correlated features—a "syndrome"—may be rapidly classified into a category that triggers an automatic action. This is an example of skill-based action with high stimulus complexity. However, when such complexity exists, the rapid action will only occur after the operator has received extensive training and experience. The skilled physician, for example, may immediately detect the pattern of symptoms indicating a certain disease and identify the appropriate treatment at once. The medical school student or intern with far less medical training, on the other hand, will evaluate the same symptoms in a much more time-consuming fashion to reach the same conclusion.

The level typified by the medical student illustrates *rule-based* behavior. Here an action is selected by bringing into working memory a hierarchy of rules, "If X occurs, then do Y." After mentally scanning these rules the decision maker will implement the appropriate action.

The situation may be familiar, but the processing is considerably less automatic and timely. The final category of action, *knowledge-based* behavior, is invoked when entirely new problems are encountered. Neither rules nor automatic mappings exist, and more general knowledge concerning the behavior of the system, the characteristics of the environment, and the goals to be obtained must be integrated in order to formulate a novel plan of action. This level is often typical of government or corporate decision making or of the troubleshooting strategies encountered by the maintenance technician in diagnosing and correcting an unfamiliar malfunction in a complex system.

Along this continuum from skill-based to knowledge-based behaviors, both the latency and the accuracy of the action selection are important. However, certain characteristics of these two measures of performance correlate with the continuum. When examining knowledge-based behavior, latency will generally be large, highly variable, and a relatively poor indicato of decision-making quality. That is, neither long nor short latency actions will necessarily indicate higher quality decisions. Accuracy, in general, will be relatively low (when compared with skill-based behavior) but will serve as a far better indicator of decision quality than latency. In the terms proposed in Chapter 1, the overall "efficiency" index of performance would be determined relatively more by accuracy than by speed. When dealing with skill-based behavior on the other hand, latency is often the critical variable, because this behavior is characteristic of highly trained operators in situations in which speed is essential and few errors are made. Latency thus becomes a relatively more reliable index of decision-making quality than it is at a knowledge-based level. Furthermore, since errors are fewer in number, they provide a less reliable source of information concerning quality.

The continuum drawn from skill-based to knowledge-based behavior is important in terms of the organization of this book. The integration of processes of perception, decision making, and action selection, which were considered in Chapter 3 within the context of decision making, will be considered again here. In that chapter, however, the context was that of knowledge- and rule-based behavior. Decision latencies were on the order of minutes, hours, or days, and their precise value was of little importance. In this chapter the same sequence of perception, decision making, and action is considered in the context of skill-based behavior. Decision latencies here are on the order of seconds and milliseconds, and the latency of response is highly important. It should be remembered, however, that most action sequences discussed in Chapter 3, if given sufficient practice and performed in a context in which speed was critical, would be discussed more appropriately in this chapter instead.

The present chapter then primarily concerns *reaction time*, or RT. What are the factors that determine the speed with which an operator can perceive a stimulus and translate that perception into a well-learned action? Latency is the critical dependent variable. Accuracy is a variable that is monitored, but it is not usually the major source of inference concerning the nature of human processing limitations. The nature of errors and the performance of tasks requiring serial responding will be considered in Chapter 10.

A large number of different variables which influence reaction time both inside and outside of the laboratory will be considered in the following pages. One of the most important of these is the degree of uncertainty about what stimulus will occur and therefore the degree of choice in the action to make. For the sprinter at the starting line of a race, there is no uncertainty about the stimulus—the sound of the starting gun—nor is there a choice of what response to make: accelerate as fast as possible. On the other hand, for the driver of an automobile, wary of potential obstacles in the road, there is both stimulus uncertainty and response choice. An obstacle could be encountered on the left, requiring a swerve to the right; on the right, requiring a swerve to the left; or perhaps at dead center requiring that the brakes be applied. The situation of the sprinter demonstrates the *simple reaction time* paradigm, the vehicle driver the paradigm of *choice reaction time.*

Examples of simple reaction time rarely occur outside of the laboratory—the sprinter's start is a rare exception. Yet the simple reaction time paradigm is nevertheless important for the following reason: All of the variables that influence reaction time can be dichotomized into those that are dependent in some way upon the choice of a response and those that are not, that is, those that only influence choice RT and those that affect all reaction times. When the simple reaction time paradigm is examined in the laboratory, it is possible to study the second class of variables more precisely because in this situation the measurement of response speed cannot be contaminated by other factors related to the degree of choice. We shall see that the influences of the latter are considerable. Hence in the following treatment we shall consider the variables that influence both choice and simple RT before discussing these variables unique to the choice paradigm.

The supplement to the chapter will then explore in detail two further aspects of reaction time and human performance. The first of these considers models that have been generated to account for the reaction time process. The second describes a particular analytical technique— the additive factors method—that has been applied to reaction time data to help understand and define the different stages in human information processing that were discussed in Chapter 1. The applications of additive factors will be discussed. Finally, we shall consider some alternative methods for defining and describing stages of information processing.

VARIABLES INFLUENCING SIMPLE AND CHOICE REACTION TIME

In the laboratory, simple reaction time is investigated by providing the subject with one response to make as soon as a stimulus occurs. The subject may or may not be warned prior to the appearance of the stimulus. Four major variables—stimulus modality, stimulus intensity, temporal uncertainty, and expectancy—appear to influence response speed in this paradigm.

Stimulus modality.   Several investigators have reported that simple reaction time to auditory stimuli is about 30 to 50 msec faster than to visual stimuli (roughly 130 msec and 170 msec, respectively; Woodworth & Schlossberg, 1965). This difference has been attributed to differences in the speed of sensory processing between the two modalities. However, Kohlberg (1971) claims that the difference is instead obtained only when experimenters fail to control for intensity differences between modalities. When Kohlberg carefully controlled for these differences, he found that auditory and visual stimuli of equal subjective intensity produced equal simple reaction times, so long as the visual stimulus was intense enough to be processed in foveal vision.

Stimulus intensity.   In a comprehensive summary of several different investigations of simple reaction time, Teichner and Krebs (1972) concluded that simple RT decreases with increases in intensity of the stimulus to an asymptotic value. This function, shown in the bottom of Figure 9.1 corresponds quite closely to the function that relates the rate of neural firing to stimulus intensity shown at the top of Figure 9.1. The striking similarity of these two functions is consistent with the view that simple RT reflects the latency of a *decision* process that something has happened (Fitts & Posner, 1967; Teichner & Krebs, 1972). This decision is based upon the aggregation over time of neural evidence in the sensory channel until a criterion is exceeded. Factors such as stimulus intensity that retard or enhance the rate of neural firing will therefore be reflected similarly in a variation of simple RT.

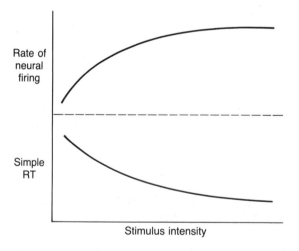

Figure 9.1   Relation between stimulus intensity, neural activity, and simple reaction time.

In this sense, the simple RT is conceived as a two-stage process as in signal detection theory. Aggregation of stimulus evidence may be fast or slow, depending on the intensity of the stimulus, and the criterion can be lowered or raised depending on the "set" of the subject. In the example of a sprinter, a lowered criterion might well induce a false start if a random noise in the environment exceeded the criterion. After one false start, the runner will raise the criterion and be slower to start on the second gun in order to guard against the possibility being disqualified. This model then attributes the only source of uncertainty in simple RT to be *temporal*.

Temporal uncertainty. The degree of predictability of when the stimulus will occur is called temporal uncertainty. This factor can be manipulated by varying the *warning interval* (WI) occurring between a warning signal and the imperative stimulus to which the subject must respond. In the case of the sprinter, two warning signals are provided: "Take your mark," and "Set." The gunshot then represents the imperative stimulus. If the warning interval (WI) is short and remains constant over a block of trials, then the imperative stimulus is highly predictable in time, and RT will be short. In fact, if the WI is always constant at around 0.5 seconds, the subject can produce a simple RT of nearly 0 seconds by synchronizing the response with the predictable imperative stimulus. On the other hand, if the warning intervals are long or variable, RT will be long. An experiment of Warrick, Kibler, Topmiller, and Bates (1964) investigated variable warning intervals as long as two and a half days! The subjects were secretaries engaged in routine typing. Occasionally they had to respond with a key press when a red light on the typewriter was illuminated. Even with this extreme degree of variability, simple RT was only prolonged to around 700 msec.

Temporal uncertainty thus results from increases in the variability and the length of the WI. When the variability of the WI is increased, this uncertainty is in the environment. When the mean length of the WI is greater, the uncertainty is localized in the subjects' internal timing mechanism, since the variability of our estimates of time intervals increases linearly with the mean duration of those intervals (Fitts & Posner, 1967).

If there is uncertainty in a stimulus, then it can be measured by information theory as described in Chapter 2. Klemmer (1957) manipulated uncertainty by varying the relation of WI variability to mean WI duration when these parameters were set for each of a series of simple RT trials. Thus a series of trials that had long and variable WIs had high temporal uncertainty and high information content. Klemmer found that the relation between the average RT on each series and the amount of temporal information in the series, as quantified by the two sources of temporal uncertainty, was linear. The linear relation between processing time and information content has important theoretical implications that will be discussed later in the chapter.

Expectancy. Klemmer's findings were based upon the *average* reaction time during a block of trials. If the average WIs within a block are long, average reaction time will be longer. On the other hand, Drazin (1961) examined individual RTs within a block of variable WIs and found that the pattern reverses. In this case, responses that follow long WIs tend to be faster than those following short WIs. This negative correlation between reaction time and warning interval is explained by the concept of *expectancy*. The longer the time since the warning signal has passed, the more the subject *expects* the imperative signal to occur. As a consequence, the response criterion is progressively lowered with the passing of time so as to produce fast RTs when the imperative stimulus finally does appear after a long wait. In terms of the sprinting example, if the delay of the starter's gun is variable from one sprint to the next, those sprinters who must wait an exceptionally long time between "Set" and the gun will produce fast RTs and fast times. On the other hand, if the gun is delayed too long there will be an increased number of false starts.

The role of expectancy and warning intervals in reaction time is critical in many real-world situations. Danaher (1980) discusses an incident in which an aircraft gaining altitude in low visibility was flying on a collision course with a second aircraft flying level and directly above. When the upper plane suddenly appeared in view of the lower one, a rapid maneuver (RT response) was required for evasion. This incident is relevant to the present discussion because a *warning signal* was provided by air traffic control just prior to the visual sighting. It was this warning signal that may well have lowered the pilots' response criterion sufficiently to initiate the timely response. (The question of why the warning was not given sooner to prevent the close pass will be discussed in Chapter 12.)

VARIABLES INFLUENCING CHOICE REACTION TIME

When actions are chosen in the face of environmental uncertainty, a host of additional variables related to the choice process itself influences the speed of action. In the terms described in Chapter 2, the operator is transmitting information from stimulus to response. This characteristic has led several investigators to use information theory to describe the effects of many of the variables on choice reaction time.

The Information Theory Model: The Hick-Hyman Law

It is intuitive that the more complex decisions or choices require longer to initiate. A straightforward example is the difference between simple RT and choice RT in which there is uncertainty about which stimulus will occur and therefore about which action to take. More than a century ago, Donders (1969) demonstrated that choice RT was longer than simple RT. The actual function that related the amount of uncertainty or degree of choice to RT was first presented by Merkel (1885). He found that RT was a negatively accelerating function of the number of stimulus-response alternatives. Each added alternative increases RT, but by a smaller amount than the previous alternative.

The theoretical importance of this function remained relatively dormant until the early 1950s when, in parallel developments, Hick (1952) and Hyman (1953) applied information theory in order to quantify the uncertainty of stimulus events (see Chapter 2). Both investigators found that choice RT increased linearly with stimulus information—$\log_2 N$, where N is the number of alternatives—in the manner shown in Figure 9.2a. Reaction time increases by a constant amount each time N is doubled or, alternatively, each time the information in the stimulus is increased by one bit. When a linear equation is fitted to the data in Figure 9.2a, then RT can be expressed by the equation $RT = a + bH_s$, a relation often referred to as the Hick-Hyman law. The constant b reflects the slope of the function—the amount of added processing time that results from each added bit of stimulus information to be processed. The constant a describes the sum of those processing latencies that are unrelated to the resolution of uncertainty. These would

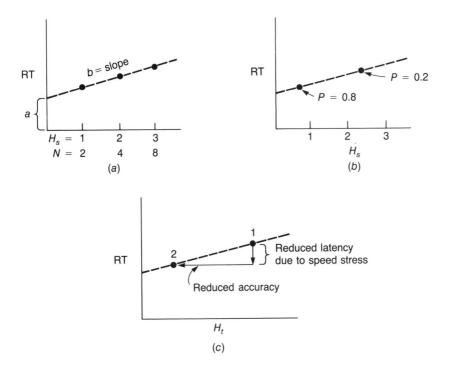

Figure 9.2 The Hick-Hyman law of choice reaction time: $RT = a + bH_s = a + bH_t$. (a) RT as a function of number of alternatives. (b) RT for two alternatives of different probabilities. (c) Change in RT and accuracy (measured by information transmission H_t) produced by different sets for speed and accuracy.

include, for example, the time taken to encode the stimulus and to execute the response. The issue of whether stimulus-response uncertainty affects perception, response selection, or both will be considered later in the chapter.

If the Hick-Hyman law is valid in a general sense, then a function similar to that shown in Figure 9.2 should be obtained when information is manipulated by various means, as described in Chapter 2. Both Hick and Hyman manipulated the number of stimulus-response alternatives N. Thus the points representing 1, 2, and 3 bits of information on the abscissa of Figure 9.2a could be replaced by the values $\log_2 2$, $\log_2 4$, and $\log_2 8$, respectively. Hyman further demonstrated that the function was still linear when the average information transmitted by stimuli during a block of trials was manipulated by varying the probability of stimuli and by varying their sequential expectancy. If probability is varied, then when N alternatives are equally likely, as described in Chapter 2, information is maximum (i.e., four alternatives yield two bits). When the probabilities are imbalanced, the average information is reduced. Hyman observed that the mean RT for a block of trials is shortened by this reduction of information in such a way that the new, faster data point still lies along the linear function of the Hick-Hyman law. When sequential constraints were imposed upon the series of stimuli, thus also increasing redundancy and reducing information conveyed, Hyman found a similar reduction in RT so that the faster point still lies along the function in Figure 9.2a.

Hyman's data measured the average reaction time for trial blocks. Fitts, Peterson, and Wolpe (1963) gave greater generality to the Hick-Hyman law by demonstrating that individual stimuli of low (or high) probability, and therefore high (or low) information content, produced RTs that also fall along the linear function describing the average RTs. Figure 9.2b presents these results for two stimuli randomly occurring in a series, one with a probability of 0.2 (high information) and the other with a probability of 0.8 (low information). Note that the response time to the low-probability, unexpected stimulus is slower, but it is slowed just enough to fall along the function at the point predicted by its higher information value, $\log_2 (0.2) = 2.2$ bits.

The Speed-Accuracy Tradeoff

In reaction time tasks, and in speeded performance in general, people often make errors. Furthermore, they tend to make more errors as they try to respond more rapidly. This reciprocity between latency and errors is referred to as the *speed-accuracy tradeoff*. According to the analysis of information transmission in Chapter 2, errors of response will reduce the information transmitted. In his original experiment, Hick (1952) examined the relation between reaction time and information transmitted as the speed-accuracy tradeoff was varied.

Hick forced subjects to respond fast (and they made more errors) or to respond accurately (and they were slow). Reaction time clearly varied across these "sets," but Hick went further and computed the information transmitted H_t from the stimulus-response matrix following the technique described in Chapter 2. He then computed the bandwidth of information transmission in bits per second by dividing decision accuracy (bits per decision) by mean RT (seconds per decision), and found that this variable (bits per second) was constant across the various speed sets. As an example, if we induce a subject to respond more rapidly in a four-choice RT task he will do so, but he will make more errors. As in Figure 9.2b, RT will be reduced, but the information transmitted will always be reduced proportionately, so that if H_t replaces H_s on the abscissa, the new data point will still lie along the original Hick-Hyman function but will now be closer to the origin. This relation is shown in Figure 9.2c. The original function is the dashed line. Point 1 is the accuracy condition, point 2 is the speed condition. Another way of describing this relation is to say that a unit of time generates a constant amount of processed information. In a number of early studies, the reciprocal of the Hick's law slope $1/b$ was described as the "bandwidth" of the subject, because this quantity estimates the amount of information transmitted per unit of time—that is, the number of bits per second that a subject is able to transmit from stimulus to response.

The results obtained by Hick actually are somewhat misleading in suggesting that performance efficiency, measured in bits per second, will be unaffected by the set for speed versus accuracy. Subsequent studies suggest that this is not the case. For example, Howell and Kreidler (1963, 1964) compared performance on both easy and complex choice RT tasks as the set for speed versus accuracy was varied by different instructions. Different subjects were told to be fast, to be accurate, or to be fast *and* accurate, and finally they were given instructions that explicitly induced them to maximize the information transmission rate in bits/sec. With both simple and complex tasks, Howell and Kreidler found that instructions changed latency and error rate in the expected directions, with the speed instructions having the largest effect on both variables. However, performance efficiency was not constant across the different sets for either task. When the choice task was easy, maximum efficiency (bits per second) was obtained by subjects instructed to maximize this quantity. When the task was complex, on the other hand, the highest level of performance efficiency (bits per second) was obtained with the speed set instructions.

Other investigations by Fitts (1966) and Rabbitt (1981), using reaction time, and by Seibel (1972), employing typing, also conclude that performance efficiency reaches a maximum value at some intermediate level of speed-accuracy set. These investigators conclude furthermore that subjects left to their own devices will seek out and select the level of set that achieves the maximum performance efficiency. This searching and maximizing behavior may be viewed as another example of optimality of human performance.

The speed-accuracy operating characteristic. Reaction time and error rate represent two dimensions of the efficiency of processing information. These dimensions are analogous in some respects to the dimensions of hit and false-alarm rate in signal detection (Chapter 2), or Task *A* and Task *B* performance in the performance operating characteristic (POC) analysis of time-sharing performance (Chapter 8). Furthermore, just as operators can adjust their response criterion in signal detection or the allocation of resources between tasks in time-sharing, they can also adjust their "set" for speed versus accuracy to various levels defining "optimal" performance under different occasions, as the experiments reported above demonstrated. The *speed-accuracy operating characteristic*, or SAOC, is a function that represents reaction time performance in a manner analogous to the receiver operating characteristic (ROC) and POC representation of detection and time-sharing performance, respectively.

Conventionally, the SAOC may be shown in one of two forms. In Figure 9.3, RT is plotted on the abscissa and some measure of accuracy (the inverse of error rate) on the ordinate (Pachella, 1974). The four different points in the figure represent mean accuracy and latency data collected on four different blocks of trials when the speed-accuracy set is shifted. The negatively accelerating function relating percent correct to RT in Figure 9.3, when considered with the observation of Howell and Kreidler (1964) and Fitts (1966) that maximum H_t occurs at intermediate error rates, has an important practical implication concerning the kind of accuracy instructions that should be given to operators in speeded tasks such as typing or keypunching: Performance efficiency will be greatest at intermediate levels of speed-accuracy set. It is reasonable to tolerate a small percentage of errors in order to obtain efficient performance, and it is probably irrational to demand "zero defects" or perfect performance. We can see why this is so by examining the speed-accuracy tradeoff plotted in Figure 9.3. Forcing the subject to commit no errors will induce impossibly long RTs.

An important warning to experimenters emphasized by Pachella (1974) and Wickelgren (1977) is also implied by the form of Figure 9.3. If experimenters instruct their subjects to "make no errors," they are forcing them to operate at a region along the SAOC in which very small

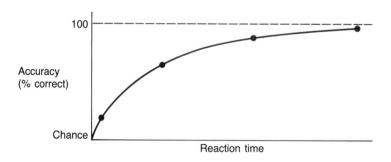

Figure 9.3 The speed-accuracy tradeoff.

fluctuations in accuracy generate very large fluctuations in latency, since the slope of the right-hand portion of Figure 9.3 is almost flat here. Hence, reaction time will be highly variable and the reliable assessment of its true value will be a difficult undertaking.

Pew (1969) has shown that when accuracy is expressed in terms of the measure log [P(correct)/P(errors)], the SAOC is typically *linear*. This relation, shown in Figure 9.4, indicates that a constant increase in time buys the operator a constant increase in the logarithm odds of being correct. In terms of the SAOC space, one should think of movement from "southeast" to "northwest" as changing performance quality, efficiency, or bandwidth (bits per second). This relationship may be captured by the mnemonic, "Northwest is best, southeast is least." Movement *along* an SAOC, on the other hand, represents different cognitive "sets" for speed versus accuracy. Two such SAOCs are shown in Figure 9.4. An excellent discussion of the speed-accuracy tradeoff in reaction time may be found in Pachella (1974) and Wickelgren (1977).

The concept of the speed-accuracy tradeoff is important when performance latency is compared between different experimental conditions or between different pieces of equipment. Three cautions here are relevant: (1) If two experimental conditions or two pieces of equipment (*A* and *B*) have equal RTs (or equal measures of processing speed), the investigator must ensure that error rate is also equal, to assure that performance is equivalent. (2) If *A* has shorter RT than *B*, the investigator needs to ensure that *A* also has an equal or lower error rate than *B*. Otherwise, the two conditions might differ in this speed-accuracy set, not in performance quality. A piece of equipment or condition that will

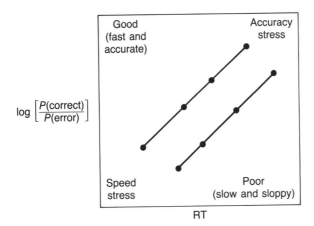

Figure 9.4 The speed-accuracy operating characteristic (SAOC). From "The Speed-Accuracy Operating Characteristic" by R. W. Pew, 1969, *Acta Psychologica, 30*, p. 18. Copyright 1969 by the North Holland Publishing Company. Reproduced by permission.

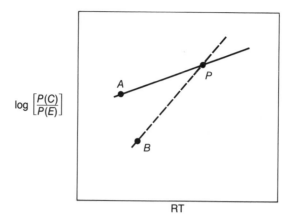

Figure 9.5 The speed-accuracy operating characteristic for two different situations, *A* and *B*. Conclusions about which is better will depend on whether speed or accuracy is stressed.

induce a higher state of arousal (see Chapter 7), for example, will be likely to shift the operator along the SAOC to more rapid but error-prone responding (Posner, 1978). (3) Even if *A* and *B* have identical RTs and error rate at one level of speed-accuracy set, in order to establish the equivalent quality of the two conditions or pieces of equipment, the speed and accuracy variables should be assessed as the speed-accuracy set is varied. This is because the two SAOCs that define performance in two conditions may not necessarily run parallel to each other as they do in Figure 9.4. Figure 9.5 shows a case in which they do not. Here *A* is better than *B* when speed is stressed, but when accuracy is stressed (point *P*) the two are roughly equivalent. Still greater stress on accuracy will make *B* superior. The lesson is that conclusions about the superiority of one system or experimental condition over the other cannot be made with certainty unless the speed-accuracy set is manipulated.

DEPARTURES FROM INFORMATION THEORY

The Hick-Hyman law has proven in general to be quite successful in accounting for the changes in RT with informational variables. The linear relation between reaction time and information, in fact, led a number of investigators to conclude that the human had a relatively constant bandwidth of information processing, provided by the inverse slope of the Hick-Hyman law function. Yet it soon became evident that information theory was not entirely adequate to describe reaction time data. We have noted already that bandwidth is not constant across wide ranges of speed and accuracy set. More crucially, five additional variables will be discussed below that influence reaction time but are not

easily quantified by information theory. These relate to subset familiarity, stimulus discriminability, the repetition effect, stimulus-response compatibility, and practice. The existence of these variables
does not invalidate the Hick-Hyman law but merely restricts somewhat
its generality and requires some degree of caution when it is applied.

Subset Familiarity

As described in Chapter 2, information measurement must assume a
given domain—that is, a set of possible alternatives from which a
stimulus may occur. To the creatively thinking operator, this set in
practice may not correspond with what is assumed by the investigator
when the information metric is applied. Garner (1974), in discussing
this issue, points out that the way we describe or think of a stimulus is
greatly influenced by our subjective belief of what its alternatives may
be. Thus when we see a line on a sheet of paper, we will not think of
describing its thickness or its blackness unless we are contrasting it
with a line that has a different width or a different shade.

An example of how the domain of possible stimuli that could occur
influences the information theory approach to reaction time is provided by an experiment of Fitts and Switzer (1962). They measured
subjects' reaction time to digit stimuli and found that choice RT to the
digits 1 and 2 ($N = 2$) was expectedly faster than RT to the digits 1
through 8 ($N = 8$). This observation is, of course, consistent with the
Hick-Hyman law, since two alternatives convey less information than
eight. However, Fitts and Switzer also found that RT to the digits 2 and
7 ($N = 2$) was much closer to the longer value ($N = 8$) than to the value
measured with the smaller set ($N = 2$), consisting of the digits 1 and 2.
Their subjects were apparently processing the digits 2 and 7 as if they
were drawn from the larger sample of 1 through 8 ($N = 8$) rather than
the smaller one. This "domain" assumed by the subjects did not correspond to that assumed by the experimenter.

Fitts and Switzer's findings have two important implications: They
suggest that applying information theory to quantify performance will
be ambiguous unless the subject is quite clear of the potential set of
alternative stimuli. The results also suggest that some attempts to
improve performance by reducing the number of possible stimuli that
could be presented to an operator may not be successful when the
stimuli are arbitrarily drawn from a larger well-learned set. In support
of this prediction, Conrad and Hull (1967) observed that the speed with
which subjects entered data into an alphabetic keyboard was not improved by reducing an alphabet from 26 to 10 letters.

Stimulus Discriminability

Reaction time is lengthened as a set of stimuli are made less discriminable from each other (Vickers, 1970). This "noninformation" factor has
some important implications. Tversky (1977) has argued that we judge

the similarity or difference between two stimuli on the basis of the ratio of shared features to total features within a stimulus, and not simply on the basis of the absolute number of shared (or different) features. Thus, the numbers 4 and 7 are quite distinct, but the numbers 721834 and 721837 are quite similar, although in each case only one digit differentiates the pair. Discriminability difficulties in RT, like confusions in memory (see Chapter 6), can be reduced by deleting shared and redundant features where possible. In the context of nuclear power plant design, Kirkpatrick and Mallory (1981) have emphasized the importance of avoiding confusability between display items by minimizing the feature similarity between separate labels.

The Repetition Effect

A number of investigators have noted that in a random stimulus series, the repetition of a stimulus-response (S–R) pair (AA) yields a faster reaction time to the second stimulus than does an alternation. For example, if the stimuli were designated A and B, the response to A following A will be faster than to A following B (e.g., Hyman, 1953; Bertelson, 1961; Kirby, 1976; Kornblum, 1969). The advantage of repetitions over alternations, referred to as the "repetition effect," appears to be enhanced by increasing N (the number of S–R alternatives), by decreasing S–R compatibility (see below), and by shortening the interval between stimuli and responses (Kornblum, 1973). Research by Bertelson (1965) and others (see Kornblum, 1973, for a summary) suggests that the response to repeated stimuli is speeded both by the repetition of the stimulus and by the repetition of the response. Thus, in a paradigm in which three stimuli (A, B, and C) are assigned to two responses (A and B to R_1, C to R_2), a stimulus repetition AA will be faster than a stimulus alternation BA, even though both have the same response. Both of these, however, will be faster than the response to A in the sequence CA in which both the stimulus and response are changed.

There are two circumstances in which the repetition effect is not observed. (1) As summarized by Kornblum (1973), the repetition effect declines with long intervals between stimuli and may sometimes be replaced by an alternation effect (faster RTs to a stimulus change). In this case, it appears that the "gambler's fallacy"—expectancy that stimuli will change—begins to come into play (see Chapter 3). Subjects do not expect a continued run of stimuli of the same sort. (2) In some tasks, such as in typing, rapid repetition of the same digit or even digits on the same hand will be slower than alternations (Sternberg, Kroll, and Wright, 1978). The typing task, however, differs from the reaction time paradigm in one fundamental respect that can explain the elimination of the repetition effect. In typing, typical of the *transcription* task which will be discussed in Chapter 10, the response sequence may be made with a considerable lag after the triggering stimulus, while subsequent stimuli are being processed. This fact allows responses to be made much closer together in time than in reaction time tasks.

Stimulus-Response Compatibility

Spatial S–R compatibility.   The issue of compatibility has been encountered before in Chapters 5 and 6. In that context the compatibility between display representation and an operator's internal model of the displayed system was discussed. Since the internal model often serves as a basis for rapid action, it is important that compatibility also be maintained between stimulus and response, the topic of interest in this chapter. The correspondence of stimuli and responses in space provides some of the clearest examples of S–R compatibility. The relations or ordering of stimuli in space should correspond with the spatial configuration of the actions required. As an example of spatial compatibility at the simplest level, Simon (1969) found that subjects have a very basic and automatic tendency to move toward the source of a light. This tendency is independent of which hand is moving and of which side of the body the light appears on (Cotton, Tzeng, & Hardyck, 1980). Therefore, if two side-by-side lights are to be responded to with two key presses, fast RTs will be observed if the left light is assigned to the left key and the right light to the right key. An incompatible right-left or left-right mapping will slow the response considerably. Spatial correspondence of stimuli should be matched by analogous spatial correspondence of responses in order to obtain maximum S–R compatibility.

Systems in the real world sometimes violate simple principles of spatial compatibility. For example, the design of most military helicopters is configured in such a way that altitude information is presented to the right of the instrument display and controlled with the left hand, while airspeed information, presented to the left, is controlled with the right hand. This configuration produces considerably worse performance than a compatible design would (Hartzell, Dunbar, Beveridge, & Cortilla, 1982).

Fitts and Seeger (1953) found that the advantages of S–R compatibility mapping also hold for more complex spatial arrays than the simple left-right mapping described above. They evaluated RT performance when each of the three patterns of light stimuli shown on the left in Figure 9.6 was assigned to one of the three response mappings (moving a lever) indicated across the top. Fitts and Seeger found that the best performance for each stimulus array was obtained when it was assigned to the spatially compatible response array: S_a to R_a, S_b to R_b, and S_c to R_c. This advantage is indicated by both faster responses and greater accuracy.

Fitts and Seeger also examined the interaction of compatibility with practice. They found that practice will benefit incompatible mappings more than compatible ones. However, incompatible performance will never truly reach the level of performance possible with higher compatibility mappings. Furthermore, under conditions of stress, there always remains a danger that the higher compatibility response tendencies will reappear (Loveless, 1963). In the example cited above in which left and right responses were assigned to right and left lights,

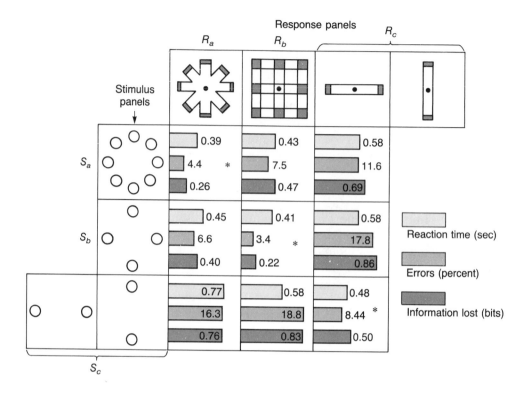

Figure 9.6   Each of the three stimulus panels on the left was assigned to one of the three response panels across the top. The natural compatibility assignments are seen down the negative diagonal and indicated by *. From "S–R Compatibility: Spatial Characteristics of Stimulus and Response Codes" by P. M. Fitts and C. M. Seeger, 1953, *Journal of Experimental Psychology, 46,* 203.

high levels of stress would induce a tendency for the operator to make the erroneous left response to the left light. A certain amount of conscious processing resources must be allocated to the response selection process in order to suppress the competing tendencies to make the dominant response to the triggering stimulus. If those resources are absent, eliminated by conditions of stress or diverted to other tasks, even a well-practiced but incompatible response assignment may lead to error.

The concept of spatial compatibility is extremely important to the system designer. An operator is often confronted by the sudden appearance of a warning light, warning tone, or movement of a meter that demands immediate action. If the action is a manual response, then compatibility relations become paramount. Thus a warning light should be served by a manual response that is close to it. If there is a spatial array of lights, then this should have its associated responses in a spatial array that is similarly oriented, preserving the same linear ordering (Alluisi, 1979). Ideally, each light should be close to the corresponding response. However, sometimes this is not possible. For example, consider an operator monitoring a fairly remote series of display indicators with a manually held keyboard. Even here, as suggested by Fitts and Seeger's results, the spatial orientation of the lights and the responses should still correspond (Hottman, 1981).

Population stereotypes. Certain compatibility relations, such as the spatial correspondence between stimulus and response and the tendency to move toward the source of stimulation, seem to be intrinsically related to the "hard wiring" of the nervous system. In contrast, others are far more determined by experience. For example, consider the relationship between the desired lighting of a room and the movement of a light switch. In North America, the compatible relation is to flip the switch up to turn the light on. In Europe, compatibility is the opposite (up is off). This difference is clearly unrelated to any difference in the physiological hardware between Americans and Europeans but rather is a function of experience. This type of preferred compatibility relation is called a "population stereotype." Smith (1981) has evaluated population stereotypes in a number of verbal-pictorial relations. For example, he asks whether the "inside lane" of a four-lane highway refers to the centermost lane on each side or to the driving lane. Smith finds that the population is equally divided on this categorization.

Population stereotypes play an important role in establishing display-control relations in instrument panels. Consider an operator who is confronted with the vertical meter shown in Figure 9.7a and wishes to adjust its position using either the knob on the right or the horizontal lever on the left. The population stereotype defines the direction of rotation or translation that the operator expects will increase the displayed quantity. In a review of compatibility and population stereotypes in display-control relations, Loveless (1963) concludes that there are a number of basic principles that govern the optimal placement and directional relations of displays and controls which will lead to the fewest inadvertent control reversals. Sometimes, as in the examples of Figure 9.7b, these relations are obvious, because all principles are in concordance. However, in other instances the panel is laid out in such a way that two principles may be in conflict. Then population stereotypes indicate which principle is the strongest and will dominate.

1. *Clockwise stereotype.* There is a basic tendency to rotate a dial clockwise for change in variables.

2. *Clockwise to increase stereotype.* There is a stereotype to rotate a dial clockwise to increase the value on a display or to move a linear display upwards or to the right (see Figure 9.7c & d).

3. *Proximity of movement stereotype.* With any rotary control the arc of the rotating element that is closest to the moving display is

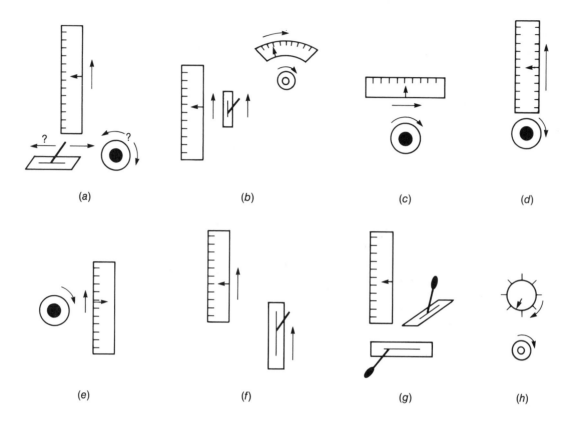

Figure 9.7 Population stereotypes in display control relations. The examples indicated by each letter are described in the text. Arrows indicate direction of change of the displayed quantity corresponding to the direction of change of the control.

assumed to move in the same direction as the display. The panel in Figure 9.7c conforms to this principle. That in Figure 9.7d is neutral, since the movement of the closest arc is at right angles to the display movement. However, the panel in Figure 9.7e violates the principle. Note here that the two arrows of motion move in opposite directions. Figure 9.7e presents an example of a conflict of principles since this relation conforms to Principle 2 (clockwise increase) but must violate Principle 3. Under such circumstances Loveless concludes that Principle 3 dominates, so that the arrangement in Figure 9.7e would be better if the displayed quantity moved downward instead of upward. Figure 9.7c of course conforms to both principles and so would be best.

4.   *Congruence.* Where possible, linear motions of control and display should be along the same axis (Figure 9.7f, not 9.7g), and rotational motions of control and display should be in the same direction (Figure 9.7b and h). However, note that in Figure 9.7h the principle of congruence (4) is pitted against that of proximity of movement (3). Here

Loveless concludes that congruence will dominate. However, the design shown in Figure 9.7b in which only the top half of the scale is read will eliminate the problems of Figure 9.7h.

5. *Congruence of location.* When there are several controls each controlling one of several displays, and each control cannot be placed close to its display, then the configuration of displays should be congruent with the configuration of controls. A square array of four displays should be associated with a square array of controls. Where this is not achieved, population stereotypes break down. The classic example of this breakdown is in the association of controls to burners on the stove (see Figure 9.8). Chapanis and Lindenbaum (1959) and Shinar and Acton (1978) report that there is a wide diversity of population stereotypes concerning which control is associated with which burner. Figure 9.8 identifies the two most compatible mappings.

Probably the most important message conveyed by the research on spatial compatibility and direction of motion stereotypes is that where possible every effort should be made to use configurations in which all principles are congruent. Where some principles are opposed, even though one may be stronger than another, the net stereotype of movement will still be weakened, and the likelihood of an inadvertent error will be increased. This might be particularly true in times of crisis or stress, when the consequences of an error are greater (Loveless, 1963).

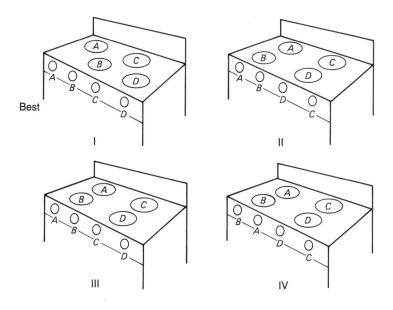

Figure 9.8 Assignment of stove control to burners. Assignment I is best, II is next, and III and IV are heavily error prone. From "A Reaction Time Study of Four Control-Display Linkages" by A. Chapanis, and L. E. Lindenbaum, 1959, *Human Factors*, 1, p. 1. Copyright 1959 by The Human Factors Society, Inc. Reproduced by permission.

Modality S–R compatibility.   Stimulus-response compatibility appears to be defined by stimulus and response modality as well as by spatial correspondence. Brainard, Irby, Fitts, and Alluisi (1962) found that if a stimulus was a light, choice RT was faster for a pointing (manual) than a voice response, but if the stimulus was an auditorily presented digit, RT was more rapid with a vocal naming response than with a manual pointing one. In a thorough review of the factors influencing choice RT, Teichner and Krebs (1974) summarized a number of studies and concluded that the four S–R combinations defined by visual and auditory input and manual and vocal response produced reaction times in the following order: A voice response to a light is slowest, a key-press response to a digit is of intermediate latency, and a manual key-press response to a light and naming of a digit are fastest.

Greenwald (1970) discusses the related concept of *ideomotor compatibility* which will occur if a stimulus matches the sensory feedback produced by the response. Under these conditions, RT will be fast and relatively automatic (Greenwald, 1979). Thus Greenwald observes fast RTs in the ideomotor compatible conditions when a written response is given to a seen letter and when a spoken response is given to a heard letter. Incompatible mappings in which a written response is made to a heard letter and a spoken response to a seen letter are slower. Ideomotor compatible mappings not only are fast, but also appear to be influenced neither by the information content of the RT task (N) nor by dual-task loading (Greenwald & Shulman, 1973). Leonard (1959) investigated reaction time in the ideomotor compatible condition when the stimuli were vibrations of keys upon which the response fingers are resting. In this compatible mapping, Leonard failed to find any effect of N on reaction time.

Wickens, Sandry, and Vidulich (1983) and Vidulich and Wickens (1982) proposed that these modality-based S–R compatibility relations may partially depend upon the central processing code (verbal-spatial) used in the task. In both the laboratory environment and in an aircraft simulator, they demonstrated that tasks which use verbal working memory are served best by auditory inputs and vocal outputs, while spatial tasks are better served by visual inputs and manual outputs. In the aircraft simulation, Wickens et al. found that these compatibility effects were enhanced when the flight task became more difficult.

The fact that certain stimulus and response modality combinations are more compatible with certain tasks than others provides useful guidelines to system designers who wish to employ speech recognition and synthesis technology in the optimum manner.

Practice

The amount of practice given to a choice RT task is closely related to S–R compatibility and is clearly the major factor defining population stereotypes. Consistent results suggest that practice, a noninformational variable, decreases the slope of the Hick-Hyman law function relating RT to information. In fact, compatibility and practice appear to trade off reciprocally in their effect on this slope. This tradeoff is nicely illustrated by comparing three studies. As discussed above, Leonard (1959) found that no practice was needed to obtain a flat slope with the highly compatible mapping of finger presses to tactile stimulation. Davis, Moray, and Treisman (1961) required a few hundred trials to obtain a flat slope with the slightly lower compatibility task of naming a heard word. Finally, Mowbray and Rhoades (1959) examined a mapping of slightly lower (but still high) compatibility. Subjects depressed keys adjacent to lights. For these unusually stoic subjects, 42,000 trials were required to produce a flat slope.

As noted above, compatibility and practice also interact within an experiment. Extensive practice may increase the speed of response to incompatible S–R assignments more than compatible ones. However, Fitts and Seeger's (1953) results demonstrated that mappings that violate spatial compatibility relations will approach but never catch up to the RT relations shown by compatible mappings.

IS INFORMATION THEORY STILL VIABLE?

Five important factors have been identified that influence RT and yet cannot be accounted for in informational terms. Teichner and Krebs (1974) assert that two of these variables, practice and S–R compatibility, along with the informational variable N, are the most potent influences on choice RT. In spite of this fact, the information metric still appears to be an important and useful concept in the reaction time paradigm. For example, investigators generally find that within a constant level of any practice, repetition, discriminability, and compatibility variable, the linearity of the Hick-Hyman law still seems to hold. Correspondingly, the effects of these five variables seems generally to be exerted on the slope of the constant function determined by RT and H_t (Fitts & Posner, 1967; Teichner & Krebs, 1974).

It is reasonable to assume that high-compatibility mappings and/or extensive practice merely allows processing to bypass the time-consuming response-selection stages (Fitts & Posner, 1967), so that the stimulus automatically activates its associated response. Furthermore, when S–R compatibility is low, then a stimulus will automatically activate an incompatible response. Then a time-consuming process of suppressing the competing response tendency is required. As described in Chapter 7, reaction time is delayed.

It seems that as long as the literal assumption is abandoned that the inverse slope of the Hick-Hyman law function reflects the bandwidth of human performance (this leads to the indefensible position that the human has infinite bandwidth when the slope is zero), then the information concept still has a lot to offer. Information theory will clearly not account for all effects on choice reaction time, but there is no reason to expect that it should. It does do a good job of describing some.

Transition

This chapter has discussed the variables that influence the speed of choosing a given response to a single stimulus. Our treatment of this topic may be expanded in one of two directions: by "zooming in" closer for a more detailed look at the choice process or by drawing back to view the choice as only one in a continuous series of actions. The chapter supplement will consider the closer look at the reaction time process. In particular, we shall consider models that describe processing operations between stimulus and response and how these operations predict the effects of variables influencing reaction time. We will then consider different approaches that have been suggested to defining the stages in reaction time. Then, in Chapter 10, we will consider the speed of action in continuous or serial performance tasks such as typing. Many of the same variables still influence processing speed, but a host of new ones are added. Chapter 10 will also consider the nature of errors in action selection, their cause, and their treatment in performance analysis.

Theories and Models
of the Reaction Time Process

MODELS OF CHOICE REACTION TIME

Considerable effort has been devoted to modeling the mechanisms in the human processing system that explain the several variables that influence reaction time. Such models are of potential value because they can predict the speed of performance in new situations and because they can parsimoniously integrate a number of findings. The Hick-Hyman law is not a model. It asserts a relation between RT and transmitted information but does not propose the underlying mechanisms or processes by which this relation comes about. The four models that will be discussed briefly below represent efforts to account for the information theory effects in terms of more specific, hypothesized processing mechanisms. More detailed treatments of these models may be found in Smith (1968), Pachella (1974), and Welford (1976).

The Serial Dichotomization Model

The serial dichotomization model proposed by Welford (1976, 1968) is intended primarily to account for the linear relation obtained between RT and $\log_2 N$. In identifying the stimulus, the brain is assumed to engage in a kind of high-speed 20-questions game, making a series of yes-no decisions (each of constant time) that successively narrow down the alternatives until the stimulus is identified. The model is thus a direct application of information theory as described in Chapter 2 and so will readily account for the linear aspects of the Hick-Hyman law. The problem is that this scheme seems plausible only when the stimulus set can be easily dichotomized. For example, if the four possible stimuli were either green or red and either a square or a circle, then a logical processing sequence might indeed be two consecutive decisions, one based on color, the next on shape. Similarly, if stimuli are lights in an array, a "split halves" technique could be employed to identify the stimulus: Is it on the left or right half of the array? Is it on the left or right side of the chosen half, etc., etc.? On the other hand, in many situations there is no simple feature that can neatly dichotomize the set of stimuli into groups. This would be the case, for example, if the stimuli were letters. Therefore, the model appears to provide a plausible description of processing only for certain kinds of stimuli.

The Repetition-Effect Model

As described earlier in this chapter, stimuli and responses that are repeated from one trial to the next yield faster reaction times than reactions to a stimulus that is different from the preceding one. Kornblum (1969, 1973) has argued that this "repetition effect" can account for all of the main effects of informational variables on RT, as manipulated in Hyman's (1953) original investigation. The reason is that the manipulation of informational variables is usually confounded with the frequency of repetition. Bearing in mind that the average RT over a block of trials is equal to the average of reaction time to a repetition and reaction time to a change or alteration weighted by their relative frequency of occurrence, then any manipulation that will increase the relative proportion of repetition trials within a series will decrease the average RT. The size of this decrease will be equal to the difference in the proportion of repetitions and alternations, multiplied by the magnitude of the repetition effect, or RE.

To provide explicit examples of how information can be confounded with P(rep), consider the three manipulations used by Hyman:

1.   An increase in N will decrease the probability of a repetition. For example, as N is increased from 2 to 4, P(rep) will decrease from 0.50 to 0.25. Thus reaction time will be lengthened by $0.25 \times$ (RE).

2.   A shift in stimulus probability, for example, in a two-choice task from [P(A) = P(B) = 0.50] to [P(A) = 0.8, P(B) = 0.2] would lead to a change from [P(rep) = P(alt) = 0.50] to [P(rep) = 0.68 and P(alt) = 0.32]. This is because the two kinds of repetitions will be an AA (P = $0.8 \times 0.8 = 0.64$) and a BB (P = $0.2 \times 0.2 = 0.04$). Thus P(rep) = 0.64 + 0.04 = 0.68. Reaction time will therefore be shortened by an amount equal to (0.68 − 0.50) (RE).

3.   The change in sequential constraints that Hyman used directly increases the frequency of repetitions, thereby shortening RT.

Thus, as Kornblum (1969) argues, all three manipulations of "information" used by Hyman inadvertently confounded this variable with the probability of a repetition in such a way that the latter variable could account for the results. Kornblum generated sequences of stimuli that differed in their information content but had equivalent values of P(rep). He found that reaction time for the different sequences was equivalent, despite the differing value of stimulus information.

Probabilistic Models of Choice Reaction Time

Probabilistic models, of which the random walk model (Fitts, 1966; Stone, 1960) and the accumulator model (Vickers, 1970) are prototypical, share many features with signal detection theory as described in Chapter 2, as well as some aspects of Bayesian decision making, discussed in Chapter 3. The two models have more features in common than features that differ between them. As an analogy to both models, consider the task of identifying whether an approaching person is a friend or not on the basis of his clothes, features, gait, and so on. The person is initially a long distance off but is progressively moving closer. As time passes, more and more evidence about the uncertain stimulus is gathered until it can be unambiguously assigned to one of the two categories. At this point a decision is made. More formally, the accumulator and random walk models share three primary common features:

1. The operator is assumed to sample relatively noisy evidence concerning the nature of the stimulus over time, evaluating this evidence as to which of the potential alternatives is most likely.

2. Following classical principles of statistical inference, the reliability of this information grows as more samples (and more time) are taken. One might envision as an analog a polaroid picture slowly developing with increased clarity over time. Correspondingly, in the example above, as the person moves closer we will have more information about his identity.

3. Because of the stress for speed, the subject imposes a response criterion, and when the amount of evidence for one stimulus or the other exceeds that criterion, the stimulus for which the evidence is greatest is selected, and the appropriate response is activated. Yet since evidence is not totally reliable, if the speed stress is too great, the wrong response may be chosen.

The accumulator, or logogen, model of reaction time (Morton, 1969; Vickers, 1970; Welford, 1976) assumes that each alternative stimulus within the possible set is associated with a separate representation in memory. Although psychologists and physiologists do not know the physical form that such a representation takes, we might for convenience assume that it is a set of neurons that communicate with each other. This representation, or *logogen* (the term used by Morton, meaning word birth), is activated when there is sensory evidence for the presence of the corresponding stimulus. It is this activation, accumulated or "counted" over time, that corresponds to the buildup of internal evidence.

As a specific example, suppose a choice RT task requires that different responses be made to the stimulus letters *A* and *K*. Each of these letters has representations in memory. When a stimulus letter *A* is presented, both representations will begin to increase in activation, as both "see" evidence for their features in the stimulus (recall the findings of stimulus confusability and feature overlap in the discussion of Neisser's visual scanning research in Chapter 4). There is a greater likelihood, however, that the *A* representation will accumulate evidence more rapidly, will exceed the criterion, and thereby will trigger the *A* response, since the representation is a direct match to the stimulus. Nevertheless, the possibility arises that because the features of the *K* or *A* stimuli are similar and because there is random activity in the perceptual system, the K representation will be driven over its criterion first, thereby resulting in an erroneous K response. This mistake will be less likely to occur if more evidence is allowed to accumulate. But more evidence of course will take more time, and so reaction time will be longer.

The major difference between the logogen and the random walk model is that in the former evidence for each alternative is accumulated independently of all others. In the random walk model, on the other hand, evidence is accumulated as an *odds* value, or difference in evidence in favor of one alternative over the other. This difference between models leads to slightly different predictions concerning the form of the reaction time distributions and the nature of errors (see Pachella, 1974). Given that the similarities between models outnumber the differences, we shall describe below only the accumulator, or logogen, model as typifying the approach of both.

The logogen model is represented schematically in Figure 9.9. In this example the noisy increase in evidence over time for each of three logogens (S, K, and A) follows presentation of the stimulus A. The criterion is indicated by a horizontal dashed line. The model accounts for the phenomena of choice reaction time in the following manner:

1. ***Probabilities.*** The subject adjusts the criteria downward for more probable alternatives (or allows the logogens of those alternatives to

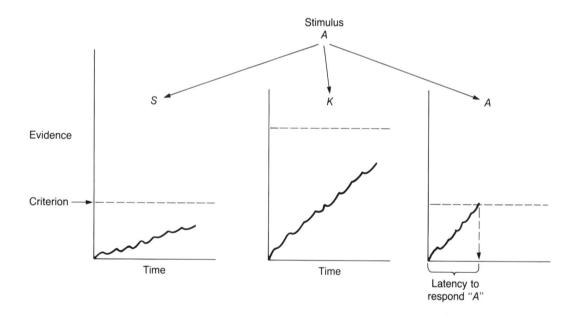

Figure 9.9   The accumulation of evidence of three hypothetical logogens on the basis of presentation of the stimulus A. Because K is less frequent than A or S, its criterion is adjusted higher. Because K shares overlapping features with A, it will show a greater accumulation of evidence than will S, which shares none.

have some "head start" or prior activation, which is equivalent). This means that evidence will accumulate to the criterion more rapidly and RT will be shorter for the more probable alternative. In Figure 9.9 the stimuli S and A are assumed to be more probable than K. Hence their criteria are lowered. Such an adjustment also implies that the more probable response will be more likely to appear as an error since it is more probable that random noise will exceed the criterion. Both of these effects are typically obtained (Broadbent, 1971).

2.   *Discriminability.* If two stimuli (and their logogens) share common features, then presentation of either stimulus will produce considerable activation of both logogens. The letters A and K share many features, as shown by the similarity of evidence-accumulation curves between the K and A logogens in Figure 9.9. In order to guard against confusion with reduced discriminability, the criteria for all logogens must therefore be raised, thereby lengthening the time for stimulus evidence to cross the criterion. Reaction time is increased for stimuli with low discriminability.

3.   *N*. As more stimulus alternatives become potentially relevant for the task at hand, the average discriminability between stimuli will be likely to decrease. To convince yourself of this, imagine that stimuli are points around the diameter of a circle of fixed size. As more stimuli are added, the mean distance between points must decrease. When discrimination is low, the difference in the rate of evidence accumulation between the logogen whose stimulus was presented and those whose stimuli were not will be reduced. Hence, RT will be raised by a cautious adjustment of the criterion, as in paragraph 2 above.

4.   *Linearity with $log_2 N$.* The amount that the criteria must be raised to keep errors constant is assumed to decrease with each added alternative because discriminability is reduced by less each time. Since there is a linear growth in evidence over time, a decreasing increment in the criterion leads to a decreasing increment in RT, thereby producing a logarithmic function.

5.   *Repetition effect.* This results from a carry-over or perseveration of activation of a logogen from the previous response. As a result, less evidence must be accumulated for a repeated stimulus. Because this perseveration dies out with the passage of time, the repetition effect itself will be smaller as the interval between stimuli is larger (Kornblum, 1973).

6.   *Stimulus-response compatibility.* The model assumes that with highly compatible S–R rules, the identification of the stimulus automatically triggers the response. When the compatibility is low, so that an incorrect response is activated, a time-consuming response suppression phase must be undertaken, which slows down the overall reaction.

7.   *The speed-accuracy tradeoff.* The speed-accuracy tradeoff results from variation in the criterion setting. Errors result when the logogen of an incorrect response will also build up activity and, because of random contributions, will actually exceed the criterion before the logogen of the correct response. This event will be more likely to occur as

alternatives are more similar to each other. It will also be more likely to occur as the criteria are lowered, requiring less evidence for a response, so that RTs will be faster. Conscious adjustment of the criterion upward and downward in this manner to produce different sets generates what is known as the speed-accuracy macrotradeoff. There is however a different form of the speed-accuracy tradeoff known as the micro-tradeoff.

8. *The speed-accuracy microtradeoff.* In the speed-accuracy mac-rotradeoff, the subject may adopt different "sets" between blocks of trials to be either slow and accurate or fast and sloppy. The speed-accuracy *microtradeoff* (Pachella, 1974; Wickelgren, 1977), on the other hand, describes the covariation of latency and accuracy *within* a block of trials of constant set. Are fast trials more likely to be correct or to be error prone? Or, phrased differently, are error trials likely to be fast or slow? The answer to this question depends upon which particu-lar characteristics of the human processing system vary from trial to trial in order to cause errors on some trials and correct responses on others. If the response criterion varies, then on trials when the criterion is lax, responses will be fast, but errors will be likely; on trials when the criterion is stringent, the responses will likely be correct but long. Therefore, trial-to-trial criterion variability induces the same negative correlation between speed and accuracy that is typical of the mac-rotradeoff. Errors will tend to be faster than correct responses. This pattern is shown in Figure 9.10*a*, in which three levels of the criterion are indicated, and the proportions of correct responses in slow, me-dium, and fast reaction times are shown in the speed-accuracy operat-ing characteristic.

On the other hand, if there is considerable variability in the quality of the signal (either the external stimulus or the internal representation of the stimulus), then on those trials in which signal quality is low, a longer time will probably be required for the appropriate logogen to reach its criterion level of activation; but because of the low stimulus quality, the response will also be more likely to be in error, since other logogens will also build up activation. When stimulus quality is high, on the other hand, responses will be fast and accurate. This relation is shown in Figure 9.10*b*, in which there is variation in stimulus quality, affecting the rate of accumulation of exclusive evidence for the correct alternative.

Evaluating results across a wide number of different experimental paradigms, negative corelations between latency and accuracy in the microtradeoff (Figure 9.10*b*) are more often observed in situations when the overall quality of the stimulus is low. Mean reaction times are therefore generally long and error rate is high (Vickers, 1970; Welford, 1976). Such situations, for example, would typify a signal detection theory experiment when some speed stress was imposed on the re-sponse (Emrich, Gray, Watson, & Tanis, 1972). Also Wickelgren (1977) reports similar findings in RT measures of recognition memory. On the other hand, more conventional reaction time paradigms with high

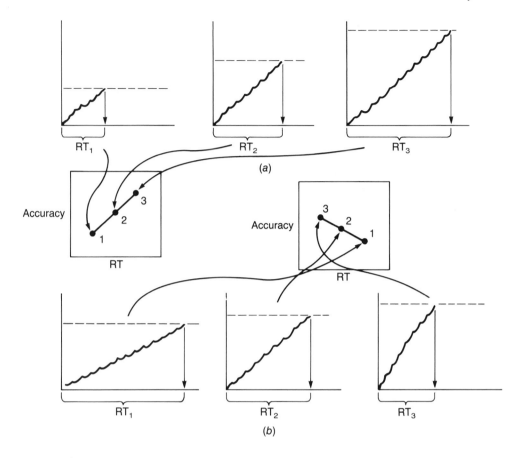

Figure 9.10 (a) Speed-accuracy microtradeoff resulting from criterion variability. (b) Speed-accuracy microtradeoff resulting from variability of stimulus evidence quality.

stimulus quality are more likely to generate the microtradeoff of Figure 9.10a similar to the macrotradeoff (Lappin & Disch, 1972). Schouten and Becker (1967) in fact found that the microtradeoff was exactly the same as the macrotradeoff. Since the latter clearly is the result of criterion variability, it may be assumed that the former is also.

The Fast Guess Model

Both the logogen model and the random walk model attribute operator strategies to the voluntary adjustment of the response criterion. By allowing the operator to control the amount of evidence acquired before a response is to be made, the speed-accuracy macrotradeoff is produced. A slightly different role of operator strategy is proposed in the *fast guess model* (Yellott, 1971). Here, under conditions of high-speed stress, it is assumed that on some portion of the trials the subject will not wait to identify the stimulus but will instead execute a random guess immediately upon stimulus detection. (With unequally probable stimuli, the guess will be that of the most probable response.) Since the response is a random guess on these trials, the accuracy will be low (at chance), but the latency will be quite short (since stimulus identification is bypassed). Therefore, the microtradeoff will be like the macrotradeoff, showing a negative speed-accuracy correlation. The macrotradeoff between these variables is accomplished in the fast guess model simply by increasing or decreasing the proportion of trials on which this fast guess strategy is employed, according to whether the set is emphasized for speed or accuracy, respectively.

STAGES IN REACTION TIME

A number of variables have been discussed that influence or prolong reaction time. The model of information processing presented in Chapter 1 also assumes that total reaction time equals the sum of the duration of a number of component processing stages (e.g., perceptual encoding, response selection, etc.). In a series of experiments and theoretical papers, psychologists have attempted to establish where these variables have their effects, and, in fact, if processing really does proceed by discrete stages or sequential mental operations. Pachella (1974) has contrasted two approaches that have been employed to justify the existence of processing stages and to examine the influences of stage duration on total RT latency. These are the subtractive method and the additive factors techniques.

The Subtractive Method

In the subtractive method an experimental manipulation is used to delete a mental operation entirely from the RT task. The decrease in RT that results is then assumed to reflect the time required to perform the absent operation. More than a century ago, Donders (1969) first used the subtractive method in RT to provide evidence for a response selection stage. He compared reaction time in which several stimuli were assigned to several responses (conventional choice RT) with reaction time in which there were several possible stimuli but only one demanded a response. This is known as the disjunctive or "go no-go" RT. The difference between these two conditions was assumed by Donders to be the time taken to select a response, since response selection does not have to be performed when only one response is required. By similar logic, when disjunctive RT was compared with the simple RT, Donders assumed that the difference reflected the latency of a stimulus discrimination stage, since only in disjunctive RT is it necessary to discriminate the stimuli. In simple RT, detection, not recognition, is sufficient to initiate the response.

The problem with the logic behind the subtractive factor method is that deleting a stage may fundamentally alter the nature of both the task and the processing at stages other than the one intended. In the above case, the disjunctive, many-stimuli-to-one response condition no longer requires that the subject discriminate *between* the various stimuli that do not require a response. These must only be discriminated from the critical stimulus. Therefore, a lower (and more rapid) level of perceptual analysis may be carried out on the critical stimulus (Pachella, 1974). For example, if the potential stimuli are the letters *A*, *K*, and *S* but only *S* is assigned to a response, then to initiate the response it is necessary only to perceive whether a letter has a curve. It is impossible to determine what extent of the RT shortening then is due to deletion of response selection and what part is due to a more rapid perceptual analysis. The subtractive technique certainly provides evidence that stages exist, but it cannot be employed to estimate their latency.

Additive Factors

An alternative to the subtractive method is the additive factors technique developed by Sternberg (1969, 1975). This technique employs converging operations in order to "home in" on the identity of stages. The procedure requires manipulating orthogonally a number of variables that prolong reaction time. In using the technique to identify processing stages, Sternberg (1969) makes three primary assumptions:

1. RT equals the sum of a series of nonoverlapping independent processing stages, each of which performs some transformation on the information. For example, perceptual encoding transforms the data from a raw sensory stimulus to an internal representation in memory.

2. The nature of the transformation (i.e., the quality of output from a stage) is independent of its own duration and also of the duration of preceding stages. If a stage has difficult processing to do—for example, when perceptual analysis is performed on a degraded stimulus—the processing may take longer, but the identity of the stimulus will be just as firmly established when passed on to the next stage as it would have been if the stimulus had been clear.

3. Two experimental manipulations that influence a common stage will produce an *interactive* effect on reaction time. That is, the effect of one manipulation on RT will be amplified at the more difficult level of the other variable. Two manipulations that influence separate stages, on the other hand, will have *additive* effects: The effect of each variable is unaffected by the level of the other variable.

An example using the additive factors method is represented in Figure 9.11. In his original demonstration of the technique, Sternberg (1969) employed the memory-search task described in Chapter 6; however, the additive factors method is really applicable to almost any

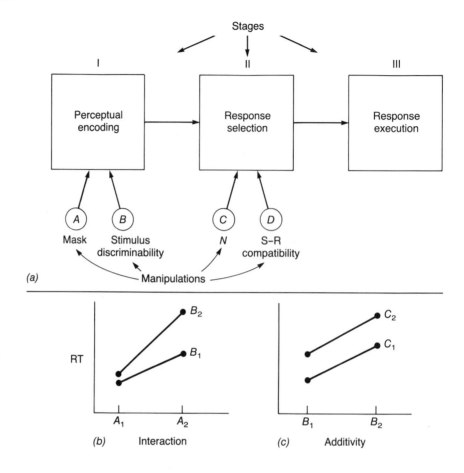

Figure 9.11 (a) Hypothetical relation between orthogonal manipulation of four variables interpreted in the framework of additive factors. (b) Variables A and B interact, effecting a common stage, as do C and D. (c) Variables B and C are additive, effecting different stages.

choice RT task. The experimenter wishes to make inferences about what manipulations influence what stages of processing, as shown in Figure 9.11a. He first identifies four experimental variables to manipulate, each identified by a letter and each run at an easy and difficult level. In our examples, these variables are the absence or presence of a mask over the stimulus (A_1 and A_2, respectively), which degrades stimulus quality; the discriminability of the stimuli in a set (B_1: high; B_2: low); the number of S–R pairs N (C_1: small; C_2: large); and stimulus-response compatibility (D_1: high; D_2: low). He then measures reaction time as each pair of variables is manipulated orthogonally. For example, RT is measured when stimuli of high and low discriminability are presented with and without a mask. The four reaction time measures in this example are shown in Figure 9.11b, indicating an interaction. The effect of the mask (A_1 versus A_2) is more pronounced when discriminability is low (B_2) than when it is high (B_1). Hence the experimenter concludes that discriminability and stimulus quality must influence the same perceptual processing stage, as in Figure 9.11a.

Next, the experimenter manipulates discriminability and set size together, by presenting stimuli of high and low discriminability in, for example, a two- and four-choice RT task. Here the results are shown in Figure 9.11c. The two variables are additive. The set-size effect is uninfluenced by the ease of discriminating the stimuli. The experimenter concludes that the two variables influence different processing stages. Since, therefore, N cannot affect perceptual encoding, it is likely instead to influence response selection. This finding would be confirmed if the experimenter manipulated N and S–R compatibility together. These variables would be found to interact, although this graph is not shown in the figure.

As experimenters have performed a large number of these orthogonal manipulations of RT difficulty, the additive factors data have provided a fairly consistent picture of processing stages. Certain pairs of variables consistently interact, and others are consistently additive. The experimenters use a certain amount of intuition to infer the stage affected by a cluster of interacting variables. For example, it is clear intuitively that S–R compatibility must influence response selection and that stimulus quality must influence perceptual processing. These "anchors" help interpret the locus of effect of other factors.

Figure 9.12 presents the pattern of additivity and interactions that have been collectively aggregated from a number of reaction time investigations. Four processing stages are portrayed across the bottom. Experimental manipulations are circled, and additive or interactive relations obtained between these manipulations are indicated by narrow or thick lines, respectively. Each line is coded by the investigation in which the manipulation was performed, and the codes are identified by the table below. The dashed arrows point to the stages inferred to be affected by the manipulation in question.

Generally, the relations that are shown in Figure 9.12 are consistent. Clusters of variables, such as S–R compatibility, N, and the repetition effect are consistently found to interact with each other by a number of investigators and to be additive with variables that logically should affect other stages. There is, however, one potential inconsistency in the table that is worthy of some discussion because of the variability it has generated in experimental results; this is the effect of probability. Some investigators (e.g., Blackman, 1975; Hawkins, MacKay, Holley, Frieden, & Cohen, 1973; Rabbitt, 1967; Sanders, 1970; Spector & Lyons, 1976) find that probability and S–R compatibility interact and thereby conclude that probability affects response selection. Yet Miller and Pachella (1973) also found that probability interacted with stimulus discriminability, clearly a perceptual variable. Finally, at least one study (Hawkins & Underhill, 1971) found additivity between probability and S–R compatibility.

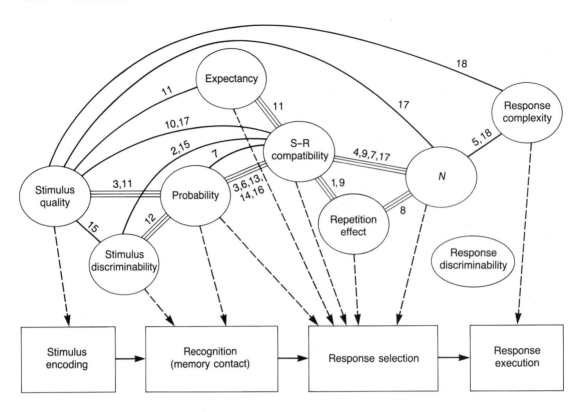

Figure 9.12 Patterns of additive factors results. $\square$ = inferred stage. $\bigcirc$ = experimental manipulation. $\equiv$ connects variables found by experimenter (number code) to interact. —— connects variables found by experimenter to be additive. --→ inference concerning the process or stage affected by the variable. Experimenter code given in table below.

1. Bertelson & Tisseyre, 1966
2. Biederman & Kaplan, 1970
3. Blackman, 1975
4. Broadbent & Gregory, 1965
5. Fitts & Peterson, 1964
6. Hawkins, MacKay, Holley, Friedin, & Cohen, 1973
7. Hawkins & Underhill, 1971
8. Hyman, 1953
9. Keele, 1969
10. McCarthy & Donchin, 1979
11. Miller & Anbar, 1981
12. Miller & Pachella, 1973
13. Rabbitt, 1967
14. Sanders, 1970
15. Schwartz, Pomerantz, & Egeth, 1977
16. Spector & Lyons, 1976
17. Sternberg, 1969
18. Whitaker, 1979
19. Wickens & Derrick, 1981

The picture is confused still further in investigations in which the probability of a stimulus and of a response are manipulated independently. This can be accomplished by assigning more than one stimulus to a given response. Then RT to two stimuli with different probabilities but assigned to the same responses can be compared. Under these circumstances some (Bertelson & Tisseyre, 1966; Hawkins, Thomas, & Drury, 1970) have found that stimulus probability is the governing factor, while others (Dillon, 1966; LaBerge & Tweedy, 1964) find that response probability is dominant.

The conclusion emerges from these investigations that probability is not a unitary effect. Hawkins et al. (1973) varied stimulus and response probability separately and found that both influence RT but only the latter interacts with compatibility. Miller and Anbar (1981) found that the component of probability that was associated with *expectancy* (manipulated through cuing, as discussed in Chapter 7) had its effect on response selection. The component related to pure stimulus frequency, on the other hand, influenced recognition.

Finally, it should be noted that the effect of stimulus probability is eliminated or greatly reduced if the various stimuli assigned to a single response class can be uniquely distinguished on the basis of a single defining feature. For example, if the two stimuli assigned to the same response are a red circle and a red square, then the role of stimulus probability will be expected to decline. Both will be responded to with equal speed even if they have different probabilities, as long as the defining feature "red" is always associated only with that response and all other stimuli are of a different color.

Applications of additive factors methodology.   While the additive factors methodology is an important theoretical method for investigating information processing, it has also served as a useful applied tool for establishing how the speed of information processing is influenced by different environmental and organistic factors such as aging (Salthouse & Sonberg, 1982), poisoning (Smith & Langolf, 1981), or mental workload as discussed in Chapter 8 (Crosby & Parkinson, 1979).

When the technique is applied for these purposes, latency in a reaction time or memory-search task is measured as the demands of different processing stages are manipulated. This is done both in the presence and absence of the factor to be investigated. This factor is then treated just like any other manipulation. If it interacts with a variable which influences a known stage, then the environment is assumed to affect the stage in question. As an example, Smith and Langolf (1981) used the Sternberg memory-search task to measure information processing of subjects who had been exposed to different amounts of mercury in industrial environments. They found an interaction between memory load and the amount of mercury poisoning in the bloodstream. Therefore, the behavioral effects of such toxins were localized at the stage of short-term memory retrieval.

Problems with Additive Factors

The additive factors technique has had a remarkable history of success in accounting for reaction time data, and it has been employed in a number of applied contexts. However, it is important to consider briefly some criticisms of the technique and at least one possible alternative methodology that has been employed to examine processing stages.

Statistical problems. Sternberg (1969) has acknowledged the difficulties that may be encountered in finding additivity. Experimental artifacts may easily produce an apparent interaction by either raising or lowering one of the points in Figure 9.11 and thereby may obscure "true" additivity. Hence, experimental results may show that there is an interaction when in fact different stages are involved.

Processes in cascade. It is probably true that assumption 2 related to stage-independence assumption (see p. 365) is not entirely valid—that is, that altered processing at one stage will not alter the quality of its output and therefore will not influence the duration of subsequent stages. In particular, recent theories of information processing presented by Eriksen and Shultz (1978) and McClelland (1979) have proposed that information is processed in a continuous flow rather than in discrete stages. In this kind of model, one stage will begin to make output available to the subsequent stage as soon as the earlier stage

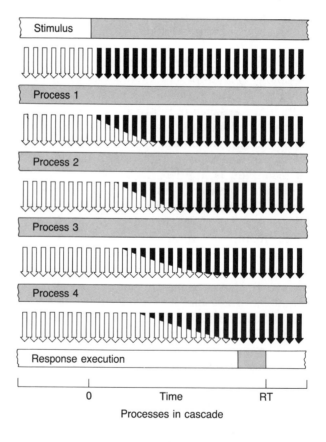

Figure 9.13 Representation of four stages or processes in cascade. From "On the Time-Relations of Mental Processes: An Examination of Processes in Cascade" by J. L. McClelland, 1979, *Psychological Review, 86*, p. 290. Copyright 1979 by the American Psychological Association. Reprinted by permission of the author.

produces any output at all. Therefore, the later stage does not wait until all processing of the earlier stages is complete before beginning its own transformations. As McClelland puts it, the processes are "in cascade." A hypothetical time graph of processing activity in each of four processes or stages in cascade taken from McClelland is shown in Figure 9.13.

A phenomenon known as *response priming* provides an explicit example of cascading processes. In a choice RT paradigm with unequal probabilities, the most probable response can be activated and begin to be initiated even before the presented stimulus has been firmly identified. This anticipation will cause some errors with the probable response (Broadbent, 1971) but will represent a sequence in which response selection begins before perceptual identification is completed. As a consequence, some of the lengthening of response selection by a manipulation that affects this stage can be "masked" or hidden by prolonging stimulus encoding (Wickens & Derrick, 1981).

A "cascade" or continuous-flow model that allows stages to operate in parallel is able to account for instances when *underadditivity* is observed, in a way that the additive factors model cannot. An underadditive relation is one in which two variables interact but the effect of each is *smaller* at the more difficult level of the other (Schwartz, Pomerantz, & Egeth, 1977). An example of such a relation is shown in the graph in Figure 9.14a. Figure 9.14b represents the duration of the two processing stages, each affected by one of two manipulations of RT

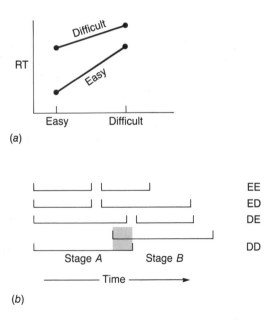

Figure 9.14 Explanation of underadditivity between two variables in terms of parallel processing. At the most difficult level of both variables, there is processing overlap, shown as shaded.

difficulty. The figure shows how the manipulations could influence the timing so as to produce the underadditive data. As long as one or the other stage is short (the level of difficulty is easy), the processing may remain serial, as shown in the top three rows. The delay imposed by each manipulation can be determined by subtracting latency at its easy level (EE) from latency at its difficult level (ED or DE). However, when both stages are prolonged at the difficult level of each variable, there is processing overlap, shown by the cross-hashings (stage B begins processing before stage A has terminated). In this case the total delay in RT will be less than that predicted by the sum of the separate delays of the two manipulations.

The Event-Related Brain Potential as an Index of Mental Chronometry

Despite some shortcomings, Sternberg's additive-factors approach provides a reasonably good approximation of the processing mechanisms involved in reaction time. In fact, the inferences concerning the influence of manipulations on stages that are made from additive and interaction data are still sound in many instances even when processes do operate in cascade (McClelland, 1979). However, no matter what theory is adopted, all reaction time investigations must make inferences about the processes between stimulus and response by looking at the final product of the response and not examining directly those intervening processes. In order to augment this mental chronometry, Donchin and his colleagues (e.g., Kutas, McCarthy, & Donchin, 1977; McCarthy & Donchin, 1979) have used the evoked brain potential to provide a direct estimate of the timing of processes up to the intermediate stage of stimulus categorization. The evoked potential was described in some detail in Chapters 7 and 8. It is a series of voltage oscillations or components that are recorded from the surface of the scalp to indicate the brain's electrical response to discrete environmental events.

An investigation by McCarthy and Donchin (1979) suggests that the late positive, or "P300," component of the evoked potential seems to covary with the duration of perceptual processing but not with the duration of response selection. Their subjects performed a reaction time task in which the displayed words "left" and "right" were responded to by pressing left and right keys. The difficulty of stimulus encoding was varied by increasing noise on the display. Response selection difficulty was manipulated by changing S–R compatibility. When RT was examined, an additive effect of both variables was observed, as would be expected by the data shown in Figure 9.12. The latency of the P300 component elicited by the stimuli was also delayed by the display mask but was *unaffected* by S–R compatibility. This finding suggested that compatibility influenced a stage *after* perceptual categorization. An interesting observation from their data was that the delay produced by the display noise was greater for RT than for P300.

This finding provides support for a cascade model, which assumes that the effects of degrading one stage (perception) can carry through to prolong processing at a subsequent stage (response selection). Longer processing at encoding will delay P300 somewhat. The final manual response will be delayed even more, because it includes slower response selection as well as slower encoding.

Conclusion: The Value of Stages

Collectively, the data from subtractive and additive factors and from the event-related potential are quite consistent with the information-processing-stage model described in Chapter 1. As we cautioned in that chapter, these data also suggest that the separation of processing stages should not be taken too literally. There undoubtedly is some overlap in time between processing in successive stages, just as the brain in general is capable of a good deal of parallel processing (see Chapters 7 and 8). However, like other models and conceptions discussed in the book, the stage concept is a useful one that is consistent with dichotomies made elsewhere between sensitivity and response bias in detection, between early and late selection in attention, and between early and late processing resources in time-sharing. The integrating value of the stage concept more than compensates for any limitations in its complete accuracy.

References

Alluisi, E. (1979). *S–R compatibility.* Paper presented to 23rd annual meeting of the Human Factors Society, Boston.

Bertelson, P. (1961). Sequential redundancy and speed in a serial two-choice responding task. *Quarterly Journal of Experimental Psychology, 12,* 90–102.

Bertelson, P. (1965). Serial choice reaction-time as a function of response versus signal-and-response repetition. *Nature, 206,* 217–218.

Bertelson, P., & Tisseyre, F. (1966). Choice reaction time as a function of stimulus versus response relative frequency of occurrence. *Nature, 212,* 1069–1070.

Biederman, I., & Kaplan, R. (1970). Stimulus discriminability and S–R compatibility: Evidence for independent effects in choice reaction time. *Journal of Experimental Psychology, 86,* 434–439.

Blackman, A. (1975). Test of the additive-factor method of choice reaction time analysis. *Perceptual Motor Skills, 41,* 607–613.

Brainard, R. W., Irby, T. S., Fitts, P. M., & Alluisi, E. (1962). Some variable influencing the rate of gain of information. *Journal of Experimental Psychology, 63,* 105–110.

Broadbent, D. E. (1971). *Decision and Stress.* London: Academic Press.

Broadbent, D. E., & Gregory, M. (1965). On the interaction of S–R compatibility with other variables affecting reaction time. *British Journal of Psychology, 56,* 61–67.

Chapanis, A., & Lindenbaum, L. E. (1959). A reaction time study of four control-display linkages. *Human Factors, 1,* 1–14.

Conrad, R., & Hull, A. J. (1967). Copying alpha and numeric codes by hand: an experimental study. *Journal of Applied Psychology, 51,* 444–448.

Cotton, W., Tzeng, O., & Hardyck, C. (1980). Role of cerebral hemispheric processing in the visual half field stimulus response compatibility effect. *Journal of Experimental Psychology: Human Perception and Performance, 6,* 13–23.

Crosby, J. V., & Parkinson, S. R. (1979). A dual task investigation of pilot's skill level. *Ergonomics, 22,* 1301–1313.

Danaher, J. W. (1980). Human error in air traffic control systems operations. *Human Factors, 22,* 535–545.

Davis, R., Moray, N., & Treisman, A. (1961). Imitation responses and the rate of gain of information. *Quarterly Journal of Experimental Psychology, 13,* 78–89.

Dillon, P. J. (1966). Stimulus versus response decisions as determinants of the relative frequency effect in disjunctive reaction time performance. *Journal of Experimental Psychology, 71,* 321–330.

Donders, F. C. (1969). On the speed of mental processes. (Translated by W. G. Koster). *Acta Psychologica, 30,* 412–431.

Drazin, D. (1961). Effects of fore-period, fore-period variability and probability of stimulus occurrence on simple reaction time. *Journal of Experimental Psychology, 62,* 43–50.

Dvorak, A. (1943). There is a better typewriter keyboard. *National Business Education Quarterly, 12,* 51–58.

Emrich, D., Gray, J. L., Watson, C. S., & Tanis, D. (1972). Response latency, confidence and ROCs in auditory signal detection. *Perception & Psychophysics, 11,* 65–72.

Eriksen, C. W., & Schultz, D. (1978). Temporal factors in visual information processing. In J. Requin (Ed.), *Attention and Performance VII.* Hillsdale, NJ: Erlbaum Associates.

Fitts, P. M. (1966). Cognitive aspects of information processing III: Set for speed versus accuracy. *Journal of Experimental Psychology, 71,* 849–857.

Fitts, P. M., & Peterson, J. R. (1964). Information capacity of discrete motor responses. *Journal of Experimental Psychology, 67,* 103–112.

Fitts, P. M., Peterson, J. R., & Wolpe, G. (1963). Cognitive aspects of information processing II: Adjustments to stimulus redundancy. *Journal of Experimental Psychology, 65,* 423–432.

Fitts, P. M., & Posner, M. A. (1967). *Human performance.* Pacific Palisades, CA: Brooks Cole.

Fitts, P. M., & Seeger, C. M. (1953). S–R compatibility: Spatial characteristics of stimulus and response codes. *Journal of Experimental Psychology, 46,* 199–210.

Fitts, P. M., & Switzer, G. (1962). Cognitive aspects of information processing I: The familiarity of S–R sets and subsets. *Journal of Experimental Psychology, 63,* 321–329.

Garner, W. (1974). *The processing of information and structure.* Hillsdale, NJ: Erlbaum Associates.

Greenwald, A. (1970). A double stimulation test of ideomotor theory with implications for selective attention. *Journal of Experimental Psychology, 84,* 392–398.

Greenwald, A. G. (1979). Time-sharing, ideomotor compatibility and automaticity. In C. Bensel (Ed.), *Proceedings, 23rd annual meeting of the Human Factors Society.* Santa Monica, CA: Human Factors.

Greenwald, H., & Shulman, H. (1973). On doing two things at once: Eliminating the psychological refractory period affect. *Journal of Experimental Psychology, 101,* 70–76.

Hartzell, E. S., Dunbar, S., Beveridge, R., & Cortilla, R. (1982, June). Helicopter pilot response latency as a function of the spatial arrangement of instruments and controls. *Proceedings of the 18th Annual Conference on Manual Control.* Dayton, OH: Wright Patterson Air Force Base.

Hawkins. H., MacKay, S., Holley, S., Friedin, B., & Cohen, S. (1973). Locus of the relative frequency effect in choice reaction time. *Journal of Experimental Psychology, 101,* 90–99.

Hawkins, H. L., Thomas, G. B., & Drury, K. B. (1970). Perception versus response bias in discrete reaction time. *Journal of Experimental Psychology, 87,* 514–517.

Hawkins, H., & Underhill, L. (1971). S–R compatibility and the relative frequency effect in choice reaction time. *Journal of Experimental Psychology, 91,* 280–286.

Hick, W. E. (1952). On the rate of gain of information. *Quarterly Journal of Experimental Psychology, 4,* 11–26.

Hottman, S. B. (1981). Selection of remotely labeled switch functions during dual task performance. In R. Sugarman (Ed.), *Proceedings, 25th annual meeting of the Human Factors Society.* Santa Monica, CA: Human Factors.

Howell, W. C., & Kreidler, D. L. (1963). Information processing under contradictory instructional sets. *Journal of Experimental Psychology, 65,* 39–46.

Howell, W. C., & Kreidler, D. L. (1964). Instructional sets and subjective criterion levels in a complex information processing task. *Journal of Experimental Psychology, 68*, 612–614.

Hyman, R. (1953). Stimulus information as a determinant of reaction time. *Journal of Experimental Psychology, 45*, 423–432.

Keele, S. W. (1969). Repetition effect: A memory dependent process. *Journal of Experimental Psychology, 80*, 243–248.

Kirby, P. H. (1976). Sequential affects in two choice reaction time: Automatic facilitation or subjective expectation: *Journal of Experimental Psychology: Human Perception and Performance, 2*, 567–577.

Kirkpatrick, M. & Mallory, K. (1981). Substitution error potential in nuclear power plant control rooms. In R. Sugarman (Ed.), *Proceedings, 25th annual meeting of the Human Factors Society*. Santa Monica, CA: Human Factors.

Klemmer, E. T. (1957). Simple reaction time as a function of time uncertainty. *Journal of Experimental Psychology, 54*, 195–200.

Kohlberg, D. L. (1971). Simple reaction time as a function of stimulus intensity in decibels of light and sound. *Journal of Experimental Psychology, 88*, 251–257.

Kornblum, S. (1969). Sequential determinants of information processing in serial and discrete choice reaction time. *Psychological Review, 76*, 113–131.

Kornblum, S. (1973). Sequential effects in choice reaction time: A tutorial review. In S. Kornblum (Ed.), *Attention and Performance IV*. New York: Academic Press.

Kutas, M., McCarthy, G., & Donchin, E. (1977). Augmenting mental chronometry: The P300 as a measure of stimulus evaluation time. *Science, 197*, 792–795.

LaBerge, D., & Tweedy, J. R. (1964). Presentation probability and choice time. *Journal of Experimental Psychology, 68*, 477–481.

Lappin, J., & Disch, K. F. (1972). The latency operating characteristic I: Effects of stimulus probability on choice reaction time. *Journal of Experimental Psychology, 92*, 419–427.

Leonard, J. A. (1959). Tactile choice reactions I. *Quarterly Journal of Experimental Psychology, 11*, 76–83.

Loveless, N. E. (1963). Direction of motion stereotypes: A review. *Ergonomics, 5*, 357–383.

McCarthy, G., & Donchin, E. (1979). Event-related potentials: Manifestation of cognitive activity. In F. Hoffmeister & C. Muller (Eds.), *Bayer Symposium VIII: Brain function in old age*. New York: Springer.

McClelland, J. L. (1979). On the time-relations of mental processes: An examination of processes in cascade. *Psychological Review, 86*, 287–330.

Merkel, J. (1885). Die zeitlichen Verhaltnisse der Willensthatigkeit. *Philosophische Studiën, 2*, 73–127.

Miller, J., & Anbar, R. (1981). Expectancy and frequency effects on perceptual and motor systems in choice reaction time. *Memory & Cognition, 9*, 631–641.

Miller, J., & Pachella, R. (1973). On the locus of the stimulus probability effect. *Journal of Experimental Psychology, 101*, 501–506.

Morton, J. (1969). Interaction of information in word recognition. *Psychological Review, 76*, 165–178.

Mowbray, G. H., & Rhoades, M. V. (1959). On the reduction of choice reaction time with practice. *Quarterly Journal of Experimental Psychology, 11*, 16–23.

Pachella, R. (1974). The use of reaction time measures in information processing research. In B. H. Kantowitz (Ed.), *Human information processing*. Hillsdale, NJ: Erlbaum Associates.

Pew, R. W. (1969). The speed-accuracy operating characteristic. *Acta Psychologica, 30*, 16–26.

Posner, M. I. (1978). *Chronometric explorations of the mind*. Hillsdale, NJ: Erlbaum Associates.

Rabbitt, P. M. A. (1967). Signal discriminability, S–R compatibility and choice reaction time. *Psychonomics Science, 7*, 419–420.

Rabbitt, P. M. A. (1981). Sequential reactions. In D. H. Holding (Ed.), *Human skills*. New York: Wiley.

Rasmussen, J. (1980). The human as a system's component. In H. T. Smith and T. R. Green (Eds.), *Human interaction with computers*. London: Academic Press.

Salthouse, T. A., & Sonberg, B. L. (1982). Isolating the age deficit of speeded performance. *Journal of Gerontology, 37*, 59–63.

Sanders, A. F. (1970). Some variables affecting the relation between relative stimulus frequency and choice reaction time. In A. F. Sanders (Ed.), *Attention and Performance III*. Amsterdam: North Holland.

Schouten, J. F., & Becker, J. A. M. (1967). Reaction time and accuracy. *Acta Psychologica, 27*, 143–153.

Schwartz, S. P., Pomerantz, S. R., & Egeth, H. E. (1977). State and process limitations in information processing. *Journal of Experimental Psychology: Human Perception and Performance, 3*, 402–422.

Seibel, R. (1972). Data entry devices and procedures. In R. G. Kinkade & H. S. Van Cott, (Eds.), *Human engineering guide to equipment design*. Washington, DC: U.S. Government Printing Office.

Shinar, D., & Acton, M. B. (1978). Control-display relationships on the four burner range: Population stereotypes versus standards. *Human Factors, 20*, 13–17.

Simon, J. R. (1969). Reaction toward the source of stimulus. *Journal of Experimental Psychology, 81*, 174–176.

Smith, E. (1968). Choice reaction time: Analysis of the major theoretical positions. *Psychological Bulletin, 69*, 77–110.

Smith, P., & Langolf, G. D. (1981). The use of Sternberg's memory-scanning paradigm in assessing effects of chemical exposure. *Human Factors, 23*, 701–708.

Smith, S. (1981). Exploring compatibility with words and pictures. *Human Factors, 23*, 305–316.

Spector, A., & Lyons, R. (1976). The locus of the stimulus probability effect in choice RT. *Bulletin of the Psychonomics Society, 7*, 519–521.

Sternberg, S. (1969). The discovery of processing stages: Extensions of Donders' method. *Acta Psychologica, 30*, 276–315.

Sternberg, S. (1975). Memory scanning: New findings and current controversies. *Quarterly Journal of Experimental Psychology, 27*, 1–32.

Sternberg, S., Kroll, R. L., & Wright, C. E. (1978). Experiments on temporal aspects of keyboard entry. In J. P. Duncanson (Ed.), *Getting it together: Research and application in human factors*. Santa Monica, CA: Human Factors.

Stone, M. (1960). Models for choice reaction time. *Psychometrika, 25*, 251–260.

Teichner, W., & Krebs, M. (1972). The laws of simple visual reaction time. *Psychological Review, 79*, 344–358.

Teichner, W., & Krebs, M. (1974). Laws of visual choice reaction time. *Psychological Review, 81*, 75–98.

Tversky, A. (1977). Features of similarity. *Psychological Review, 84*, 327–352.

Vickers, D. (1970). Evidence for an accumulator model of psychophysical discrimination. *Ergonomics, 13*, 37–58.

Vidulich, M., & Wickens, C. D. (1982). The influence of S–C–R compatability and resource competition on performance of threat evaluation and fault diagnosis. In R. E. Edwards (Ed.), *Proceedings, 26th annual meeting of the Human Factors Society*. Santa Monica, CA: Human Factors.

Warrick, M. S., Kibler, A., Topmiller, D. H., & Bates, C. (1964). Response time to unexpected stimuli. *American Psychologist, 19*, 528.

Welford, A. T. (1968). *Fundamentals of skill*. London: Metheun.

Welford, A. T. (1976). *Skilled performance: Perceptual and motor skills*. Glenview, IL: Scott, Foresman.

Whitaker, L. A. (1979). Dual-task interference as a function of cognitive processing load. *Acta Psychologica, 43*, 71–84.

Wickelgren, W. (1977). Speed accuracy tradeoff and information processing dynamics. *Acta Psychologica, 41*, 67–85.

Wickens, C. D., & Derrick, W. (1981, March). *The processing demands of higher order manual control: Application of additive factors methodology*. (Technical Report EPL-80-1/ONR-80-1). Champaign, IL: University of Illinois, Engineering Psychology Research Laboratory.

Wickens, C. D., Sandry, D., & Vidulich, M. (1983). Compatibility and resource competition between modalities of input, central processing, and output: Testing a model of complex task performance. *Human Factors, 25*, 227–248.

Woodworth, R. S., & Schlossberg, H. (1965). *Experimental psychology*. New York: Holt, Rinehart, & Winston.

Yellott, J. T. (1971). Correction for fast guessing and the speed accuracy trade-off. *Journal of Mathematical Psychology, 8*, 159–199.

Serial Reaction Time, Transcriptions, and Errors

OVERVIEW

The analysis in Chapter 9 dealt primarily with the selection of a single discrete action in the reaction time task. We considered there the factors that influenced reaction time, as well as some of the models that have been formulated to account for the process. Many tasks in the real world, however, call for not just one but a series of repetitive actions. Typing and assembly-line work are two examples. The factors described in Chapter 9 that influence single reaction time are just as important in influencing the speed of repetitive performance. However, the fact that several stimuli must be processed in sequence brings into play a set of additional influences that relate to the timing and pacing of sequential stimuli and responses. These timing variables also interact in some important respects with the variables discussed in Chapter 9.

This chapter will deal with serial or repeated responses. Initially we shall focus on the simplest case: only two stimuli presented in rapid succession. This is the paradigm of the psychological refractory period, and its discussion, while somewhat theoretical, integrates characteristics of both Chapters 8 and 9. Next we will examine response times to several stimuli in rapid succession, the serial reaction time paradigm. This discussion will lead us to an analysis of transcription skills such as typing. Finally, we will conclude our overall treatment of action, both single and serial, by considering the nature of errors in action.

THE PSYCHOLOGICAL REFRACTORY PERIOD

The psychological refractory period, or PRP, paradigm (Telford, 1931) is one in which two reaction time tasks are presented close together in time and compete for common processing resources. The temporal separation between the two stimuli is called the *interstimulus interval*, or ISI. In this paradigm, even when the stimuli associated with the two tasks (S_1 for RT_1 and S_2 for RT_2) are in different modalities and the responses are different, the reaction time to the second stimulus (RT_2) is consistently delayed from a single-task control condition. Normally, RT_1 is unaffected by the presence of RT_2. Suppose, for example, a subject is to press a key (R_1) as soon as a tone (S_1) is heard and speak

(R_2) as soon as a light (S_2) is seen. If the light is presented a fifth of a second or so after the tone, the subject will be slowed in responding to the light (RT_2) because of the response to the tone. However, reaction time to the tone (RT_1) will be unaffected.

In his original investigation of the paradigm, Telford (1931) proposed a sort of refractoriness in the brain's information-processing mechanisms to account for this slowing, similar to the refractoriness in neural firing. When a single neuron is stimulated twice in succession at very short intervals, the response to the second stimulus will be delayed. The analogous effect in the "psychological" domain of reaction time led to the name "psychological refractory period." Since Telford's investigation, two theories have been proposed to account for the specific mechanisms underlying the RT_2 delay (Smith, 1967).

Expectancy Theory

Elithorn and Lawrence (1955) proposed that the subject simply does not *expect* S_2 to occur so soon after S_1. Therefore, following the principles outlined in Chapter 9, the subject will be slow in responding. The stimulus S_1 in this case serves a kind of warning signal for S_2. If lowered expectancy serves as the basis for the RT_2 delay, then within a block of reaction time trials in which the ISI remains *constant*, the subject should eventually build up maximum expectancy for S_2 at the time it occurs. Therefore, RT_2 should be fast when the ISI is constant. Creamer (1963) performed such an experiment and found that the PRP is still observed (RT_2 is still delayed) even when ISI is held constant within a block of trials. These results suggest that there must be another mechanism besides lowered expectancy that is operating to produce the RT_2 delay. This second mechanism appears to be related to the human's inability to select responses to two stimuli in parallel. It is formally explained by the *single-channel theory*. Some evidence as to the nature of this alternative mechanism was provided by Reynolds (1966), who observed that the nature of the processing of S_1 influences the magnitude of the RT_2 delay. Thus, RT_2 will be more delayed if RT_1 is choice reaction time than if it is simple reaction time. This effect would not be predicted if the function of S_1 is simply to provide a warning signal for S_2. The finding, however, is quite compatible with the single-channel theory, which we shall now consider.

Single-Channel Theory

The single-channel theory of the PRP was originally proposed by Craik (1947) and has subsequently been expressed and elaborated by Davis (1965), Bertelson (1966), and Welford (1967). It is compatible with Broadbent's (1958) conception of attention as an information-processing bottleneck that can only process one stimulus or piece of information at a time (see Chapter 7). In explaining the PRP effect, single-channel theory assumes that the processing of S_1 temporarily "captures" the single-channel bottleneck of the decision-making/response-selection stage. Thus until R_1 has been released (the single channel has finished processing S_1), the processor cannot begin to deal with S_2. The stimulus S_2 must therefore wait at the "gates" of this single-channel bottleneck until they open. This "waiting time" is what prolongs RT_2. According to this view, anything that prolongs the processing of S_1 will increase the PRP delay of RT_2. Reynolds' (1966) experiment confirmed this prediction. The PRP delay was lengthened if reaction time involved a choice rather than a simple response.

This bottleneck in the sequence of information processing activities does not appear to be located at the peripheral sensory end of the processing sequence (like blinders over the eyes that are not removed until R_1 has occurred). If this were the case, then no processing of S_2 whatsoever could begin until RT_1 is complete. However, as described in Chapter 7, much of perception is relatively "automatic." Therefore the basic *perceptual* analysis of S_2 can proceed even as the processor is fully occupied with selecting the response to S_1 (Karlin & Kestinbaum, 1968; Keele, 1973). Only after its perceptual processing is completed does S_2 have to wait for the bottleneck to dispense with R_1. These relations are shown in Figure 10.1.

Imagine as an analogy a kindergarten teacher (the bottleneck) who must get two children ready for recess (S_1 and S_2). Both are able to put on their coats by themselves (this is the automatic "early" processing that does not require the teacher—attention—to function), but both need the teacher to zip or button the coats (select and execute a response). Therefore, how long child$_2$ will have to wait is a joint function of how soon she arrives after child$_1$ (the ISI) and how long it takes the

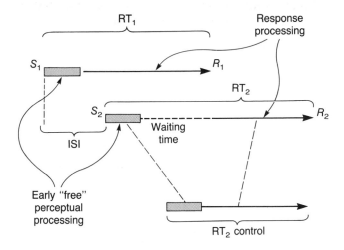

Figure 10.1   Single-channel theory explanation of the psychological refractory period. The figure shows the delay (waiting time) imposed on RT_2 by the presence of RT_1.

teacher to button child$_1$. But this waiting time will not include the time it takes child$_2$ to put on her coat, since this can be done in parallel with the buttoning of child$_1$. Only once child$_2$'s coat is on, must she wait. Therefore, the total time required for child$_2$ to get the coat on (analogous to RT$_2$) is equal to the time it normally takes to put the coat on and have the teacher button it, plus the waiting time. The latter may be predicted by the ISI.

Returning to the PRP paradigm, the *delay* in RT$_2$, beyond its single-task base line, will increase linearly (on a one-to-one basis) with a decrease in ISI and an increase in the complexity of response selection of RT$_1$, since both of these increase the waiting time. This relation is depicted in Figure 10.1. Assuming that the single-channel bottleneck is perfect (that is, postperceptual processing of S$_2$ will not start at all until R$_1$ is released), then the relation between ISI and RT$_2$ will look like that shown in Figure 10.2. When ISI is long (much greater than RT$_1$), then RT$_2$ is not delayed at all. When ISI is shortened to about the length of RT$_1$ (actually a little less, because of the "free perceptual processing" of S$_2$), some temporal overlap will occur and RT$_2$ will be prolonged because of a waiting period. This waiting time will then increase linearly as ISI is shortened further.

The relation between ISI and RT$_2$ shown in Figure 10.2 is not always entirely accurate (Kahneman, 1973), but generally does describe rather successfully a large amount of the PRP data (Bertelson, 1966; Kantowitz, 1974; Smith, 1967). There are, however, three important qualifications to the general single-channel model as it has been presented so far.

1. When the ISI is sufficiently short (less than about 100 msec), a qualitatively different processing sequence occurs; both responses are emitted *together* (grouping) and *both* are delayed (Kantowitz, 1974). It is as if the two stimuli are occurring so close together in time that S$_2$

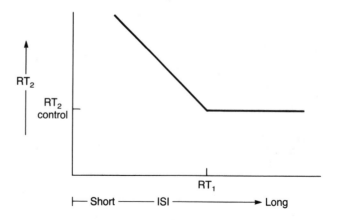

Figure 10.2 Relation between ISI and RT$_2$ predicted by single-channel theory.

"gets through" the gate while it is still accepting S_1 (Welford, 1952; Kantowitz, 1974).

2. Sometimes RT_2 suffers a PRP delay even when the ISI is greater than RT_1. This delay occurs when the subject is monitoring the feedback from the response of RT_1 as it is executed (Welford, 1967).

3. The 45 degree angle of the function determined by RT_2 and ISI is, of course, predicted by the strict serial processing view of the bottleneck which assumes that only one response can be selected at a time. Kahneman (1973) asserts that this relation may be somewhat less than 45° and argues on this basis that some processing associated with the second response occurs in parallel with the selection of R_1. This conclusion is, of course, more compatible with the capacity or resource conception of attention as described in Chapter 8, since the concept of resources is operationally defined as a sharable commodity. The issue of whether there is strict seriality or some degree of parallel processing, while of interest from a theoretical perspective, is probably not very important in application. What is important is the general finding that there is delay in processing successive signals, and this delay will occur even when there is no temporal overlap (that is, ISI > RT_1).

SERIAL REACTION TIME

In the real world, operators are more likely to encounter a series of stimuli that must be rapidly processed than a simple pair of such stimuli. In the laboratory the former situation is realized in the serial reaction time paradigm. Here a series of RT trials occur sufficiently close to each other in time so that each RT is affected by the processing of the temporally adjacent stimuli in the manner described by the single-channel theory. A large number of factors influence performance in this paradigm, typical of tasks ranging from quality control inspection to typewriting to assembly-line manufacturing. Many of these variables were considered in Chapter 9. For example, S–R compatibility, stimulus discriminability, and practice influence serial RT just as they do single-trial-choice RT. However, some of these variables interact in important ways with the variables that describe the sequential timing of the successive stimuli. Other variables describe that timing itself. Six of these factors will be considered below: decision complexity, number of information sources, pacing, response complexity, preview, and feedback.

Decision Complexity

The decision complexity advantage. Chapter 9 described how the linear relation between choice reaction time and the amount of information transmitted—the Hick-Hyman law—was seen to reflect a capacity limit of the human operator. The slope of this function expressed as seconds per bit could be inverted and expressed as bits per second. Early interpretations of the Hick-Hyman law assumed that the latter figure provided an estimate of the "bandwidth" of the human processing system. As decisions become more complex, decision rate slows proportionately.

Research indicates, however, that the *latency* with which a *single* response is made in the discrete RT trial does not predict the *rate* at which a *series* of responses can be made in the serial RT paradigm. That is, the two reciprocal ways of expressing capacity limitations—bits per second and seconds per bit—are not interchangeable. According to the assumptions underlying this inversion, the information-transmission capacity is a constant, fixed rate. This assumption would imply that in a serial RT paradigm, the complexity of a decision (bits per decision) and decision rate (decisions per second) should trade off reciprocally. For example, if we could make one 6-bit decision/sec, we should also be able to make two 3-bit decisions/sec, three 2-bit decisions/sec, or six 1-bit decisions/sec.

In fact, this tradeoff does not appear to hold. The most restricting limit in human performance appears to relate to the absolute number of decisions that can be made per second rather than the number of bits that can be processed per second. Humans are better able to process information delivered in the format of one 6-bit decision per second than in the format of six 1-bit decisions per second (Alluisi, Muller, & Fitts, 1957; Broadbent, 1971). Thus the frequency of decisions and their complexity do *not* trade off reciprocally. The advantage of a few complex decisions over several simple ones may be defined as a *decision complexity advantage*. This finding suggests that there is some fundamental limit to the central-processing or decision-making rate, independent of decision complexity, that constrains the speed of other stages of processing. Such a limit might well explain why our motor output often "outruns" our decision-making competence. The "uhs" or "uhms" that we sometimes interject into rapid speech provide examples of how our motor system fills in the noninformative responses while the decision system is slowed by its limits in selecting the appropriate response (Welford, 1976).

Debecker and Desmedt (1970) performed an experiment to determine the upper limit of decision-making speed when decisions were of the simplest possible kind in which event information was still transmitted. Subjects performed a 1-bit "go-no-go" serial RT task, in which only one of two stimuli was responded to. The rate of stimulus presentation was gradually increased until the information transmission rate (bits transmitted per second) no longer increased. This critical rate was found to be at 2.5 decisions/sec, or one decision/400 msec. This value then seems to provide an upper bound on decision-making speed in serial performance.

The most general implication of the decision complexity advantage is that greater gains in transmission performance may be achieved by calling for a few complex decisions than by calling for many simple decisions. Several investigators suggest that this is a reasonable guideline. For example, Deininger, Billington, and Riesz (1966) evaluated push-button dialing. A sequence of 5, 6, 8, or 11 letters to be dialed was drawn from a vocabulary of 22, 13, 7, and 4 alternatives, respectively (22.5 bits each). The total dialing time was lowest using the shortest number of units (5 letters), each delivering the greatest information content per letter. Thus, again, frequency of decisions and complexity of decisions did not trade off reciprocally. In a somewhat different application, D. P. Miller (1981) examined the time it took operators to proceed through a 64-item hierarchical computer data menu in order to choose a single one of the 64 items. Item identification time was more rapid when they made two 5-bit selections than when they made three 4-bit or four 3-bit choices.

A property of working memory that was discussed in Chapter 6 is seemingly related to the decision complexity advantage. This was Yntema's (1963) finding that memory capacity is expanded when subjects must retain a few items with several attributes (i.e., items of greater complexity), as opposed to several items with a few attributes.

Implications for keyboard design. The decision complexity advantage has implications for the task of typewriting. For example, Seibel (1972) concluded that making text more redundant (less information per key stroke) will increase somewhat the rate at which key responses can be made (decisions per second) but will decrease the overall information transmission rate (bits per second). It follows from these data that processing efficiency could be increased by allowing each key press to convey more information than the 4.5 bits provided on the average by each letter (see Chapter 2). One possibility is to allow separate keys to indicate certain words or common sequences such as *and, ing,* or *th.* This "rapid type" technique has indeed proven to be more efficient than conventional typing, given that the operator receives a minimal level of training (Seibel, 1963). However, if there are too many of these high-information units, the keyboard itself will become overly large. In this case efficiency may decrease because the sheer size of the keyboard will increase the time it takes to locate keys and to move the fingers from one key to another (see Chapter 11).

One obvious solution to this motor limitation is to allow "chording," in which simultaneous rather than sequential key presses are required. This approach would increase the number of possible "strokes" without imposing a proportional increase in the number of keys. Thus, with only a five-finger keyboard, it is possible to produce $2^5 - 1$, or 31, possible chords without requiring any finger movement to different keys. With ten fingers resting on ten keys the possibilities are $2^{10} - 1$, or 1,023.

A number of studies indeed suggest that the greater information available per key stroke in chording provides a more efficient means of transmitting information. For example, Seibel (1963) found that increasing the number of possible chords beyond five had little effect on processing speed (chords per minute) but increased overall information transmission. Another study by Seibel (1964) examined skilled court typists using a chording "court writer," or stenotype. Although these operators responded at a third of the rate of skilled typists in terms of the number of keys/unit time, they also succeeded in transmitting twice as much information (bits per second), again consistent with the decision complexity advantage. Lockhead and Klemmer (1959) observed successful performance with a chord typewriter, while Conrad and Longman (1965), found that chording was better than conventional typing as a means of sorting letters in automatic mail-sorting consoles. In fact, most automated mail-sorting consoles are equipped with a chording scheme in which critical categories of postal information are indicated by chording different subsets of eight keys using the four fingers on each hand.

The preceding description of chording has emphasized the greater information transmission potential of the chord. A study by Gopher and Eilam (1979; Gopher, 1983) demonstrates the advantage of the chord in terms of its reduced motor demand and provides an interesting example of the use of spatial compatibility in system design. Gopher and Eilam developed a three-fingered chord writer for Hebrew characters in which two successive depressions of the three fingers "fill in" the top and bottom row of a 2 × 3 matrix. The spatial form of the filled matrix then corresponds directly to that of the Hebrew character. This patterning is shown for three characters in Figure 10.3. Because of the natural spatial compatibility between the manual response and the visual shape of the letter, their keyboard has proven to be remarkably

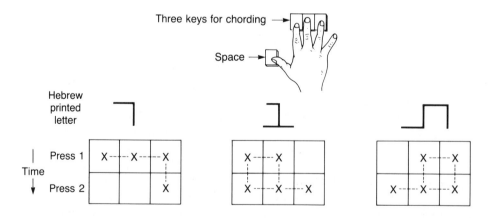

Figure 10.3   Hebrew-letter keyboard. From "Development of the Letter Shape Keyboard" by D. Gopher and Z. Eilam, 1979, in C. Bensel (Ed.), *Proceedings, 23rd Annual Meeting of the Human Factors Society* (p. 43), Santa Monica, CA: Human Factors. Copyright 1979 by the Human Factors Society, Inc. Reproduced by permission.

easy to master. A further advantage of this and all other chording keyboards is that responses do not depend on visual feedback. Since the fingers need never be removed from their "home" keys, these keyboards can be readily used without error, even as the eye is continuously fixated elsewhere.

Load Stress

Conrad (1951) used the term "load stress" to refer to the number of sources (or channels as defined in Chapter 7) along which stimuli may appear. It follows from our discussion of divided attention in Chapter 7 that a performance decrement will probably occur if the operator is required to monitor or process more than one channel at a time. If increasing the number of channels produces a concomitant increase in the number of signals to be processed per unit time, then the decrement is not surprising. An investigation by Goldstein and Dorfman (1978) determined that there is also a decrease in performance resulting from load stress when the rate of information delivery remains constant. That is, although an operator can successfully monitor or respond to one channel that provides X stimuli/min, that operator may not be able to monitor N channels each with X/N stimuli/min.

In their experiment Goldstein and Dorfman required subjects to monitor 1, 2, or 3 meters that had to be responded to with a key press if the indicators entered a danger zone. These "events" could vary in their frequency. The subjects' response reversed the indicated movement, but only after some inertial delay. This inertia required subjects to anticipate and reverse an indicator before it entered the danger zone. Goldstein and Dorfman found that the effect of *load* was prepotent over the effect of *speed* in degrading performance. That is, when the number of events/unit time was constant, performance was better on one source than several. More dramatically, subjects did a better job of processing 72 events/min on one meter than only 24 events/min distributed across three meters, even though the latter condition generated a mere 8 events/meter.

Goldstein and Dorfman's results are probably not surprising, in light of the findings of Tulga and Sheridan (1980) reported in Chapter 7. These investigators found that attention allocation becomes increasingly nonoptimal with an increase in the number of information sources. The results also explain perhaps why the frequency of visual scanning is considered to be a major contributor to operator workload (Moray et al., 1979; Simmons, 1979; see Chapter 8).

Pacing

The pacing factor, probably one of the most important influences on performance in serial RT tasks, affects the circumstances under which the operator proceeds from one stimulus to the next. Pacing can be defined in terms of two dimensions, one dichotomous and one continuous, that generally describe the degree of time constraints placed upon the human operator (Gottsdanker & Senders, 1980). The dichotomous dimension contrasts a *force-paced* with a *self-paced* stimulus rate. In the force-paced schedule each stimulus follows the preceding stimulus at a constant interval. The critical parameter is the interstimulus interval, or ISI. The frequency of stimulus presentation in the force-paced schedule is thus independent of the operator's responses. Work on an assembly line in which the items continuously move past the operator on a conveyor belt is force-paced. The speed of the belt defines the ISI. In the self-paced schedule each stimulus follows the previous *response* by a constant interval known as the *response-stimulus interval,* or RSI. In this case, the frequency with which stimuli appear depends upon the latency of the operator's response.

The continuous dimension in pacing defines the value of the timing parameters. Either self-paced or force-paced schedules may be perceived as leisurely if the RSI or ISI, respectively, is long, so that a long time passes between a response and a subsequent stimulus. If, on the other hand, the RSI is reduced to near zero or the ISI to a value near the average reaction time, then the speed stress can be quite intense indeed. In fact, the self-paced schedule with a zero RSI can seem just as "forced" as the force-paced schedule if the operator is responding rapidly. No matter how short the response, the subsequent stimulus will always be waiting.

The differences between these two schedules and the ease with which one or the other may be implemented in such systems as automatic postal sorting, assembly-line work, or industrial inspection has led investigators repeatedly to ask which is better. Welford (1968, 1976) argues on intuitive grounds in favor of the self-paced schedule. This is because as long as there is any variability in the RT response (because of fluctuations in decision complexity, stimulus quality, or operator efficiency), the self-paced schedule will allow long responses at one time to be compensated for by shorter responses at another. Welford argues that the force-paced schedule allows no such flexibility. Either the ISI must be set longer than the longest expected RT to avoid temporal overlap, in which case the number of stimuli processed will be fewer than in the self-paced rate, or the ISI can be set shorter (i.e., at the mean response rate of the self-paced condition). In this case a long RT to one stimulus will cause processing to overlap in time with the subsequent stimulus. A PRP effect will then result. This effect would not be damaging if the subsequent stimulus were permanently displayed and required a short latency response. However, if the stimulus was transient and disappeared before the process of S_1 was dispensed with, an error might result. If the processing of S_2 was also lengthy, the PRP effect would propagate to S_3, and so on.

Empirical data that argue decisively in favor of one schedule over another are not easy to obtain, as they depend critically upon the experimental choice of the RSI and ISI parameters. An appropriate comparison would require that the following steps be taken: (1) collect data from the self-paced condition with a short RSI, (2) examine the distribution of RTs in this condition, (3) choose an ISI that would curtail the longest N percentile of the self-paced RT responses, (4) collect data from the force-paced condition at this ISI, and (5) contrast error rates and response latencies between conditions, assuming that N% of the RTs in the self-paced conditions would have been responded to with chance accuracy.

When this procedure was employed by Brown (1957) and by Wagenaar and Stakenberg (1975), the latter using a four-choice visual-manual RT task, both found that there were fewer errors in the force-paced condition than would be predicted by assuming that all responses beyond the ISI cutoff would be chance guesses. In fact, Wagenaar and Stakenberg actually found superior performance (both more rapid and more accurate) in the force-paced condition. Furthermore, increases in pacing rate in this condition did not induce corresponding depressions in accuracy. The reason for this apparent improvement with force-pacing is not readily apparent. Broadbent (1971) attributes the improvement to an increase in *arousal* and a corresponding increase in efficiency that occurs in the force-paced condition (see Chapter 7), although Wagenaar and Stakenberg argue against this hypothesis.

An alternative explanation attributes the difference between schedules to the length of the RSI that is chosen in the self-paced condition. Data provided by Welford (1967) suggest that if the RSI is less than 200 msec, then monitoring feedback of the previous response prevents processing of the subsequent stimulus from beginning until the critical period has passed. In Wagenaar and Stakenberg's experiment, the RSI value was set at 100 msec, a level that would produce this sort of disruption of every self-paced stimulus. This fact would then lower the accuracy of processing on all stimuli in the self-paced schedule. For the force-paced schedule, accuracy would be influenced only on those trials following an RT with a particularly great latency.

Like those results in the laboratory, comparative results of pacing schedules in more real-world environments have also generated somewhat contradictory conclusions (see Salviendi & Smith, 1981, for a good discussion of these issues). McFarling and Heimstra (1975) compared self- and force-paced schedules of quality control inspection as subjects detected faulty wirings in circuit boards. They found that the self-paced schedule produced slower but more accurate inspection than did the force-paced schedule. Accuracy of both schedules declined with increased board complexity. The speed of the self-paced schedule (number of items inspected per unit time) declined with complexity as well, while the speed of the force-paced schedule was of course unchanged, since this was the parameter fixed by the experimenter. However, subjective ratings on a number of bipolar adjective scales generally favored the self-paced schedule. Unfortunately, the two critical intervals (ISI and RSI) were not tightly controlled or equated in this study. However, Drury and Coury (1981) concluded that in inspection tasks (see Chapter 7), because of the high variability of search time, self-paced schedules will always produce a higher expected accuracy.

In the context of automatic postal sorting, Conrad (1960) concluded that machine pacing will not produce as great speed of sorting as will a self-paced schedule. Salviendi and his co-workers have examined pacing in an industrial environment and have observed that self-paced schedules, although producing more accurate performance, also tend to induce a greater level of stress as measured by heart rate (Manenica, 1979) and non-work-related movements (Basila & Salviendi, 1979). Knight and Salviendi (1981), however, found that the increased stress is a consequence of the greater scheduling and time-keeping chores required of the self-paced subject. If these chores can be reduced by presenting better informational cues, then the heart-rate-related measure of stress also declines.

At the present time, then, the data appear to be sufficiently ambiguous that firm conclusions are difficult to draw concerning which schedule is better. One reason is that differences between schedules depend so much upon the timing parameters, the RSI and ISI, that are chosen. If these are short, for example, then with long-duration tasks, subjects in a force-paced schedule will be unable to rest, unless these too are built into the system. With long RSIs, on the other hand, rest will occur often. It seems safe to assert, however, that as the variability of processing latency increases, for whatever reasons, the relative merits of self-paced as opposed to force-paced schedules will improve. The exact level of variability at which a superiority of self- over force-paced schedules will be observed cannot be stated with confidence.

Response Complexity

In a serial RT task, increasing the time or precision constraints of the response exerts three specific effects that all serve to slow down performance. As a trivial example, consider the difference between depressing a key upon which one's finger is resting and moving the finger to a distant key before pressing. (1) Most obviously, a more complex response will probably take longer to execute. In the present example, this is the added time to move from the home key to the target key, an issue that will be discussed in more detail in Chapter 11. (2) A number of investigations have established that it takes longer to select responses of greater complexity in choice RT paradigms. For example, Klapp and Irwin (1976) have shown that the time to initiate a vocal or manual response is directly related to the duration of the response. Sternberg, Kroll, and Wright (1978) found that it takes progressively longer to initiate the response of typing a string of characters as the number of characters in the string is increased.

While the first two factors (movement time and initiation time) influence the discrete trial reaction time as well as the serial, the third factor relating to response complexity is unique to the serial RT paradigm. This concerns (3) the feedback demands of response monitoring.

In our discussion of the PRP paradigm, we saw there is good experimental evidence that delays in the second reaction time could occur even after the response to the first stimulus was completed. Welford (1976) and Davis (1965) have attributed this refractory delay to the operator's relatively automatic tendencies to monitor the execution of each response. This processing demand temporarily precludes processing a second stimulus. As the response is made more complex, this time during which the bottleneck is occupied monitoring the response of RT_1 will be raised accordingly.

A factor closely related to response complexity is that of response repetition. As the repetition effect was described in Chapter 9, we saw that a response that repeated itself was more rapid than if it followed a different response (Kornblum, 1973). As we indicated in that discussion, however, there is a trend in many serial RT skills such as typewriting for the opposite effect to occur, in which a response is slowed by repetition.

The differing effects of response repetition between the discrete trial RT paradigm described in Chapter 9 (in which repetition was good) and the typing task described here (in which it is harmful) is worthy of note. In Chapter 9 the repetition effect was considered to be a kind of "shortcut" which eliminated the repeated engagement of the response-selection stage. If a stimulus repeated, then the same response was activated as before, and time was saved. Reaction times in this case were relatively long (around 200–300 msec), however, compared to the interresponse times in typing, which may be less than half that amount. This produces a speed of 5–8 responses/sec. In light of Debecker and Desmedt's (1970) data, which showed a limit to serial response speed of 2–3 decisions/sec, the high speeds in typing indicate that the separate responses may be selected without engaging a higher-level decision process. As a consequence there is no longer any benefit to repetitions by shortcutting this process. In fact, with the shorter interresponse times, processing instead begins to impose upon the refractoriness of the motor system when repeated commands are issued to the same muscles (Fitts & Posner, 1967). In this case, it becomes advantageous to employ separate muscle groups for successive responses. Hence the advantage for alternations.

The slowing of repeated responses is greatest when a single finger is repeated. This is particularly true when the repeated finger must strike two different keys because movement time is now required. However, slowing is also evident when two different digits on the same hand are repeated (Rummelhart & Norman, 1982). For example, Sternberg, Kroll, and Wright (1978) observed that words which used key strokes of alternate hands were typed faster than those in which all letters were typed with one hand. One characteristic of the keyboard layout on the conventional Sholes, or "QWERTY," keyboard is that keys are placed in such a way that common sequences of letters will be struck with keys that are far apart and therefore, on the average, likely to be struck with different hands.

Unfortunately, the QWERTY design fails to include some other characteristics that would lead to more rapid performance. For example, more letters are typed on the row above than on the home keys, thus adding extra movement. Also the amount of effort between the two hands is not balanced. Dvorak (1943) designed a keyboard that reflects these two considerations and that attempts as well to maximize between-hand alternations. There is some consensus that with proper training, the Dvorak keyboard could lead to improved typing performance. However, the Dvorak board will probably never prove to be practical because the 5–10% estimated improvement (Alden, Daniels, & Kanarick, 1972; Norman & Fisher, 1982) is small relative to the familiarity that users have with the conventional keyboard and to the great inertia against change.

Preview and Transcription

In the previous section an important difference between serial-reaction-time task and typing was identified in terms of the effect of repetitions. Repetitions helped in serial RT performance but hindered in typing. There are clearly other important differences between these two tasks in terms of the maximum speed of continuous performance. As established by Debecker and Desmedt, the serial RT task cannot proceed more rapidly than 2.5 decisions/sec even under the most optimal conditions. Yet skilled typists can execute key strokes at a rate of more than 15 per second for short bursts (Rummelhart & Norman, 1982). The major difference here is in the way in which typing and, more generally, the class of *transcription tasks* (e.g., typing, reading aloud, musical performance) are structured to allow the operator to make use of *preview*, *lag*, and *parallel processing*. These are characteristics that allow more than one stimulus to be displayed at a time (preview is available) and therefore allow the operator to lag the response behind perception. Thus at any point in time the response executed is not necessarily relevant to the stimulus that was most recently encoded but is more likely to be a stimulus encoded earlier in the sequence. Therefore perception and response are occurring in parallel. Whether one speaks of this as "preview" (seeing into the future) or "lag" (responding behind the present) obviously depends upon the somewhat arbitrary frame of reference one chooses to define the present.

Preview and lag are both possible in either self-paced or force-paced paradigms. Thus, in typing (a self-paced activity), the typist typically encodes letters (as judged from visual fixation) approximately one second before they are entered into the keyboard. Similarly, in reading aloud, the voice will lag well behind the eye fixations. Oral translation is a force-paced task because the auditory speech flows at a rate that is not determined by the translator. Here there is a lag of a few words between when a word is heard and when it is spoken. In general, this lag in transcription is beneficial to performance. Yet the physical constraints of the task determine the extent to which a lag is possible. In a self-paced task a lag can be created only if a preview of two or more stimuli is provided (e.g., two or more stimuli are displayed simultaneously). This typifies the cases of typing or reading aloud in which preview is essentially unlimited and is subject only to the constraints of visual fixation (see Chapter 4). In a force-paced paradigm in which a stimulus does not wait for a response to occur—for example, taking oral dictation or translating spoken languages—the operator need only build up a slight lag or queue before responding.

When the operator uses preview and lag, he must maintain a running "buffer" memory of encoded stimuli that have not yet been executed as responses. It is therefore interesting to consider why the lag is beneficial in transcription, in light of the fact that task-induced lags between input and response were shown to be harmful in the running memory tasks discussed in Chapter 6. A major difference between the two cases concerns the size of the lag involved. The lag typically observed in transcription is short—around one second—relative to the decay of working memory; so the contents of memory are readily available at the time that output is called for. Resource-demanding rehearsal processes need not be invoked. The problems of running memory are typically observed when the delay is greater.

Since the costs encountered in running memory tasks are not present in the lags of transcription, then it is possible for the transcriber to realize the two important benefits of these lags: allowance for variables and for chunking (Shaffer, 1973; Shaffer & Hardwick, 1970). Each of these will be discussed in turn.

1. *Allowance for variability.* In a nonlagged system, if an input is encountered that is particularly difficult to encode, this delay will be shown at the response as well. Correspondingly, a particularly difficult response will slow down processing of the subsequent input in a PRP-like fashion. On the other hand, if there is a lag between input and output, accounted for by a buffer of three or four items, then a steady stream of output at a constant rate can proceed (for at least a short while) even if input is temporarily slowed. The buffer is just "emptied" at a constant rate. A prolonged input will, of course, temporarily reduce the size of the buffer until it may be "refilled" with a more rapidly encoded input. If easily encoded stimuli do not appear, then the operator may eventually be forced to adopt a less efficient nonlagged mode of processing. This is what occurs when the text on a page being typed becomes degraded.

2. *Allowance for chunking.* There is good evidence in typing that inputs are encoded in chunks, so that the letters within each chunk are processed more or less in parallel. (Recall the "whole-word effect" studied by Reicher, 1969, and discussed in Chapter 4.) The output, however, must be serial (assuming that a chord writer is not employed). The creation of a lag, therefore, allows the steady flow of serial output to proceed even as the buffer is suddenly increased in the number of required output units (key presses stored) as a result of the "parallel" perception of a chunk. Furthermore, factors affecting the speed of individual responses (i.e., the reach time for letters on a keyboard) are totally unrelated to encoding difficulty. A buffer will prevent slow responses from disrupting subsequent encoding. Figure 10.4 provides a schematic example in which a relatively constant output stream (response per unit time) is maintained in spite of variations in encoding speed and buffer contents caused by variations in input chunks and input quality. These variations will cause the size of the buffer to vary yet allow an even, "rhythmic" flow of responses.

Time-sharing in transcription.   The ability of skilled operators to use preview as efficiently as they do is testimony both to the accuracy of the running "buffer" memory, which holds inputs until they are ready to be output, and to our ability to efficiently time-share the various stages of processing involved. Detailed analysis of the transcription task indicates that the response to one letter is not disrupted

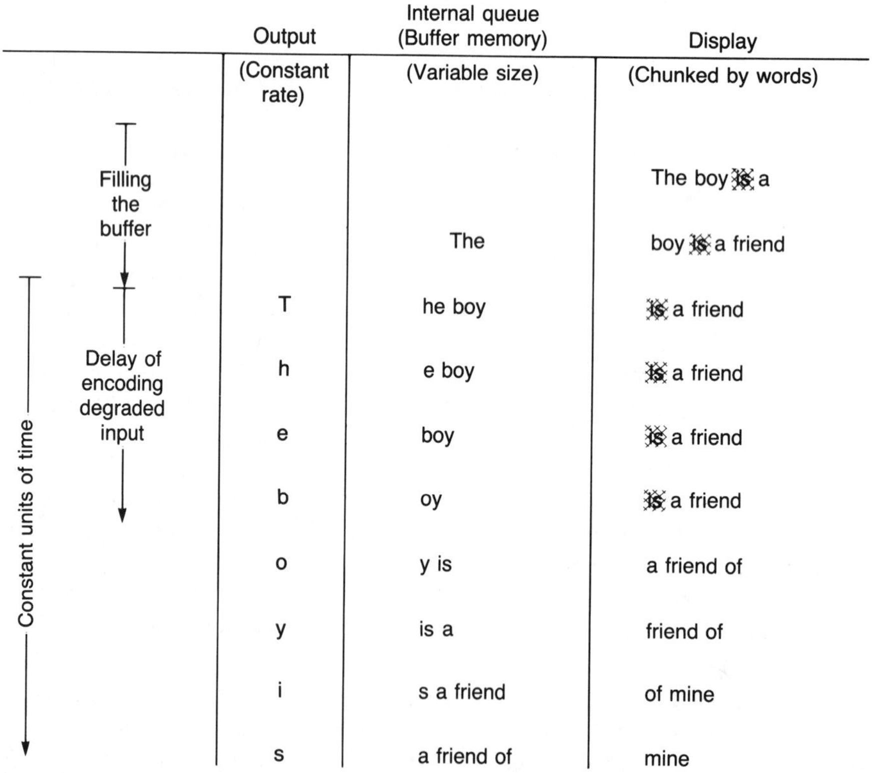

	Output	Internal queue (Buffer memory)	Display
	(Constant rate)	(Variable size)	(Chunked by words)
Filling the buffer			The boy 🖾 a
		The	boy 🖾 a friend
	T	he boy	🖾 a friend
Delay of encoding degraded input	h	e boy	🖾 a friend
	e	boy	🖾 a friend
	b	oy	🖾 a friend
	o	y is	a friend of
	y	is a	friend of
	i	s a friend	of mine
	s	a friend of	mine

Constant units of time (along left vertical axis)

Figure 10.4   Schematic representation of transcription, showing variable input rate (caused by chunking and stimulus quality) providing constant output rate. Note the long delay when encoding the degraded word "is."

by encoding of a future letter (Shaffer, 1973). In addition to encoding one stimulus and responding to a different one, a third parallel activity involves maintaining material in the buffer memory (the buffer is, of course, short enough in its temporal characteristics so that it may properly be labeled as a preattentive code). A fourth process carried out in parallel is that of continuously monitoring for errors of action. Rabbitt (1981) argues on the basis of his experimental results that this is a parallel process. These results show that experimental variables which influence the speed of responding leave error-correction time unaffected. Finally, through high-speed photographic analysis, Rummelhart and Norman (1982) have argued that two or more of a typist's responses themselves are executed in parallel. Movement of a finger toward a key that will be needed one or two strokes in the future proceeds even as key strokes for present letters are executed.

The efficiency with which all of these activities are performed in parallel becomes all the more remarkable in light of Shaffer's (1975) finding, described in Chapter 8, that *two* transcription tasks—typing and auditory shadowing, both demanding a verbal central processing code and neither involving predictable input—can nevertheless be efficiently time-shared. The efficiency of performance of Shaffer's subjects seems to depend not only upon their high skill levels as typists, but also upon the "natural" compatibility between input and output modalities as discussed in Chapter 9. Thus, while typing (visual-manual) and shadowing (auditory-vocal) were efficiently shared, a greater disruption of time-sharing resulted when the modalities were reversed and typing of an auditory message was time-shared with reading.

The use of preview.   The availability of preview does not, of course, mean that preview will necessarily be used. The unskilled typist will still type one letter at a time and will go no faster if preview is available than it if it is not. The skilled typist depends heavily upon preview for efficient transcription. Investigations by Shaffer and Hardwick (1970), Shaffer (1973), and Hershon and Hillix (1965) suggest that preview helps performance. These data make clear the joint benefits of preview: making available more advance information and giving the operator an opportunity to perceive chunks (see Chapter 6). Hershon and Hillix (1965) had subjects type a written message, of which various numbers of letters could be displayed in advance. They increased the number of preview letters from 1 to 2, then to 3, then to 6, and finally to an unlimited number. Greater preview provided some benefits to typing of random-letter strings, but much greater benefits to the typing of random words. The greater benefit in the latter condition, of course, resulted because of the "chunkability" of words.

Shaffer (1973) conducted a systematic investigation of preview effects using one highly skilled typist. Shaffer examined the differences in interresponse times (IRT) as the subject typed varying kinds of text with different amounts of preview. Slowest typing was obtained when typing random letters with no preview, a condition equivalent to a self-paced serial RT task. Here the IRT was 500 msec. Progressively shorter IRTs (faster typing) were obtained when typing random letters with unlimited preview (IRT = 200 msec) and typing random words with preview (IRT = 100 msec). In a fourth condition, words in coherent text were typed with unlimited preview. These results were interesting because no further gain in typing speed was observed over the random-word condition. This finding suggests that the benefits of preview are *not* related to the semantic level of processing, but rather to the fact that preview allows the letters within chunk-sized units (i.e., words) to be processed in parallel (see Chapter 4).

Further support for the conclusion that word chunks and not semantic content is the critical factor in preview is provided by another of Shaffer's findings. When typing either coherent prose or random words, eight letters of preview are sufficient to accomplish all necessary gains in performance. Eight letters would be sufficient to encompass the great majority of words but not generally enough to extract coherent semantic meaning. The absence of heavy semantic involvement in transcription would thereby explain how subjects in Shaffer's (1975) experiment were able so successfully to perform two concurrent transcription tasks. As long as neither was competing for semantic central-processing codes, both tasks employing separate input-output modes with "natural" compatible visual-manual and auditory-speech linkages could proceed in parallel (Wickens, Sandry, & Vidulich, 1983).

A second conclusion of Shaffer's research is that the benefits of chunking are primarily perceptual and that they may be manifested in terms of storage but not in terms of response. Groups of letters are perceived as a unit, perhaps are stored in the buffer memory as a unit, but are rarely output to the keyboard as an integrated motor program; that is, they are not a highly overlearned response pattern, such as one's signature, that is executed as an open-loop motor "chunk" or motor program (Keele, 1973; Summers, 1981; see Chapter 11). Shaffer's arguments are based primarily upon an analysis of the interresponse times between letters within a word. If these are, in fact, parts of a "motor chunk" or motor program, then the sum of the IRTs within a chunked word should be less than a similar sum of mean IRTs derived from all words. For example, since the IRT averaged across typing of words with preview is 100 msec, then, to be typed as a chunk, the total IRT in the word "cat" should be *less* than 200 msec (i.e., less than two 100-msec IRTs). Shaffer observes that this "motor packaging" effect holds true only for a few extremely common sequences (*and*, *the*, -*ing*) but not for the general class of common words. This conclusion is supported by more recent investigations of typing performed by Sternberg et al. (1978) and Gentner (1982). Gentner, for example, performed a detailed analysis of finger movement in typing and concluded that these movements were determined entirely by constraints of the hands and keyboard interacting with the letter sequence. Higher-level word units played no role in response timing.

Feedback

The quality and timing of feedback in transcription has an important influence upon performance. The potential *negative* effects of feedback are probably most pronounced in shadowing or translation when the feedback from one's own voice will tend to mask the signal to be processed. Welford (1976) notes that skilled translators have acquired the ability to filter out the feedback from their own voice. It is this absence of the natural feedback processing that may account for the somewhat unnatural sounds and inflections of the translator's voice. Delays of feedback can also have a degrading effect on transcription performance, just as they disrupt performance in general. The influence of artificially delayed feedback on the performance of such generally automated skills as speaking or handwriting has been well documented by Smith (1962).

A major concern in this regard is in the area of computer-based text editing (Embley & Nagy, 1981), in which a heavily loaded time-sharing machine or a transmission line of limited bandwidth can make the delay between a key entry and a displayed response perceptually noticeable. The amount of disruption caused by delay varies considerably as a function of the task performed. When there is considerable cognitive involvement on the part of the operator, minor delays in computer response to operator entries are not serious since this delay period usually constitutes operator "think time." However, even in this case, it is important that delays be relatively constant since this predictability allows the operator to anticipate the occurrence of the system response and engage in cognitive activity during the interim (Embley & Nagy, 1981). Considerably more constraining and disruptive are delays in the "echo" of key-stroke entries as presented visually on the display. Here, delays longer than 100 msec appear to disrupt input of transcription performance (R. B. Miller, 1968). However, Long (1976) found a difference between skilled and novice operators in this regard. Skilled transcribers, after brief training, were not hampered by variable feedback delays averaging 160 msec in length. Like the professional translators, they were apparently successful in tuning out feedback from their own responses. In typing, this filtering is accomplished simply by not looking at the display.

In summary, humans are remarkably efficient at performing transcription tasks, overcoming many of the "limits" of reaction time and time-sharing described earlier. Only relatively small margins for improvement are probably feasible in the redesign of the typewriter (Norman & Fisher, 1982). Furthermore, it is likely that future developments in automatic speech recognition devices may eventually make the human's manual function in typing obsolete. The read or created message may simply be spoken to the computer and a typed text prepared from the automatically recognized speech. The major benefit of research on typing would not then appear to be on issues of keyboard redesign. Instead, the value of this research seems to be in terms of how models of "optimal" typing performance, such as those of Shaffer and Hardwick (1970), Rummelhart and Norman (1982), Gentner (1982), and Card, Moran, and Newell (1983; see Chapter 1), can be applied to design interfaces for other transcription and data-entry tasks such as those involved in more complex human-computer interfaces.

ERRORS IN PERFORMANCE

Most of our treatment of action in Chapters 9 and 10 has emphasized the speed of human performance. In most real-world systems, however, the designer is more concerned with the existence of errors or mistakes than with the latency of performance. The consequences of a delay in bringing a nuclear reactor on line, for example, are certainly far less serious than those of a mistake in the start-up process. When entering data into a keyboard, one is often more concerned with accuracy than with speed.

In Chapter 9 accuracy was considered in the context of the speed-accuracy tradeoff and of the criterion model of reaction time. If it is assumed that operators can voluntarily trade off speed for accuracy, then this assumption allows the investigator to focus on variables that affect the *latency* of processing, because any variable which increases processing latency (except for an accuracy "set") will also be likely to induce more errors of performance if the subject is forced to perform more rapidly. Furthermore, since processing latency (a mean) is usually a more sensitive and reliable measure than is processing error rate (a proportion, which may often be very low), it is often convenient and advisable for the system designer to study influences on processing speed. The designer may then assume that differences in RT between two system configurations of perhaps only a few tens of milliseconds will ultimately translate into an increased likelihood of error. That is, systems that are slower for operators to use will probably be more likely to induce errors. This assumption is probably reasonable in most cases, although Santee and Egeth (1982) caution that there may be exceptions in tasks where the quality of perceptual data is extremely low. Under these circumstances they report that latency and accuracy data may not always give equivalent results.

While much can be gained then from studying latency, there is some reason to focus explicitly on the causes and sources of human error, particularly as some of these are not necessarily factors that would induce increased latency. For example, it is interesting to consider that sometimes errors are most likely to occur when we are performing a skill in a highly practiced, skill-based, automated mode—the very conditions under which speed might be expected to be most rapid.

There are also some experimental variables which decrease latency but appear to do so not by increasing the overall efficiency of information processing but rather by shifting the operator toward the speed end of a speed-accuracy tradeoff (see Figures 9.4 and 9.5). Preparation or expectancy for a stimulus-response pair is one such variable (Posner, 1978). If one expects a stimulus to occur, it will be responded to more rapidly, but this response will also be more likely to be given if the unexpected stimulus occurs. Another example is provided by the use of auditory (speech) displays in aircraft. When used for noncritical advisory warnings, the greater alerting of the auditory modality (see Chapter 7) may cause the pilot to view the signal with greater urgency than necessary and hence be more likely to make an erroneous response. As a consequence of this speed-accuracy bias, a decision has been made not to use speech displays for nonpriority warnings on the new Boeing 757 and 767 aircraft.

Given then that errors are of some interest by themselves and not just as a correlated variable to latency, the following sections consider some of the kinds of errors of action that are made. We will also discuss the issue of using human-error data to predict system reliability.

Categories of Errors

Norman (1981) has identified a number of potential causes for human error in complex task performance. Norman makes a distinction between two categories of errors that corresponds with the continuum of action proposed by Rasmussen (1980), discussed in Chapter 9, that runs from skill-based to rule-based to knowledge-based behavior. Norman identifies errors of *misperception* as those that are more likely to occur when the operator is functioning at a rule- or knowledge-based level relying upon conscious decision making. These errors, which Norman calls *mistakes*, typify many of the errors in decision making described in Chapter 3: An operator integrates or processes cues poorly, has a faulty understanding of the nature of the situation at hand, and so formulates an incorrect action. Mistakes are less likely to occur at an automated level of performance because well-learned perceptual processes conventionally take place with high accuracy.

A second class of errors called *mode errors* occur more frequently in skill- or rule-based behavior. These result when a particular action that is highly appropriate in one mode of operation is performed in a different, inappropriate mode because the operator has not correctly perceived the appropriate context. A potentially disastrous example of a mode error would be a pilot who attempts to raise the landing gear while still on the runway, having misperceived that he has left the runway. Mode errors are of increasing concern in human-computer interactions when the operator occasionally must deal with keys that serve very different functions dependent upon the setting of another part of the system. Even with the simple typewriter I have often experienced the frustrations of intending to type a string of digits (e.g., 1 9 6 5), when mistakenly I have left the case setting in the uppercase mode and so produce the string ! (¢ %. Mode errors may occur in computer text editing, in which a command that is intended to delete a line of text may instead delete an entire page (or data file!) because the command was executed in the wrong mode. Certain computer text-editing systems are particularly unforgiving in this respect because mode errors are quite likely to occur, and their consequences are drastic (Norman, 1981b).

Mode errors are a joint consequence of relatively automated performance—when the operator fails to be aware of which mode is in operation—and of improperly conceived system design in which such mode confusions can have major consequences. The reason, of course, that mode errors can occur is that a single action may be made in both appropriate and inappropriate circumstances.

Ideally, certain safeguards or lockout logic should be built into intelligent systems to prevent erroneous actions from occurring in inappropriate circumstances when the consequences of error are serious. However, in some situations of multimode operation, such as text editing, the lockout option may not be possible, for the computer may have no reason for assuming that the operator should be in one mode or the other. An appropriate solution in this case would be to make actions with more severe erroneous consequences (i.e., deleting an entire data file), more difficult options to select. This is sometimes done by interrogation: "Are you sure you wish to perform option X?" Such an interrogation would be designed to interrupt the level of automated performance in which the mode error is likely to occur. An additional safeguard against mode errors would be to ensure that salient stimulus or contextual conditions are quite different when operating in different modes. For example, a computer display could be formatted in a background of a different color as a function of the current mode of operation. This ever-present cue could serve as a constant reminder of the appropriate mode.

Capture errors, or *slips*, are prominent when the operator is functioning at a skill-based level of behavior. They occur when three conditions are present in a sequence of action: (1) An inappropriate action is selected and performed because the tendencies for this action may already be high. (2) The stimuli that trigger the action are quite similar to the stimuli that would trigger the appropriate action. Thus, when the stream of behavior "passes too close" to the inappropriate action-triggering stimulus, the latter is selected. (3) The operator is functioning in a skill-based automatic mode in which the selection of the action is not closely monitored.

Examples of such slips abound in everyday experience. When typing a word such as "practicum," we may find ourselves typing "practice" or "practical" instead, as the sequence of letters "p r a c t i c" will provide a context that "captures" the more familiar final string ("a l," or "e"). Classic slips are often slips of the tongue. We might introduce an acquaintance as "Colin MacDonald," when his name is Colin Mac-Farland, if we had already had a friend of the former name.

Examples of slips caused by faulty instrument design are notorious when two controls are placed close together. When an operator who is functioning at an automated mode needs to use the less frequent one, the fact that the two stimuli (the controls) are close together may cause the more frequent response to be made. We have all probably confused windshield wipers and car lights in this fashion in poorly designed dashboards. A more serious example is provided by the close proximity of the control of landing gear and flaps in some propeller-driven aircraft. The use of controls coded by different shapes is an example of a tactile display that helps to discriminate the controls and make slips less likely to occur.

The interesting characteristic of capture errors is that they often tend to occur when one is performing at the most automatic and highly skilled level of well-learned tasks. As a result, heavy processing resources are not invested in the selection of successive actions. This diversion of attention from a more normal action sequence may lead to the highly skilled execution of the inappropriate action sequence triggered by similar stimulus conditions. Capture errors will therefore probably be the inevitable consequence of operation in an otherwise desirable, skill-based mode of performance. To guard against their occurrence, system designers should try to guarantee that a task is configured so that a habit of greater activation or strength is not available in a context when one of lesser strength is called for. A classic example described in Chapter 9 is a configuration in which a low S–R compatibility response is called for (e.g., activating the lower of two switches when the upper of two lights appears). When the operator is functioning at a consciously controlled level, the appropriate action will be likely to be performed. When, however, performance reaches an automated mode (e.g., as attention is diverted elsewhere), a capture error of the more compatible—but inappropriate—response will become likely.

Human Reliability

An important reason for studying human error is, of course, to identify error causes and to suggest possible corrective solutions. A different reason for investigating human error relates to the analysis of human reliability (Swain, 1977; Swain & Guttman, 1980). A fairly precise analytic technique exists to predict the reliability (probability of failure or mean time between failure) of a complex *mechanical* or *electrical* system with components of known reliabilities that are configured serially or in parallel. For example, consider a system consisting of two components, each with a reliability of 0.9 (i.e., a 10% chance of failure during a specified time period). If the components are arranged in series, so that if either fails the total system fails, then the probability of system failure is precisely $1 - (0.9 \times 0.9) = 1 - 0.81 = 0.19$. If they are arranged in parallel (redundantly), so that the system will fail only if both fail, then the probability of system failure is $0.1 \times 0.1 = 0.01$.

Swain (1977; Swain & Guttman, 1980) has argued that a corresponding calculation of the reliability of complex man-machine systems such as the nuclear power reactor can be made by mathematically integrating human and machine error information in the common calculations. The results of this analysis, if accurate, would be quite valuable in predicting the circumstances in which performance breakdowns might occur. It is clear, however, that some caution is required in human reliability analysis because human error is fundamentally different from machine error, and so it cannot necessarily be treated mathematically in the same fashion (Adams, 1982; Sheridan, 1981). Adams has cited four specific ways in which human and machine error may differ. These pertain to the nonindependence of human error, error monitoring, parallel human components, and the integration of human and machine reliability.

Nonindependence of human errors.   When analyzing machine errors in reliability analysis, the assumption is sometimes made that the probability that one component will fail is independent of the failure probability of another. While this assumption is dubious when dealing with equipment, with humans it is particularly untenable. Often if we make one mistake, our resulting frustrations may increase the likelihood that we will make a subsequent bungle. Alternatively, the first mistake may increase our care and caution in subsequent operations and make future errors less likely. Whichever is the case, it is impossible to claim that the probability of making an error at time T is independent of whether an error was made at time $T - N$. The actuarial data base on human-error probability which is used to predict reliability will not easily capture these dependencies because they are determined by mood, caution, personality, and other uniquely human properties.

Error monitoring.   When machine components fail, they require outside repair or replacement. Yet humans normally have the capability to monitor their own performance, even when operating at a relatively automated level. As a result, they often correct errors before they ultimately affect system performance. This is particularly true with capture errors or action slips. The operator who accidentally activates the wrong switch may be able to shut it off quickly and activate the right one before any damage is done. Rabbitt (1981), for example, notes that slips of response selection in typing are detected with very high frequency. Even errors of perception may be monitored and corrected a reasonable proportion of the time. Thus it is quite difficult to associate the probability of a *human* error with the probability that it will induce a *system* error.

Parallel components.   When machine reliability is considered, the operation of two parallel (or redundant) components is assumed to be independent. For example, three redundant autopilots are often used on an aircraft so that if one fails, the indication of the two in agreement may be employed as the "true" guidance input. In this case, none of the autopilots will influence the other's operation (unless they are all affected by a superordinate factor such as a total loss of power). This independence cannot, however, be said to hold true of human behavior. In the control room of a power station, for example, it is unlikely that the diagnosis made by one operator in the face of a malfunction will be independent of that made by another. Thus it is unlikely that there will be independent probabilities of error shown by the two operators. Social factors may make the two operators relatively more likely to agree in their diagnosis than had they been processing independently, particularly if one is in a position of greater authority.

Integrating human and machine reliabilities. Adams (1982) argues that it is difficult to justify mathematically combining human-error data with machine-reliability data, derived independently, to come up with joint reliability measures of the total system. Here again a nonindependence issue is encountered. When a machine component fails (or is perceived as being more likely to fail), it will probably alter the probability of human failure in ways that cannot be precisely specified. It is likely, for example, that the operator will become far more cautious, trustworthy, and reliable when interacting with a system that has a higher likelihood of failure or with a component that itself has just failed than when interacting with a system that is assumed to be infallible. This is a point that will be considered again in the context of our discussion of automation in Chapter 12.

The important message here, as stated succinctly by both Sheridan (1981) and Adams (1982), is that a considerable challenge is required to integrate actuarial data of human error with machine data in order to estimate system reliability. Unlike some other domains of human performance (see particularly manual control in Chapter 11), precise mathematical modeling of *human* performance does not appear to allow accurate prediction of *total system* performance. While the potential benefits of accurate human reliability theory are great, it seems more likely that more immediate human-factors benefits will be realized if effort is focused upon case studies of individual errors in performance (Pew, Miller, and Feehrer, 1981). These case studies can be used to diagnose the resulting causes of errors and to recommend the corrective system modification.

Transition

The major class of tasks that have been considered in this chapter are those in which a repeated series of responses is required. We have been most concerned with the timing of each separate response and a categorical classification of whether each response was correct or not. The size of an error did not matter. In the following chapter we will consider again tasks in which several manual actions are carried out repetitively over time. However, in this domain of continuous manual control, considering tasks such as driving and flying, some errors are always expected, and it is the *size* of those errors that is the critical measure of performance efficiency. Processing time is again important, but response latency is not directly examined as a dependent variable. Instead, what is important is how this latency translates to error. Finally, the tasks are continuous, and it is not easy to specify when a response starts and stops. What is most important is the analog form of that response.

References

Adams, J. A. (1982). Issues in human reliability. *Human Factors, 24,* 1–10.

Alden, D. G., Daniels, R. W., & Kanarick, A. F. (1972). Keyboard design and operations: A review of the major issues. *Human Factors, 14,* 275–293.

Alluisi, E., Muller, P. I., & Fitts, P. M. (1957). An information analysis of verbal and motor response in a force-paced serial task. *Journal of Experimental Psychology, 53,* 153–158.

Basila, B., & Salviendi, G. (1979). Non-work related movements in machine-paced and self-paced work. In C. Bensel (Ed.), *Proceedings, 23rd meeting of the Human Factory Society,* Santa Monica, CA: Human Factors.

Bertelson, P. (1966). Central intermittency twenty years later. *Quarterly Journal of Experimental Psychology, 18,* 153–163.

Brown, R. A. (1957). Age and "paced" work. *Occupational Psychology, 31,* 11–20.

Broadbent, D. (1958). *Perception and communications.* London: Pergamon Press.

Broadbent, D. (1971). *Decision and stress.* New York: Academic Press.

Card, S., Moran, T. P., & Newel, A. (1983). *The psychology of human-computer interactions.* Hillsdale, NJ: Erlbaum Associates.

Conrad, R. (1951). Speed and load stress in sensori-motor skill. *British Journal of Industrial Medicine, 8,* 1–7.

Conrad, R. (1960). Letter sorting machines—paced, "lagged," or unpaced. *Ergonomics, 3,* 149–157.

Conrad, R., & Longman, D. S. A. (1965). Standard typewriter vs. chord keyboard—an experimental comparison. *Ergonomics, 8,* 77–88.

Creamer, L. R. (1963). Event uncertainty, psychological refractory period, and human data processing. *Journal of Experimental Psychology, 66,* 187–194.

Craik, K. W. J. (1947). Theory of the human operator in control systems I: The operator as an engineering system. *British Journal of Psychology, 38,* 56–61.

Danaher, J. W. (1980). Human error in ATC system operations. *Human Factors, 22,* 535–545.

Davis, R. (1965). Expectancy and intermittency. *Quarterly Journal of Experimental Psychology, 17,* 75–78.

Debecker, J., & Desmedt, R. (1970). Maximum capacity for sequential one-bit auditory decisions. *Journal of Experimental Psychology, 83,* 366–373.

Deininger, R. L., Billington, M. J., & Riesz, R. R. (1966). The display mode and the combination of sequence length and alphabet size as factors of speed and accuracy. *IEEE Transactions on Human Factors in Electronics, 7,* 110–115.

Drury, C., & Coury, B. G. (1981). Stress, pacing, and inspection. In G. Salviendi & M. J. Smith (Eds.), *Machine pacing and operational stress.* London: Taylor & Francis.

Dvorak, A. (1943). There is a better typewriter keyboard. *National Business Education Quarterly, 12,* 51–58.

Elithorn, A., & Lawrence, C. (1955). Central inhibition—Some refractory observations. *Quarterly Journal of Experimental Psychology, 7,* 116–127.

Embley, D. W., & Nagy, G. (1981). Behavioral aspects of text editors. *Computing Surveys, 13,* 33–70.

Fitts, P. M., & Posner, M. A. (1967). *Human performance.* Pacific Palisades, CA: Brooks Cole.

Gentner, C. R. (1982). Evidence against a central control model of timing in typing. *Journal of Experimental Psychology: Human Perception and Performance, 9,* 793–810.

Goldstein, I. L., & Dorfman, P. W. (1978). Speed stress and load stress as determinants of performance in a time-sharing task. *Human Factors, 20,* 603–610.

Gopher, D. (1983). On the contribution of vision-based imagery to the acquisition and operation of a transcription skill. In W. Printz & A. Sanders (Eds.), *Cognition and motor processes.* New York: Springer.

Gopher, D., & Eilam, Z. (1979). Development of the letter shape keyboard. In C. Bensel (Ed.), *Proceedings, 23rd annual meeting of the Human Factors Society.* Santa Monica, CA: Human Factors.

Gottsdanker, R. M., & Senders, J. W. (1980, November). *On the estimation of mental load* (Technical Report No. AFOSR-79-0122). Santa Barbara: University of California, Psychology Department.

Hershon, R. L., & Hillix, W. A. (1965). Data processing in typing: Typing rate as a function of kind of material and amount exposed. *Human Factors, 7,* 483–492.

Kahneman, D. (1973). *Attention and effort.* Englewood Cliffs, NJ: Prentice-Hall.

Kantowitz, B. H. (1974). Double stimulation. In B. H. Kantowitz (Ed.), *Human information processing.* Hillsdale, NJ: Erlbaum Associates.

Karlin, L., & Kestinbaum, R. (1968). Effects of number of alternatives on the psychological refractory period. *Quarterly Journal of Experimental Psychology, 20,* 160–178.

Keele, S. W. (1973). *Attention and human performance.* Pacific Palisades, CA: Goodyear.

Klapp, S. T., & Irwin, C. I. (1976). Relation between programming time and duration of response being programmed. *Journal of Experimental Psychology: Human Perception and Performance, 2,* 591–598.

Knight, J., & Salviendi, G. (1981). Effects of task stringency of external pacing on mental load and work performance. *Ergonomics, 24,* 757–764.

Kornblum, S. (1973). Sequential effects in choice reaction time: A tutorial review. In S. Kornblum (Ed.), *Attention & Performance IV.* New York: Academic Press.

Lockhead, G. R., & Klemmer, E. T. (1959, November). *An evaluation of an 8-key word-writing typewriter* (IBM Research Report RC-180). Yorktown Heights, NY: IBM Research Center.

Long, J. (1976). Effects of delayed irregular feedback on unskilled and skilled keying performance. *Ergonomics, 19,* 183–202.

McFarling, L. H., & Heimstra, N. W. (1975). Pacing, product complexity, and task perception in simulated inspection. *Human Factors, 17,* 361–367.

Manenica, I. (1979). Comparison of some physiological indices during paced and unpaced work. *International Journal of Production Research, 15,* 261–275.

Miller, D. P. (1981). The depth/breadth trade-off in hierarchical computer menus. In R. Sugarman (Ed.), *Proceedings, 25th annual meeting of the Human Factors Society.* Santa Monica, CA: Human Factors.

Miller, R. B. (1968). Response time in non-computer conversational transactions. In *Proceedings, 1968 Fall Joint Computer Conference.* Arlington, VA: AFIPS Press.

Moray, N., Johanssen, G., Pew, R. W., Rasmussen, J., Sanders, A., & Wickens, C. D. (1979). Report of the experimental psychology group. In N. Moray (Ed.), *Mental workload: Its theory and measurement.* New York: Plenum Press.

Norman, D. A. (1981a). Categorization of action slips. *Psychological Review, 88,* 1–15.

Norman, D. A. (1981b). The trouble with Unix. *Datamation, 27,* 139–150.

Norman, D. A., & Fisher, D. (1982). Why alphabetic keyboards are not easy to use: Keyboard layout doesn't matter much. *Human Factors, 24,* 509–520.

Pew, R. W., Miller, D. C., & Feehrer, C. E. (1981). Evaluating nuclear control room improvements through analysis of critical operator decisions. In R. Sugarman (Ed.), *Proceedings, 25th annual meeting of the Human Factors Society.* Santa Monica, CA: Human Factors.

Posner, M. I. (1978). *Chronometric explorations of the mind.* Hillsdale, NJ: Erlbaum Associates.

Rabbitt, P. M. A. (1967). Time to detect errors as a function of factors affecting choice-response time. *Acta Psychologica, 27,* 131–142.

Rabbitt, P. M. A. (1981). Sequential reactions. In D. Holding (Ed.), *Human skills.* New York: Wiley.

Rasmussen, J. (1980). The human as a system component. In H. T. Smith and T. R. G. Green (Eds.), *Human interaction with computers.* London: Academic Press.

Reicher, G. M. (1969). Perceptual recognition as a function of meaningfulness of stimulus material. *Journal of Experimental Psychology, 81,* 275–280.

Reynolds, D. (1966). Time and event uncertainty in unisensory reaction time. *Journal of Experimental Psychology, 71,* 286–293.

Rumelhart, D., & Norman, D. (1982). Simulating a skilled typist: A study of skilled cognitive-motor performance. *Cognitive Science, 6,* 1–36.

Salviendi, G., & Smith, M. J. (Eds.). (1981). *Machine pacing and occupational stress.* London: Taylor & Francis.

Santee, S. L., & Egeth, H. E. (1982). Do reaction time and accuracy measure the same aspects of letter recognition. *Journal of Experimental Psychology: Human Perception and Performance, 8,* 489–501.

Seibel, R. (1963). Discrimination reaction time for a 1,023-alternative task. *Journal of Experimental Psychology, 66,* 215–226.

Seibel, R. (1964). Data entry through chord, parallel entry devices. *Human Factors, 6,* 189–192.

Seibel, R. (1972). Data entry devices. In H. S. Van Cott & R. G. Kinkade (Eds.), *Human engineering guide to equipment design.* Washington, DC: U.S. Government Printing Office.

Shaffer, L. H. (1973). Latency mechanisms in transcription. In S. Kornblum (Ed.), *Attention and Performance IV.* New York: Academic Press.

Shaffer, L. H. (1975). Multiple attention in continuous verbal tasks. In S. Dornic (Ed.), *Attention and Performance V.* New York: Academic Press.

Shaffer. L. H., & Hardwick, J. (1970). The basis of transcription skill. *Journal of Experimental Psychology, 84,* 424–440.

Sheridan, T. (1981). Understanding human error and aiding human diagnostic behavior in nuclear power plants. In J. Rasmussen and W. B. Rouse (Eds.), *Human detection and diagnosis of system failures.* New York: Plenum Press.

Simmons, R. (1979). Methodological considerations of visual workload of helicopter pilots. *Human Factors, 21,* 353–368.

Smith, K. V. (1962). *Delayed sensory feedback and balance.* Philadelphia: Saunders.

Smith, M. (1967). Theories of the psychological refractory period. *Psychological Bulletin, 19,* 352–359.

Sternberg, S., Kroll, R. L., & Wright, C. E. (1978). Experiments on temporal aspects of keyboard entry. In J. P. Duncanson (Ed.), *Getting it together: Research and application in human factors.* Santa Monica, CA: Human Factors.

Summers, J. J. (1981). Motor programs. In D. H. Holding (Ed.), *Human skills.* New York: Wiley.

Swain, A. D. (1977). Error and reliability in human engineering. In B. Wolman (Ed.), *International encyclopedia of psychiatry, psychology, psychoanalysis, and neurology* (Vol. 4, pp. 371–373). New York: Von Nostrand Reinhold.

Swain, A. D., & Guttman, H. E. (1980). *Handbook of human reliability analysis with emphasis on nuclear power plant application* (NUREG/CR–1278). Washington, DC: U.S. Government Printing Office.

Telford, C. W. (1931). Refractory phase of voluntary and associative response. *Journal of Experimental Psychology, 14,* 1–35.

Tulga, M. K., & Sheridan, T. B. (1980). Dynamic decisions and workload in multitask supervisory control. *IEEE Transactions on Systems, Man, and Cybernetics, SMC–10,* 217–232.

Waganaar, W. A., & Stakenburg, H. (1975). Paced and self-paced continuous reaction time. *Quarterly Journal of Experimental Psychology, 27,* 559–563.

Welford, A. T. (1952). The psychological refractory period and the timing of high speed performance. *British Journal of Psychology, 43,* 2–19.

Welford, A. T. (1967). Single channel operation in the brain. *Acta Psychologica, 27,* 5–21.

Welford, A. T. (1968). *Fundamentals of skill.* London: Methuen.

Welford, A. T. (1976). *Skilled performance.* Glenview, IL: Scott, Foresman.

Wickens, C. D., Sandry, D., & Vidulich, M. (1983). Compatibility and resource competition between modalities of input, central processing, and output: Testing a model of complex task performance. *Human Factors, 25,* 227–248.

Yntema, D. (1963). Keeping track of several things at once. *Human Factors, 6,* 7–17.

Continuous
Manual Control

OVERVIEW

In the performance of most tasks, the information that is encoded and the decisions that are made must be translated into action. The preceding chapters of this book have assumed for convenience that the form of this action is relatively simple compared to perceptual or central-processing activities. Little attention was given to the analog form or the time-space trajectory of the response. In this chapter we will consider the class of tasks in which this analog time-space trajectory is critical. This is the domain of continuous manual control.

Human performance in manual control has been considered from two quite different perspectives: the skills approach and the tracking approach. Each has used different paradigms and different analytical tasks, and each has been generalized to different applied environments. The skills approach has primarily considered analog motor behavior when the operator must produce or reproduce a movement pattern from memory, in circumstances when there is little environmental uncertainty. The gymnast performs such a skill when executing a complex maneuver; so too does the assembly-line worker who coordinates a smooth integrated series of actions around a set of environmental stimuli—the products to be assembled—that are highly predictable from one instance to the next. Because there is little environmental uncertainty, such skills in theory may be performed perfectly and identically from trial to trial. In terms of Figure 1.1, the behavior is described as "open loop," since there is little need to process the visual feedback from the response. The emphasis of experiments on skills has focused heavily on the time course of skill acquisition and the optimal conditions of practice, whether addressed from the traditional learning point of view (Bilodeau & Bilodeau, 1969) or from a more recent information-processing perspective (Holding, 1981; Stelmach, 1978).

In contrast to the skills approach, the tracking approach examines human abilities in controlling dynamic systems to make them conform with certain time-space trajectories in the face of environmental uncertainty (Kelly, 1968; Poulton, 1974). Most forms of vehicle control fall into this category. The research on tracking has been oriented primarily toward engineering, focusing on mathematical representations of the human's analog response when processing uncertainty. Unlike the skills approach, which focuses on issues of learning and practice, the tracking approach generally addresses the behavior of the well-trained operator.

The general treatment of manual control presented below reflects this dichotomy but acknowledges that, as so often is the case in psychology, the dichotomy is really more of a continuum. The basketball player who executes a skilled, highly practiced maneuver by herself may be engaging in a pure "open-loop skill," but when she does so in the middle of a game with a defender providing some degree of environmental uncertainty, the response becomes more of a compromise between open-loop skills and tracking. Correspondingly, while much of the research on tracking does assume that the operator is well trained, some tracking research has considered the role of practice as well (Fuchs, 1962; Briggs, Fitts, & Bahrick, 1957; Pew, 1974).

We shall begin by considering at an atomistic level the simplest form of analog response—the minimum time taken to move from a starting point to a target with constrained accuracy. The data describing this skill, it turns out, are quite well captured by a basic "law" of motor control, whose principles seem to underlie both open-loop skills and tracking. More complex forms of open-loop skills will then be discussed before turning to an extensive treatment of manual control in tracking and vehicle control.

Finally, the chapter supplement will consider in some detail the efforts that have been made to model the human operator in tracking and will provide an introduction to the mathematical language of frequency-domain analysis which has been used in these modeling efforts.

OPEN-LOOP MOTOR SKILLS

Discrete Movement Time

Pioneering investigations by Woodworth (1899) and Brown and Slater-Hammel (1949) found that the time required to move the hand or stylus from a starting point to a target obeys the basic principles of the speed-accuracy tradeoff. Quite intuitively, faster movements terminate less accurately in a target, while targets of high constraints (small area), requiring increased accuracy, are reached with slower responses. The amplitude of a movement also influences this speed-accuracy relation. It takes a longer time to move a greater distance into a target of fixed area. However, if precision is allowed to decline with longer movements, then movement time is essentially unchanged with length.

Fitts (1954) investigated the relation between the three variables of time, accuracy, and distance in the paradigm shown in Figure 11.1. Here the subject is to move the stylus as rapidly as possible from the start to the target area. Fitts found that when movement amplitude (A) and target width (W) were manipulated, the joint effects of these variables were summarized by a simple equation that has subsequently been known as Fitts' law:

$$\text{Movement time } (MT) = a + b \log_2 \left(\frac{2A}{W}\right)$$

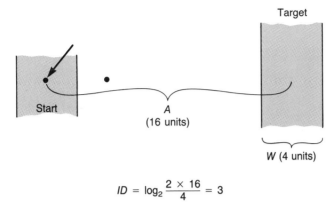

$$ID = \log_2 \frac{2 \times 16}{4} = 3$$

Figure 11.1 The Fitts movement-time paradigm. The movement may be either a single discrete movement from start to target or a series of alternating taps between the two targets.

where a and b are constants. This equation describes formally the speed-accuracy tradeoff in movement: Movement time and accuracy (target width W) are reciprocally related. Longer movements can be made (increasing A), but if their time is to be kept constant, then accuracy must suffer proportionately. That is, the target into which the movement will terminate, W, must be widened.

Fitts described the specific quantity $\log_2 (2A/W)$ as the *index of difficulty* (*ID*) of the movement. In Figure 11.1, the *ID* = 3. Movements of the same index of difficulty can be created from different combinations of A and W but will require the same time to complete. Figure 11.2 shows the linear relation between *MT* and *ID* obtained by Fitts when different combinations of amplitude and target width were manipulated. Each *ID* value shows the similar movement time created by two or three different amplitude/width combinations. The high degree of linearity is evident for all but the lowest condition, in which the linear relation slightly underpredicts *MT*.

Several more recent investigations have demonstrated the generality of Fitts' law. For example, Fitts and Peterson (1964) found that the law is equally accurate for describing single discrete movements or reciprocal tapping between two targets. Langolf, Chaffin, and Foulke (1976) observed that the relation accurately describes data for manipulating parts under a microscope. Drury (1975) found that the law accurately described movement of the foot to pedals of varying diameter and distance, while Card (1981) has employed the law as a basic predictive element of key-reaching time in keyboard tasks. Jagacinski, Repperger, Ward, and Moran (1980) have extended the model to predicting performance in a dynamic target-acquisition task.

Other investigators have examined more theoretical properties of the basic relation. For example, Fitts and Peterson (1964) studied the relation between movement time and the time to initiate the movement and

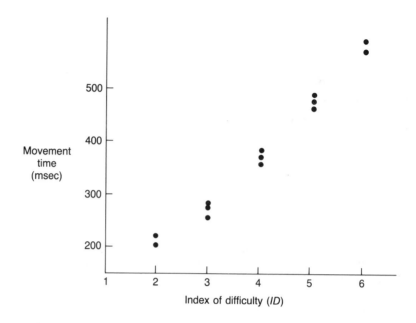

Figure 11.2   Data on movement time as a function of the index of difficulty. From "The Information Capacity of the Human Motor System in Controlling the Amplitude of Movement" by P. M. Fitts, 1954, *Journal of Experimental Psychology, 47,* 385.

found that these two were relatively independent of each other. Increasing the index of difficulty which made movement time longer left the reaction time to initiate the movement unaffected. Conditions that varied reaction time (i.e., single versus choice) had no effect on subsequent movement time. Kelso, Southard, and Goodman (1979) examined movement times for simultaneous two-handed movements to two targets of varying index of difficulty. They observed that the two movements were not independent of each other but that movement time of the easier (lower *ID*) hand was slowed down so as to be in synchrony with the time taken for the more difficult and therefore slower movement.

Models of discrete movement.   Figure 11.3*a* shows a typical trajectory or time history recorded as the stylus approaches the target in the paradigm of Figure 11.1. Two important characteristics of this pattern are apparent. (1) The general form of the movement is that of an exponential approach to the target, with an initial high-velocity approach followed by a smooth final "homing" phase. In the earliest research in this area, Woodworth (1899) distinguished between these two phases, labeling the first as the *initial ballistic* and the second as *current control*. (2) The velocity profile of the movement shown in Figure 11.3*b* reveals that control is not continuous but appears to

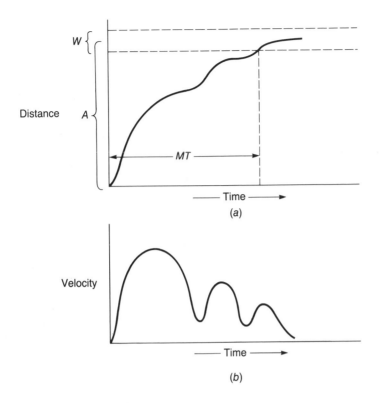

Figure 11.3 Typical position (a) and velocity (b) profile of Fitts' law movement.

consist of a number of discrete corrections, each involving an acceleration and a deceleration.

Several investigators have attempted to provide precise models describing the processes that generate the typical time histories shown in Figure 11.3 (e.g., Crossman & Goodeve, 1963; Fitts, 1954; Howarth & Beggs, 1981; Keele, 1973; Schmidt, Zelaznik, & Frank, 1977), and some have argued that alternative formulations of the model provide a slightly better account for the data (e.g., Howarth & Beggs, 1981). Two characteristics however transcend most of these modeling and experimental approaches. (1) All agree that the basic form provides a reasonably close and parsimonious approximation to the data even if it is not entirely accurate. (2) Most are based upon a feedback processing assumption. As the operator approaches the target he samples, either continuously or intermittently, the remaining error to the target and implements proportional (to the error) corrections to nullify the error. This sample-and-correct process is continued until the target boundary is crossed. Such behavior will produce the generally exponential approach shown in Figure 11.3 in which stylus velocity is roughly proportional to momentary error. This characteristic of target-aiming responses, as we shall see in the following section, is an important one in describing continuous tracking skill, as well as the discrete control discussed here.

Motor Schema

Visually guided responses such as those described by Fitts' law are, of course, critical components in a wide variety of real-world skills such as those required in assembly-line work, target acquisition, or performance on complex or unfamiliar keyboards. Yet with highly learned skills performed under conditions of minimal environmental uncertainty, it is evident that visual feedback is not necessary. We say that these skills may be performed in *open-loop* fashion. In fact, as a consequence of the visual dominance discussed in Chapter 7, this visual feedback may actually be harmful. Shoe tying, touch-typing, or the performance of a skilled pianist are good examples of skilled performance that does not require visual feedback.

Psychologists and motor-learning theorists have identified two general characteristics of such well-learned motor skills: (1) They may well be dependent upon feedback, but the feedback is *proprioceptive*. Information from the joints and muscles is relayed back to central movement-control centers in order to guide the execution of the movement in accordance with centrally stored goals or "templates" of the ideal time-space trajectory (Adams, 1971, 1976). (2) The pattern of desired muscular innervation may be stored centrally in long-term memory and executed as an open-loop *motor program* without benefit of feedback correction and guidance (Keele, 1968; Schmidt, 1975; Summers, 1981). The relative importance of these two sources of information—proprioceptive feedback versus centrally stored "programs"—in the control of highly skilled motor sequences has been hotly debated (for different perspectives, see Adams, 1971, 1976; Keele, 1968, 1973; Schmidt, 1975; Summers, 1981). While the difference between these two positions is of considerable theoretical importance, the distinction is less important to human engineering concerns.

The terms *motor program* and *motor schema* have been used to label highly overlearned skills that, as a consequence of this learning, are not dependent upon guidance from visual feedback (Keele, 1968; Schmidt, 1975; Shapiro & Schmidt, 1982; Summers, 1981). The concept of the motor program or schema may be best defined in terms of four correlated attributes: high levels of practice, low attention demand, single response selection, and consistency of outcome.

High levels of practice.   Extensive practice is perhaps the most critical defining attribute of the motor program. Skills that are not highly practiced are unlikely to possess the further attributes defined below.

Low attention demand.   The motor program tends to be *automated* in the terms described in Chapter 8. In that discussion, task practice was assumed have a major influence on resource demand. Here the limited resource demand is a major criterion for defining the motor program. A well-learned complex sequence of responses may be executed while disrupting only slightly the performance of a concurrent task (Bahrick & Shelley, 1958).

Single response selection.   Within the framework of the information-processing model presented in Chapter 1, it is assumed that a single response selection is required to activate or "load" a single motor program, even though the program itself may contain a number of separate discrete responses. Thus resources are demanded only once, at the point of initiation when the program is selected.

Programs may vary in their complexity. As discussed in Chapter 10, investigators such as Klapp and Erwin (1976), Shulman, Jagacinski, and Burke (1978), and Martenuik and MacKenzie (1980) have argued that motor programs of greater complexity will take longer to "load." Therefore, choice reaction times will be longer when responses of greater complexity are chosen. For example, Shulman et al. found that reaction time to initiate a double key press was longer than for a single press. The relation between program complexity and reaction time only appears to hold, however, so long as the program cannot be "loaded" in advance. In a simple RT task, for example, it is possible to "load" or activate the entire response sequence in advance of the imperative signal, since that response is the only one possible. In this case, the relation between program complexity and response latency is no longer observed (Klapp and Erwin, 1976; Martenuik & MacKenzie, 1980).

Consistency of outcome: Programs versus schemata.   A motor program is assumed to generate very similar space-time trajectories from one replication to another. Keele (1968) initially described the program as a relatively invariant pattern of muscular innervations that were centrally stored and executed. However, a number of investigators pointed out that what is consistent is not the *process* of muscular innervation (and therefore the specific pattern of neural commands) but the *product* of the response (Bartlett, 1932; Pew, 1970, 1974; Schmidt, 1975; Shapiro & Schmidt, 1982). Thus the signature of one's name meets the criteria of a motor program. Yet the signature may vary drastically in terms of the actual muscular commands used (or even total muscle groups) depending upon the context in which one's name is signed—whether, for example, the name is signed on a small horizontal piece of paper or on a large vertical blackboard. In his original discussion, Bartlett (1932) pointed out that the specific muscular patterns involved in a tennis player's swing were quite different in different circumstances. MacNielage (1970) argued that the articulation of familiar words is an example of motor programs. The product of an articulation is roughly the same whether the speaker speaks normally or through clenched teeth. Yet the process of muscular innervation is totally changed between these two conditions.

In both of these examples, drastic changes in motor patterns have occurred. Yet certain characteristics of the time-space trajectory have remained invariant across the modification. Thus whatever is learned and stored in long-term memory cannot be a specific set of motor commands but must represent a more generic or general set of specifications of how to reach the desired goal. These specifications were labeled by Bartlett (1932) and Schmidt (1975) as a *motor schema*. Once a schema is selected, the process of loading requires that the specific instance parameters be specified in order to meet the immediate goals at hand (Pew, 1970, 1974). Two major attributes appear to be preserved in the final output: (1) the *relative* timing of highlights (directional changes) in the movement, which may be slowed down or speeded up in this absolute value, and (2) the relative positioning of these highlights in x, y, and z coordinates of space, even as the absolute extent of the movement may be expanded or shrunk along any of these three dimensions.

The concept of the motor schema has done a great deal to further the understanding and description of processes in skilled motor performance. Its practical implications are less relevant to the areas of performance, however, than to those of learning and training. The implications to training have a direct analogy to the implications of the perceptual schema concept discussed in Chapter 5. If schemata are based upon variability, then training of those schemata should emphasize exposure to a variety of instances. In the case of the motor schema, different variations of the desired time-space trajectory should be practiced. This hypothesis has received some limited experimental support with relatively precisely defined laboratory tasks (Shapiro & Schmidt, 1982) but awaits strong empirical variation with more complex skills.

TRACKING OF DYNAMIC SYSTEMS

In performing manual skills, as described in the last section, we often guide our hands through a coordinated time-space trajectory. Yet at other times we use our hands to guide the position of some other analog system or quantity. At the simplest level, the hand may merely guide a pointer on a blackboard or a light pen on a video display. The hand may also be used to control the steering wheel and thereby position a vehicle on the highway, or it may be used to adjust the temperature of a burner or the closure of a valve in order to move the parameters of a chemical process through a predefined trajectory of values over time. When describing human operator control of physical systems, as typified by the two more complex examples provided above, research moves from the domain of perceptual motor skills and motor behavior to the more engineering domain of tracking. This shift in domain results primarily because of the great influence of three inanimate elements on the performance of the operator who must make a system respond in correspondence to a desired goal. These are (1) the *dynamics* of the system itself: how it responds in time to the forces applied; (2) the *input*, or desired trajectory of the system; and (3) the *display*, the means whereby the operator views or hears the information concerning the desired and actual state of the system. These three elements interact with many of the human operator's limitations discussed in previous chapters in such a way as to present difficulties to a human's tracking in the real world.

Real-world tracking is demonstrated in practically all aspects of vehicle control, ranging from bicycles to aircraft, ships, and space vehicles. It also characterizes many of the tasks performed in complex chemical and energy process control industries, when flow, pressure, and temperature must be controlled and regulated. In the experimental laboratory, the tracking paradigm is typically implemented on a computer in which the subject controls a system whose dynamics are computer simulated by manipulating a control stick and observing the response as a moving symbol on a visual display.

The Tracking Loop: Basic Elements

Figure 11.4 presents the basic elements of a tracking task. These elements will be described within the context of automobile driving, although the reader should realize that the elements may generalize to any number of different tracking tasks.

When driving an automobile, *the human operator* perceives a discrepancy between the desired state of his or her vehicle and its actual

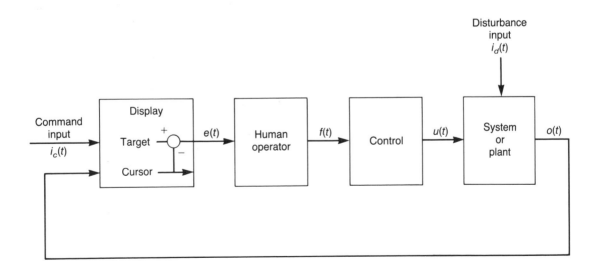

Figure 11.4 The tracking loop.

state. The car may have deviated from the center of the lane or may be pointing in a direction away from the road. The driver wishes to reduce this discrepancy, the function of time error $e(t)$. To do so, a force (actually a torque) $f(t)$, measured in pounds, is applied to the steering wheel or control. This force in turn produces a rotation $u(t)$ of the wheel itself. The relation between the force applied and the steering wheel movement is defined as the *control dynamics*. Movement of the wheel or control by a given time function $u(t)$ in turn causes the vehicle's actual position to move laterally on the highway. This movement is the *system output* $o(t)$. The relation between control position $u(t)$ and system response $o(t)$ is defined as the *plant dynamics*. When presented on a display, the representation of this output position is called the *cursor*. If the operator is successful in the correction, then it will serve to reduce the discrepancy between vehicle position on the highway $o(t)$ and the desired, or "commanded," position at the center of the lane $i(t)$. On a display, the symbol representing the input is called the *target*. The difference between the output and input signals then is the error $e(t)$, the starting point of our discussion. The good driver will respond in such a way as to keep $o(t) = i(t)$, or $e(t) = 0$. It should be clear from the course of this discussion and from the form of Figure 11.4 why tracking is often called closed-loop behavior.

Because error occurrences in tracking stimulate the need for corrective responses, the operator need never respond at all as long as there is no error. However, errors typically arise from one of two sources. *Command inputs*, $i_c(t)$, are changes in the *target* that must be tracked. For example, if the road curves, it will generate an error for a vehicle traveling in a straight line and so will necessitate a response. *Disturbance inputs*, $i_d(t)$, are those applied directly to the system. For example, a wind gust that buffets the car off the highway is a disturbance input. So also is an inadvertent movement of the steering wheel by the driver. Either kind of input may be *transient*, such as a step displacement or a gradual shift. In the first case, imagine that the crosswind on a highway suddenly shifts. In the second case, imagine that the crosswind gradually increases as a car goes around a curve. Alternatively, the input may be *continuous*, in which case it may be described as either predictable and periodic or random. Examples of these four different kinds of inputs are shown in Figure 11.5. As described in. Chapter 4, either random or periodic inputs may be represented in the frequency domain as spectra. The representation of tracking signals in the frequency domain will be discussed more fully in the chapter supplement.

The source of all information necessary to implement the corrective response is the *display*. For the automobile driver, the display is simply the field of view through the windshield, but for the aircraft pilot making an instrument landing, the display is represented by the instruments of pitch, roll, and altitude themselves. An important distinction may be drawn between *pursuit* and *compensatory* displays. A pursuit display presents independent movement of both the target and the cursor. Thus the driver of a vehicle views a pursuit display, since

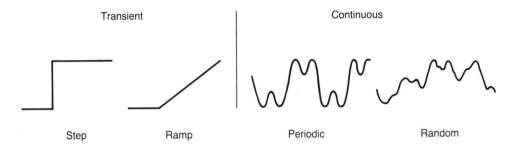

Transient   Continuous

Step   Ramp   Periodic   Random

Figure 11.5 Tracking inputs.

movement of the automobile can be distinguished and viewed independently from the curvature of the road. A compensatory display presents only movement of the error relative to a fixed reference on the display. The display provides no indication of whether this error arose from a change in system output or command input. Most compensatory displays are artificial means of depicting real-world conditions, and these will be discussed in some detail later in the chapter.

Finally, tracking performance is typically measured in terms of error. It is calculated at each point in time and then cumulated and averaged over the duration of the tracking trial. Kelly (1968) gives a good discussion of different means of calculating tracking performance.

Transfer Functions

Figure 11.4 presents three examples in which a time-varying input to a system produces a time-varying response: The operator's force $f(t)$ applied to the control produced control displacement $u(t)$. The displacement $u(t)$ produced a change in system position $o(t)$. Finally, in the same terms, we may think of the error $e(t)$ "applied to," or viewed by, the human as generating the force $f(t)$. In all cases, the *transfer function* represents the mathematical relation between the input and output of a system. The transfer function may be expressed either in terms of a mathematical equation or graphically by showing the time-varying output produced by a given time-varying input. When systems are said to be *linear*, their transfer functions may be conceived as built from the transformations of a number of fundamental, atomistic dynamic elements. Because the limits of human tracking performance depend in important ways upon the transfer function of the system being controlled, and because many of the models of tracking behavior have used transfer functions to describe human performance,[1] it is important to describe these fundamental dynamic elements.

---

[1]When describing human behavior in terms of transfer functions, the output of the human is usually considered to be the control position $f(t)$. It is assumed that the human "intends" to produce a given position. The force $f(t)$ used to achieve this position is achieved fairly automatically.

Figure 11.6 shows the dynamic response of eight of these basic elements to the step input shown in Figure 11.5; this response is sometimes called the *step response*. In addition, the mathematical equation that reflects output to input is presented in two different formats: One is in terms of differential equations in the time domain and the second in terms of the Laplace transform in the frequency domain. The reader should not be concerned with the Laplace function at this point, as it will be covered in more detail in the chapter supplement. Each of these elements will be considered briefly below.

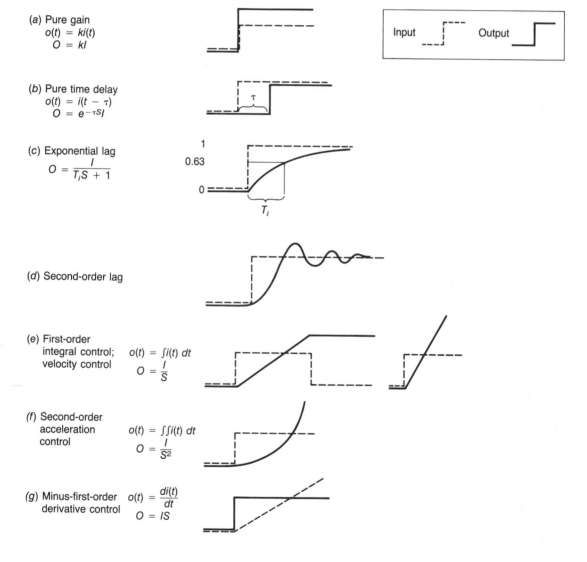

(a) Pure gain
$o(t) = ki(t)$
$O = kI$

Input $\quad$ Output

(b) Pure time delay
$o(t) = i(t - \tau)$
$O = e^{-\tau S}I$
τ

(c) Exponential lag
$O = \dfrac{I}{T_iS + 1}$
1
0.63
0
T_i

(d) Second-order lag

(e) First-order integral control; velocity control
$o(t) = \int i(t)\, dt$
$O = \dfrac{I}{S}$

(f) Second-order acceleration control
$o(t) = \int\int i(t)\, dt$
$O = \dfrac{I}{S^2}$

(g) Minus-first-order derivative control
$o(t) = \dfrac{di(t)}{dt}$
$O = IS$

Figure 11.6 Basic dynamic elements in tracking. The transfer functions for most elements are presented in both the time domain (lower-case letters) and the Laplace domain (capital letters).

Pure gain.   A pure-gain element describes the ratio of the amplitude of the output to that of the input. The element in Figure 11.6*a* has a gain of 2, since the output is twice the size of the input. High-gain systems, like the steering mechanism of a sports car, are highly responsive to inputs. However, they may sometimes lead to instability, as will be seen below. Low-gain systems tend to be described as "sluggish," since large inputs produce only small outputs.

Pure time delay.   The pure time delay, or transmission lag, delays the input but reproduces it in identical form T seconds later (Figure 11.6*b*). This would describe the response of a remotely located robot equipped with a television camera on the surface of the moon, controlled from a master on earth. Feedback from control signals sent to the robot would be reproduced precisely at the operator's terminal, but only after the four-second delay necessary for the signal to be relayed to the robot and for the picture to be relayed back to earth. A pure time delay has no effect on gain, nor does gain have any effect on time delay.

Exponential lag.   Some lags do not reproduce the input identically but instead gradually "home in" or stabilize on the target input. The *exponential lag*, shown in Figure 11.6*c*, is defined by its *time constant* T_I, which is the time that the output takes to reach 63% of its final value. The response shown in Figure 11.6*c* bears considerable resemblance to the human target-acquisition response described by Fitts' law in Figure 11.3*a*. In a more general sense, it describes the response of many systems with a built-in negative feedback loop to ensure that an output is reached. A similar response would be shown by the tires on a car with a hydraulic power-steering system following the command indicated by steering-wheel position or by the temperature in a thermostatically regulated room following a discrete input adjustment to the thermostat.

Second-order lags.   Like the lags in Figure 11.6*b* and *c*, the step response shown in Figure 11.6*d* also eventually reproduces the commanded step input; however, it does so only after considerable oscillation. This response is typical of many dynamic physical systems with mass and spring constants. Imagine, for example, the response of a large weight attached to a spring if the spring is suddenly displaced to a new position (Figure 11.7). Eventually the output will stabilize on its new commanded location, but only after the oscillations induced by the spring have died down. If the mass is heavy or the spring constant is small, the frequency of oscillations will be low. Thus the ratio of spring constant to mass determines this critical *resonant frequency* of oscillation. If the oscillation takes place in a viscous atmosphere (i.e., underwater), then its amplitude will be greatly attenuated and the oscillation will die out rapidly. In fact, if there is enough viscosity, it may not oscillate at all. This attenuation is defined by the *damping ratio* of the lag.

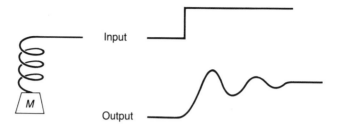

Figure 11.7   The mass-spring system providing an example of a second-order lag.

Integrator, velocity-control, or first-order system.   The step response shown in Figure 11.6e is a constant velocity (change in position per unit time) with a magnitude proportional to the step size. In calculus this response is defined by the time integral of the input. Notice that if the input is withdrawn, the velocity returns to zero, but the output is at a new location. Such systems are frequently encountered in manual control. An example is the relation between the angle of steering wheel deflection and the heading of a vehicle. A constantly held wheel position will produce a constant rate of change of heading, or rate of turn. Any first-order or velocity-control system must also be defined by its gain. A low-gain system is shown on the left of Figure 11.6e. A high-gain system is on the right. The term "order" refers to the number of time integrations in the transfer function. Therefore, with one integration, the system shown in Figure 11.6e is a first-order system. The first-order dynamic response is closely related to the exponential lag shown in Figures 11.3 and 11.6c. If a system makes a first-order response to the *error* rather than to the command input, the result will be an exponential lag. Since the response velocity to correct the error is proportional to the size of the error, then as the error is reduced, response velocity is reduced proportionately. In Figure 11.6c the velocity approaches zero, on the right side, as the error approaches zero.

Double-integrator, acceleration-control, or second-order system. A second-order system combines two integrators in series. The step response shown in Figure 11.6f is therefore the constant acceleration that would be obtained if the velocity response in Figure 11.6e were integrated a second time. Like the second-order lag, the pure second-order system is typical of any physical system with large mass and therefore great inertia when a force is applied. It is *sluggish* in that it will not respond immediately, particularly if the gain is low. When tracked, second-order systems also tend to be *unstable,* or difficult to control, because once the system does begin to respond, its high inertia will tend to keep it going in the same direction and cause it to "overshoot" its destination. The operator will have to make a series of reverse corrections. The oscillatory behavior of a human operator tracking a second-order system, in fact, often looks quite similar to the step response of the second-order lag, shown in Figure 11.6d. As an intuitive example of a pure second-order system, try to imagine moving a large smooth stone to a precise location on a surface of very smooth ice. Second-order systems are very prevalent in aviation, seagoing vehicles, and chemical processes.

Differentiator.   The minus-first-order, or differentiator, control system, shown in Figure 11.6g, will produce an output of a value equal to the rate of change of the input. The step response of the differentiator is theoretically a spike of infinite height and 0 width, since the "step" is an instantaneous change in position (and so has infinite velocity). As a result, the step response is not shown in Figure 11.6g. Instead, the *ramp response* is depicted. The system response is therefore just the opposite of the first-order system shown in Figure 11.6e. In the calculus representation these two are also opposite. If a time function is differentiated and then integrated, the original function will be recovered. In isolation, differential control systems are not frequently observed. An example might be an electrical generator in which the output (current) is proportional to the rate of turn of the input coils. However, differential control systems are of critical importance when they are placed in series with systems of higher order. They can reduce the "effective" order of the system by "canceling" one of the integrators and so make it easier to control. For example, a differentiator placed in series with the second-order system of Figure 11.6g will produce a first-order system. This point will be considered later.

Frequency-domain response.   The transient dynamic response of the elements shown in Figure 11.6 is in the time domain. Yet engineers are often more concerned about the response of these elements to periodic or random inputs. Indeed, most tracking studies involve continuous inputs. For some of the dynamic elements of Figure 11.6, the response to random or periodic inputs is intuitively quite predictable from the step response. A pure gain, for example, will reproduce a periodic signal perfectly but at a higher or lower amplitude, given by the value of the gain. For other elements, however, the response to periodic inputs is considerably more complex. In the chapter supplement we shall consider the frequency-domain response of different elements as they are used in human operator modeling.

Human Operator Limits in Tracking

The previous chapters of this book have identified a number of limitations in human information processing. Five of these limits in particular influence the operator's ability to track. These are the limits of processing time, information transmission rate, predictive capabilities, processing resources, and compatibility. Each of these will be described briefly in the context of manual control. Each will then be considered in more detail as they influence specific aspects of the manual-control task.

Processing time. The discussion of reaction time in Chapter 9 suggested that humans do not process information instantaneously. In tracking, a perceived error will be translated to a control response only after a lag, referred to as the effective time delay (McRuer & Jex, 1967; McRuer & Krendel, 1959). Its absolute magnitude seems to depend somewhat upon the order of the system being controlled. Zero and first-order systems are tracked with time delays from 150 to 300 msec. For a second-order system, the delay is longer, about 400–500 msec, reflecting the more complex decisions that need to be made (McRuer & Jex, 1967).

Time delays, whether the result of human processing or system lag, are harmful to tracking for two reasons: (1) Obviously, any lag will cause output to no longer line up with input. The error thus resulting, shown as the shaded region in Figure 11.8, will grow with the magnitude of the delay. (2) Often more seriously, when periodic or random inputs are tracked, delays will induce problems of *instability*, producing oscillatory behavior. These problems will be considered in greater detail below.

Bandwidth. Tracking involves the transmission of information, whether this is displayed as a command or as a disturbance-induced error. As discussed in Chapter 3, time-varying input and output signals may be quantified using continuous information theory, and it is not surprising that the same limitations of information transmission in discrete tasks are evident in continuous tracking as well. Thus Elkind and Sprague (1961), Crossman (1960), and Baty (1971) found that the limit of information transmission in tracking is between 4 and 10 bits/sec, depending upon the particular conditions of display. Crossman (1960), for example, found that maximum transmission rate is considerably greater with pursuit than with compensatory displays. Elkind observed that the rate increases if a preview of the input is available

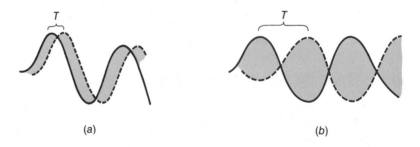

(a)   (b)

Figure 11.8 Error resulting from time delay: (*a*) small, (*b*) large.

before it is tracked. This is like an automobile driver previewing segments of the road ahead before the vehicle actually reaches them.

In Chapter 10 it was argued that the limits of serial reaction time were defined by the frequency of decisions, not by the complexity of those decisions. In tracking too there appears to be an upper limit in the frequency with which corrective decisions can be made that is more restrictive than the limit imposed by their complexity. This frequency limit in turn determines the maximum *bandwidth* of random inputs that can be tracked successfully; it is normally found to be between 0.5 and 1.0 Hz (Elkind & Sprague, 1961). This value corresponds quite closely with estimates that the maximum frequency with which corrections are exerted in tracking is roughly 2 times/sec (Craik, 1948; Fitts & Posner, 1967). Since two corrections are required for each cycle, this limit corresponds to a bandwidth of 1 cycle/sec. This limit appears to be a central one, related to processing uncertainty in the tracking signal, rather than a motor one, because operators have no difficulty in tracking *predictable* signals as high as 2–3 Hz (Pew, 1974; Pew, Duffenback, & Fensch, 1967). The limit of two corrections/per second in continuous tracking is slightly more restrictive than the maximum decision-making speed in the serial RT paradigm of 2.5 decisions/per second (Debecker & Desmedt, 1970; see Chapter 10). The reason for this difference can be explained by the fact that Debecker and Desmedt used the simplest possible 1-bit decision, whereas in random-input tracking there is typically a greater amount of uncertainty/decision.

Prediction and anticipation.   Fortunately, human operators are rarely placed in real-world environments in which they must track inputs at bandwidths so high that the limits on processing rate become restrictive. The more serious limits instead appear to be imposed when operators track systems like ships and aircraft that have lags. Here the operator must *anticipate* future errors on the basis of present values in order to make control corrections that will only be realized by the system output after a considerable lag. Consider the pilot of a supertanker sailing in a channel who realizes the vessel is off course and wishes to correct this. Because of the high inertia of the ship and its higher-order control characteristics, a correction delivered to the rudder will not substantially alter the ship's course for a matter of tens of seconds. Therefore, to stay within the limits of the channel effectively, the operator must base corrections upon future error and not present error. Corrections based on present error will be too late.

Future error, of course, equals the difference between future input and future output. In the case of ship control, future command input can easily be *previewed* (this is the view of the channel or path to be negotiated). But future output must be derived and anticipated, a function, as we have noted in Chapter 3 and again in Chapter 7, that humans do not perform effectively. In tracking, this limitation occurs in part because higher derivatives (velocity and acceleration) of the error signal must be perceived. Where a signal *is* at the moment is best indicated, of course, by where it is. But where a signal *will be* in the future is best indicated by its velocity and acceleration. Ample data are available to suggest that humans perceive position changes more precisely than velocity changes, and both velocity and position changes more precisely than acceleration (Fuchs, 1962; Gottsdanker, 1952; Kelly, 1968; McRuer et al., 1968; Runeson, 1975). Thus, when tracking slow, sluggish systems, the operator's perceptual mechanisms are called upon to perform functions for which they are relatively ill-equipped.

Processing resources.   Another source of difficulty in anticipation relates to the resource demands of spatial working memory, as discussed in Chapters 6 and 8. When anticipating where a sluggish, higher-order system like a supertanker will be in the future, it helps to be able to perceive its acceleration, but it is also important to be able to perform calculations and estimations of where that system will be in the future given an *internal model* of the system's dynamics (Gill, Wickens, Donchin, & Reid, 1982a; Moray, 1981; Pew & Baron, 1978). For the operator who is not highly trained, the operations based on this internal model demand the processing of working memory. Tracking thus is readily disrupted by concurrent tasks. The limits of human resources also account for tracking limitations when the operator must perform more than one tracking task at once, that is, in dual-axis tracking. This particular situation will be considered in more detail at the end of the chapter.

Compatibility.   The discussion of S–R compatibility in Chapter 9 emphasized that certain spatial compatibility relationships were relatively "natural". Because tracking is primarily a spatial task, it is apparent that these compatibility relations should affect tracking performance. The research on control and display relations in tracking suggests that indeed they do.

Effect of System Dynamics on Tracking Performance

The interface between human limitations and the dynamic properties of the system to be controlled determines the level of tracking performance. We shall consider the effects on performance of three important characteristics of those system dynamics: gains, time delay, and order. In certain combinations these variables produce problems of stability, a factor which will be considered separately.

Gain.   Both tracking performance and subjective ratings of effort appear to follow an inverted U-shaped function of system gain (Gibbs, 1962; Hess, 1973; McRuer & Jex, 1967). Whether tracking steps or compensating for random disturbances, systems with intermediate levels of gain are easiest to track. The advantage of middle-gain systems results from the tradeoff between the costs and benefits of more extreme gains. When gain is high, minimal control effort is required to produce large corrections. For example, the steering wheel on a sports car need be turned only slightly to round a curve. Thus, in a sense, high gain is economical of effort, and this economy is quite valuable when continuous corrections are required to track random input. On the other hand, gain that is too high can lead to overcorrections, oscillations, and instability if there are lags in the system (see below). This too is undesirable and can be eliminated by reducing gain. The crossover point of the two functions of instability and effort determines the "optimal" level of gain. The precise level of gain that is optimum cannot be specified in a general sense because it is determined by the extent to which effort at too low a level of gain and instability at high gain are the more important criteria.

Time delay.   Pure time delays are universally harmful in tracking, and tracking performance gets progressively worse with greater delays. The reason is apparent from the discussion on page 420. If a control input will not be realized until some point in the future, then the corrective input must be based upon the future value of error rather than its present value. Such anticipation, as noted, is imperfectly done.

The effects of exponential lags, on the other hand, are often less harmful. An exponential lag is, in a sense, a combination of a zero-order, or position, control and a first-order, or velocity, control. In Figure 11.6c note that immediately following the step input, the response of the exponential lag looks very much like that of the velocity control system in Figure 11.6e. Only later does it look like the response of the time delay in Figure 11.6b. In fact, when controlled with high-frequency corrections, a system with an exponential lag behaves very much like the first-order system. We shall see below that first-order systems have some substantial advantages over systems of zero order. It is these advantages that prevent exponential lags controlled at higher frequencies from exerting the kinds of harmful effects that the pure time delay does.

System order.   The effects of system order on all aspects of performance may be best described in the following terms: Zero-order and first-order systems are roughly equivalent, each having its costs and benefits. Both are also equivalent to exponential lags, which, as we have seen, are a sort of combination of zero- and first-order. However, with orders above first, both error and subjective workload increase dramatically. The reason that zero- and first-order systems are nearly equivalent may be appreciated by realizing that successful tracking requires that both position and velocity be matched (Poulton, 1957). Under some circumstances, matching one of these quantities might be more important than the other. Compare the two functions in Figure 11.9. In Figure 11.9a position error is reduced quite frequently to zero, but the velocities of input and output are rarely matched. In Figure

(a)   (b)

Figure 11.9   Styles of control: (a) minimized position error but high velocity
error, (b) minimized velocity error but high position error.

11.9b, while velocity is closely matched, the positions of input and
output rarely agree. Which form of tracking is superior? Clearly the
answer to this question depends upon the circumstances. If one were a
passenger in an aircraft, the response in Figure 11.9a would not suggest
a comfortable ride compared to that of Figure 11.9b, but if the aircraft
were flying at a low level, following a precise course with a minimum
margin for error, the performance in Figure 11.9b might be disastrous.

Viewing performance thus as a mixture of position matching and
velocity matching, the fact that the input to a zero-order control system
directly accomplishes the former and that input to a first-order velocity
system accomplishes the latter indicates why neither is unequivocally
preferable to the other. The intermediate level between zero- and first-
order control can be created either by linearly combining the outputs of
the two pure orders ("rate-aided" systems) or by varying the time
constant of an exponential lag (Wickens, in press). These systems
generate performance that is also equivalent to either pure first- or pure
zero-order control, depending upon the relative importance of making
a position or velocity match (Chernikoff & Taylor, 1957).

In contrast, control systems of the second order and higher are
unequivocally worse than either zero- or first-order systems (Kelly,
1968). The problems with second-order control are manyfold. As noted
above, in order to control any higher-order system effectively, one must
anticipate its future state from its present. To do so requires that higher-
error derivatives be perceived as a basis for correction, a process known
as "generating lead," and humans, as we have discussed, do not perform this function well. As McRuer and Jex (1967) observed, the operator's effective time delay is also longer when higher derivatives must be
processed under second-order control. This increased lag contributes
an additional penalty to performance.

Second-order systems may be controlled by two strategies. The one
described above requires the operator to perceive the higher error
derivatives continuously and respond smoothly on the basis of this
information. An alternative strategy of second-order control is sometimes referred to as "bang-bang," double-impulse, or time-optimal con-

trol (Gill et al., 1982a; Hess, 1979; Young & Miery, 1965). Here the operator perceives an error and reduces it in the minimum time possible with an open-loop "bang-bang" correction. As shown in Figure 11.10, this is accomplished by throwing the stick "hard over" in one direction to generate maximum acceleration for half the interval, and then reversing the stick to produce maximum deceleration for the other half, monitoring and checking the result at the end. Because the double-impulse strategy reduces large errors in the shortest possible time, it is referred to as a form of "optimal control" (Young, 1969).

While the double-impulse control eliminates the need for continuous perception of error derivatives of smooth analog control, it does not necessarily reduce the total processing burden. More precise timing of the responses is now required, and an accurate "internal model" of the state of the system must be maintained in working memory, in order to apply the midcourse reversal at the appropriate moment (Jagacinski & Miller, 1978). Also as suggested by Figure 11.10, the bang-bang strategy will produce high velocities. Like the reponse shown in Figure 11.9a, there are conditions when a lower-velocity "smooth ride" is preferable. The *optimal control models* of tracking behavior discussed in the chapter supplement dictate the appropriate strategy, given the importance of a smooth ride versus low error.

Instability. A major concern in the control of real-world dynamic systems is whether or not control will be *stable*—that is, whether the output will follow the input and eventually stabilize without the excessive oscillations shown by the second-order lag in Figure 11.6d. Oscillatory and unstable behavior can result from two quite different causes: positive and negative feedback.

To illustrate *positive feedback systems*, imagine two people sleeping under an electric blanket with dual temperature controls, one for each side. Suppose that the controls inadvertently become switched. If A now feels cold, she will turn the heat up. This will lead to an increase

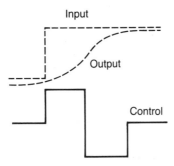

Figure 11.10 Required "bang-bang" response to reduce an error with a second-order system in minimum time.

in heat on the other side causing B to turn the heat down, and thereby leading A to feel still colder. Person A will then adjust the heat still higher and the resulting chain of events is evident. This is an example of a positive feedback system: An error once in existence is magnified.

While the above example is unlikely to occur in real-world systems except if switches are inadvertently misconnected, there is a second kind of positive feedback which is not unusual in aviation systems. An intuitive example of this kind of positive feedback occurs when one must balance an inverted pendulum. A computer analog of this task was built as the *critical instability tracking task*, described as a workload assessment technique in Chapter 8 (Allen & Jex, 1968; Jex, McDonnel, & Phatac, 1966). The feedback loop of the "critical task," as it is commonly called, is shown in Figure 11.11. An error e, once detected by the system, will generate a proportional output velocity o, whose value is determined by the gain λ. However, unlike "purposeful" human control in which negative feedback subtracts this velocity from the existing error, the positive feedback loop in the critical task adds the velocity to the error, thereby increasing the rate of error movement away from the center. This response is similar to the dynamics of a balanced stick. If there is a small error (from the vertical) it will begin to fall, and its rate of falling (increase in error) will, to a point, increase as it falls farther. The overall rate of falling will furthermore be proportional to the shortness of the stick, analogous to the value of λ.

Positive feedback loops characterize a number of complex dynamic vehicles. An example is the control of a booster rocket with its guidance controlled by swiveling tail-mounted engines. Positive feedback loops may also be found in certain aspects of the control of helicopters and other complex aircraft. Like second-order systems, those with positive feedback are universally harmful for the obvious reason that they cannot be left unattended. If control is not exercised they will eventually diverge, just as the inverted pendulum balanced on the

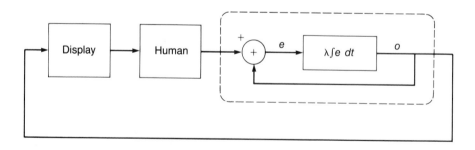

Figure 11.11   Dynamics of critical instability tracking task. From "A 'Critical' Tracking Task for Manual Control Research" by H. R. Jex, J. P. McDonnel, and A. V. Phatac, 1966, *IEEE Transactions on Human Factors in Electronics, HFE–7*, p. 139. Copyright 1966 by the Institute of Electrical and Electronics Engineers, Inc. Reprinted by permission.

fingertip must eventually fall unless the finger is moved back under the top of the stick.

Negative feedback systems are more typical. Humans and most well-designed systems function in such a way as to reduce rather than increase detected errors. This is the property of a negative feedback system, described clearly in treatments by Moray (1981), Kelly (1968), Jagacinski (1977), and Toates (1975). Good stability normally results from such "purposeful" control action. However, there are certain occasions when even a negative feedback loop with the best error-correcting intentions produces oscillatory or even unstable behavior. A potentially disastrous example in aircraft control occurs when *pilot-induced oscillations* are produced. Here the vertical path of the aircraft swings violently up and down, with growing amplitude (Hess, 1981).

Instability caused by negative feedback results from high gain coupled with large phase lags. For example, the second-order system is "sluggish" with a long lag in its response. Hence control of second-order systems tends to be unstable. However, instability may also result with lower-order systems when there are long delays in system response.

The reason why high gain and long phase lag collectively produce instability may be appreciated by the following example taken from Jagacinski (1977). Imagine that you are adjusting the temperature of shower water to your "ideal" reference value (a command input). You are controlling to reduce the error and so acting as a negative feedback system. However, because of the plumbing, there is a lag between your adjustment of the water faucet and the change in water temperature—the source of the perceived error used to guide correction. If your gain is high, then when you feel initially cold, you increase the hot water by a large amount and will continue to increase it so long as you feel cold. As a consequence, you will probably overshoot, scald yourself, and the error will now be on the "hot" side. If your gain remains high, then the compensatory cooling correction will also be "overapplied," and the water will, after a lag, become too cold. The eventual temperature-time history will be a series of growing oscillations.

Clearly in these circumstances you must reduce your gain in order to avoid the unstable behavior (and scalded backside!) resulting from the time lag. In the shower example a gain reduction involves applying a smaller corrective turn of the faucet in response to the detected error—tolerating a mild discomfort now in anticipation of an eventual stable response.

The difference between high- and low-gain systems and their association with unstable and sluggish behavior, respectively, is an important one in human performance, because gain may be conceived as a "bias parameter," like the response criterion in signal detection or the speed-accuracy set in reaction time. It is a parameter, then, which can be strategically adjusted to different values according to different environmental conditions or strategic goals. The difference in tracking performance of the two systems shown in Figure 11.9, for example, could be attributed in part to a difference in the gain of the feedback system that is tracking the error: high in Figure 11.9a, low in Figure 11.9b.

The question of stability and its dependence upon gain and lag is, of course, critically important in the design and testing of piloted vehicles and represents a major application of manual-control research. However, engineering-oriented research on stability is normally performed using the language of the *frequency* domain. We described this language briefly in Chapter 4 when discussing how the speech signal was represented. The supplement to the present chapter covers the frequency domain analysis of tracking in more detail, with the goal of making the sometimes mystical language of this area more understandable to the psychologist.

There is an alternative control strategy for making corrections when the lag is long. This strategy is to base your control correction on the trend of the error rather than its absolute level. Thus, in the shower example, if you feel that the water is getting warmer even if you are still too cold in an absolute sense, this trend can serve as signal to stop increasing the heat.

This control strategy, based on error rate rather than error value, should by now be familiar. On pages 422 and 424 we suggested that it formed the basis of anticipation when controlling systems with long lags. Here we see that this strategy is essential, because it is often necessary to avoid instability. Also when describing the control elements in Figure 11.6, we suggested that a differentiator could "cancel" an integration in controlling higher-order systems. When the human responds predictively on the basis of trends, he is effectively becoming a differentiator and so is canceling one of the integrators of second-order dynamics. In the following section we will see how certain tracking displays have been modified in order to make this prediction easier and induce humans to control like differentiators.

Displays

Input prediction and preview. Several times in this chapter we described the problem associated with prediction and anticipation in tracking. These, in fact, can be divided into problems of predicting the command input and problems of predicting the future trajectory of the plant. The future of the plant, in turn, is determined by operator corrections or by disturbance inputs. When there are lags in the tracking loop of T seconds caused either by the human operator's time delay or by system lags, then it is essential to know what the error will be T seconds into the future, so that corrections can be formulated on the basis of this predicted error, not the present information. Thus, if a future error is perceived and corrected now, the system will realize its appropriate correction the T seconds later that will be appropriate.

Clearly the future input will be most accurately available when there is *preview*. The automobile driver, for example, has preview available of the course of the road ahead. Figure 11.12 shows preview as it might be presented on a typical tracking display with the future course shown on the top of the display. The large benefits of preview (Crossman, 1960; Elkind & Sprague, 1961; Reid & Drewell, 1972) result primarily because it enables the operator to compensate for processing lags in the tracking loop. The fact that the operator's time delay is only in the order of 200–500 msec suggests that half a second of preview should be all that is needed when one is tracking systems that have no lags of their own. Reid and Drewell (1972) varied the amount of preview available while subjects tracked a first-order system. The results shown for one subject in Figure 11.13 suggested that performance did indeed benefit greatly for the first half second of preview offered but improved minimally with greater amounts. On the other hand, when there are longer system lags, preview is used farther into the future. The supertanker pilot is "tracking" the channel several hundred yards ahead of the bow of the ship. This is a command input signal that will not be traversed by the ship until minutes later.

In the absence of preview, the human operator must use whatever information and computational facilities are available to *predict* the future course of the input. To some extent this prediction may be based upon the statistical properties of the input. For example, if the bandwidth of a random input is low, the present position and velocity provide some constraints upon future position. For example, even if we had no preview of the input function shown in Figure 11.12, we would

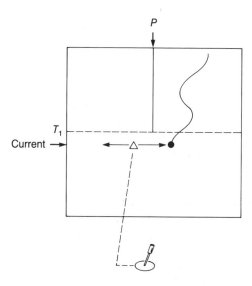

Figure 11.12   Tracking with preview.

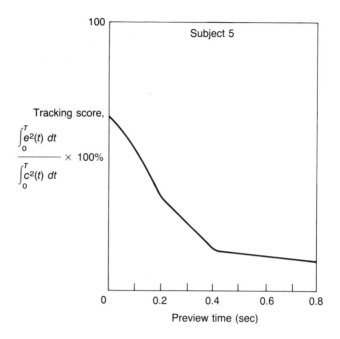

Figure 11.13   Effect of varied preview interval on tracking performance. From "A Pilot Model for Tracking with Preview" by D. Reid and N. Drewell, 1972, *Proceedings, 8th Annual Conference on Manual Control* (Wright Patterson AFB Ohio Flight Dynamics Laboratory Technical Report AFFDL–TR–72, 92) (p. 200), Washington DC: Government Printing Office.

consider it unlikely that the input would be to the left of point P at time T_1. Our past experience with this input tells us that it just doesn't change that rapidly.

To the extent that the input is nonrandom or contains periodicities, the constraints on future input are increased considerably and prediction becomes easier. When this occurs, the operator can track using what Krendel and McRuer (1968) described as a *precognitive mode*. Poulton (1957) has carefully described the manner in which both statistical and repetitive or stored properties of the input signal can facilitate tracking behavior.

Output prediction and quickening.   As we have described previously, the future trajectory of the output may be predicted with some assurance given its current position, velocity, and acceleration. The best estimate of where a higher-order system with some mass and inertia will be in the future is provided by a combination of its present position and its higher derivatives. This kind of prediction, as we have seen, is not readily done and extracts a heavy toll on operator resources. In trying to reduce this burden, engineering psychologists have developed displays in which a computer estimates error (or output) derivatives and explicitly presents these as predicted symbols of future position (Kelly, 1968; Gallagher, Hunt, & Williges, 1977). This format is called a *predictive display*. A typical two-dimensional predictive display for aircraft control investigated by Jensen (1981) is shown in Figure 11.14. The plane's current position in the sky is shown as the wide markers in the middle of the display designated "fixed airplane." Its estimated position four seconds into the future is shown by the predictor element. The display indicates that the future will find the aircraft closer to horizontal and a little farther to the left than its current position.

The predictive elements of Figure 11.14 may be "driven" by directly computing the position, velocity, and acceleration of the present system state and adding these values together with appropriate weights. Alternatively, these values may be inferred by directly measuring different internal states of the system (Gallagher, Hunt, & Williges, 1977). To provide an example of these two computational procedures, a predictive display of a car's future lateral position on the highway could be driven by computing the present values of its lateral position, velocity, and acceleration. Alternatively, since these three values are roughly equivalent to the current position of the car, the heading of the car, and

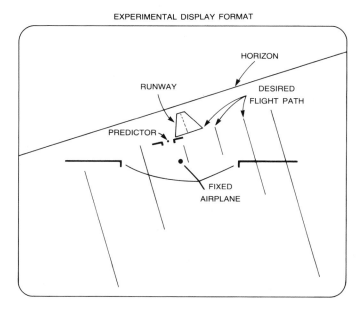

EXPERIMENTAL DISPLAY FORMAT

Figure 11.14 Typical predictor display in flight control. From "Prediction and Quickening in Prospective Flight Displays for Curved Landing and Approaches" by R. J. Jensen, 1981, *Human Factors, 23,* p. 358. Copyright 1981 by The Human Factors Society, Inc. Reproduced by permission.

the deflection of the steering wheel, respectively, then the latter three variables could be directly measured, weighted, and summed to provide an accurate predictive display.

Every predictive display must make some assumptions about the future forces acting on the system (either from operator control inputs or from disturbances). For example, is it assumed in predicting the future trajectory of the system that the operator will apply an optimal correction, a suboptimal correction, or no correction at all? The nature of these assumptions may differ somewhat from display to display and will determine the accuracy and therefore the effectiveness of the display. It is also apparent that the accuracy of any prediction will decline into the future. Just how far into the future a prediction is valid depends on such things as the sluggishness of the system and the frequency of control or disturbance inputs. Predictive accuracy in the future will increase with more sluggish systems and with lower-frequency inputs.

No matter how the predictive information is derived, predictive displays have proven to be of great assistance to operators' tracking of higher-order systems. This theme appeared in Chapters 3 and 7 and will be reiterated in Chapter 12 as well. Jensen (1981) demonstrated the value of prediction based upon both velocity and acceleration information in the cockpit display shown in Figure 11.14. In an excellent discussion of the topic, Kelly (1968) described the tremendous benefit of predictive displays for submarine depth control, an example of very sluggish third-order dynamics.

In 1954, Birmingham and Taylor proposed a technique known as *quickening* that was closely related to the predictive display. A quickened display presents only a single indicator of "quickened" tracking error which is calculated by combining the present error position, velocity, and acceleration. Like the prediction element in a predictive display, the quickened element indicates where the system will be likely to be in the future if it is not controlled. Unlike the predictive display, a quickened display has no indication of the current error. The justification for this absence is that the current error provides no information that is useful for correction. This, of course, has the disadvantage that there are certainly times when you want to know where you are and not just where you will be.

An important advantage of quickening over the predictive display is that the former contains just one element and so is more economical of space. A predictive display always needs to include a time dimension. In order to provide this benefit of display economy without incurring the cost of an inaccurate picture of the present, Gill et al. (1982b) developed what they called a *pseudoquickened* display. Like the quickened display, only a single symbol was presented, but this symbol accurately corresponded to true position. The predicted or quickened error was indicated by intensity changes on one side or the other of the error signal. These intensity changes, integral to the main error cursor, "commanded" the operator to respond at a time and in a direction necessary to realize optimal control.

Pursuit versus compensatory displays. The goal of tracking is to match the output to the input, or to minimize the error. These two seemingly equivalent statements define the *pursuit* and *compensatory* display formats, respectively. As described earlier in the chapter, the operator with a pursuit display views the command input and the system output separately (as influenced by disturbance inputs and purposeful control responses). In the compensatory display, only the difference between these two—the error—is portrayed. Pursuit displays generally provide superior performance to compensatory (Poulton, 1974) for two major reasons. These relate to the ambiguity of compensatory information and the compatibility of pursuit displays.

On the compensatory display, the operator is unable to distinguish between the three potential causes of error: command input, disturbance input, and the operator's own control action. As a consequence of this ambiguity, control is more difficult than in the pursuit display when command and disturbance inputs can be distinguished. If, however, there is only one source of input, then the advantage of the pursuit display decreases, since errors on the compensatory display are now less ambiguous. Flying an aircraft toward a fixed runway, for example, is a tracking task in which there is information only in the disturbance input. The runway, representing the command input, does not move. On the other hand, flying an aircraft toward the runway on an aircraft carrier now has a command input added if the carrier is moving or rolling in the sea swells. Flying a space vehicle toward a moving target is a task with only command input, since there will be no environmental disturbances acting on the vehicle (this assumes that the operator "knows" the influence of all commands delivered by herself to the vehicle).

Even when there is only a command input present, however, the pursuit display will still provide some advantage because its stimulus response compatibility is greater (see Chapter 9). If the command input is suddenly displaced to the left, this input will require a leftward correction on the pursuit display. A left-moving stimulus thereby is corrected with a leftward response. This is an inherent motion compatibility that is consistent with the operator's tendency to move toward the source of stimulation (Roscoe, 1968; Roscoe, Corl, & Jensen, 1981; Simon, 1969). On the other hand, in the compensatory display the left-moving command input will be displayed as a right-moving error. In this case, a right-moving stimulus now requires an incompatible leftward response. Whenever some portion of tracking input is command, this compatibility factor will benefit the pursuit display.

Pursuit versus compensatory behavior. Previous discussion in this chapter emphasized the importance of the closed, negative feedback loop in tracking. We assumed that operators continuously process the difference between where they are and where they would like to be and respond appropriately. That is, they *compensate* for an error. This style of tracking is referred to as closed loop, or *compensatory* behavior. The operator need not, however, process the error directly. When tracking with a pursuit display, the operator may attend only to the input. If there are no disturbances acting on the plant and the operator knows the plant dynamics, then he may ignore the output and assume that it will follow the commands that he has provided for the system to track the input. If he ignores the output, then he *cannot* be processing the error, since the error is by definition the difference between input and output. This strategy is referred to as open-loop or *pursuit* behavior. It leads to more efficient tracking because, unlike compensatory behavior, pursuit behavior does not require that an error be present in the first place in order to generate a corrective response.

The contrast between pursuit and compensatory behavior is demonstrated by the two mechanisms which the human eyeball uses to track moving targets in order to keep them in foveal vision (Young & Stark, 1963). The *saccadic* mode is compensatory. Based upon a discrepancy between where the target is located and where the eye is fixated, a discrete jump, or saccade, is programmed to reduce the error. (This is essentially a zero-order system. The size of the movement is proportional to the size of the error.) There is, however, a *pursuit* mode that generates a constant velocity of eyeball rotation whenever the target shows constant velocity motion. The two mechanisms work fundamentally independently and in parallel.

Although it appears logical that pursuit behavior will occur with pursuit displays and compensatory behavior with compensatory displays, this association is not necessary in a pursuit display because the operator may focus attention either on error or input. Pew (1974) has described much human tracking behavior as a combination of the two behaviors. The operator will track in the pursuit mode when possible, but he will supplement it with compensatory corrections when errors build up. Conversely, pursuit behavior is possible even with a compensatory display when there are no disturbances, although this is more difficult. An operator who knows precisely the effects of control manipulations on the plant response can mentally "subtract" this contribution from the perceived error. The difference is the command input, and this can be tracked directly with pursuit behavior.

Krendel and McRuer (1968) have contrasted compensatory and pursuit behavior to represent the progression of skill acquisition in tracking. In their *successive organization of perception* (SOP) model of tracking, the authors propose that three modes of tracking, as shown in Figure 11.15, describe the progression of tracking behavior with practice. In the compensatory mode (Y_c), the operator acts as an error corrector with a corresponding time lag which leads to poorer performance. With greater skill, the operator becomes progressively more able to track in the pursuit mode (Y_p), thereby responding directly to

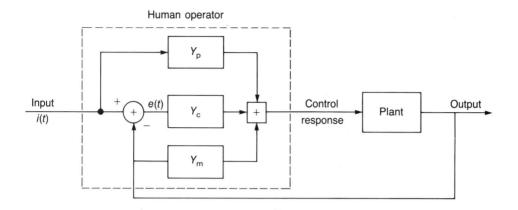

Figure 11.15 Successive operation of perception (SOP) model of Krendel and
 McRuer (1968) shows compensatory loop (Y_c), pursuit loop (Y_p), and
 memory or precognitive loop (Y_m). From "A Review of Quasi-linear Pilot
 Models" by D. T. McRuer and H. R. Jex, 1967, *IEEE Transactions on
 Human Factors in Electronics, 8*, p. 240. Copyright 1967 by the Institute of
 Electrical and Electronics Engineers, Inc. Reprinted by permission.

the input and nullifying some of the existing lag. Finally, a third mode
of tracking is identified. This is known as the *precognitive* mode (Y_m),
which is possible only when the input is nonrandom. In this case, the
operator can store the input patterns in long-term memory and respond
on the basis of this stored information. Since the inputs can be predicted in advance, tracking can be carried out with no lag whatsoever
(Poulton, 1957). This behavior is truly open loop, since it may take
place in the total absence of feedback. For example, as we traverse the
curves of a familiar driveway, we may do so if we choose with our eyes
closed. This is an example of precognitive tracking.

Precognitive, pursuit, and compensatory behavior may be experimentally dissociated from each other by the following means: If the
total display can be blanked for periods of time, and the operator
continues to reproduce the input successfully, behavior must be precognitive. If just the output symbol is eliminated and tracking is successful, then behavior is pursuit. If not, then behavior must have been
compensatory. Unfortunately these manipulations somewhat distort
the tracking task in their effort to determine how the operator can track.
A more ingenious way of naturally distinguishing the characteristics of
pursuit and compensatory behavior without artificially disrupting portions of the display is to exploit the fact that disturbance inputs can
only be corrected with compensatory behavior. Since these are not
shown as command inputs they cannot be pursued. If the operator
tracks an input function that is made up of several independent components and if these are distributed between command and disturbance,
then two separate transfer functions of the human operator simultaneously tracking the two sets of inputs can be compared (Allen & Jex,

1968). If the two transfer functions are identical, then all behavior must be compensatory as this is the "lowest common denominator" of tracking behavior. If they differ, the differences will reflect the better pursuit behavior used to track the command input.

MULTIAXIS CONTROL

Humans must often perform more than one tracking task simultaneously. The aircraft pilot, for example, controls both the pitch and roll of the aircraft; when we drive we track lateral position, heading, and velocity. Even riding a bicycle involves tracking lateral position while also stabilizing the vertical orientation of the bike. In general, as discussed in Chapter 8, there is a cost to multiaxis control that results from the division of processing resources between tasks. However, the severity of this cost is greatly influenced by the nature of the relation between the two (or more) variables that are controlled and the way in which they are physically configured.

Cross-Coupled and Hierarchical Systems

A major distinction can be drawn between multiaxis systems in which the two variables to be controlled as well as their inputs are essentially independent of each other, and those in which there is cross-coupling so that the state of the system or variable on one axis partially constrains or determines the state of the other. An example of two basically independent axes is provided by the control of attitude and elevation in a gunnery task. On the other hand, there is a small degree of cross-coupling between the pitch and roll axes of aircraft control. What the pilot does to control the roll (rotation around the axis of the aircraft fusilage) and resulting change in aircraft heading has a small effect on the pitch (nose up, nose down) and therefore the altitude change of the aircraft. At the far extreme of cross-coupling, control of the heading and lateral position of an automobile on the highway are highly cross-coupled axes. In this case, the two cross-coupled tasks are considered to be *hierarchical*. That is, lateral position cannot be changed independently of a control of heading. The steering wheel, which directly controls heading, is *used* to obtain a change in lateral position.

Many higher-order control systems in fact possess similar hierarchical relationships. Lower-order variables must be controlled in order to regulate or track higher-order variables. Figure 11.16 shows analogous representations of three such hierarchically organized control systems: automobile control, aircraft heading control, and submarine depth control. In each case the operator controls the variable on the far left, with the final goal of tracking the variable on the far right. For driving, this is a second-order task, since there are two integrals in the control loop. For flying and submarine control, it is a third-order task. Variables on the left are said to be "inner loop," and those on the right define "outer-

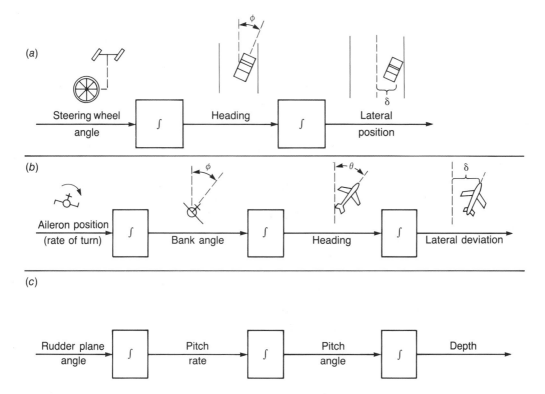

Figure 11.16   Three examples of hierarchical control systems: (a) automobile driving, (b) aircraft heading control, (c) submarine depth control.

loop" variables. Kelly (1968), Roscoe (1968), and Roscoe et al. (1981) provide good explanations of hierarchically organized control systems.

While hierarchical systems may often have a number of *displayed* elements (e.g., steering-wheel angle, vehicle heading, lateral position), there is normally only one control element which gives control over the inner-loop variable (i.e., the steering wheel). Kelly (1968) and Roscoe (1968) have argued strongly that it is important to organize the displayed elements coherently in a manner that conforms to the operator's internal model of the system being controlled. Thus, if three variables bear an ordered relation to each other by virtue of three increases in system order, as in the submarine depth control shown in Figure 11.16, then displays of the three variables should be presented in a format such that the coherent ordering is preserved.

The strategy of hierarchical loop control is one in which the operator typically sets goals for the highest-order, or "outer-loop," variable (i.e., a change in the aircraft lateral position from a desired path). To accomplish this, he must, in turn, control the variable of the next lower level (a change in aircraft heading, which is equal to lateral velocity). This, in turn, places constraints on the variables of the next inner loop (bank angle equal to rate of change of heading) whose rate of change must then be controlled by the innermost loop, aileron control. Hierarchical control thus involves the parallel efforts to control "outer loops" through regulation of inner loops. At any given time the operator may be focusing attention on error of inner-loop variables, outer-loop variables, or both. Because the inner-loop control is often rapid and is somewhat "mindless" in pursuit of the more purposeful, cognitive outer-loop goals, systems are being designed with automation of the inner-loop control so that the operator has a direct means of controlling outer-loop variables. For example, the automated cockpit will allow the pilot to "dial in" a desired heading on the autopilot, and the automated control system will accomplish the necessary tracking of inner-loop variables (ailerons, bank angle) to attain the goal. The issue of automated tracking control will be discussed further in Chapter 12.

Factors That Influence the Efficiency of Multiaxis Control

Display separation.   Whether hierarchical, cross-coupled, or independent, multiaxis control will obviously be harmed if the error or output indicators are increased in separation across the visual field. When the separation is sufficiently great so that indicators may not be simultaneously in foveal vision, then, as discussed in Chapter 7, operator scan patterns may provide a useful index of the sequence of information extraction from the display. In some of the earliest classic work on formatting of aviation displays, Fitts, Jones, and Milton (1950) provided fundamental data on the importance of various sources of information in flight control derived from instrument scan patterns. The principles of display formatting derived from scan information were summarized in Chapter 7. More recently, Harris and Christhilf (1980) have been able to discriminate between momentary fixations on instruments that extract information necessary for tracking control (minimum dwell time of 500 msec) and those that are basic check reading for the purpose of monitoring instrument position (200–300 msec). It appears that the shorter duration is sufficient to obtain an approximate "ball park" estimate of position; the longer fixations include an added amount of time necessary for the precise assessment of position and velocity.

The problems associated with visual scanning and sampling strategies were discussed in some detail in Chapter 7. The resulting loss of performance was attributed to "peripheral interference." When the eye is fixated upon one display, the other will be in peripheral vision and therefore generate data of lower quality. In tracking this does not mean, however, that a display can only be tracked if it is fixated. There is good evidence that a considerable amount of parafoveal information concerning both position and velocity may be used to attain effective control (Allen, Clement, & Jex, 1970; Levison, Elkind, & Ward, 1971; Wickens, in press). Wickens, Sandry, and Hightower (1982) have found, furthermore, that a peripheral axis in a multiaxis display is tracked considerably better when it falls in the left visual field than in the right. This finding makes sense in terms of the structure of the visual system and cerebral hemispheres discussed in Chapter 4. The left visual field has direct access to the right hemisphere. The right hemisphere processes spatial information and so would play a dominant role in tracking.

The obvious solution to the problems of degraded performance with the separation of tracking displays is the same as it was with the discrete tasks discussed in Chapter 7: Minimize display separation by bringing the displayed axes closer together. In the extreme, two control dimensions may be represented by motion of a single variable in the x and y axes of space—an integrated display. In this case, peripheral interference no longer contributes a cost to multiaxis tracking, and other sources of diminished efficiency may be identified. Three such sources, related to resource demand, control similarity, and display/control integration will now be considered.

Resource demand.   Navon et al. (1982) note that the cost to multiaxis tracking with a single display and control is surprisingly small. It seems that the integration of the two axes as two integral dimensions of a single object fosters a certain amount of "cooperation" between the tasks. Nevertheless, to the extent that some interference between tracking tasks results from resource competition, as discussed in Chapter 8, it is not surprising to find that the cost of dual-axis control increases as the resource demands of a single axis are increased. As an example, Baty (1971) had subjects time-share tracking of two zero-order, two first-order, and two second-order systems. He found little evidence for a difference in interference between zero- and first-order control—the two were described earlier in this chapter to be substantially similar in their demands when performed singly. However, the magnitude of interference imposed by time-shared second-order control was considerably greater.

Similarity of control.   An experiment performed by Chernikoff, Duey, and Taylor (1960) indicates that the increasing resource demand of higher-order control may under certain circumstances be lessened by making the control dynamics more similar between the two axes. Different orders of control require the operator to adopt different control strategies, a situation that will be discussed further in the chapter supplement. Therefore, it is apparently more difficult for the operator to time-share control with two different strategies than to maintain a single strategy for both axes.

In their experiment, Chernikoff et al. asked subjects to perform dual-axis tracking with all three control orders (zero, first, and second) in all pairwise combinations. It was therefore possible to evaluate error on a given axis as a function of the order of control on the time-shared axis. The data are shown in Figure 11.17. For zero-order tracking, error increases

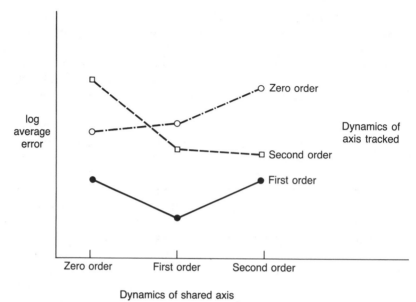

Figure 11.17 Effect of time-shared control order on tracking performance at different orders. Note that the absolute height of the three curves is arbitrary. What is important is the change in error within each curve. From "Two-Dimensional Tracking with Identical and Different Control Dynamics in Each Coordinate" by R. Chernikoff, J. W. Duey, and F. V. Taylor, 1960, *Journal of Experimental Psychology, 60*, p. 320. Copyright 1960 by the American Psychological Association. Adapted by permission of the authors.

with higher order on the paired axis. However, for first-order tracking, error is lower when the time-shared axis is also of first order than it is when time-shared with the lower, but different, zero-order system. Likewise, for second-order tracking, performance is no worse when the shared axis is also second order than when shared with the lower but different first- and zero-order controls. In fact, it is a good bit higher when shared with zero order. The advantage of the lower resource demands of zero- and first-order tracking is nullified by the greater interference resulting from the fact that separate dynamics must be controlled. Wickens, Tsang, and Benel (1979) have found that the requirement to share different dynamics is also a contributor to increased subjective mental load, as well as reduced time-sharing efficiency.

The findings of Chernikoff et al. can be placed in the context of the causes of dual-task interference described in Chapter 8. In that discussion, task interference was generally assumed to be an increasing function of task similarity. An exception to this principle was noted, however, when similarity also allows some degree of *cooperation* between tasks. In Chapter 7, this cooperation was seen in the phenomenon of redundancy gain. In the present example, the cooperation is manifest when a single control strategy can be activated in working memory and applied to both axes simultaneously.

Display and control integration.   When two axes are tracked, the degree of display or the control integration may be varied independently. The four quadrants in Figure 11.18 show the four different display-control combinations that can be generated by integrating or separating the axes on both displays and controls. Further options are available when the separate-separate display is employed (quadrant IV), for now it is possible to present the two axes either in parallel or at right angles to each other. Comparing the latter two configurations, Levison, Elkind, and Ward (1971) found that the right-angle placement is superior. In this case the dissimilarity (between the two axes of motion) is helpful because the control responses on the two axes are independent. To the extent that the two hands are moving in a common plane, it will be more difficult to maintain independence of their separate responses. Kelso, Southard, and Goodman (1979) reported a similar difficulty in maintaining independent timing of responses by two hands moving in similar trajectories.

Investigations by Chernikoff and Lemay (1963) and Baty (1971) compared the four quadrants of Figure 11.18. These studies concluded that when two axes with like dynamics are shared, there is an advantage to integrating display and control (DI, CI), over separating them (DS, CS) (quadrant I is better than quadrant IV). Chernikoff and Lemay also compared conditions of quadrant II (DI, CS) with those of quadrant III (DS, CI) and found performance in quadrant II to be superior. That is, the effect of integrating displays was generally more beneficial than that of integrating controls. This finding is to be expected. Integrating displays produces a clear reduction in visual scanning and allows a

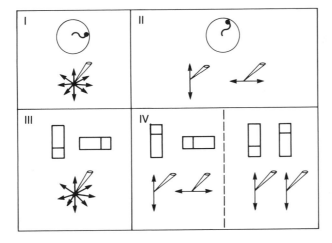

Figure 11.18   Different configurations of display and control integration.

holistic object perception (see Chapter 5). Integrating controls reduces total motor activity but increases the possibility of response interference, since motor control for both tasks must now be generated by the common hemisphere (see Chapter 8).

The most interesting results of Chernikoff and Lemay's study occurred when the same set of display-control configurations were employed with two axes of different dynamics (zero and second order). Under these conditions, as noted in the previous section, there is an overall cost to performance. Chernikoff and Lemay found, however, that this "cost of dissimilarity" was least when the display and control were both separate, as in quadrant IV (the condition that had generated greatest cost with similar dynamics), and was greatest when the control was integrated, as in quadrants I and II.

These results are again in keeping with the competition/cooperation factor in dual-task performance discussed in Chapter 8. When cooperation is possible between some aspects of two tasks, there is a benefit to be gained by maintaining high proximity in their configuration; in this case the proximity is achieved by display and control integration. When there is competition—different control strategies required—proximity should be minimized by separating the tasks. Furthermore, this separation will be most beneficial when the separated elements are those that would be most likely to compete. In the case of tracking, the competition is presumably highest at the motor end, where humans have a difficult time executing different independent responses that are not highly practiced. Therefore, the greatest savings to the cost of mixed dynamics are realized by separating the controls.

Auditory Displays

As we have described in the preceding section, a major difficulty with multiaxis control occurs when the operator must scan between displays. As noted, this difficulty may be reduced by display integration, but this procedure can also impose some cost if the dynamics differ. Furthermore, there are clearly limits to how tightly information can be condensed in the visual field without encountering problems of display clutter (see Chapter 7). One potential solution is to present tracking-error information through the auditory modality. The discussions of multiple resource theory in Chapter 8 suggested that the auditory modality was indeed capable of processing spatial information. After all, we do this whenever we turn our head to the source of sound. It is important to realize, however, that audition is probably less intrinsically compatible with spatial processing than is vision (see Chapter 9). These properties suggest that single-task auditory tracking will never be superior and will probably be inferior to single-axis visual tracking, but it may provide benefits in environments with a heavy visual load.

Generally speaking, all of these hypotheses have been confirmed by the experimental data (Wickens, in press). The earliest investigation of auditory tracking, carried out in World War II, examined if the auditory modality could be used to convey turn, bank angle, and airspeed information in an aircraft simulator (Forbes, 1946). The display, known as "flybar," indicated turn by a "sweeping" tone from one ear to the other, whose sweep rate was proportional to turn rate. Bank angle was indicated by pitch changes in one ear or the other, and airspeed by the frequency of interruption of a single pitch. With sufficient training on this display, Forbes found that both pilots and nonpilots could fly the simulator as well as with the visual display.

More recently, with more precise performance measurement, both Isreal (1980) and Vinge (1971) have found that auditory tracking is nearly, but not quite, equivalent to visual. In the single-axis display used by these investigators, error was represented by the apparent spatial location of a tone. This was adjusted by playing tones of different relative intensity through stereo headphones. In addition, the absolute value of the error was represented redundantly by tone pitch. Low error was indicated by a low pitch. Vinge (1971) found that control over two independent tracking axes was superior when one was presented auditorily and the other visually, as compared to a visual-visual condition. These results replicate the findings of within-modality resource competition discussed in the dual-task literature in Chapter 8. Although Isreal (1980) found that single-task auditory and visual tracking was nearly equivalent, he also found that auditory tracking was more disrupted by a secondary task displayed in *either* modality. These results suggest that in terms of Figure 8.3 auditory tracking is more "resource-limited" than visual, a condition which probably reflects the fact that we rarely use our sense of hearing to make fine manual adjustments in space. As discussed in Chapter 9, the auditory modality is less spatially compatible than the visual.

Mané and Wickens (in press) hypothesized that the greatest benefit to auditory displays would probably occur when they presented error information that was redundant with information already delivered along visual channels. The benefits of redundancy gain were, of course, discussed previously in the context of signal detection theory and absolute judgment in Chapter 2 and parallel processing in attention in Chapter 7. The usefulness of this variable was again shown by Mané and Wickens in tracking.

In their experiment, a flight-control task similar to stabilizing one axis of a helicopter was presented in three ways: visually, auditorily using the redundant pitch and location display, or in a redundant bimodal display. This task was tracked either alone or in the presence of a demanding visual side task analogous to what the helicopter pilot might face while scanning the outside field of view. While single-task auditory performance was considerably worse than performance with the visual or redundant display, it did not suffer as much from the task loading as did the visual display. Furthermore, the redundant display showed the least disruption of all by the requirements of the concurrent visual side task.

In spite of its early success in the flybar experiment and subsequent studies showing near equivalence with vision, the use of auditory displays in tracking has received only minimal investigation over the past four decades. This neglect results in part from the fact that the auditory channel is more intrinsically tuned to the processing of verbal (speech) information. Hence the feeling is that the auditory channel should be dedicated to speech processing. Also, the auditory modality is hampered somewhat because it does not have spatial reference points that are precisely defined, as vision does. Nevertheless, it does appear that under certain conditions auditory spatial displays could provide valuable supplementary information, particularly if this information were presented along channels that do not peripherally mask the comprehension of speech input.

MODELING THE HUMAN OPERATOR IN MANUAL CONTROL

There are many circumstances in vehicle control in which the need to keep error low but also to maintain stability greatly constrains the controller's freedom of action to engage in different strategies of control. These constraints have one great advantage. They allow human performance in manual control to be modeled and predicted with a far greater degree of precision than is possible in many other tasks. In fact, the mathematical models of tracking performance that have been derived have been some of the most accurate, successful, and useful of any of the models of human performance that we have discussed. Knowing before it is built whether an aircraft with a given set of dynamics is flyable, by combining the model of vehicle dynamics with the transfer function of a pilot, can lead the systems designer to realize a tremendous savings in engineering cost. The chapter supplement will describe two such models—the crossover model (McRuer & Krendel, 1959) and the optimal control model (Kleinman, Baron, & Levison, 1971). Both of these share certain characteristics with previous models discussed in the book. They specify optimal behavior and allow for some tradeoff of operator strategies. Before presenting these models, however, the supplement will describe in some detail the "engineering" language of the frequency domain, used primarily to discuss and interpret these models.

Transition

Motor and manual control is often difficult, time consuming, and heavily loading. At the same time, as noted on page 438, many aspects of such control reflect lower-level, noncognitive processes that in the context of Chapter 1 might readily be assigned to machines in a systems analysis. This approach is clearly being adopted to some extent. Robots in industrial assembly tasks are performing tracking, and so also are autopilots and stability augmentation devices in aircraft control. Unfortunately, however, the trend toward automation is not without a number of problems, and this fact has led some to conclude that automation in certain contexts may have proceeded far enough already (Weiner & Curry, 1980). The final chapter of this book will examine process control as a broader extention of manual control. It will then examine automation in the context of manual and process control functions, and finally it will consider automation in the broader context of other decision-making and cognitive tasks that have been discussed in previous chapters.

Engineering Models
of Manual Control

As noted previously, it is possible to think of the human operator perceiving an error and translating it to a response in the same conceptual terms as we think of any other dynamic element responding to an input to produce an output. Figure 11.6 showed examples of several such elements. The objective of human-operator models in tracking is to describe the human in similar terms to these dynamic elements.

FREQUENCY-DOMAIN REPRESENTATION

Figure 11.6 showed the step response of the different dynamic elements as functions of time—that is, the response in the time domain. The mathematical expression of these elements—a differential or integral equation—was also a function of time. Each different element was *uniquely* described by its step response and its time-domain transfer function.

In most manual-control research in engineering psychology, it is preferable to represent transfer functions in the frequency domain in terms of spectra. This procedure was described briefly in our discussion of the speech signal in Chapter 4, and the reader is referred to that discussion to see how continuous signals are defined as spectra. In tracking, we assume that the subject tracks an error, which is a continuous signal varying in time (see Figure 11.5), to produce a response, also a continuous time-varying signal. Spectral analysis breaks each of the signals down into its component frequencies of oscillation as shown in Figure 4.9. Then the transfer function between the two signals, input and output, is specified by two fundamental relations that exist between two signals specified at each frequency of oscillation. These relations are the *gain*, or *amplitude ratio*, and the *lag*. The gain is the ratio of output to input amplitude. The lag is the amount by which the output trails the input. Lag is normally expressed in degrees of a cycle rather than in units of time. Thus, a 1-second time delay will be a 1-cycle lag at a frequency of l cycle/sec (1 Hz), but will be a 2-cycle lag at a 2-Hz frequency.

These two properties of the relation between input and output are shown at the top of Figure 11.19. On the left, at a frequency of oscillation of 1 Hz, the output amplitude is twice the magnitude of the input.

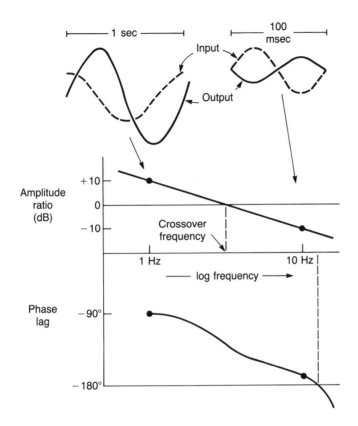

Figure 11.19   Bode plot of system $O(t) = K\int i(t - \tau)\,dt$

Hence the amplitude ratio or gain is 2/1, or 2. The output also lags behind the input by one-fourth of a cycle: The output reaches its peak when the input is already halfway to its trough. Hence, the phase lag is said to be 90°. To the right is shown a higher frequency of 10 Hz. Here the gain is less than 1, since the input amplitude is larger than the output, while the phase lag is 180°: Peaks of the inputs occur at troughs of the output and vice versa. The two signals are "out of phase."

Typically the frequency-domain representation of a transfer function is depicted in a *Bode plot*, an example of which is shown in the middle of Figure 11.19. Each Bode plot actually consists of two functions, plotting gain and phase on the same frequency axis. Therefore, at the top of the Bode plot, the amplitude ratio expressed in *decibels* is shown as a function of log frequency. Across the bottom, phase lag in degrees is expressed again as a function of log frequency. The particular gain and phase relationships of the 1 and 10 Hz frequencies at the top of the figure are shown as four points in the Bode plot. In fact, these points have been connected with solid lines in order to show the gain and phase values that would have been observed if the dynamic system whose input and output we were measuring had the time-domain transfer function

$$Y = K\!\int\!e(t - \tau_e)\,dt$$

This function is a first-order system with gain and time delay, a combination of the dynamic elements shown in Figure 11.6a, b, and e.

In the Bode plot, the value of K determines the "intercept" of the gain function. Changes in K will shift the curve up and down but will leave its slope unchanged. The constant time delay causes a greater phase lag at higher frequencies. Hence there is an exponential drop-off with the logarithm of frequency. The phase lag would increase linearly if frequency were plotted on a linear rather than a logarithm scale. Finally, on a Bode plot the single integral in the transfer function contributes an amplitude ratio that becomes progressively smaller at high frequencies and is attenuated at a rate of -20 dB for each decade increase of frequency. The phase lag produced by a first-order system is always a constant 90° at all frequencies. In Figure 11.19 the lag caused by the integrator and the lag caused by the time delay are simply added together. At very low frequencies the time delay lag is negligible, and so the constant integrator lag of 90° is the only one seen.

Often two transfer functions are placed in series. For example, in the tracking loop shown in Figure 11.4 the transfer function of the human operator, the control, and the plant are all in series. In this case, the components of the combined Bode plot are simply added. Thus, since a second-order system (Figure 11.6f) is just two first-order systems in series (Figure 11.6e), the second-order Bode plot would correspond to the added components of two first-order Bode plots. The result would be a plot with an amplitude ratio slope of -40 dB/decade (gains are multiplied, so their logarithmic values in decibels are added) and phase lag of 180°.

At this point it is possible to see why higher-order systems lead to instability in some closed-loop negative-feedback systems. Recall from our earlier discussion that instability was caused by a combination of high phase lag and high gain. The long phase lag in responding to periodic signals (180° for second order) is present at every frequency. The other element that leads to instability—high gain—is present only at lower frequencies. These two characteristics jointly lead to a critical principle in the analysis of system stability: *If the gain is greater than 1 (0 dB) at frequencies at which the phase lag is also greater than 180°, the system will be unstable when responding to frequencies of that value.* The reason is that when the phase lag is 180°, a correction intended to reduce an error at that frequency will, by the time it is realized by the system response (one half-cycle or 180° later), be *added to* rather than subtracted from the error, since the error will now have reversed in polarity. If the gain is greater than 1, this counterproductive correction will increase the error, leading to the kinds of oscillations described earlier in the chapter.

According to this principle of closed-loop stability, the system shown in Figure 11.19 is stable. The frequency at which the phase lag becomes greater than 180° (about 12 Hz) is higher than the frequency at which the gain curve dips below 1 (above 4 Hz). This latter frequency measure—critical for stability analysis—is called the *crossover frequency*. Figure 11.20 shows a Bode plot of the same system in Figure 11.19 responding with higher gain, raising the amplitude ratio curve. This change moves crossover frequency to a higher value, which is now greater than the critical frequency at which the phase lag becomes greater than 180°. The system in Figure 11.20 will therefore be unstable.

First- and Second-Order Lags in the Frequency Domain

Figure 11.21 shows Bode plots for "pure" zero-, first-, and second-order systems. Each order adds a phase lag of 90° and increases the slope of the gain function by -20 dB/decade. Earlier we suggested that the first-order, or exponential, lag shown in Figure 11.6c represented something of a compromise between a zero- and first-order system. Examination of the Bode plot of the first-order lag, presented in Figure 11.22, shows how this compromise is realized in the frequency domain. When a first-order lag is driven by low-frequency inputs, the system responds as a zero-order system. There is no phase lag, and the gain is constant at all

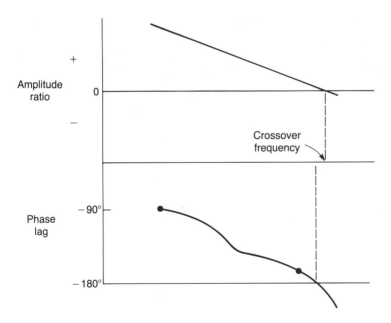

Figure 11.20   Bode plot of the system in Figure 11.19 now unstable because of a higher value of K.

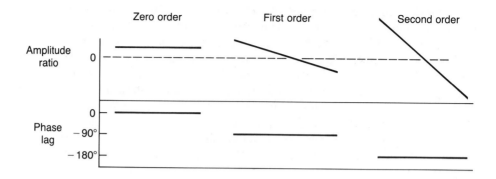

Figure 11.21 Bode plot of zero-, first-, and second-order system.

frequencies. When driven by high frequencies, on the other hand, the system responds as the first-order system of Figure 11.21. The "break point" transitioning from zero- to first-order is given by the frequency $1/T_I$. A corresponding mixture of orders is observed with the second-order lag, whose Bode plot is shown in Figure 11.23. At low frequencies the response is again zero order. At very high frequencies it is second order, similar to that shown in Figure 11.21. At the critical midfrequency range the system responds with increased amplitude. This is the "resonant frequency" of the system and is the same as the frequency of oscillation shown in the step response in Figure 11.6d. If the input to the second-order lag, mass-spring system shown in Figure 11.7 were oscillated at gradually increasing frequencies, it too would show a critical resonant frequency with very high gain. Only small oscillations would be necessary to produce large movements of the mass.

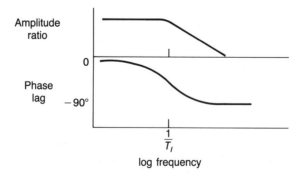

Figure 11.22 Bode plot of a first-order lag, showing zero-order behavior at low frequency and first-order behavior at high frequency.

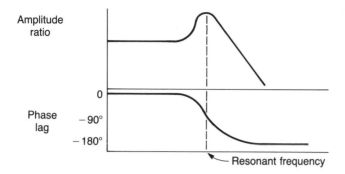

Figure 11.23 Bode plot of second-order lag.

Operations in the Frequency and Laplace Domain

In Figure 11.6, the dynamic elements were described by differential equations in the time domain. The functions were of the form $f(t)$. When dealing with systems in the frequency domain, dynamic elements are expressed in the form $F(jw)$, where jw represents a particular frequency characteristic. Alternatively, they may be represented in the Laplace domain in the form $F(S)$. The Laplace operator S is one that accounts for both frequency-domain and time-domain characteristics, and so it is the most general way of describing dynamic systems and their associated inputs and outputs. A description of the mathematical nature of the frequency-domain and Laplace-domain operations is beyond the scope of this chapter, and the reader is referred to good treatments by Toates (1975), Licklider (1960), and Moray (1981) for intuitive discussions of their derivations.

In this discussion we shall focus on one major benefit of the Laplace representation, the reason for its attractiveness to engineering analysis. Both signals and dynamic elements can be represented by Laplace transforms. To determine the output of a dynamic element from its input, all that needs to be done is to multiply the Laplace transform of the input by the transform of the dynamic element, and the product will be the transform of the output. Reciprocally, if we know the Laplace transform of the input and the output, we can compute the transfer function simply by dividing output by input. This turns out to be a lot simpler than performing differential and integral calculus in the time domain. The Laplace-domain transfer functions of some of the dynamic elements are shown in Figure 11.6. Here the simplicity is evident. The Laplace representation of an integrator is K/S. That of a differentiator is KS. Hence placing the two transfer functions in series produces $K/S \times KS = K^2$; that is, a pure gain. In the discussion of models of human-operator tracking presented below, both the time-domain and the Laplace-domain transfer functions will be considered:

the time domain because it is perhaps slightly more intuitive, the frequency domain or Laplace domain because it provides a bridge to the engineering literature where these models are often used.

MODELS OF HUMAN OPERATOR TRACKING

The initial efforts to model human tracking behavior in the late 1940s and the 1950s were described as *quasi-linear* models (Licklider, 1960; McRuer, 1980; McRuer & Jex, 1967; McRuer & Krendel, 1959). The term *quasi-linear* derives from the engineer's assumption that the operator's control behavior in perceiving an error and translating it to a response can be modeled as a linear transfer function like those dynamic elements shown in Figure 11.6. However, they acknowledge that this representation is indeed only an approximation to linear behavior. That is why the modifier "quasi" is attached. Because the human response is not truly linear, it is referred to as a *describing function* rather than a transfer function. Quasi-linear models have been applied with greatest success to describing tracking behavior in the frequency domain.

The Crossover Model

Early efforts to discover the invariant characteristics of the human operator as a transfer function relating perceived error e(t) to control response u(t) encountered considerable frustration (see Licklider, 1960, for an excellent discussion of these models). The most successful of these early approaches, the crossover model developed by McRuer and Krendel (1959; McRuer & Jex, 1967), was successful because it departed from previous efforts in one important respect, which may be appreciated by reviewing Figure 11.4. Rather than looking for an invariant relation between error and operator control u(t), McRuer and Jex examined the relation between error and *system* response o(t). In this form, their model allows the operator's describing function to be flexible and change with the system transfer function in order to achieve the characteristics of a "good" control system.

As described earlier, the two primary characteristics of good control are low error and a high degree of system stability. In order to meet these two criteria, the crossover model asserts that the human responds in such a way as to make the total "open-loop" transfer function—the function that relates perceived error to system output—behave as a first-order system with gain and effective time delay. That is,

$$o(t) = K\!\int e(t - \tau_e)\, dt$$

or, in the Laplace domain,

$$O(S) = KE^{-\tau_e S}/S$$

This transfer function is the simple crossover model, and its frequency-domain representation was in fact that which was shown in Figure 11.19. The way in which the crossover model describes the human and the plant is shown in Figure 11.24. As we have noted, in the Laplace domain the transfer function of two components in series can simply be multiplied together. That is why the Laplace domain function is equal to *HG*.

The crossover model is described by two parameters: the gain and the effective time delay. The gain is of course the ratio of output velocity to perceived error. (Output is expressed as velocity rather than amplitude because the transfer function is an integrator which produces velocity output from position input.) Humans adapt very nicely so that they adjust their own gain compensatorily to increases or decreases in system gain, in order to maintain the total open-loop gain, the ratio O/E, at a constant value (McRuer & Jex, 1967). The variable τ_e is the effective time delay, described above as the continuous analog of the human operator's discrete reaction time. Unlike gain, for which there are advantages and disadvantages are both high and low levels, long time delays are invariably harmful. Thus τ_e represents one of the class of "efficiency" parameters discussed in Chapter 1. There is not much that the human operator can do to shorten this parameter.

The third element of the model is its first-order characteristics—the single time integration. It is important in understanding the crossover model to consider why the operator chooses to behave in a way that makes the operator-plant team respond as a first-order system. The answer may be given in terms of two compromises—one expressed intuitively, one formally. At an intuitive level it is possible to view a first-order response as a compromise between the costs and benefits of

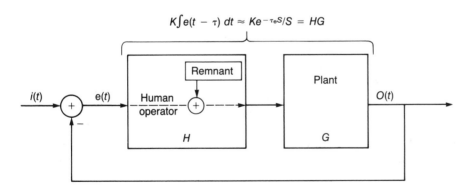

$$K \int e(t - \tau)\, dt \approx K e^{-\tau_e S}/S = HG$$

Figure 11.24   The crossover model of McRuer and Krendel (1959). From "A Review of Quasi-linear Pilot Models" by D. T. McRuer and H. R. Jex, 1967, *IEEE Transactions on Human Factors in Electronics, 8,* p. 240. Copyright 1967 by the Institute of Electrical and Electronics Engineers, Inc. Adapted by permission.

zero- and second-order controls. Zero-order controls are tight and, as shown in Figure 11.6a, instantly correct detected errors. However, if the errors are sudden steps as in Figure 11.6, then an instant response of this kind might be quite unpleasant if, for example, one were riding in a vehicle corrected in this manner. Greater "smoothness" of correction is thereby obtained by control at higher order. However, second-order control as we have noted is so sluggish that it becomes unstable. Hence, first-order control makes an appropriate compromise between jerkiness of control and sluggish instability.

The formal compromise expresses the rationale for adopting a first-order control system in terms of an attempt to meet the two criteria of a "good" control system: low error and stability (McRuer & Jex, 1967). Considering the system depicted in Figure 11.24, it is evident that low error (at least as spatially defined) will be accomplished by a system in which the gain of the entire *closed*-loop transfer function (relating command input I to system output O) is equal to 1. In this case, inputs will be directly matched in amplitude by outputs, and barring any phase lags, the error will be zero. However, in order to accomplish a unity gain of the closed-loop transfer function, the open-loop function, describing the relation of error to output, must have an infinite (or very high) gain (i.e., small errors should be corrected with large corrective responses in the opposite direction). This property can be shown formally (see Jagacinski, 1977, and Wickens, in press).

While "tight" closed-loop control and low error can therefore be obtained by making the open-loop gain very high, as we have seen, a high-gain response strategy will generate stability problems whenever system phase lags are greater than 180° of the frequencies being corrected. These problems could occur any time an operator with effective time delay enters the control loop or whenever the plant itself has phase lags. As a consequence, at high frequencies good control must maintain the open-loop gain to a value of less than 1 at a frequency, the crossover frequency, below that where the phase lag is greater than 180°. The first-order system with its downsloping amplitude ratio in the frequency domain, shown in Figures 11.19 and 11.21, nicely accomplishes this function.

The human operator, the "flexible" element of the open-loop function HG, meets these two criteria by responding in such a way as to make HG behave like a first-order system, with the unavoidable time delay. Thus the form of the crossover model, $o(t) = K\int e(t - \tau_e)\, dt$, is adopted. Referring to Figure 11.19, it is apparent that such a function will produce high gain at low frequencies and low gain at high frequencies, thereby jointly meeting the criteria of minimizing low-frequency error and maintaining high-frequency stability.

In order to achieve this form of the open-loop transfer function $O/E = GH$, the ideal human operator must adapt to changes in the system transfer function. This adaptation is accomplished by changing the form of H (the operator's control response to an error) to first order when the system is zero order, to zero order when the system is first order, and to a minus-first-order or derivative control system when the system is second order. In the Laplace domain, the reason for this adaptation is very easy to see. The total transfer function *HG* must be first order, *K/S*. Thus when the plant is zero order (*K*), the human is 1/*S* and $K \times 1/S = K/S$. When the plant is first order (1/*S*), the human is zero order (*K*) and $1/S \times K = K/S$. When the plant is second order (1/S^2), the human is a derivative controller *KS* and $KS \times 1/S^2 = K/S$. With the second-order control dynamics, the fact that the human becomes a derivative controller agrees with our earlier discussion of second-order control. In that discussion we said that perceiving the derivative of the error signal (behaving as a *KS* controller) was an aid to anticipation and prediction. Here we see that it is not only desirable, but also may be essential to maintain stability. If third-order dynamics are to be controlled, the crossover model is less applicable because this requires the human operator to adopt a minus-second-order, KS^2, or double-derivative control function in order to maintain stability, responding directly to error acceleration. As noted previously, human abilities to perceive acceleration are limited (Fuchs, 1962; McRuer, 1980).

In an extensive series of validation studies, McRuer and his colleagues (McRuer, 1980; McRuer & Jex, 1967; McRuer & Krendel, 1959; McRuer et al., 1968) found that the human behaves quite closely according to the crossover model when performing compensatory tracking tasks with random input. When Bode plots of the human transfer function between error and output are constructed in a compensatory task, more than 90 percent of the variance of well-trained operators can be accounted for by the simple two-parameter model. The model, therefore, compares favorably with other models of human behavior that we have discussed. Figure 11.25, taken from McRuer and Jex (1967), shows the unchanging form of the crossover model Bode plot *HG* as humans track zero-, first-, and second-order systems. The figure then indicates how the human operator adapts to compensate for the changing dynamics of the system. Only when the system becomes second order does the fit of data to the model begin to deteriorate somewhat. This is because, as noted on page 424, with second-order systems operators are likely to engage in some nonlinear "bang-bang" behavior that cannot be captured by the linear transfer function.

The mathematical equation of the crossover model describing functions cannot predict all of the output that will be observed when the human tracks a given error signal. The remaining variance in system response that is not accounted for by the linear describing function is referred to as the *remnant*. Some of the remnant results from nonlinear forms of behavior like the impulse control shown in second-order tracking. In addition, a remnant is also caused by such factors as time variations in the describing function parameters or random "noise" in human behavior. It is often depicted as a quantity injected into the human processing signal as shown in Figure 11.24 (Levison, Baron, & Kleinman, 1969; McRuer & Jex, 1967).

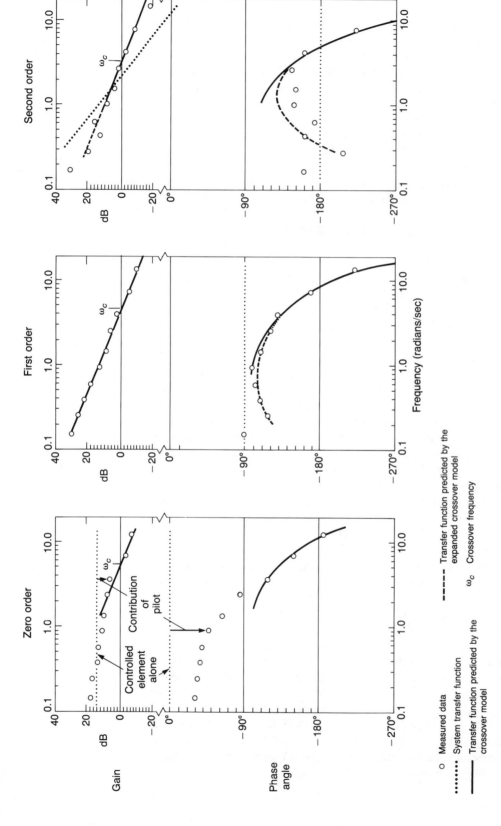

Figure 11.25 Human operator adaptation to plant dynamics with the crossover model. From "A Review of Quasi-linear Pilot Models" by D. T. McRuer and H. R. Jex, 1967, *IEEE Transactions on Human Factors in Electronics*, 8, p. 235. Copyright 1967 by the Institute of Electrical and Electronics Engineers, Inc. Adapted by permission.

The crossover model has proven to be quite successful in accounting for human behavior in dynamic systems. It has allowed design engineers to predict the closed-loop stability of piloted aircraft by combining the transfer function of the aircraft provided by aeronautical engineers with the crossover model of its pilot. It has also provided a useful means of predicting the mental workload encountered by aircraft pilots from the amount of lead or derivative control that the pilot must generate to compensate for higher-order control lags (Hess, 1977). Finally, the time delay, gain, and remnant measures of the model provide a convenient means of capturing the changes in the frequency domain that occur as a result of such factors as stress, fatigue, dual-task loading, practice, or supplemental display cues (Wickens & Gopher, 1977). To the extent that these three parameters capture fundamental changes in human processing mechanisms, their expression forms a more economical means of representation than does the raw Bode plot.

THE OPTIMAL CONTROL MODEL

Despite the great degree of success which the crossover model has had in accounting for tracking behavior, it is not without some limitations (McRuer, 1980; Pew & Baron, 1978). (1) It is essentially a frequency-domain model and so does not readily account for time-domain behavior. (2) The form of the model and its parameters are based purely upon fits of the equations to the input-output relations of tracking; they are not derived from consideration of the processing mechanisms actually used by the human operator. (3) Unlike models of reaction time, signal detection, or dual-task performance, the crossover model does not readily account for different operator strategies of performance. While the operator is assumed to be flexible—adapting the order of the describing function or compensatorily adjusting gain to the form of the plant—these adjustments are dictated by characteristics of the system. They are not chosen according to "styles" of tracking or instructional sets in the same way that beta was adjusted in signal detection theory. On the other hand, the optimal control model, in contrast to the crossover model, incorporates an explicit mechanism to account for this sort of strategic adjustment.

The basic components of the optimal control model are shown in Figure 11.26. The human operator perceives a set of displayed quantities and must exercise control in order to minimize a quantity J known as the quadratic *cost functional* and shown in the middle of the figure. This quantity J, a critical component of the optimal control model, is often expressed in the form

$$J = \int (A\dot{u}^2 + Be^2)\, dt$$

The integrated quantity within parentheses that is to be minimized is a weighted combination of squared error and squared control ve-

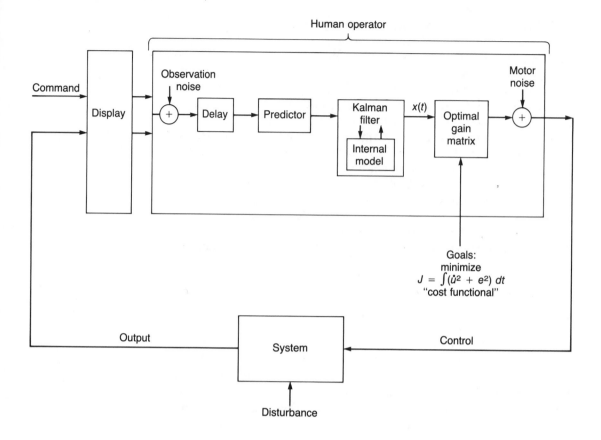

Figure 11.26   The optimal control model.

locity. The relative constants of the two terms will depend upon the relative importance of control precision e, as opposed to control effort $\dot{u}$. These may vary from occasion to occasion or from operator to operator. As shown in the comparison of the two signals in Figure 11.9, effort to minimize error at all costs may require fairly rapid control, but efforts to make a ride comfortable, or to save fuel if controlling a space vehicle, require less control action and therefore increase error. Thus as the operator trades off between tracking with low error and tracking with smooth control, the quantities u^2 and e^2 will also trade off accordingly.

The assumption that the operator attempts to minimize the cost function represents one aspect of the "optimal" characteristic of the optimal control model. In order to minimize this quantity, the operator adjusts the gains that are used to translate the various perceived quantities (i.e., error and error velocity) into control. These gains are represented by the *optimal gain matrix* in Figure 11.26, a set of rules whose formulations are beyond the scope of the present treatment (see Baron, Kleinman, & Levison, 1970).

Optimal control is not perfect control. The human operator suffers two kinds of limitations: time delay and disturbance. Typically all of the time delays in operator response are "lumped" into one delay, shown in Figure 11.26. The sources of disturbance are attributable either to external forces or to the remnant of the operator. In Figure 11.26, these are shown as a noise in perceiving (observation noise) and a noise in responding (motor noise). In order to provide the most precise and current estimate of the current state of the system x upon which to base the optimal control, the operator must engage in two further processing operations: (1) *Optimal prediction* is done to compensate for the time delay, taking into account the manner in which future states may be predicted from past states. (2) *Estimation* of the true state of the system from the noisy state perturbed by the remnant is accomplished by applying a *Kalman filter*. The Kalman filter is an optimization technique similar in some respects to the Bayesian decision-making procedures described in Chapter 3. It combines estimates of system noise, operator remnant, and an internal model of the system to make a "best guess" of true system state. Application of the Kalman filter represents a second manner in which the human is assumed to behave optimally, since the estimation of system state is based upon an optimal weighting of system noise and operator remnant. The final current, best guess of the system state $x(t)$ is then used as the input to the control gains that were selected by the cost function J, and from this process the appropriate motor response is generated, although perturbed by motor noise.

Like the crossover model, the optimal control model has also proven quite successful in accounting for much of the variance of human performance in tracking. However, its computational complexities, as well as the greater number of parameters in the model that must be specified to "fit" the data, make it somewhat more difficult to apply. Nevertheless, the ability of the model to account for shifts in operator strategies gives it a desirable degree of flexibility that the crossover model does not possess. Discussions by Curry, Kleinman, and Hoffman (1977), Baron and Levison (1977), Hess (1977), and Levison, Elkind, and Ward (1971) provide examples of the manner in which the model has been applied to optimize the design of aviation systems and to assess operator workload and attention allocation in a quantitative model of attention. Levison et al. (1971) have directly related the level of remnant or observation noise, shown in Figure 11.26, to the fraction of resources allocated to a tracking axis.

Summary

Models of human tracking have come about as close as any to providing useful predictive information that can assist in system design. As we have pointed out, however, these models are accurate precisely because they describe behavior that is fairly tightly constrained. The pilot of certain aircraft simply cannot depart from the error and stability prescriptions of the crossover model or else the plane will crash. Yet, ironically, it is these very constraints on flexibility that have made manual controls some of easiest functions to replace by automation, thereby potentially making the models obsolete. However, the extent to which humans will be replaced by automatic controllers is determined by a host of other factors above and beyond the simple ability of automatic controllers to perform as well as humans. These issues will be addressed in Chapter 12.

References

Adams, J. A. (1971). A closed loop theory of motor learning. *Journal of Motor Behavior, 3,* 111–150.

Adams, J. A. (1976). Issues for a closed loop theory of motor learning. In G. E. Stelmach (Ed.), *Motor control: Issues and trends.* New York: Academic Press.

Allen, R. W., Clement, W. F., & Jex, H. R. (1970, July). *Research on display scanning, sampling, and reconstruction using separate main and secondary tracking tasks* (NASA CR-1569). Washington, DC: NASA.

Allen, R. W., & Jex, H. R. (1968, June). *An experimental investigation of compensatory and pursuit tracking displays with rate and acceleration control dynamics and a disturbance input* (NASA CR-1082). Washington, DC: NASA.

Anderson, R. O. (1970). *A new approach to the specification and evaluation of flying qualities* (AFFDL-TR-69-120). Wright Patterson Air Force Base, OH: Air Force Flight Dynamics Laboratory.

Bahrick, H. P., & Shelly, C. (1958). Time-sharing as an index of automatization. *Journal of Experimental Psychology, 56,* 288–293.

Baron, S., Kleinman, D., & Levison, W. (1970). An optimal control model of human response. *Automatica, 5,* 337–369.

Baron, S., & Levison, W. H. (1977). Display analysis with the optimal control model of the human operator. *Human Factors, 19,* 437–457.

Bartlett, F. C. (1932). *Remembering: A study in experimental social psychology.* London: Cambridge University Press.

Baty, D. L. (1971). Human transinformation rates during one-to-four axis tracking. *Proceedings of the 7th Annual Conference on Manual Control* (NASA SP-281). Washington, DC: U.S. Government Printing Office.

Bilodeau, E. A., & Bilodeau, I. McD. (Eds.). (1969). *Principles of skill acquisition.* New York: Academic Press.

Birmingham, H. P., & Taylor, F. V. (1954). A design philosophy for man-machine control systems. *Proceedings, I. R. E., 42,* 1748–1758.

Briggs, G. E., Fitts, P. M., & Bahrick, H. P. (1957). Learning and performance in a complex tracking task as a function of visual noise. *Journal of Experimental Psychology, 53,* 329–387.

Brown, J. S., & Slater-Hammel, A. T. (1949). Discrete movements in a horizontal plane as a function of their length and direction. *Journal of Experimental Psychology, 10,* 12–21 (147).

Card, S. K. (1981). The model human processor: A model for making engineering calculations of human performance. In R. Sugarman (Ed.), *Proceedings, 25th Annual meeting of the Human Factors Society.* Santa Monica, CA: Human Factors.

Chernikoff, R., Duey, J. W., & Taylor, F. V. (1960). Two-dimensional tracking with identical and different control dynamics in each coordinate. *Journal of Experimental Psychology, 60,* 318–322.

Chernikoff, R., & Lemay, M. (1963). Effect of various display-control configurations on tracking with identical and different coordinate dynamics. *Journal of Experimental Psychology, 66,* 95–99.

Chernikoff, R., & Taylor, F. V. (1957). Effects of course frequency and aided time constant on pursuit and compensatory tracking. *Journal of Experimental Psychology, 53,* 285–292.

Craik, K. W. J. (1948). Theory of the human operator in control systems II: Man as an element in a control system. *British Journal of Psychology, 38,* 142–148.

Crossman, E. R. F. W. (1960). The information capacity of the human motor system in pursuit tracking. *Quarterly Journal of Experimental Psychology, 12,* 1–16.

Crossman, E. R. F. W., & Goodeve, P. T. (1963). *Feedback control of hand movement and Fitts' law.* Paper presented to the Experimental Psychology Society, Oxford, U.K.

Curry, R. E., Kleinman, D. L., & Hoffman, W. C. (1977). A design procedure for control/display systems. *Human Factors, 19,* 421–436.

Debecker, J., & Desmedt, R. (1970). Maximum capacity for sequential one-bit auditory decisions. *Journal of Experimental Psychology, 83,* 366–373.

Drury, C. (1975). Application of Fitts' Law to foot pedal design. *Human Factors, 17,* 368–373.

Elkind, J. I., & Sprague, L. T. (1961). Transmission of information in simple manual control systems. *IRE Transactions on Human Factors in Electronics, HFE-2,* 58–60.

Eriksen, B. A., & Eriksen, C. W. (1974). Effects of noise letters upon the identification of a target letter in a non-search task. *Perception & Psychophysics, 16,* 143–149.

Fitts, P. M. (1954). The information capacity of the human motor system in controlling the amplitude of movement. *Journal of Experimental Psychology, 47,* 381–391.

Fitts, P. M., & Jones, R. E. (1961). Psychological aspects of instrument display: Analysis of 270 "pilot-error" experiences in reading and interpreting aircraft instruments. In H. W. Sinnaiko (Ed.), *Selected papers on human factors in the design and use of control systems* (pp. 359–396). New York: Dover.

Fitts, P. M., Jones, R. E., & Milton, J. L. (1950). Eye movements of aircraft pilots during instrument-landing approaches. *Aeronautical Engineering Review, 9*(2), 24–29.

Fitts, P. M., & Peterson, J. R. (1964). Information capacity of discrete motor responses. *Journal of Experimental Psychology, 67,* 103–112.

Fitts, P. M., & Posner, M. A. (1967). *Human performance.* Pacific Palisades, CA: Brooks/Cole.

Forbes, T. W. (1946). Auditory signals for instrument flying. *Journal of Aeronautical Science, 13,* 255–258.

Fuchs, A. (1962). The progression regression hypothesis in perceptual-motor skill learning. *Journal of Experimental Psychology, 63,* 177–192.

Gallagher, P. D., Hunt, R. A., & Williges, R. C. (1977). A regression approach to generate aircraft predictive information. *Human Factors, 19,* 549–566.

Gibbs, C. B. (1962). Controller design: Interactions of controlling limbs, time-lags, and gains in positional and velocity systems. *Ergonomics, 5,* 385–402.

Gill, R., Wickens, C. D., Donchin, E., & Reid, R. (1982a). The internal model: A means of analyzing manual control in dynamic systems. *Proceedings, IEEE Conference on Systems, Man, and Cybernetics.* New York: Institute of Electrical and Electronics Engineers, Inc.

Gill, R., Wickens, C. D., Donchin, E., & Reid, R. (1982b). Pseudo quickening: A new display technique for the control of higher order systems. In R. E. Edwards (Ed.), *Proceedings, 27th annual meeting of the Human Factors Society.* Santa Monica, CA: Human Factors.

Gottsdanker, R. M. (1952). Prediction-motion with and without vision. *American Journal of Psychology, 65,* 533–543.

Harris, R. L., & Christhilf, D. M. (1980). What do pilots see in displays? In G. Corrick, E. Hazeltine, & R. Durst (Eds.), *Proceedings, 24th annual meeting of the Human Factors Society.* Santa Monica, CA: Human Factors.

Hess, R. A. (1973). Nonadjectival rating scales in human response experiments. *Human Factors, 15,* 275–280.

Hess, R. A. (1977). Prediction of pilot opinion ratings using an optimal pilot model. *Human Factors, 19,* 459–475.

Hess, R. A. (1979). A rationale for human operator pulsive control behavior. *Journal of Guidance and Control, 2,* 221–227.

Hess, R. A. (1981). An analytical approach to predicting pilot-induced oscillations. *Proceedings, 17th Annual Conference on Manual Control.* Pasadena, CA: Jet Propulsion Laboratory.

Holding, D. H. (Ed.). (1981). *Human skills.* New York: Wiley.

Howarth, C. I., & Beggs, W. D. A. (1981). Discrete movements. In D. H. Holding (Ed.), *Human skills.* New York: Wiley.

Howarth, C. I., Beggs, W. D. A., & Bowden, J. M. (1971). The relationship between speed and accuracy of movement aimed at a target. *Acta Psychologica, 35,* 207–218.

Isreal, J. (1980). *Structural interference in dual task performance: Behavioral and electrophysiological data.* Unpublished Ph.D. dissertation, University of Illinois, Champaign.

Jagacinski, R. J. (1977). A qualitative look at feedback control theory as a style of describing behavior. *Human Factors, 19,* 331–347.

Jagacinski, R. J., & Miller, D. (1978). Describing the human operator's internal model of a dynamic system. *Human Factors, 20,* 425–434.

Jagacinski, R. J., Miller, D. P., & Gilson, R. D. (1979). A comparison of kinesthetic-tactual and visual displays via a critical tracking task. *Human Factors, 21,* 753–761.

Jagacinski, R. J., Repperger, D. W., Ward, S. L., & Moran, M. S. (1980). A test of Fitts' law with moving targets. *Human Factors, 22,* 225–233.

Jensen, R. J. (1981). Prediction and quickening in prospective flight displays for curved landing and approaches. *Human Factors, 23,* 333–364.

Jex, H. R. (1979). A proposed set of standardized sub-critical tasks for tracking workload calibration. In N. Moray (Ed.), *Mental workload: Its theory and measurement.* New York: Plenum Press.

Jex, H. R., McDonnel, J. P., & Phatac, A. V. (1966). A "critical" tracking task for manual control research. *IEEE Transactions on Human Factors in Electronics, HFE–7,* 138–144.

Keele, S. W. (1968). Movement control in skilled motor performance. *Psychological Bulletin, 70,* 387–403.

Keele, S. W. (1973). *Attention and human performance.* Pacific Palisades, CA: Goodyear.

Kelly, C. R. (1968). *Manual and automatic control.* New York: Wiley.

Kelso, J. A., Southard, D. L., & Goodman, D. (1979). On the coordination of two-handed movements. *Journal of Experimental Psychology: Human Perception and Performance, 5,* 229–259.

Klapp, S. T., & Erwin, C. I. (1976). Relation between programming time and duration of response being programmed. *Journal of Experimental Psychology: Human Perception and Performance, 2,* 591–598.

Kleinman, D. L., Baron, S., & Levison, W. H. (1971). A control theoretic approach to manned-vehicle systems analysis. *IEEE Transactions in Automatic Control, AC–16,* 824–832.

Krendel, E. S., & McRuer, D. T. (1968). Psychological and physiological skill development. *Proceedings, 4th Annual NASA Conference on Manual Control* (NASA SP–182). Washington, DC: U.S. Government Printing Office.

Langolf, C. D., Chaffin, D. B., & Foulke, S. A. (1976). An investigation of Fitts' law using a wide range of movement amplitudes. *Journal of Motor Behavior, 8,* 113–128.

Levison, W. H., Baron, S., & Kleinman, D. L. (1969). A model for human controller remnant. *IEEE Transactions in Man-Machine Systems, MMS–10,* 101–108.

Levison, W. H., Elkind, J. I., & Ward, J. L. (1971, May). *Studies of multi-variable manual control systems: A model for task interference* (NASA CR–1746). Washington, DC: NASA.

Licklider, J. C. R. (1960). Quasilinear operator models in the study of manual tracking. In R. D. Luce (Ed.), *Mathematical psychology.* Glencoe, IL: Free Press.

MacNielage, P. F. (1970). Motor control of serial ordering of speech. *Psychological Review, 77,* 182–196.

Mané, A., & Wickens, C. D. (in press). *Redundant auditory presentation of the tracking signal in a dual task environment.* Unpublished paper, Department of Psychology, University of Illinois, Champaign, IL.

Martenuik, R. G., & MacKenzie, C. L. (1980). Information processing in movement organization and execution. In R. S. Nickerson (Ed.), *Attention and Performance VIII.* Hillsdale, NJ: Erlbaum Associates.

McRuer, D. T. (1980). Human dynamics in man-machine systems. *Automatica, 16,* 237–253.

McRuer, D. T., Allen, R. W., Weir, D. H., & Klein, R. H. (1977). New results in driver steering control models. *Human Factors, 19,* 381.

McRuer, D. T., & Jex, H. R. (1967). A review of quasi-linear pilot models. *IEEE Transactions on Human Factors in Electronics, 8,* 231.

McRuer, D. T., & Klein, R. H. (1975, May). Comparison of human driver dynamics in simulators with complex and simple visual displays and in an automobile on the road. *Proceedings, 11th Annual Conference on Manual Control* (Technical Memorandum NASA TM X-62, 464). Washington, DC: U.S. Government Printing Office.

McRuer, D. T., & Krendel, E. S. (1959). The human operator as a servo system element. *Journal of the Franklin Institute, 267,* 381–403, 511–536.

McRuer, D. T., et al. (1968). *New approaches to human-pilot/vehicle dynamic analysis* (AFFDL-TR-67-150). Dayton, OH: Wright Patterson Air Force Base.

Moray, N. (1981). The role of attention in the detection of errors and the diagnosis of errors in man-machine systems. In J. Rasmussen & W. Rouse (Eds.), *Human detection and diagnosis of system failures.* New York: Plenum Press.

Navon, D., Gopher, D., Chillag, W., & Spitz, G. (1982, August). *On separability and interference between tracking dimensions in dual axis tracking* (Technical Report). Haifa, Israel: Technion Israeli Institute of Technology.

Pew, R. W. (1970). Toward a process-oriented theory of human skilled performance. *Journal of Motor Behavior, 11,* 8–24.

Pew, R. W. (1974). Human perceptual-motor performance. In B. Kantowitz (Ed.), *Human information processing.* Hillsdale, NJ: Erlbaum Associates.

Pew, R. W., & Baron, S. (1978). The components of an information processing theory of skilled performance based on an optimal control perspective. In G. E. Stelmach (Ed.), *Information processing in motor control and learning.* New York: Academic Press.

Pew, R. W., Duffenback, J. C., & Fensch, L. K. (1967). Sine-wave tracking revisited. *IEEE Transactions on Human Factors in Electronics, HFE-8,* 130–134.

Poulton, E. C. (1957). Learning the statistical properties of the input in pursuit tracking. *Journal of Experimental Psychology, 54,* 28–32.

Poulton, E. C. (1974). *Tracking skills and manual control.* New York: Academic Press.

Reid, D., & Drewell, N. (1972). A pilot model for tracking with preview. *Proceedings, 8th Annual Conference on Manual Control* (Wright Patterson AFB Ohio Flight Dynamics Laboratory Technical Report AFFDL-TR-72, 92). Washington, DC: U.S. Government Printing Office.

Roscoe, S. N. (1968). Airborne displays for flight and navigation. *Human Factors, 10,* 321–322.

Roscoe, S. N. (1980). *Aviation psychology.* Ames, IA: Iowa State University Press.

Roscoe, S. N., Corl, L., & Jensen, R. S. (1981). Flight display dynamics revisited. *Human Factors, 23,* 341–353.

Runeson, S. (1975). Visual predictions of collisions with natural and non-natural motion functions. *Perception & Psychophysics, 18,* 261–266.

Schmidt, R. A. (1975). A schema theory of discrete motor skill learning. *Psychological Review, 82,* 225–260.

Schmidt, R., Zelaznik, H., & Frank, J. S. (1977). *Motor output variability: An alternative interpretation to Fitts' law.* Paper presented to Big Ten Symposium on Information Processing in Motor Control, Madison, WI.

Shapiro, D. C., & Schmidt, R. A. (1982). The schema theory: Recent evidence and development of implications. In J. A. S. Kelso & J. E. Clark (Eds.), *The development of movement control and coordination.* Norwich, U.K.: Wiley.

Shulman, H. G., Jagacinski, R. J., & Burke, M. W. (1978). *The time course of motor preparation.* Paper presented at the 19th Annual Meeting of the Psychonomics Society, San Antonio, TX.

Simon, J. R. (1969). Reactions toward the source of stimulation. *Journal of Experimental Psychology, 81,* 174–176.

Stelmach, G. E. (Ed.). (1978). *Information processing in motor control and learning.* New York: Academic Press.

Summers, J. J. (1981). Motor programs. In D. H. Holding (Ed.), *Human skills.* New York: Wiley.

Toates, F. (1975). *Control theory in biology and experimental psychology.* London: Hutchinson Education.

Vehemans, P. J. (1981). Prediction and state in simple dynamic systems: What is learned? In H. Stassen (Ed.), *First European Annual Conference on Human Decision Making and Manual Control.* (From p. 700) New York: Plenum Press.

Vidulich, M. D., & Wickens, C. D. (1981, December). *Time-sharing manual control and memory search: The effects of input and output modality competition, priorities, and control order* (Technical Report EPL–81–4/ONR–81–4). Champaign, IL: University of Illinois, Engineering Psychology Research Laboratory.

Vinge, E. (1971). Human operator for aural compensatory tracking. *Proceedings, 7th Annual Conference on Manual Control* (NASA SP–281). Washington, DC: U.S. Government Printing Office.

Weiner, E. L., & Curry, E. R. (1980). Flight deck automation: Problems and promises. *Ergonomics, 23,* 995–1012.

Wickens, C. D. (1976). The effects of divided attention on information processing in manual tracking. *Journal of Experimental Psychology: Human Perception and Performance, 2,* 1–17.

Wickens, C. D. (in press). The effects of control dynamics on performance. In K. Boff & L. Kaufman (Eds.), *Handbook of perception and performance.* New York: Wiley.

Wickens, C. D. & Derrick, W. (1981, January). *The processing demands of second order manual control: Application of additive factors in methodology* (Technical Report EPL–81–1/ONR–81–1). Champaign, IL: University of Illinois, Engineering Psychology Research Laboratory.

Wickens, C. D., Gill, R., Kramer, A., Ross, W., & Donchin, E. (1981, October). The processing demands of higher order manual control. In J. Lyman and A. Bejczy (Eds.), *17th Annual Conference on Manual Control.* La Canada, CA: Jet Propulsion Laboratory.

Wickens, C. D., & Gopher, D. (1977). Control theory measures of tracking as indices of attention allocation strategies. *Human Factors, 19,* 349–365.

Wickens, C. D., Sandry, D. C., & Hightower, R. (1982, October). *Display location of verbal and spatial material: The joint effects of task-hemispheric integrity and processing strategy* (Technical Report EPL–82–2/ONR–82–2). Champaign, IL: University of Illinois, Engineering Psychology Research Laboratory.

Wickens, C. D., Tsang, P., & Benel, R. (1979). The dynamics of resource allocation. In C. Bensel (Ed.), *Proceedings, 23rd annual meeting of the Human Factors Society.* Santa Monica, CA: Human Factors.

Woodworth, R. S. (1899). The accuracy of voluntary movement. *Psychological Review, 3,* Monograph Supplement No. 2.

Young, L. R. (1969). On adaptive manual control. *Ergonomics, 12,* 635–675.

Young, L. R., & Meiry, J. L. (1965). Bang-bang aspects of manual control in higher-order systems. *IEEE Transactions on Automatic Control, 6,* 336–340.

Young, L. R., & Stark, L. (1963). Variable feedback experiments testing a sampled data model for eye tracking movements. *IEEE Transactions on Human Factors in Electronics, HFE–4,* 38–51.

Process Control
and Automation

12

PROCESS CONTROL

In Chapter 11 we discussed manual control and tracking of dynamic systems. In most of this treatment we assumed either implicity or explicity that the system was a vehicle. Yet this need not be the case. The dynamics of many industrial, chemical, and energy conversion processes must also be "tracked," by manipulating controls to compensate for disturbances and to follow prescribed inputs. However, we will treat process control in a separate chapter for four reasons: (1) These processes are generally more complex than vehicle dynamics, with a greater number of interacting variables. (2) The system responses are often sufficiently slow so that human manual control is more discrete than analog. (3) As a consequence of this slower control it is an area which is less constrained by motor limitations, but one that makes issues of decision making, perception, and memory, discussed in Chapters 3–6, much more relevant. (4) Finally, the process control task is closely tied to concepts of automation, discussed at the end of this chapter.

Overview

Process control is certainly not synonymous with automation. Much of the control, as we shall see, is carried out manually. However, because of the tremendous complexities of processes involved and because hazardous environments and toxic materials are often employed, the process control environment is one in which automation is an inevitable companion. Regulating and controlling large chemical, energy, or thermal processes impose many demands that are simply beyond the capabilities of the human operator. Some of these limitations result from obvious physical constraints. Humans, for example, cannot readily manipulate the chemicals that are involved in many industrial processes; they cannot handle the radioactive fuel in nuclear processes; nor can they come in physical contact with elements at the extreme temperatures of many energy-conversion systems. Other constraints are directly attributable to the complexity of the process variables involved, which impose limits on human cognitive processes. For whichever reason, many components of process control have been automated, and increasing computer technology makes it inevitable that automation will probably proceed a good bit further. It is essential therefore that the nature of human involvement in process control be clearly specified so that the automation that is implemented will be appropriate and functional.

Our discussion of process control will provide a number of links to treatments of basic human processes and limitations covered in earlier chapters. Indeed, process control is a domain which at one time or another can impose upon all of the limitations of human performance that have been discussed, and so it makes a fitting conclusion to the book. The reader is referred to excellent treatments by Edwards and Lees (1974), Sheridan and Johannsen (1976), Umbers (1979), and Rasmussen and Rouse (1981) for discussions of the process control field that are more detailed than that provided below.

Characteristics of Process Control

The specific kinds of processes of concern to the human operator in this domain are diverse. They include the management of cake-baking ovens (Beishon, 1969) and the control of blast furnaces (Bainbridge, 1974), paper mills (Beishon, 1966), distribution of electrical power (Umbers, 1976; Williams, Seidenstein, & Goddard, 1980), distillation of gas (Queinnec, DeTerssac, & Thon, 1981), and, of course, nuclear power (Rasmussen, 1979, 1981; Sheridan, 1981). These diverse examples possess three general characteristics that allow them to be classified together and discussed somewhat apart from other systems discussed in earlier chapters.

First, the process variables that are controlled and regulated are slow. They have long time constants, relative to the systems that were controlled or tracked in Chapter 11. Thus the control delivered by an operator in process control may not produce a visible system response for seconds and sometimes even minutes. Using the terminology introduced in Chapter 11, the operator is more often controlling *outer-loop variables*, while automated adjustment and feedback loops handle the inner-loop control. Thus the operator of a blast furnace may choose a "set point" of desired temperature, and automated inner-loop control will provide the amount of fuel and energy to the furnace necessary to achieve that temperature some minutes later.

Second, although controls are often adjusted in discrete fashion, the variables that are being controlled are essentially analog, continuous processes. This means, in terms of concepts discussed in Chapters 5 and 6, that the operator's ideal internal model of the processes should also be analog and continuous rather than discrete and symbolic.

Third, the processes typically consist of a large number of interrelated variables. Some are hierarchically organized as in Figure 11.16; others are cross-coupled, so that changes in one variable will influence several other variables simultaneously. The staggering magnitude of this complexity is demonstrated by the display confronting the typical supervisor of a nuclear power plant control room (Figure 12.1). Grimm (1976), for example, has noted that the complexity of power station control rooms, as measured by the number of controls and displays,

Figure 12.1 Typical nuclear power plant control room. Controls are placed on the benchboard or just above, quantitative displays are placed on the vertical segment of the boards, and qualitative annunciator displays are placed in the uppermost segments of the boards. From *Human Factors Review of Nuclear Power Plant Design: Final Report* (Project 501, NP-309-SY) by Electrical Power Research Institute, March 1977, Sunnyvale, CA: Lockhead Missiles and Space Co.

grew geometrically from under 500 in 1950 to around 1,500 in 1970 to more than 3,000 in 1975. This complexity of interactions can severely tax the operator's mental model of the status of the plant, a model whose level of accuracy is important for normal control and critical for the response in the face of malfunctions or failures.

Control Versus Diagnosis

The process control operator's task has typically been described as hours of intolerable boredom punctuated by a few minutes of pure hell. This dichotomy, although perhaps somewhat overstated, serves nicely to discriminate between the two major functions of the process controller: the normal control and regulation of the process and timely detection, diagnosis, and corrective action in the face of the very infrequent malfunctions that may occur. As the incident at Three Mile Island described in Chapter 1 indicates, the low frequency at which these failures occur must not diminish concerns for their accurate detection and diagnosis.

The duties of the process controller normally begin with the start-up of the process—for example, bringing a nuclear reactor on line. While a fairly standardized set of procedures is followed during start-up, the cognitive demands of this phase of operation are heavy, and the task is complex. As described later in the chapter, a large number of warning signals, designed to be appropriate for steady-state operation, may be flashing during the transient conditions of start-up. Once the system is in a steady state, the operator typically engages in periodic adjustments or "trimming" of process variables, in order to keep certain critical process parameters within designated bounds and meet changing production criteria. In the framework of the tracking paradigm, these adjustments may be made because of disturbances (i.e., a reduction in water quality will require more purification) or because of command inputs. For example, the need for electrical power output will increase at peak-use times.

While the control aspects of process control are thus somewhat allied with the discussion of manual control in Chapter 11, the detection and diagnosis aspects are more closely related to the discussions of those topics in Chapters 2 and 3, as well as to the treatment of mental representations in Chapters 5 and 6. The dichotomy that is drawn here between normal control and abnormal detection and diagnosis is further justified by a number of other experimental results from process control research which suggest the independence of the two functions.

As one example of this dichotomization, White (1981) investigated subjects' ability to control a set of three simulated energy-generation processes. Control was required in order to maintain certain output values within critical bounds. Disturbance inputs applied to the system tended to drive the values out of bounds, a condition indicated by auditory warnings. White found that variables which influenced subjects' display-sampling procedure had little influence on their control behavior, while those variables influencing control behavior exerted little effect on display sampling. This dichotomy, reminiscent of similar dichotomies between detection sensitivity and response criterion (Chapter 2), early and late processing resources (Chapter 8), and early and late processing stages (Chapter 9), indicates that a similar independence may be operating between the processes of perceptual sampling, detection, and diagnosis and the control task.

If the two phases are truly independent, it would suggest that those who are good controllers may not necessarily be effective at detection and diagnosis and, conversely, that good diagnosticians may not be good controllers. A study by Landeweerd (1979) suggested that individual differences in ability of these two components were indeed independent. Landeweerd assessed the accuracy of operators' visual-spatial images of a chemical distillation process (drawing a schematic diagram) and the accuracy of their verbal-causal models (statements of what caused what in the process). He then assessed the operators' abilities both to control the process under normal operating conditions and to diagnose failures. Landeweerd found that the accuracy of the visual-spatial image correlated reliably with detection and diagnosis but not with control; causal-model accuracy, on the other hand, correlated with control ability but not with diagnosis. Drawing a similar conclusion in the realm of training, rather than individual differences, Kessel and Wickens (1982) found that operators who had prior training in the detection of dynamic system failures were not helped by this training in the ability to control these systems.

Rouse and Morris (1981) observed a different manifestation of the contrast between detection and control in the performance of operators engaged in a chemical production process. They found that operator performance was dictated in part by a tradeoff between priorities. Control strategies which yielded high production (and therefore high economic payoff) were not conducive to accurate fault detection. On the other hand, more cautious strategies, designed to detect faults when they occurred, were exercised at the expense of production. This strategic tradeoff, reminiscent of tradeoffs encountered in earlier discussions of signal detection, reaction time, and manual control, is one that must be considered in the analysis of such systems. The two phases may be independent of each other at some levels but may actually compete on others.

Yet another manifestation of this competition between control and diagnosis was discussed in Chapter 7 in the context of sampling and attention control. There Moray (1981) argued that the scanning strategies appropriate for normal operations—sequentially focusing only on one of a cluster of tightly coupled variables—were inappropriate for abnormal conditions. Under abnormal conditions, normal scanning strategies will not allow the operator to detect a *drop* in the normal correlation between variables within a cluster that would be expected to occur as a consequence of a malfunction. Rasmussen and Lind (1981) discuss the potentially disastrous consequences that could result when, in the presence of a malfunction, an operator continues to sample with inappropriate routines, based upon the assumption that the system is functioning normally.

Finally, Landeweerd (1979), in articulating the differences between control and diagnosis, notes that the controller in normal circumstances must focus attention upon the forward flow of events: what causes what. The diagnostician in times of failure, on the other hand, must often reverse the entire pattern: what was caused by what. The difficulties that people have dealing with diagnosis rather than prediction—in reversing the normal causal sequence of events—was discussed in some detail in Chapter 3. Also, the contrast between the two directions is somewhat analogous to the distinction made between route and survey knowledge in our discussion of navigation in Chapter 5. Route knowledge, which is frame-dependent, is like the forward causal chain of control. Survey knowledge, requiring a more neutral frame of reference equipped to handle novel situations, is more closely related to the frame of reference required for diagnosis, in which the operator must be prepared to consider alternative perspectives.

The preceding discussion has emphasized that control and detection/diagnosis, both integral aspects of the overall process control task, may be independent in terms of operator abilities and experimental variables but are somewhat incompatible in terms of economically imposed strategies and the performance of routines common to both (i.e., scanning and sampling behavior). Because this distinction between control and detection/diagnosis is manifest in a number of different ways, the two aspects of the process controller's task will be considered separately in the treatment that follows.

Control

Performance strategies.   The operator's primary responsibilities during normal process control are to monitor system instruments and periodically adjust or "trim" control settings in order to maintain production quantities within certain bounds. While the task thereby bears some similarity to tracking, a major contrast between the two domains is that tracking normally involves closed-loop behavior while much of the responding in process control is optimally performed using open-loop strategies. This strategy, however, becomes available only after the process is well learned. In a simulated process control task, Crossman and Cooke (1962) and McLeod (1976) found that subjects employed both modes of control but performed better when the open-loop mode was used. Kragt and Landeweerd (1974) found that operators' performance evolved from closed- to open-loop strategies when they received training in a simulated process control task. Beishon (1966) reported that skilled paper mill operators would initiate their adjustments with open-loop control action, followed later by discrete closed-loop adjustment.

The reason open-loop is preferable to closed-loop behavior in process control is that the long time constants typically involved when the process responds to control manipulations cause a closed-loop strategy to be an inefficient and potentially unstable mode of behavior. We saw in Chapter 11 that long time delays contributed to closed-loop instability. Process variables often change so slowly that operators cannot readily perceive their higher derivatives in a manner that makes for stable closed-loop control (McLeod, 1976). Patternote and Verhagen (1979) supported this conclusion when observing the behavior of operators controlling a simulated distillation process. They found that behavior changed from open- to closed-loop strategies as the time constants of the simulated process were shortened.

When the task of the process controller is carefully analyzed, successful control like successful tracking requires that three important components be present: (1) a clear specification and understanding of the future goals of production, that is, a command input; (2) an accurate mental representation of the current state of the process which, because of the sluggish nature of the variables involved, must be used to predict future state; and (3) an accurate "internal model" of the dynamics of the process. This third element is essential if open-loop control is to be employed, since the internal model represents the means by which a plan of control action is formulated to bring the process to a specified future goal. If there is no internal model, the operator must respond, wait to see what happens, and then respond again. That is, he must engage in slow and inefficient closed-loop control.

These three components dictate that certain conditions should be met for optimum control. The following two sections will consider these conditions as they are manifest in the role of future process information (describing the first and second components) and the concept of the internal model (the third component).

The future in process control.   As we have discussed, the process control task is made difficult by the heavy reliance upon prediction and planning that is required to guide activities around future states. In Chapter 11 we showed how the planning of future activities in control may be partitioned into two components: planning and anticipating future goals (these may be command inputs or internally specified goals) and anticipating or predicting the future system responses. Since the system in process control responds slowly, actions implemented now must be based upon future, not present, states. The importance of both future goals and future states has implications for the design of process control environments.

Verbal protocols taken from process and industrial control operators indicate that a significant portion of their time is spent in planning operations (Bainbridge, 1974), anticipating future demands or slacks in the process. In some environments the planning demands can become quite substantial indeed. Typical, for example, is the dispatcher who must schedule the disposition of certain resources over time. Here, as might be anticipated, *preview* displays, such as that shown in Figure 7.2, have been found to provide considerable benefits to performance (Shackel, 1976). Smith and Crabtree (1975) have proposed that computer aiding may be used with great benefits in these circumstances by providing the operator with the flexibility to "try out" different scheduling solutions in response to the anticipated arrival of events. With sufficient computing power, these solutions can be played out in fast time, and the optimal one may be selected.

The use of computer aiding can potentially achieve even greater value when a *prediction* of the system response is displayed. Here the same advantages to predictive display are present as those that were described in Chapter 11; however, these advantages are typically even greater in process control because of the longer time constants involved

(Moray, 1981). Long time constants usually mean that there is little motion observable in display indicators (McLeod, 1976). As a consequence, status indicators that are based only upon present time do not provide adequate trend information. Such trend information may, however, be derived from one of three sources: (1) *Historical displays*, typically presented by strip charts, give some estimate of future trends, to the extent that the future may be mentally extrapolated from the past. An example of a historical display used in nuclear power is shown in Figure 12.2. (2) *Predictive displays*, as described in Chapter 11, generate predicted outputs based upon computer models of the plant. West and Clark (1974) found that considerable assistance was provided by the predictive display but little by the historical strip chart. Sheridan (1981) has advocated the use of fast-time computer models of the plant that can be played with different inputs to evaluate and compare alternate control strategies. Whitefield, Ball, and Ord (1980) have suggested a similar fast-time predictor display to help air traffic controllers determine the best way to resolve potential conflict situations. Sheridan has also advocated the use of off-line predictive models as a technique in fault diagnosis. In this technique, the operator would be allowed to replay different scenarios in an effort to find one that matches the pattern of events actually produced by the malfunction.

(3) The third source of predictive information, found to be of use by Brigham and Liaos (1975), is derived from the display of *intervening process variables* between the control manipulations and the actual

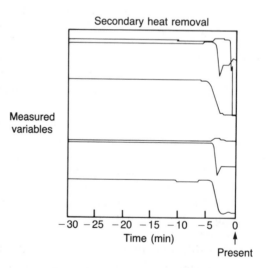

Figure 12.2 Historical strip-chart display. From "An Evaluation of Nuclear Power Plant Safety Parameter Display Systems" by D. Woods, J. Wise, and L. Hanes, 1981, *Proceedings, 25th Annual Meeting of the Human Factors Society* (p. 111), Santa Monica, CA: Human Factors. Copyright 1981 by the Human Factors Society, Inc. Reproduced by permission.

output changes. In the terms described in Chapter 11 and shown in Figure 11.16, this is analogous to the information that the controller of higher-order hierarchical systems may derive from a display of lower-order variables—the information, for example, that the automobile driver controlling lateral position derives from vehicle heading and steering-wheel angle, both derivative predictors of future lateral position.

Because the future state of a system continuously unfolds in analog fashion, there would seem to be an advantage to analog over digital predictive displays. Research in the area seems to agree with this conclusion (e.g., Pikaar & Twente, 1981; West & Clark, 1974). Trends, either historical or predictive, are compatibly presented as analog displays. On the other hand, the trends require a symbolic transformation if they are to be assessed from digital readings. Pikaar and Twente (1981), for example, found that subjects could identify the existence of a complex multi-input, multi-output relationship more rapidly and easily with an analog than with a digital display.

The advantage of analog displays in process control (supplemented when needed by digital check reading) does not, however, imply that information should be presented on long banks of meters, with one meter for each variable (Goodstein, 1981; Moray, 1981). The presentation of information in this nonintegral format, although conventionally used in many process control displays, fails to provide the conceptual integration and relation between variables that can be best achieved through the judicious use of computer technology. The advantage of integral representation was discussed in Chapters 3 and 5 and will be emphasized in greater detail below.

The internal model in process control.   The internal model is a hypothetical construct that is assumed to account for three aspects of process control behavior: (1) It guides display sampling and scanning of the multivariate systems as discussed in Chapter 7 (Moray, 1981). In the context of Chapter 4, the internal model acts as a framework that contributes some top-down organization to the operator's effort to interpret the mass of bottom-up data (dials, meters) typical of the process control environment. (2) It is used to formulate plans of action and translate intended goals (future system responses) into present control actions. The internal model in this context is the basis for open-loop control. (3) The internal model also forms the source of the operator's expectancies of which variables will covary with certain others. Consequently, it is used to provide a basis for failure detection when the system response does not correspond with those expectancies, as well as for fault diagnosis to the extent that the operator can identify the qualitative nature of the departure from expectancy.

The internal model is but a hypothetical construct employed to account for these different sets of behavior. As a result, considerable recent research has been directed toward precisely defining its properties (Bainbridge, 1974; Gill, Wickens, Donchin, & Reid, 1982; Jagacinski

& Miller, 1978; Kessel & Wickens, 1982; Rasmussen, 1979; Veldhuyzen & Stassen, 1977). There is by now sufficient experimental evidence to allow some general statements to be made about the specificity of an internal model and its code of representation.

1.   *Specificity.* A given internal model appears to be reasonably specific to a given process. Thus, Umbers (1979) concludes, "Once a control skill has been learned the operator may not always find it easy to transfer to another control task, even if it is similar"[1] (Umbers, 1979, p. 266). Kessel and Wickens (1982) found that subjects' ability to manually control a dynamic system was not assisted by previous experience monitoring and detecting failures in the same system under autopilot control. Both Crossman and Cooke (1962) and Kragt and Landeweerd (1974) report that an understanding of the *general* scientific principles underlying the process provides no assistance in the operator's ability to control it. This finding also suggests that an internal model is not a highly general concept. In contrast, however, understanding the *specific* principles underlying the given system to be controlled does assist the operator in process control (Attwood, 1970; Brigham & Liaos, 1975). Landeweerd, Seegers, and Praageman (1981), for example, found that operators with two hours instruction on the dynamics of a complex chemical process learned to control the process more rapidly than those who received information on the input/output relations for the same duration of study. The findings of these investigators reinforce the validity of the internal model concept. They suggest that effective control is not learned merely by acquiring a series of stimulus-response mappings but is based instead upon an understanding of the structure of casual sequencing in the plant. The importance of this understanding will be emphasized further when we discuss failure detection and fault diagnosis below.

2.   *Code of representation.* The discussion presented in Chapters 4 and 5 contrasted verbal and analog forms of mental representation and suggested that concepts could be associated with a continuum between these two end points. Several investigators have speculated as to the code of representation of the internal model. Bainbridge (1981) titles her detailed article on the internal model concept, "Mathematical Equations or Processing Routines?" Indeed the assumption that the internal model is either verbal or expressible in verbal form is implicit in the use of verbal protocols to model operator control behavior. Bainbridge (1974) has employed this technique with some success in order to understand the planning routines of blast furnace operators. Yet as several investigators have pointed out, verbal protocols have certain limitations (Rasmussen & Lind, 1981; Umbers, 1979).

---

[1]From "Models of the Process Operator" by G. Umbers, 1979, *International Journal of Man-Machine Studies, 11,* p. 226. Copyright 1979 by Academic Press, Inc. (London) Ltd. Reprinted by permission.

Bainbridge (1979, 1981) cautions that operators may not say what they actually do, particularly if they are interviewed, and the actions that they actually take run contrary to standardized prescribed procedures. In a different view, Broadbent (1977) points out that much of the understanding of the analog dynamics of complex systems may be inaccesible to verbal explanations. This is similar to the child who may be highly skilled at riding a bicycle, thereby showing an understanding of its dynamics, but be unable to provide a coherent description of those dynamics. This, Broadbent suggests, may occur when highly skilled routines become automatized and designated to lower hierarchical levels of processing. Brigham and Liaos (1975) have provided some direct support for the verbal inaccessibility of the internal model. As discussed on page 472, in the simulated dynamic process investigated by Brigham and Liaos, operators performed better when they viewed the response of intermediate variables in the process controlled; yet they were unable to provide any detailed description of the manner in which they used that information.

In contrast to the studies of analog process control, when electronics troubleshooting was investigated Rasmussen and Jensen (1974) found that electronic technicians were able to provide accurate verbal protocols of their troubleshooting strategies. It is possible that this contrast in ability may be accounted for in terms of the continuous visual-spatial analog characteristics of process control, as opposed to the discrete, logical steps followed in diagnostic electronics troubleshooting (Rouse, 1981; see also Chapter 4).

The contrast between analog-spatial and verbal modes in process control has been pursued in three further investigations. The study of individual differences in process control and supervision conducted by Landeweerd (1979) was discussed earlier in this chapter. He found that subjects who had a better visual-spatial *image* of the distillation process—that is, who were able to draw a visual schematic—performed better in failure diagnosis; those who possessed a better verbal causal *model*—what causes what in the process—made better controllers. Consistent with this finding, Landeweerd et al. (1981) found that a high degree of visual imagery did *not* assist operators in the control of a complex chemical distillation process. In an unpublished study of failure detection, Weingartner (1982) ascertained the spatial processing underlying this aspect of process control. She employed the logic of within-code interference outlined in Chapters 6 and 8. Two tasks sharing a common code of working memory should interfere more than two tasks using separate codes. Weingartner observed greater interference between a multielement failure-detection task and a spatial, as opposed to a verbal, working memory task. Both memory tasks involved auditory input so that neither competed with failure detection for peripheral visual input channels.

In summary, the results seem to provide clear evidence that the concept of an internal model is viable. A given model is relatively specific to the dynamic process under consideration, is not totally accessible to verbalization (Brigham & Liaos, 1975; Broadbent, 1977; Rasmussen, 1981), and is spatial in its characteristics when detection or diagnosis is required (Landeweerd, 1979; Weingartner, 1982). However, certain components clearly are verbal (Bainbridge, 1974, 1981), and these are useful when control is exercised (Landeweerd, 1979). The possible differences between internal models for detection and diagnosis and those for control is consistent with the dichotomy drawn between these two phases in the previous section and is an area worthy of further investigation.

Implications.   The internal model has implications for both training and display design. The absence of a high level of model generality for a particular system suggests that such training should emphasize neither specific stimulus-response associations nor highly general principles but should focus instead on the causal relations between variables within the specific system in question. The fact that the model is not completely verbal suggests that such training should include a heavy dose of exposure to spatial/analog formats. Kessel and Wickens (1982) recommend that a considerable degree of hands-on control of variables should be provided in training, even if these variables will not actually be controlled manually in real-time supervision of the system in question. For operators who have the opportunity to enter the control loop, the relations between variables may be appreciated better and stored more permanently than if the system were merely monitored from an "out of the loop" perspective or studied abstractly.

In terms of task configuration, two implications either directly or indirectly derive from the internal model concept: (1) Since an internal model should optimally guide sampling, Moray (1981) has emphasized that an important function of computer assistance is to remind the operator when display elements should be sampled (see Chapter 7). This reminder should be based upon a true model of the system. Consequently, it will serve to bring operator sampling behavior, and therefore the internal model, into correspondence with the ideal. Such sampling reminders are reasonably important in conventional large panel displays such as those shown in Figure 12.1, with one instrument for each variable. Moreover, for centralized computer-driven displays where the operator sits in front of one central monitor and must "call up" each variable to be monitored, a sampling guide becomes even more important. With the centralized display the operator's only guide of what to sample (or how long it has been since he last sampled a given variable) is his own memory. The fallibility of memory was well documented in Chapter 7.

(2) Rasmussen and Lind (1981) point out that an internal model may have several different levels of representation. An operator may think about a process in terms of physical variables (steam pressure, valve positions, temperature, flow) or more abstract higher-level variables

(energy conservation, heat exchange, fuel economy). The system may also be represented in different causal "molecules" with certain variables conceptually integrated and others separated. In order to provide information that is compatible with the level and format of representation, displays should be flexible and able to present information in the most compatible format (see Chapter 5). While this may not be possible with conventional mechanical instrument displays, it is clearly a feasible alternative with computerized displays. Research in process control must therefore address the issue of internal model representation and display formatting in order to provide guidelines for building appropriate computer-driven displays. This point will be addressed more directly in the following section.

Fault Detection

When discussing fault detection and diagnosis in process control, our treatment comes full circle, touching base again with the concepts discussed in Chapter 2 (vigilance and detection) and Chapter 3 (decision making and diagnosis). However, we now address these concepts distinctly from the point of view of the process monitor's task. In spite of the formal equivalence between failure detection in the process control environment and the concepts of detection and vigilance discussed in Chapter 2, there are two characteristics of process control that emphasize its differences from the simulated environment that is most often employed in the laboratory vigilance paradigms:

1. In process control the operator is not typically waiting passively between failures but is intermittently engaged in moderate levels of control adjustments, parameter checking, and log keeping necessary to maintain a current internal model of the system. These are the sort of intervening activities that maintain at least a modest level of arousal and eliminate this source of vigilance decrement.

2. When a failure does occur it is normally indicated with a visual and/or auditory alarm. The latter will be sufficiently salient to call attention to itself so that the probability of a miss (in the vigilance sense) is, in fact, quite low. Nevertheless, the consequences of a miss may still be drastic.

Yet these two qualifications do not entirely eliminate the concern about failure detection in process control. In the design of any discrete warning device it is necessary to assume a threshold for the indicated variable that is sufficiently tolerant so that random variations in the process variables do not trigger false alarms. (See Chapter 2 for a discussion of the consequence of false-alarm-prone warning indicators.)

In order to attain this level of tolerance, it is likely that true failures which occur gradually rather than catastrophically will be displayed by trends in the variables that are visible prior to the activation of the warning. A sensitive, alert operator with a well-formulated internal model will be able to use this "advance information" to prepare for the upcoming event and possibly take corrective action before the alarm sounds. As one means of enhancing the sensitivity of operators to variables approaching abnormal regions, Moray (1981) has suggested that the detection process can be facilitated by providing operators with displays that are scaled in terms of their probability of normal value, rather than absolute physical units, and by providing more integrated configural representation of variables.

In this regard, several investigators have voiced concern over the one variable for one display concept that governs the display philosophy of many process control operations (Goodstein, 1981; Moray, 1981). These investigators argue instead for integrated displays, a concept discussed in some detail in Chapter 5. These may simply depict schematic representations of the causal relationships between process variables. Alternatively, the integrated display concept might be carried to a further level of abstraction as shown in Figure 12.3. The example shown is the spoke or polar display of 12 primary safety variables in nuclear power monitoring developed originally by Coekin (1969). Shown in Figure 12.3a is the prototype investigated by Woods, Wise, and Hanes (1981). The different radial scales are adjusted so that normal operation is represented by a perfect circle, while departures indicate a distortion. The distortion shown in Figure 12.3b is that resulting from a loss of coolant supplied to a nuclear reactor.

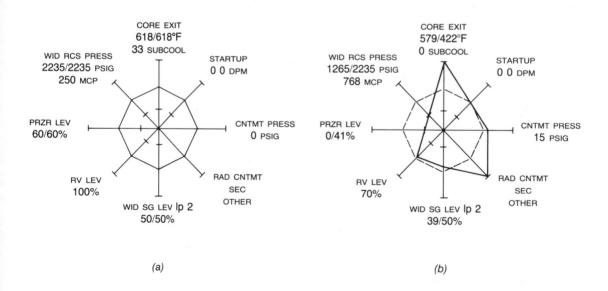

(a)   (b)

Figure 12.3 Integrated spoke or polar display for monitoring critical safety parameters in nuclear power. (a) Normal operation; (b) wide-range iconic display during loss-of-coolant accident. From "An Evaluation of Nuclear Power Plant Safety Parameter Display Systems" by D. Woods, J. Wise, and L. Hanes, 1981, *Proceedings, 25th Annual Meeting of the Human Factors Society* (p. 111), Santa Monica, CA: Human Factors. Copyright 1981 by the Human Factors Society, Inc. Reproduced by permission.

The integration of the 12 indicators into a single object should enhance the ability of operators to detect abnormalities, to the extent that these parameters correlate or covary in real-world occurrences. One reason that an integrated format is advantageous is provided in a study by Medin, Altom, Edelson, and Freko (1982), who have shown that subjects tend to be particularly sensitive to configural or correlated values when making medical diagnoses. As described in Chapter 2, moreover, Garner and Fefoldy (1970) have advocated the use of integral displays to enhance the processing of *correlated* information between variables. The integral representation of the 12 parameters shown in Figure 12.3, like the object-display concept investigated by Jacob, Egeth, and Bevon (1976) and discussed in Chapter 5, emphasizes this integral configuration.

In summary, either a vigilant operator or one equipped with well-configured displays will be likely to suspect or detact a malfunction prior to the activation of annunciator systems. However, even if this is not the case, detection will nevertheless be fairly well guaranteed by the alarms and lights of the annunciation system. Then *diagnosis* becomes the next phase of operator response. As we shall see below, however, there is a certain paradoxical relation between detection and diagnosis: The very characteristics of annunciators that may guarantee timely detection—salience and prominence—can severely inhibit the effective use of those annunciators in diagnosis.

Fault Diagnosis

Once a failure or abnormality is detected, the operator is faced with a choice of what actions to take in order to meet three criteria. These criteria, in decreasing order of importance, are as follows: (1) actions that will ensure plant safety, (2) actions that will not jeopardize plant economy and efficiency, and (3) diagnostic actions that will localize and correct the fault. Unfortunately, these criteria are not always compatible with each other, and operators who must function in a probabilistic world in which outcomes of actions cannot be predicted with certainty should at least be certain about the relative importance of each. This importance must be emphasized, for example, when a choice must be made to take a turbine or plant off the line, thereby ensuring safety (criterion 1) but potentially sacrificing economy (criterion 2). The status of the third criterion, diagnosis, is also somewhat ambivalent. In the nuclear industry, written procedures are being altered to emphasize that control actions that will restore critical system variables to their normal range must take priority over diagnosis of a fault (Zach, 1980). The intent of this shift in priorities is to emphasize that the operator need not worry about what caused the failure until the system has been "stabilized." On the other hand, diagnosis is often a necessary precursor to restoring or maintaining plant safety. As the incident at Three Mile Island revealed (see Chapter 1), misdiagnosis can potentially lead to disastrous circumstances.

The problems in fault diagnosis encountered by the human monitor of process control are an inherent consequence of the ambiguity of fault symptoms. Given a particular fault, it may be possible to predict precisely the resulting symptoms that would result. However, the converse is not the case; a given set of observed symptoms—particularly as filtered by limitations of the operator's working memory and attention (see Chapter 3)—may have resulted from a large number of different causes. As we discussed in Chapter 3, diagnosis is far more difficult than prediction. The following sections will consider two alternative sources of diagnostic information that may be available to the process control supervisor after a failure.

Annunciator and alarm information.   A good deal of preliminary information concerning the nature of a failure is *potentially* available in the alarms and annunciators that first indicate its existence. The word "potentially" is emphasized, however, because from the operator's point of view the annunciator information is often essentially uninterpretable. This unfortunate state of affairs results because the vast interconnectedness of the modern process control or nuclear power plant often means that one "primal" failure will drive conditions at other parts of the plant out of their normal operating range so rapidly that within minutes or even seconds scores of annunciator lights or buzzers will create a buzzing-flashing confusion. At one loss-of-coolant incident at a nuclear reactor, more than 500 annunciators changed status within the first minute and more than 800 within the first two (Sheridan, 1981). Operators on the scene of the Three Mile Island incident complained that this rapid growth prevented them from obtaining good information concerning the initial "primal" conditions that led to the other secondary failures (Goodstein, 1981). The problem can be exacerbated by a logic which is sometimes incorporated whereby a central master alarm (usually auditory) will sound whenever a fault appears anywhere in the system. Difficulties potentially occur when this alarm, often very annoying, is turned off. When this happens, some systems are configured to turn off the causal alarms, or "freeze" their state so that even if the variables return to normal conditions this information is not registered (Goodstein, 1981; Sheridan, 1981). Thus the very annoyance of the master alarm that guarantees failure *detection* induces a form of behavior that may be counterproductive to effective diagnosis.

Much has been written concerning the poor design of warning and annunciator systems in both nuclear power plants (e.g., Benel, McCafferty, Neal, & Mallory, 1981; Hopkins et al., 1982; Osborne, Barsam, & Burgy, 1981; Sheridan, 1981) and aviation (Cooper, 1977; Thompson, 1981). In a comprehensive evaluation of human factors goals in the nuclear industry, a report by Hopkins et al. (1982) concludes that reseach in improving the effectiveness of alarm and annunciator systems should be of the highest importance. These analyses are in agreement in their conclusions that the confusing nature of annunciator systems could be greatly reduced by implementing certain basic principles. These principles would also increase the diagnostic value of the annunciators. Four categories of general innovations appear to be particularly useful: sequencing, grouping and prioritizing, color, and informativeness.

1. *Sequencing.* This procedure allows the operator to recover an accurate picture of the unfolding of a progressive series of annunciators and therefore be better able to deduce the "primal fault." A simple variant of sequencing is the concept of a "first out panel" in which the first alarm to appear in a series is somehow distinctively identified for a prolonged period (i.e., by a flashing light; Benel et al., 1981). The more complex form of sequencing, dependent upon computer services, requires a buffer memory that could replay on a schematized display—at a speed chosen by the operator—the sequence of annunciator appearances. Lees (1981) discusses the role of the computer in analyzing the alarm sequence and in using its own intelligent logic to help identify the primal fault. Sequencing is thus another example of how "smart" displays can assist the fallable records of human memory.

2. *Grouping and prioritizing.* Of course any computer-driven replay of an event sequence should ideally be organized spatially in such a manner that displays of systems that are closely related functionally are also closely spaced physically. This recommendation is equally important for the placement of annunciators themselves. Annunciators for functionally related systems should be physically close. In this way, it is likely that a given fault will trigger annunciation of a "primal cluster" of indicators and lead to a pattern that is far easier to interpret than would be a random appearance of lights distributed all across the control panel.

Of course there are tradeoffs that may have to be made. The concept of "functional proximity," for example, may be defined in several dimensions. The indicators of the same pressure valve on two different reactors are, in one sense, highly similar. Yet in another sense, their similarity is low when compared to the similarity between the valve indicator and the indicator of pressure on its input side. The latter two, belonging to a single system, are more likely to be causally related in a failure and therefore belong to the same fault cluster. In addition to functional similarity, the similarity of *priority* is yet another dimension that, as some have argued, should guide annunciator proximity. Certain warnings are of high priority (i.e., potential exposure of the radioactive core in a nuclear reactor or excessive steam pressure approaching explosion point) whereas others are less so. Analysts have argued that one physical dimension of annunciator displays should define priority. Benel et al. (1981) suggest a matrix array in which the vertical dimension defines priority and the horizontal defines some functional proximity. Alternatively, priority may be defined by color. In commercial aviation, system designers are careful to let priority guide the selection of speech displays that talk to the pilot (Hanson et al., 1982). In summary, in agreement with the general concept of stimulus/central-processing compatibility put forth in Chapters 5 and 6, the principles of grouping here assert that the conceptual "cognitive" dimensions that define annunciators (functional proximity, urgency) should correspond with the physical dimensions that define their placement and format.

3. *Color*. As discussed in Chapters 2 and 7, color, if properly and consistently used, has the potential to aid greatly in interpreting the diagnostic information available from alarms. Yet again, following the principle of physical-conceptual compatibility, it is extremely important that display colors be consonant with conceptual population stereotypes (e.g., red = danger, yellow = caution, green = normal) and that these stereotypes be consistently adhered to (Osborne, et al., 1981). A classic example of the lack of consistency has been pointed out by Osborne et al. in process plants where the convention is followed that green indicates the status of an open, flowing valve and red the status of a closed one. Unfortunately, under normal plant operations some valves will be open and some will be closed. The resulting "Christmas tree" depiction of normal operation may be indistinguishable at a glance from the appearance of a failure. Osborne et al. contrast this coding with a higher conceptual level of coding in which green indicates the normal valve position (whether open or closed) and red the abnormal. In this "Greenboard" display, abnormalities will thereby show up with great salience as any red indicator in a sea of green.

The complex issues involved in good display design are revealed, however, when it is seen that even this concept has certain limitations. On one hand, it runs contrary to stereotypes in the electronics industry in which green means "circuit not live" and red means "circuit hot, do not touch." This would place alternate coding interpretations side by side where electrical and hydraulic systems are mixed. On the other hand, the same valve positions that are normal during one phase of operation—the steady running of a nuclear plant—may be abnormal during alternate phases such as start-up and shutdown. This means that the Christmas tree pattern will reappear during these transient phases and provide an uninterpretable array that cannot readily distinguish normal from abnormal conditions. This is particularly dangerous because start-up and shutdown are periods of the highest cognitive load. The only solution here is to incorporate computer logic to make the assignment of color code to valve position dependent not only by criteria of normality, but also by the operational phase.

4. *Informativeness*. The spirit of providing useful annunciator information is reinforced by Thompson's (1981) suggestion that annunciators should be made to supply a fair degree of information, in the formal sense of the word discussed in Chapter 2. We concluded in Chapter 10 that humans can process a smaller number of information-rich stimuli more efficiently than a large number of stimuli of small information content. This characteristic of processing reinforces the value of more informative "higher level" annunciator systems that can provide a single piece of evidence for a set of symptoms that covary, rather than one annunciator for each. Further economy is gained if the single physical annunciator may be configured to provide more than one bit of information.

Diagnostic search.   In the absence of explicit information from the sequence of alarms, the process control operator will normally be required to engage in a series of question-asking diagnostic troubleshooting operations. In Chapter 3 many of the limitations of human operator performance in diagnosis and troubleshooting were discussed in a variety of contexts. This discussion will not be repeated in this chapter except to reemphasize the three major causes of nonoptimal behavior: limitations on working memory and attention (both exacerbated by display formats that require a serial scan) and shortcomings associated with certain logical operations (the use of negative evidence and deductions from the absence of symptoms). Instead, the present section will present two descriptive approaches that have been applied specifically to the process control fault-diagnosis environment. Both are taken from the analytic work of Jens Rasmussen (1979, 1980, 1981), and both involve the concept of hierarchical representation. In one case, this hierarchy is defined in terms of level of abstraction and understanding of the process dynamics. In the other case, it is defined in terms of the automation of action.

1.   *Level of abstraction.* As Rasmussen and Lind (1981) have emphasized, the analysis and understanding of a process may be carried out at several different levels of abstraction: in terms of the direct physical variables involved (i.e., pressure, temperature), in terms of more abstract variables (energy-mass exchanges), or at still higher levels of abstraction (economic costs and benefits). The diagnostic process, like perceptual categorization discussed in Chapters 4 and 5, involves an interplay of bottom-up and top-down processing moving up and down this hierarchy of abstraction. Thus failure symptoms (like sensory evidence) are only displayed at the lowest level of abstraction—typically in terms of pressure and temperature readings—but for these to be interpreted, higher levels may need to be brought play. In particular, many action selections need to be based upon information at the highest levels of abstraction—for example, what are the consequences of a shutdown of a unit? What alternative energy sources are available?

Rasmussen and Lind (1981) argue that operators must move up and down the abstraction hierarchy, at times "zooming in" for closer analysis and scrutiny at particular levels. The existence of this sort of processing dictates that operators will be well served by flexible computerized displays that present diagnostic information at the level of abstraction requested by the operator at any given moment. Rasmussen and Lind make a further recommendation as to the format of such displays. In order to allow a common invariant scheme of representation format at all levels, they propose that all levels possess certain invariant features that can be highlighted in the displays. This invariance at all levels will reduce the cognitive complexity involved when several different levels are viewed in sequence, as they might be during diagnosis. The three features they propose are the following:

a. The *process* observed. This they refer to as the "what" level.

b. The *implementation* of the process by units at a lower level, or the "how" level. This may be described in terms of the physical components themselves if it is at a lowest level, or in terms of lower hierarchical levels if the relevant process is higher.

c. The *purpose* of the process for the next higher level (the "why" level). In turn, the purpose of a process may be assigned to one of three categories. A process may *implement* part of a higher-level function, as where a unit may serve to bring the temperature of a quantity through a certain time trajectory; it may serve to *supply* a product to the higher-level function, as in the role of a fuel pump; or it may serve to *maintain* a condition within proper limits, the role, for example, of a homeostatic regulator or pressure-relief valve.

The conceptual analyses carried out by Rasmussen then represents an important converging link between cognitive models of thinking and problem solving on the one hand and human computer interactions on the other. It is clear, however, that research in this area has not gone far beyond the stage of conceptual modeling. Given the tremendous complexities of both human cognition in problem solving and of the system itself, there would appear to be a great potential for gains to be made by developments in this area.

2.   *Level of automatization.* In Chapter 9, three levels of automatization of action were contrasted: skill-based, rule-based, and knowledge-based behavior. While the continuum was shown there to be relevant to a wide variety of skills, it is important to note that Rasmussen (1980, 1981) originally developed this categorization to be applied in the context of the process control environment. In an effort to bring Rasmussen's conceptual analysis into application, Pew, Miller, and Feehrer (1981) have used the skill-, rule-, and knowledge-based trilogy as a framework for analyzing four "critical incidents" that have actually occurred in the nuclear power industry. None of these cases approached the magnitude of the Three Mile Island incident, but all of them presented the potential for disaster.

Figure 12.4 presents Pew et al.'s modification of Rasmussen's hierarchical representation, in the context of the incidents that they investigated. Perceptual information variables flow from automated skill-based perceptions on the lower left to knowledge-based problem solving and interpretation at the top. Correspondingly, the column at the right flows from knowledge-based actions at the top to rule-based and skill-based actions toward the bottom. Linkages between and across any levels are possible. For example, automatic acknowledgment of an alarm light reflects skill-based perception and response, as characterized by the arrow across the bottom of the ladder. If the

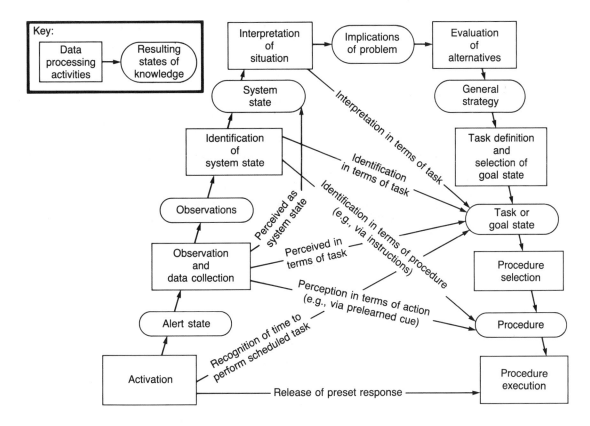

Figure 12.4 Hierarchy of perception, decision, and action adapted from Rasmussen (1980). From "Evaluating Nuclear Control Room Improvements Through Analysis of Critical Operator Decisions" by R. W. Pew, D. C. Miller, and C. E. Feehrer, 1981, *Proceedings, 25th Annual Meeting of the Human Factors Society* (p. 101), Santa Monica, CA: Human Factors. Copyright 1981 by the Human Factors Society, Inc. Reproduced by permission.

operator follows a standardized set of procedures in response to a failure that was only diagnosed after a lengthy analysis, this sequence represents rule-based action following knowledge-based perception.

On the basis of their analysis, Pew et al. concluded that the operators in these incidents were required to work at all different levels of the hierarchy. They concluded further that once operators proceeded beyond the skill-based level, one of the greatest hindrances to timely diagnosis and action was time stress. Reiterating points made in Chapter 3 and emphasized by Sheridan (1981) in his discussion of the nuclear power operator's job, Pew et al. point to the effects of stress in inducing cognitive "tunnel vision" (see Chapter 7). The undersirable effects of stress resulting from limited time can be most easily reduced by maximizing the efficiency with which diagnostic information at any level of automatization can be encoded. To do so requires again that the system designer carefully consider the nature of operator's information needs and the flexible display formats that are most compatible with those needs.

Conclusion

Human factors research on process control is a relatively young domain. Many of the recommendations concerning display layout, anthropometry, control location, lighting and legibility, and clearly written procedures translate directly from the general practice of good human factors (Bailey, 1982; McCormick & Sanders, 1982), and this approach has been implemented with some degree of success. Yet it is equally clear that great improvements are possible by combining theories of cognitive psychology as these are manifest in process control understanding and fault diagnosis with increasingly available computer/display technology. This is an area in which relatively little engineering psychology research has been conducted to assess the utility of the recommendations made by such researchers as Rasmussen and Umbers. The potential payoffs of this research appear to be quite high, in light of the fact that systems are evolving toward greater complexity with increasing levels of automation. In the following section we shall conclude the book by considering the issues in automation in more detail, emphasizing the particular shortcomings that may result.

AUTOMATION

The call for increased automation of complex systems has been a natural response to instances in which human error has contributed to disasters or near disasters. The importance of automation was articulated in the early 1950s by Birmingham and Taylor (1954), who proposed automation in manual control and stated that "speaking mathematically he (man) is best when doing least." The trend toward automation has been encouraged both by the increasing availability of sophisticated technology and by the fact that in many situations, particularly in aviation, increasing automation leads to more economical, fuel-efficient operation (Lerner, 1983; Weiner & Curry, 1981). It is also a trend that may ultimately reduce personnel costs, as operators—often expensive to train and employ—are replaced by automated components. Despite the apparent attractions of automation, however, there are a number of less obvious shortcomings. These will be dealt with at length after we consider some examples of different kinds of automation.

Examples and Purposes of Automation

Automation varies from that which totally replaces the human operator by computer or machine to computer-driven "assists" that provide some unburdening of an overloaded operator. The different purposes of automation may be assigned to three general categories:

1. ***Performing functions that the human operator cannot perform because of inherent limitations.*** This category describes many of the complex mathematical operations performed by computers (e.g., those involved in statistical analysis). In the realm of dynamic systems, examples of this class of automation would include control guidance in a manned booster rocket in which the time delay of a human operator is sufficiently long to make the rocket go unstable (see Chapter 11), aspects of control in complex nuclear reactions in which the dynamic processes are too complex for the human operator to respond to "on line," or robots for the manipulation of materials in hazardous or toxic environments. In these and similar circumstances, automation appears essential and unavoidable whatever its costs may be.

2. ***Performing functions that the human operator can do but performs poorly or at the cost of high workload.*** Examples here would include the autopilots that control many aspects of flight on commercial jet aircraft (e.g., Boeing 747, 757, 767, DC–10, and L–1011; Lerner, 1983) and the automation of certain complex monitoring functions such as the ground proximity warning systems (GPWS) (Danaher, 1980; Weiner & Curry, 1981). For pilots, the GPWS integrates information about altitude, location, sink rate, and other factors and decides whether the aircraft is approaching dangerously close to the ground in the wrong circumstances. If it is, an auditory alert is sounded. A third example could be the automating of fault-diagnosis procedures. Although not formally implemented, an automated diagnosis system, suggested for operation in nuclear plants, would integrate symptoms in case of a malfunction and would make an inference concerning their possible causes. Finally, human operators could be aided by computer offering of optimal information for decision making and diagnosis. This technique would select information on those attributes that are most diagnostic for a given decision-making problem, rather than letting the operator select attributes freely. Chapter 3 provided evidence that operator selections in information seeking were not always optimal and discussed this kind of automation.

3. ***Augmenting or assisting performance in areas where humans show limitations.*** This category is similar to the preceding one but is distinct in that automation is intended not as a replacement for integral aspects of the task but as an assist on peripheral tasks necessary to accomplish the main task. As we have seen, there are major bottlenecks in performance because of limitations in human short-term memory and prediction/anticipation. The following examples of automation would provide useful means of unburdening the operator of these limitations. An automated display or "visual echo" of auditory messages is one such example. In aviation this might involve a speech recognition device that interprets a command received from air traffic control and prints the command on a visual display. This procedure would eliminate the need for the pilot to rehearse the information in

working memory until it is entered. It also would relieve the operator of the requirement to jot down such information manually. A second example would be the identification or tagging of aircraft symbols with their flight number, altitude, and other factors on computer-generated air traffic control displays (Danaher, 1980). A reduction of working memory load and/or manual writing is accomplished by this procedure. Another example would be a computer displayed "scratch pad" of the output of diagnostic tests in fault diagnosis of the chemical, nuclear, or process control industries. As suggested in Chapter 3, this procedure would greatly reduce memory load. Or computer displayed "checklists" of required procedures, coupled with a voice recognition system, would allow operators to indicate orally when each procedure had been completed. The automated display could then "check off" each step accordingly. Another way a computer could assist is in the optimal scheduling of activities projected into future time horizons (Whitefield, Ball, & Ord, 1980; see Chapter 7). Finally, as described in Chapter 11, the predictive display in tracking provides a great assist, again by reducing the anticipation requirement.

Costs and Benefits of Automation

The potential benefits of the classes of automation described above can be easily demonstrated. In almost every incident, such as Three Mile Island, in which human error has led to a disaster, subsequent analyses often conclude, "If only such and such a function had been automated, then the human operator would not have had the opportunity to make the blunder that produced the event." On the other hand, as we shall describe below, there are a number of incidents and accidents which may be potentially attributed to the direct or indirect effects of automation itself. The problem is that neither examples where human error is obvious nor those in which automation is a problem can by themselves provide objective evidence of whether automation is, on the whole, a good or a bad thing. This uncertainty is reflected in Figure 12.5, which may be interpreted within the framework of the discussion of program evaluation by Einhorn and Hogarth (1978) presented in Chapter 3. The visible examples of failures described above both fall in the right cells of Figure 12.5. However, as described in Chapter 3, an objective assessment of the overall merits of automation must take into account data in the left cells—data that are so routine in nature that they are rarely tabulated. How many times have automated systems or human operators worked successfully? Furthermore, to provide more reliable data in these two cells, one would have to know how automated and manual systems would have performed doing equivalent kinds of jobs. While it is possible to establish tight control of job equivalency in laboratory investigations, these data are not readily available when assessing real-world performance.

A specific example of the lack of complete data necessary to evaluate automation is discussed by Weiner and Curry (1981) and pertains to the

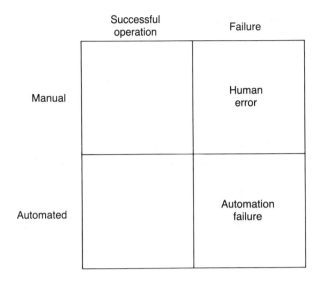

Figure 12.5   Failure in manual and automated systems.

ground proximity warning signal in aviation. The GPWS, it will be recalled, is an alarm system that warns the pilot if the plane is flying too close to the ground or is approaching the ground too rapidly. Pilots have complained about false alarms and other difficulties with the system, thereby highlighting the failures of automation (Lerner, 1983). Yet it is impossible to document how many potential crashes have been avoided because of the presence of the GPWS in the modern cockpit. Statistics on the reduced accident rate since its mandatory incorporation in 1974, however, suggest that these advantages exist and that the failures of the system should be tolerated. Occasional failures are not sufficient grounds for eliminating the system.

Given the absence of data concerning normal operation of both manual and automated systems, it is therefore somewhat difficult to make empirically based statements that automation is better or worse than manual performance in certain environments. What is possible, however, is to provide a more accurate subjective evaluation of the kinds of problems that will be encountered by automation, thereby emphasizing the guidelines for implementing corrective solutions. Such evaluations, however, still would not allow an explicit statement to be made of whether manual performance could have done better in the equivalent task. Bearing these qualifications in mind, the reader should be aware that the discussion of the problems of automation presented in the following section is certainly not an indictment of automation. Complete automation of many chemical and process plants may proceed for years without failure. The discussion merely serves to emphasize the reasons for caution whenever the implementation of automation is considered.

Problems with Automation

Increased monitoring load.   As described in Chapter 2, the automation of functions once carried out by humans moves the human operator to the higher level of a "supervisory controller" (Sheridan & Johannsen, 1976) or "supervisor-sentinel" (Pew, 1969). The operator now has less to do but, as a consequence of the increased number of system components in the automated system, has many more indicators of component status to monitor. The data discussed in Chapter 2 suggested furthermore that humans do not make terribly effective monitors. An intuitive solution that could be adopted in light of this shortcoming is to automate the monitoring function as well. This was the solution described in the previous section, implemented when "smart" or centralized alarm systems were incorporated in process control systems (Lees, 1981) or aviation (Hansen et al., 1982). Yet automated monitoring functions too have their problems: they add still more components which are also subject to failure, and they leave the operator even further removed from the ultimate process under his responsibility. This increased "distance" can be quite harmful to successful control and diagnosis in case a failure occurs (Lerner, 1983).

Proliferation of components.   The proliferation of components brought on by increased system complexity and automation has two associated problems. The first problem, as noted previously, is that more components dictate that there is a greater likelihood that something somewhere in the system will fail. This is a consequence of the basic reliability equation: The probability of any one thing failing is equal to one minus the product of the reliability of all components. Since component reliabilities are less than one, the product will decrease (and one minus the product will increase) as more components are included (Adams, 1982). The proliferation of components requires more displayed elements—some 886 separate annunciators on the Lockheed L–1011 aircraft. This increase in display complexity, particularly if the elements are not carefully configured, will magnify the problems associated with fault diagnosis described in the previous section.

The second problem is that with more system components, each one having its own annunciator indicative of its operating state, there is an increased likelihood that at least one system will show a false alarm (e.g., the indicator will falsely say that the system is malfunctioning). This is what happened when the Eastern Airlines L–1011 crashed in the Everglades in 1972, the incident described in Chapter 1. A warning light incorrectly indicated the status of the plane's landing gear. That problem diverted the crew's attention from guidance of the aircraft and ultimately led to the disaster. Furthermore, as we saw in Chapter 2, such false alarms may cause the operator to distrust the indicators and produce the "voluntary vigilance decrement"—a conscious conservative adjustment of the detection criterion. The ultimate effect of system false-alarm rate on operator strategies is one that requires careful evaluation. In this regard, Weiner and Curry (1981) argue that it is important for annunciator systems to be designed so that the operator is able to check the validity of an alarm where possible.

False sense of security.   The previous section pointed out a problem that could occur if the operator distrusts an automated component. Weiner and Curry (1981) and Danaher (1980) identified a problem associated with the opposite state of affairs, when the operator places too much trust in the automated system. In this case, total belief in the infallibility of the system can lull the operator into a false sense of security, not making checks that would otherwise be advisable. Two examples cited by Danaher (1980) illustrate this tendency.

The first concerns a near collision that was described in Chapter 9 in the context of warning signals for reaction time. In 1975, near Detroit, a DC–10 was climbing to altitude and approaching from underneath a Lockheed L–1011 that was flying directly overhead. The two aircraft were on a collision course. The air traffic controller noted the developing situation on his display. However, because the display was automated with appropriate tags and warnings, he decided not to take the immediate warning precautions, but turned to handling other aircraft, *knowing he would be alerted when and if a collision became imminent.* He was relieved by a second operator before the situation developed further and did not inform his replacement. When the second operator at last noted the danger on the display, he quickly alerted the DC–10 to descend, only one second before the DC–10 pilot visually sighted the L–1011 through the clouds a short distance away. A potential disaster was averted through some fast maneuvering by the DC–10 pilot—resulting in a number of injuries aboard the aircraft. The point here is that had the air traffic control display not been equipped with the automated equipment, the controller would have been less conservative in issuing his warning. The message of this example is not to suggest that such automated warning devices are bad. As noted above they have probably prevented far more incidents than they have caused. The important point rather is to emphasize that the system designer must think of ways of preventing the human operator from placing too much trust in an automated device.

In the second illustration, the 1972 Eastern L–1011 crash into the Everglades described above, disaster was not averted. Here the flight-deck personnel, while in holding pattern, were attempting to diagnose the cause of the malfunctioning landing-gear indicator. A major contributing cause of disaster was their unwarranted faith that the autopilot would hold them at altitude while their full attention was diverted to diagnosis. The autopilot did not hold, and the plane gradually descended into the Everglades with 99 fatalities. Clearly with no autopilot on board, the pilot would have had to continue to fly and hold altitude, delegating the fault diagnosis to other members of the flight deck.

Out-of-the-loop familiarity. When the operator is replaced by an automatic controller, he reduces his level of interaction or familiarity with the state of the system. There is some evidence that when a malfunction of the system does occur (for example, an engine failure), the operator will be slower to detect it and will require a longer time to "jump into" the loop and exert the appropriate corrective action if he is not integrally part of the loop. The increased latency and reduced accuracy of failure detection when out of the loop was demonstrated by Wickens and Kessel (1979, 1980), Kessel and Wickens (1982), Young (1969), and Ephrath and Young (1981). In all of these experiments operators were required to detect changes in the dynamics of a tracking system that was either controlled manually by the operator or controlled by an autopilot.

There is, on the other hand, at least one experiment indicating the opposite conclusion, that detection performance is better by a monitor than by a controller (Ephrath & Young, 1981, experiment 2). Ephrath and Young suggested that this difference in results was probably caused by the increased level of workload in the experiment that showed monitor superiority. However, another possibility is that the difference results from the use of qualitatively different kinds of failures in the two experiments. Ephrath and Young defined their failure as a gradual bias in the directional navigation rather than a change in the plant dynamics themselves. Wickens and Kessel (1981) concluded that the results are consistent in suggesting that, when monitoring for changes in the system dynamics themselves, performance advantages will be best when the operator remains in the loop, particularly when concurrent task workload is relatively low.

The role of system familiarity as described in the preceding paragraph applies to the immediate real-time loss of information regarding the momentary state of the system. There also appear to be long-term consequences of being removed from the control loop, evident to the extent that pilots or controllers may lose proficiency as they receive less and less "hands on" experience. This skill remains one of critical importance as long as the potential remains for them to intervene, for example, in cases of autopilot failure. The loss-of-familiarity problem is particularly evident in civil aviation when copilots gain experience flying heavily automated wide-bodied jets (e.g., Boeing 747, DC–10), where there is the potential to lose proficiency, and then transfer to be pilots of the less automated narrow-body jets, where they must "revive" their proficient manual control skills (Weiner & Curry, 1981). The fact that familiarity may be gradually lost from "out-of-the-loop" experience argues strongly for the importance of frequent retraining periods. During these sessions, the operator of complex systems may enter the loop in simulators and experience the dynamic relationship between system variables.

Higher-level operator error. As described above, automation does not eliminate the possibility of human error. It merely relocates the sources of human error to a different level. This may be manifest in several different situations. For example, there are "set up" errors in which an automated system is programmed incorrectly prior to its use. As a specific example, automatic inertial navigation systems that make corrective changes in midflight require that pilots enter the coordinates of these changes into the system prior to takeoff. A mistake in this entry process will be a potential source of error that may go undiagnosed even at the critical turn point (Weiner & Curry, 1981).

There are also sources of human error in the manufacture and maintenance of automated equipment. Errors at any of these levels can be potentially disastrous. One classic example is provided by an event that was a major contributing cause to the Three Mile Island incident. As described in Chapter 1, a pressure relief valve indicator was designed to display to the operator what the valve was commanded to do and not what it actually did. When the two did not correspond, the operators formed the wrong impression of the state of the system and initiated actions that worsened the situation.

As automated systems become more and more complex, greater responsibility often shifts to the computer programmer, who must design programs to foresee possible combinations of events that might require automated responses. Yet, as noted in Chapter 1, it is just this sort of "creative problem solving" in response to low-probability events that benefits from the unique contributions that the human operator at the scene has to offer. Programming cannot take into account all possible eventualities. If an automated fault-diagnosis system is not equipped to interpret a certain event that actually does occur (involving, for example, difficult-to-foresee compound failures), what should be its appropriate response? If it makes a best (but wrong) guess, this choice might be far worse than making no response at all because of the operator's tendency to trust the system. If the operator does believe the system and so maintains an incorrect hypothesis, this trust would thereby induce all of the consequences of cognitive vision in hypothesis testing described in Chapter 3 (Sheridan, 1981).

The answers to these issues concerning the fallibility of automated diagnostic systems are not known, although human factors engineers have acknowledged the heavy responsibility that is placed upon programmers and system designers as a result (Goodstein, 1981; Lees, 1981). It would seem advisable for computerized systems to err on the side of caution in attaching confidence ratings to their own decision-making or diagnostic performance, particularly when confronting ill-structured problems where the rigidity of computer programmed solutions is likely to lead to nonoptimal performance (Buck & Hancock, 1978; Laughery & Drury, 1979).

Loss of cooperation between human operators.   Automation by definition removes some functions from human operator control. In his excellent treatment of issues in automation, Danaher (1980) points out the sometimes intangible benefits that result from the interaction of two thinking cooperative human beings in a complex system, in contrast to the impersonal communications between human and computer. In the air transport system, in particular, this level of flexibility and cooperation between air traffic controller and pilot may help to resolve ambiguities that could otherwise prove difficult.

In spite of the greater degree of cooperation and flexibility inherent in a human, as opposed to a machine component, it is possible nevertheless to follow certain guidelines that will "humanize" elements in the automated systems. Rouse (1977, 1981) and Steeb, Weltman, and Freedy (1976) have discussed these issues at some length, in the course of their investigations of *cooperative* man-machine systems. That is, systems in which a given task may be performed either by the human operator or by the computer, depending upon the particular circumstances present at any instant in time. The concept of "cooperative," tolerant computer systems is closely related to the discussion in Chapter 6 of human-computer interactions. On the basis of the research of Rouse, Steeb, and their colleagues, the following four guidelines would appear to provide better performance:

1. *Forgiving or fault-tolerant computer.* The computer system should be designed to allow common human errors to occur without greatly disrupting the task processing.

2. *Who's in charge?* When there is a task that may be performed either by the human or by the machine, the question arises as to who (human or computer) should have the final decision. It perhaps is logical to argue that the human—the more sophisticated component—be provided with the final choice, and many responsible human operators would demand that choice. Yet humans also tend to become insensitive evaluators of the degraded quality of their own performance precisely at the time when machine assistance is needed most. As a solution to this dilemma, Steeb et al. argue that, while humans can continue to have final choice, executive mechanism on the machine side should have a continually available record of performance to use as a basis for displaying a continuous *recommendation* of who should be in control.

3. *Feedback.* Clear, unambiguous feedback should be provided concerning who *is* in charge at any moment in time.

4. *Training and understanding the processing of automated equipment.* The operator's understanding of the mechanism by which the component does the job is a major determinant of how well an operator trusts an automated component to assist and cooperate. If the component is viewed as a magic black box that generates outputs from inputs through some unknown algorithm, it will be far less trusted than if the operator understands the procedures followed by the component, particularly if the operator perceives the component to be going through the same procedures as he or she would in deriving an output (Weiner & Curry, 1981). This demand normally requires some investment into training the operator to understand, trust, and use the automated process.

User Acceptance

The criterion for evaluating automation from the perspective of job satisfaction moves well beyond the domain of engineering psychology to that of sociology and industrial psychology. There is probably as much diversity in the acceptance of automated systems as there is in the personality of the users and the capabilities of the systems themselves. For example, some pilots prefer hands-on control at all phases of flight while others prefer to use the autopilot at every opportunity. It is clear, however, that acceptance by the user is a critical factor in determining the economic viability of an automated system. An optimal automated system which is available but is never deployed might as well never have been developed in the first place. One which is mandatory and potentially effective but disliked could well be worse than no system at all, if employed in the service of a hostile user.

In their discussion of flight-deck automation, Weiner and Curry (1980) present certain guidelines that would seem to produce more acceptable automated systems. These, to some degree, parallel the guidelines for cooperative systems discussed above.

1. Design the automated system to perform the task in a manner consistent with how the user would perform it. In this manner the operator's internal model of the automated system's performance is consonant with his own.

2. The desires and needs for automation may vary across occasions and across individuals. Therefore automation should be flexibly available and not mandatory.

3. In a similar vein, automated decision systems should provide guidelines and recommendations, not commands. An accepting, "forgiving" system will be better accepted by the user than an autocratic one.

4. Extensive training in the use of the automated system should be provided so that the operator may truly appreciate its potential benefits and mode of operation and so that failure and malfunctions may be better understood.

The list of guidelines for automation goes on, and the reader is referred to Weiner and Curry's article, along with that of Danaher (1980) and many of the chapters in Rasmussen and Rouse's (1981) book for further guidelines. Suffice it to say that at this point the computer revolution is upon us and the inevitable consequence will be increased automation of function. Human demands as programmers, supervisors, and diagnosticians will expand even as their role as active continuous controllers might diminish. Such an evolutionary trend is consonant with the theme of this book: Engineering psychology will never diminish as an area of importance for society but will only shift its areas of concern to parallel the changes in systems with which humans interact.

Summary

The discussion of process control and automation in this final chapter has alluded to every previous chapter in the book, while also referring to issues of learning and training—major components of engineering psychology not covered here. This "global" characteristic emphasizes an important general point: The best design of the system must take into account human limitations and strengths, but these are not isolated components as the somewhat arbitrary division of chapters might suggest. The human brain is complex. Perception, memory, attention, and action all interact. A well-designed system must consider the way in which these processes interrelate, as well as the limitations of each. A system that is so designed will never eliminate the possibility of human error. But it can at least remove the system designer as a contributing factor, and that will be a major step.

References

Adams, J. A. (1982). Issues in human reliability. *Human Factors, 24,* 1–10.

Attwood, D. D. (1970). The interaction between human and automatic control. In F. Bolam (Ed.), *Paper making systems and their control.* London: British Paper and Board Makers Association.

Bailey, R. W. (1982). *Human performance engineering: A guide for system designers.* Englewood Cliffs, NJ: Prentice-Hall.

Bainbridge, L. (1974). Analysis of verbal protocols from a process control task. In E. Edwards & F. P. Lees (Eds.), *The human operator in process control.* London: Taylor & Francis.

Bainbridge, L. (1979). Verbal reports as evidence of the process operator's knowledge. *International Journal of Man-Machine Studies, 11,* 411.

Bainbridge, L. (1981). Mathematical equations or processing routines? In J. Rasmussen & W. B. Rouse (Eds.), *Human detection and diagnosis of system failure.* New York: Plenum Press.

Beishon, R. J. (1966). *A study of some aspects of laboratory and industrial tasks.* Unpublished D.Phil. thesis, University of Oxford.

Beishon, R. J. (1969). An analysis and simulation of an operator's behavior in controlling continuous baking ovens. In F. Bressen & M. deMontmollen (Eds.), *The simulation of human behavior.* Paris: Durod.

Benel, D. C. R., McCafferty, D. B., Neal, V., & Mallory, K. M. Issues in the design of annunciator systems. In R. C. Sugarman (Ed.), *Proceedings, 25th annual meeting of the Human Factors Society.* Santa Monica, CA: Human Factors.

Birmingham, H. P., & Taylor, F. V. (1954). *A human engineering approach to the design of man-operated continuous control systems* (Report No. 433). Washington, DC: U.S. Naval Research Laboratory.

Brigham, F. R., & Liaos, L. (1975). Operator performance in the control of a laboratory process plant. *Ergonomics, 29,* 181–201.

Broadbent, D. E. (1977). Levels, hierarchies, and the locus of control. *Quarterly Journal of Experimental Psychology, 29,* 181–201.

Buck, J. R., & Hancock, W. M. (1978). Manual optimization of ill-structured problems. *International Journal of Man-Machine Studies, 10,* 95–111.

Coekin, J. A. (1969). A versatile presentation of parameters for rapid recognition of total state. *Proceedings, International Symposium on Man-Machine Systems* (IEEE Conference Record 69 C59-mms). New York: IEEE.

Cooper, G. E. (1977). *A summary of the status and philosophies relating to cockpit warning systems* (NASA Contractor Report NAS2-9117). Washington, DC: NASA.

Crossman, E. R. F. W., & Cooke, J. E. (1962). *Manual control of slow response systems.* Paper presented at the International Congress on Human Factors in Electronics, Long Beach, CA.

Danaher, J. W. (1980). Human error in ATC systems operations. *Human Factors, 22,* 535-545.

Edwards, E., & Lees, F. P. (1974). *The human operator in process control.* London: Taylor & Francis.

Einhorn, H. J., & Hogarth, R. M. (1978). Confidence in judgment: Persistence of the illusion of validity. *Psychological Review, 85,* 396-416.

Einhorn, H. J., & Hogarth, R. M. (1981). Behavioral decision theory: Processes of judgment and choice. *Annual Review of Psychology, 32,* 53-88.

Electrical Power Research Institute. (1977, March). *Human factors review of nuclear power plant design* (Project 501, NP-309-SY). Sunnyvale, CA: Lockheed Missiles and Space Co.

Ephrath, A. R., & Young, L. R. (1981). Monitoring vs. man-in-the-loop detection of aircraft control failures. In J. Rasmussen & W. B. Rouse (Eds.), *Human detection and diagnosis of system failures.* New York: Plenum Press.

Garner, W. R., & Fefoldy, G. (1970). Integrality of stimulus dimensions in various types of information processing. *Cognitive Psychology, 1,* 225-241.

Gill, R., Wickens, C. D., Donchin, E., & Reid, R. (1982). The internal model: A means of analyzing manual control in dynamic systems. *Proceedings, IEEE Conference on Systems, Man, and Cybernetics.* New York: Institute of Electrical and Electronics Engineers, Inc.

Goodstein, L. P. (1981). Discriminative display support for process operators. In J. Rasmussen & W. B. Rouse (Eds.), *Human detection and diagnosis of system failures.* New York: Plenum Press.

Grimm, R. (1976). Autonomous I/O-colour-screen-system for process-control with virtual keyboards adapted to the actual task. In T. B. Sheridan & G. Johannsen (Eds.), *Monitoring behavior and supervisory control.* New York: Plenum Press.

Hanson, D., Boucek, G., Smith, W., Hendrickson, J., Chickos, S., Howison, W., & Berson, B. (1982). Flight operations safety monitoring effects on the crew alerting system. In R. Edwards (Ed.), *Proceedings, 26th annual meeting of the Human Factors Society.* Santa Monica, CA: Human Factors.

Hopkins, C. D., Snyder, H., Price, H. E., Hornick, R., Mackie, R., Smillie, R., & Sugarman, R. C. (1982). *Critical human factor issues in nuclear power regulation and a recommended comprehensive human factors long-range plan* (NUREG/CR-2833; Vols. 1-3). Washington, DC: U.S. Nuclear Regulatory Commission.

Jacob, R. J. K., Egeth, H. E., & Bevon, W. (1976). The face as a data display. *Human Factors, 18,* 189-200.

Jagacinski, R. J., & Miller, R. A. (1978). Describing the human operator's internal model of a dynamic system. *Human Factors, 20,* 425-439.

Kessel, C. J., & Wickens, C. D. (1982). The transfer of failure-detection skills between monitoring and controlling dynamic systems. *Human Factors, 24,* 49-60.

Kragt, H., & Landeweerd, J. A. (1974). Mental skills in process control. In E. Edwards & F. P. Lees, *The human operator in process control.* London: Taylor & Francis.

Landeweerd, J. A. (1979). Internal representation of a process fault diagnosis and fault correction. *Ergonomics, 22,* 1343-1351.

Landeweerd, J. A., Seegers, H. J., and Praageman, J. (1981). Effects of instruction, visual imagery, and educational background on process control performance. *Ergonomics, 24,* 133-141.

Laughery, K. R., Jr., & Drury, C. J. (1979). Human performance and strategy in a two-variable optimization task. *Ergonomics, 22,* 1325-1336.

Lees, F. P. (1981). Computer support for diagnostic tasks in the process industries. In J. Rasmussen and W. B. Rouse (Eds.), *Human detection and diagnosis of system failures.* New York: Plenum Press.

Lerner, E. J. (1983). The automated cockpit. *IEEE Spectrum, 20,* 57-62.

McCormick E. J., & Sanders, M. S. (1982). *Human factors in engineering and design.* New York: McGraw-Hill.

McLeod, P. (1976). Control strategies of novice and experienced controllers with a slow response system (a zero-energy nuclear reactor). In T. B. Sheridan & G. Johannsen (Eds.), *Monitoring behavior and supervisory control.* New York: Plenum Press.

Medin, D. L., Altom, M. W., Edelson, S. M., & Freko, D. (1982). Correlated symptoms and simulated medical classification. *Journal of Experimental Psychology: Learning, Memory, and Cognition, 37–50.*

Moray, N. (1981). The role of attention in the detection of errors and the diagnosis of failures in man-machine systems. In J. Rasmussen & W. B. Rouse (Eds.), *Human detection and diagnosis of system failures.* New York: Plenum Press.

Osborne, P. D., Barsam, H. F., & Burgy, D. C. (1981). Human factors considerations for implementation of a "green board" concept in an existing "red/green" power plant control room. In R. C. Sugarman (Ed.), *Proceedings, 25th annual meeting of the Human Factors Society.* Santa Monica, CA: Human Factors.

Patternote, P. H., & Verhagen, L. H. J. M. (1979). Human operator research with a simulated distillation process. *Ergonomics, 22,* 19.

Pew, R. W. (1969). Comments on 'promotion of man': Challenges in socio-technical systems; designs for the individual operator. In E. O. Attinger (Ed.), *Global Systems Dynamics.* Munich/New York: Karger.

Pew, R. W., Miller, D. C., & Feehrer, C. E. (1981). Evaluating nuclear control room improvements through analysis of critical operator decisions. In R. C. Sugarman (Ed.), *Proceedings, 25th annual meeting of the Human Factors Society.* Santa Monica, CA: Human Factors.

Pikaar, R. N., & Twente, T. H. (1981). How people discover input/output relationships. In H. G. Stassen (Ed.), *First European Annual Conference on Human Decision Making and Manual Control.* New York: Plenum Press.

Queinnec, Y., DeTerssac, G., & Thon, P. (1981). Field study of the activities of process controllers. In H. G. Stassen (Ed.), *First European Annual Conference on Human Decision Making and Manual Control.* New York: Plenum Press.

Rasmussen, J. (1979). Reflections on the concept of operator workload. In N. Moray (Ed.), *Mental workload: Its theory and measurement.* New York: Plenum Press.

Rasmussen, J. The human as a system component. In H. Smith & T. Green (Eds.), *Human interaction with computers.* London: Academic Press.

Rasmussen, J. (1981). Models of mental strategies in process plant diagnosis. In J. Rasmussen & W. B. Rouse (Eds.), *Human detection and diagnosis of system failures.* New York: Plenum Press.

Rasmussen, J., & Jensen, A. (1974). Mental procedures in real life tasks: A case study of electronic trouble shooting. *Ergonomics, 17,* 193–207.

Rasmussen, J., & Lind, M. (1981). Coping with complexity. In H. G. Stassen (Ed.), *First European Annual Conference on Human Decision Making and Manual Control.* New York: Plenum Press.

Rasmussen, J., & Rouse, W. B. (1981). *Human detection and diagnosis of system failures.* New York: Plenum Press.

Rouse, W. B. (1977). Human-computer interactions in multitask situations. *IEEE Transactions on Systems, Man, and Cybernetics, SMC-7,* 384–392.

Rouse, W. B. (1981). Experimental studies and mathematical models of human problem solving performance in fault diagnosis tasks. In J. Rasmussen & W. B. Rouse (Eds.), *Human detection and diagnosis of system failures.* New York: Plenum Press.

Rouse, W. B., & Morris, N. M. (1981). Human problem solving in a process control task. In H. G. Stassen (Ed.), *First European Annual Conference on Human Decision Making and Manual Control.* New York: Plenum Press.

Shackel, B. (1976). Process control—simple and sophisticated display devices as decision aids. In T. B. Sheridan & G. Johannsen (Eds.), *Monitoring behavior and supervisory control.* New York: Plenum Press.

Sheridan, T. B. (1981). Understanding human error and aiding human diagnostic behavior in nuclear power plants. In J. Rasmussen & W. B. Rouse (Eds.), *Human detection and diagnosis of system failures.* New York: Plenum Press.

Sheridan, T. B., & Johannsen, G. (Eds.). (1976). *Monitoring behavior and supervisory control.* New York: Plenum Press.

Smith, H. T., & Crabtree, R. (1975). Interactive planning. *International Journal of Man-Machine Studies, 7,* 213.

Steeb, R., Weltman, G., & Freedy, A. (1976). *Man/machine interaction in adaptive computer-aided control: Human factors guidelines* (Technical Report PTR-1008-76-1/31). Woodland Hills, CA: Perceptronics.

Thompson, D. A. (1981). Commercial aircrew detection of system failures: State of the art and future trends. In J. Rasmussen & W. B. Rouse (Eds.), *Human detection and diagnosis of system failures.* New York: Plenum Press.

Umbers, I. G. (1976). *A study of cognitive skills in complex systems.* Unpublished Ph.D. dissertation, University of Aston, England.

Umbers, I. G. (1979). Models of the process operator. *International Journal of Man-Machine Studies, 11,* 263–284.

Veldhuyzen, W., & Stassen, H. G. (1977). The internal model concept: An application to modeling human control of large ships. *Human Factors, 19,* 367–380.

Weiner, E. L., & Curry, R. E. (1980). Flight deck automation: Promises and problems. *Ergonomics, 23,* 995–1012.

Weingartner, A. (1982). *The internal model of dynamic systems: An investigation of its mode of representation.* Undergraduate honors thesis, University of Illinois, Department of Psychology, Champaign, IL.

West, B., & Clark, J. A. (1974). Operator interaction with a computer controlled distillation column. In E. Edwards & F. P. Lees (Eds.), *The human operator in process control.* London: Taylor & Francis.

White, T. N. (1981). Modeling the human operator's supervisory behavior. In H. G. Stassen (Ed.), *First European Annual Conference on Human Decision Making & Manual Control.* New York: Plenum Press.

Whitefield, D., Ball, R., & Ord, G. (1980). Some human factors aspects of computer-aiding concepts for air traffic controllers. *Human Factors, 22,* 569–580.

Wickens, C. D., & Kessel, C. (1979). The effect of participtory mode and task workload on the detection of dynamic system failures. *IEEE Transactions on Systems, Man, and Cybernetics, 13,* 21–31

Wickens, C. D., & Kessel, C. (1980). The processing resource demands of failure detection in dynamic systems. *Journal of Experimental Psychology: Human Perception and Performance, 6,* 564–577.

Wickens, C. D., & Kessel, C. (1981). Failure detection in dynamic systems. In J. Rasmussen & W. B. Rouse (Eds.), *Human detection and diagnosis of system failures.* New York: Plenum Press.

Williams, A. R., Seidenstein, S., & Goddard, C. J. (1980). Human factors survey of electrical power control centers. In G. Corrick, E. Haseltine, & R. Durst (Eds.), *Proceedings, 24th annual meeting of the Human Factors Society.* Santa Monica, CA: Human Factors.

Woods, D., Wise, J., and Hanes, L. (1981). An evaluation of nuclear power plant safety parameter display systems. In R. C. Sugarman (Ed.), *Proceedings, 25th annual meeting of the Human Factors Society.* Santa Monica, CA: Human Factors.

Young, L. R. A. (1969). On adaptive manual control. *Ergonomics, 12,* 635–657.

Zach, S. E. (1980). Control room operating procedures: Content and format. In G. E. Corrick, E. C. Haseltine, & R. T. Durst, Jr. (Eds.), *Proceedings, 24th annual meeting of the Human Factors Society.* Santa Monica, CA: Human Factors.

APPENDIX A

Some Values of d'

P(hit)	P(false alarm)					
	0.01	0.02	0.05	0.10	0.20	0.30
0.51	2.34	2.08	1.66	1.30	0.86	0.55
0.60	2.58	2.30	1.90	1.54	1.10	0.78
0.70	2.84	2.58	2.16	1.80	1.36	1.05
0.80	3.16	2.89	2.48	2.12	1.68	1.36
0.90	3.60	3.33	2.92	2.56	2.12	1.80
0.95	3.96	3.69	3.28	2.92	2.48	2.16
0.99	4.64	4.37	3.96	3.60	3.16	2.84

Selected values from *Signal Detection and Recognition by Human Observers* (Appendix 1, Table 1) by J. A. Swets, 1969, New York: Wiley. Copyright 1969 by John Wiley & Sons, Inc. Reproduced by permission.

APPENDIX B

Values Computed from the Formula

$$A' = 1 - \frac{1}{4}\left\{\frac{P(FA)}{P(H)} + \frac{[1 - P(H)]}{[1 - P(FA)]}\right\}$$

P(hit)	P(false alarm)						
	0.00	0.05	0.10	0.15	0.20	0.25	0.30
0.05	0.762	0.500					
0.10	0.775	0.638	0.500				
0.15	0.787	0.693	0.597	0.500			
0.20	0.800	0.727	0.653	0.577	0.500		
0.25	0.812	0.752	0.692	0.629	0.566	0.500	
0.30	0.825	0.774	0.722	0.669	0.615	0.558	0.500
0.35	0.837	0.793	0.748	0.702	0.654	0.605	0.553
0.40	0.850	0.811	0.771	0.730	0.687	0.644	0.598
0.45	0.862	0.827	0.792	0.755	0.717	0.678	0.637
0.50	0.875	0.843	0.811	0.778	0.744	0.708	0.671
0.55	0.887	0.859	0.829	0.799	0.768	0.736	0.703
0.60	0.900	0.874	0.847	0.820	0.792	0.762	0.732
0.65	0.912	0.889	0.864	0.840	0.814	0.787	0.760
0.70	0.925	0.903	0.881	0.858	0.835	0.811	0.786
0.75	0.937	0.918	0.900	0.876	0.855	0.833	0.811
0.80	0.950	0.932	0.913	0.894	0.875	0.855	0.835
0.85	0.962	0.946	0.930	0.912	0.894	0.876	0.858
0.90	0.975	0.960	0.944	0.930	0.913	0.897	0.881
0.95	0.987	0.974	0.960	0.946	0.932	0.917	0.903
1.00	1.00	0.987	0.975	0.962	0.950	0.937	0.925

Author Index

Subject Index